Optical Fiber Communications

This textbook is intended for undergraduate and postgraduate students of electronics and communication engineering. Providing an overview of the historical development of optical fiber communications, the book introduces readers to the basics of optical power and then discusses optical receivers, optical transmitters and optical amplifiers. It has discussions on attenuation, transmission losses, and optical sources like semiconductor light emitting diodes and lasers. The book also elaborates several dispersion-management schemes that restore an amplified signal to its original state. Besides this, the author clearly explains the concepts and applications of wavelength division multiplexing using optical fibres and different optical components. Finally, an overview of measurement techniques is presented to motivate students to perform lab activities and develop working projects.

Theoretical concepts have been elaborated starting from the basics and are well-supported by illustrations, numerical problems and step-by-step solved examples. Each topic is discussed in an interactive manner which includes its definition, its interpretation, example data, relevant illustrations and features such as facts-to-know. Adequate mathematical derivation and geometrical representation are included, wherever necessary. Other useful features include learning objectives, points to remember, important equations, key-terms with definitions, short answer type questions with answers, review questions and multiple choice questions.

T. L. Singal is a Professor at the Department of Electronics and Communication Engineering at Chitkara University, Chandigarh. Besides his teaching experience, he has 22 years industrial experience with leading telecom companies in India, Germany and the USA. His interest areas include digital communication systems, future generation digital cellular networks and wireless and internet technologies.

Optical Fiber Communications

Principles and Applications

T. L. Singal

CAMBRIDGE
UNIVERSITY PRESS

CAMBRIDGE
UNIVERSITY PRESS

4843/24, 2nd Floor, Ansari Road, Daryaganj, Delhi 110002, India

Cambridge University Press is part of the University of Cambridge.

It furthers the University's mission by disseminating knowledge in the pursuit of education, learning and research at the highest international levels of excellence.

www.cambridge.org
Information on this title: www.cambridge.org/9781316610046

First published 2016

Printed in India by Magic International Pvt. Ltd., Noida

A catalogue record for this publication is available from the British Library

Library of Congress Cataloging-in-Publication Data
Names: Singal, Tarsem Lal, author.
Title: Optical fiber communications : principles and applications / Tarsem Lal Singal.
Description: Delhi, India : Cambridge University Press is part of the University of Cambridge, [2016] | Includes bibliographical references and index.
Identifiers: LCCN 2016012701 | ISBN 9781316610046 (pbk.)
Subjects: LCSH: Optical fiber communication.
Classification: LCC TK5103.592.F52 S56 2016 | DDC 621.382/75--dc23 LC record available at https://lccn.loc.gov/2016012701

ISBN 978-1-316-61004-6 Paperback

Dedicated to

my school teacher Shri Kasturi Lal Jindal who
taught me how to learn, how to teach
and how to make others learn

Contents

Figures

Tables

Acknowledgments

Writing this book in tune with the mandatory requirements of outcome-based education, involved extensive research and efforts. I am grateful to all those who directly or indirectly provided me guidance and support. At the outset, I would like to express my gratitude for the encouragement and inspiration received from Dr Ashok Chitkara, Chancellor Chitkara University; Dr Madhu Chitkara, Vice-Chancellor Chitkara University; Dr Archana Mantri, Pro-Vice Chancellor Chitkara University; my colleagues and students of Chitkara University.

I would like to thank the editorial team at Cambridge University Press for bringing out this book in its present form.

The dream of my beloved parents, who wished me to be mentor for aspiring young engineering students, is fulfilled through the publication of this book. Their blessings are similar to that bestowed by the Almighty. I remain indebted to my wife Pinki, my daughter Ritu and my son Pankaj, for their continuous support.

Special thanks to the reviewers for taking out time and providing encouraging comments and valuable suggestions regarding improvement of the manuscript.

I am sure that every student will find this book rich in content along with its unique pedagogical features, which fully justify objective-oriented learning. This will certainly give the book a clear-cut advantage over other books on a similar course. The academic community as a whole will enjoy the simplified yet extensive and elaborate approach to every topic covered in the book. All efforts have been made to make this book error-free, but I believe there is always scope for improvement in our efforts. Your valuable suggestions/feedback are most welcome at *tarsemsingal@gmail.com*

CHAPTER

Introduction 1

Chapter Objectives

After studying this chapter, you should be able to

- ✓ get a historical overview of optical fibers and optical fiber communications;
- ✓ give reasons for the use of optical fiber in preference to wire cable and suggest suitable applications for fiber–optics;
- ✓ describe essential elements of optical fiber communications link;
- ✓ know advantages and disadvantages of optical fibers.

Light wave at higher frequency range of electromagnetic spectrum (3×10^{11}–3×10^{16} Hz) is used for transmission of information through fibers as transmitting medium in optical fiber communications. It can offer a large bandwidth (more than 50 THz) for data transmission. The emergence of fiber optics as a dominant technology for long-distance broadband services is discussed in this chapter. The basic configuration of optical fiber communication system comprises of an information source, a voltage-to-current converter, an optical source, a channel coupler, an optical fiber channel, an optical repeater, an optoelectronic detector, an electronic receiver, and the output device. The need, advantages, disadvantages, and wide range of applications of optical fiber communication are covered so as to generate interest to know more about the subject in subsequent chapters.

1.1 Historical Development

Optical fiber communication has been developed over the last two centuries. The first optical communication system, known as the *'optical telegraph'*, was invented in the 1790s by French engineer Claude Chappe. In 1880, Alexander Graham Bell patented the photophone—an optical telephone device for transmission of speech using a beam of light. During the 1920s, an experiment was conducted by J. L. Baird of England and C. W. Hansell of the USA, for transmission of images for TV/Facsimile systems using arrays of uncoated fiber cables. In the 1950s, A. V. Heel and H. Hopkins protected a bare glass fiber by covering it with a transparent cladding having lower index of refraction. As a result, crosstalk between fibers was greatly reduced in addition to providing protection from contamination. This led to the development of the *flexible fiberscope*, which is widely used in the medical field.

> ***Note:*** In 1956, N. S. Kapany of England coined the term 'fiber optics'. The initial applications of optical fiber were not in communications at all, because the early fibers were too lossy. Bundles of fibers were used for medical imaging to view inaccessible parts of the human body.

By 1960, attenuation of the order of 1 dB/m was achieved with glass-clad fibers. This was acceptable for medical imaging applications but not for voice/data transmissions. The invention of the lasers in the 1960s marked the beginning of a new era in modern optics, called *Photonics.* Maiman developed an experimental optical amplifier by using lasers in the electromagnetic spectrum. However, the reliability of long-distance laser links operating in the millimeter-wave region was limited mainly due to various atmospheric turbulences like clouds, rains, and fog.

> ***Note:*** The invention of the laser greatly accelerated research efforts in fiber–optic communications. A laser can operate at higher frequency, offer relatively high output optical power, and carry an extremely wideband signal. Thus, it is ideally suitable for use in high-capacity optical fiber communications networks.

In 1970, Maurer, Keck, and Schultz developed single-mode fused silica fibers, with attenuation less than 2 dB/km at the operating wavelength of 633 nm, which paved the way for fiber–optics technology for long distance optical communication. In 1977, the development mainly focused on multi-mode fibers with core diameters of 50 nm or 62.5 mm, and having a refractive index gradient between fiber core and cladding. Such fibers having attenuation of about 2 dB/km were used to transmit optical signals at 850 nm wavelength from GaAlAs laser diodes up to several kilometers without the use of signal regenerators. This was followed by the use of InGaAsP lasers at 1300 nm wavelength having fiber attenuation of 0.5 dB/km only, and reduced pulse dispersion as compared to that at 850 nm. In the early 1980s, the first long-distance transatlantic backbone networks were developed for telecommunication purpose using single-mode fiber as communication medium and optical sources at 1300 nm wavelength. This technology is followed as one of the standards for optical fiber communication networks even today.

> ***Note:*** Bell Laboratories successfully transmitted 1 billion bps through a fiber cable for 600 miles without the use of any regenerator.

1.1.1 Advances in Optical Fiber Communication

One inherent challenge to provide higher data rate in optical fiber communication is managing the dispersion effect. There are various methods employed to enhance the data rate of the fiber system. One of them is the wavelength division multiplexing (WDM) method. However, for any optical fiber system increase in the bit rate will increase the dispersion effect on the system leading to pulse broadening, causing incorrect reception of data at the receiver.

- A new generation optical fiber system operating at 1550 nm wavelength region using single-mode fibers having fiber loss of about 0.2–0.3 dB/km finds widespread applications in high-capacity submarine systems.
- Low-loss single-mode fibers enable larger repeater spacings. Moreover, submarine cables employing Erbium-doped fiber amplifiers can avoid the use of electro–optic regenerators and provide data rate capacity up to 5 Gbps.

- With the recent development of the Dense Wavelength Division Multiplexing (DWDM) technology, multiple optical signals generated by different sources (each at 10 Gbps data rate) can be transmitted simultaneously up to physical distance of approximately 400 km.

Facts to Know

Optical fiber communication has only been practical since about 1970, when glass fiber was finally made with low enough loss to be useful. The invention of the laser diode, at about the same time, helped to make optical fiber communication practical.

1.1.2 Generation of Light Wave Systems

Optical fiber communication systems or light wave systems were developed over several years in a series of generations, based on its operating wavelength and improved performance.

1. *First Generation.* In the 1970s, the earliest optical fiber communication systems were developed using infrared LED and GaAs semiconductor lasers as optical sources, a silica-based optical fiber as transmission medium and low-cost photodetectors at operating wavelength of near 850-nm region. These systems provided 50–100 Mbps data transmission rates with repeater spacings of the order of 10 km. But due to its relatively high attenuation (≈ 3 dB/km), it became less attractive subsequently.
2. *Second Generation.* In the early 1980s, optical fiber systems were developed to operate near 1300 nm wavelength region, with lower fiber loss (less than 1 dB/km, typically 0.5 dB/km). The development of InGaAsP semiconductor lasers with simultaneous oscillation of several longitudinal modes and detectors alongwith single-mode fibers, exhibiting low dispersion, offered 1–2 Gbps transmission data rates with repeater spacings higher than that of 40–50 km.
3. *Third Generation.* In the 1990s, the silica fibers were developed at 1550 nm wavelength which offered theoretical minimum attenuation of approximately 0.2 dB/km. These optical fiber communication systems offered data speeds over 2.4 Gbps with repeater spacings of 100 km or more. Systems using InGaAsP lasers operating in a single longitudinal mode and dispersion-shifted fibers could operate at 10 Gbps date rate.
4. *Fourth Generation.* With the advent of the wavelength-division multiplexing (WDM) technique for increased data rate capability and of optical amplification methods for employing greater repeater spacings, a revolution began in the development of optical fiber communication in the spectral region extending from 1450–1620 nm. By the year 2001, the light wave WDM systems used in-line erbium-doped fiber amplifiers with 60–80 km spacing for compensation of fiber losses and operation at 10 Tbps data rate.
5. *Fifth Generation.* Subsequent availability of dry fibers (single-mode dispersion-shifted), Raman amplification techniques, and optical solitons (very short optical pulses that counteract the dispersion effect due to fiber nonlinearity and thereby preserve their shape) enabled to extend the wavelength region from 1300–1650 nm for simultaneous working of thousands of WDM channels at the rate of 40–160 Gbps.

Table 1.1 depicts the important aspects of five generations of light wave systems.

Table 1.1 Five generations of light wave systems

Generation	*Wavelength (μm)*	*Fiber Type*	*Bit Rate*	*Fiber Losses (dB/km)*	*Repeater Spacings*
1st (1970s)	0.85	Multimode (graded core)	2–45 Mbps	≥ 1	≈ 10 km
2nd (Early 80s)	1.3	Multimode (graded core)	45–90 Mbps	0.5–1.0	≈ 40 km
3rd (Late 80s)	1.55	Single mode	≥1.7 Gbps	≈0.3	≈ 60–70 km
4th (Early 90s)	1.45–1.62 (Typical 1.55)	Single mode (dispersion-shifted)	2.4 Gbps	≈ 0.2	≈ 80 km
5th (In 2000s)	1.50–1.57 (Typical 1.55)	Single mode (dispersion-shifted/soliton) + Fiber Amplifier	≥2.4 Gbps	0.1–0.2	≥ 100 km

From the above data, it is observed that from the 1st generation to the 5th generation, there has been continuous improvement in the performance of optical fiber communication systems and networks. First generation optical fiber systems operated at relatively low bit rates and were essentially based on multimode fibers. Since the commercial introduction of single-mode fiber systems in public telecommunication networks, there has been an ever growing improvement in the overall performance of the optical fiber systems in all segments of telecommunications with high transmission data rate capabilities.

Facts to Know

Today, in addition to short-distance 660 nm systems, 850 nm, 1310 nm, and 1550 nm systems (for longest link lengths) are being extensively manufactured and deployed for telecommunication purposes. Longer is the operating wavelength, better is the system performance, but at a much higher cost.

1.1.3 Recent Developments in Fiber–Optics

More recent developments in the field of fiber–optics can be summarized as below:

1. Dense Wavelength Division Multiplexing (DWDM) links for enhanced capacity utilization
2. Erbium-doped Optical Fiber Amplifiers (EDFAs)
3. Dispersion compensating soliton-based optical fiber transmission techniques
4. Dispersion compensating fibers
5. Integrated optics which deals with the miniaturization and integration of various optical components such as electro–optic modulators, directional couplers, splitters, combiners, etc.
6. Use of photonic switching architectures which use integrated optical switches
7. Optical MEMs which provide data-rate transparent switching services
8. The 1625-nm based fibers, comparable with that of 1550-nm fibers

All these developments aim at achieving fiber attenuation as small as 0.16 dB/km (at 1550 nm), data transmission rates in excess of 2 Gbps, and repeater spacings of more than 200 km, and laser with lifetime of over a million hours.

Note: As a result of significant improvements in data transmission speed and repeater spacings in optical fiber systems, newer standards such as Dual Queue Distributed Bus (DQDB), Fiber Distributed Data Interface (FDDI), Synchronous Optical Networks (SONET) and Synchronous Digital Hierarchy (SDH) have also emerged.

Microwave photonics deals with optical generation, processing, distribution and photonic analog-to-digital conversion of microwave signals. It is based on the principle that a high frequency microwave signal can be generated by beating two different optical waves from two free-running laser diodes at a photodetector. The resultant frequency of the microwave signal will be equal to the difference in frequency between the two input optical waves. Thus, it is possible to generate an electrical signal having a microwave frequency in THz range, and having a high phase noise. An opto–electronic oscillator is nowadays used to generate a low phase noise microwave signal. Microwave photonics finds applications in sensor networks, radar communications, instrumentation and electronic warfare systems.

1.2 Electromagnetic Spectrum

The propagation of an optical signal (or even an electrical signal) through any transmitting medium takes place in the form of electromagnetic waves or signals.

- In a *wireline medium*, electromagnetic signals propagate along a metallic cable in the form of voltage (or current) waveforms.
- In a *wireless medium* through free space, electromagnetic signals propagate in the form of radio waves, usually termed as electromagnetic waves.
- In an *optical fiber medium*, the information signals propagate as electromagnetic light waves.

Definition of electromagnetic wave: The analog combination of electrical voltage and magnetic field propagates through air or space, and is called an electromagnetic wave or simply an 'em wave'.

By nature, radio signal transmissions take place on one radio frequency or with a very narrow bandwidth. Electromagnetic signal is distributed throughout an almost infinite range of frequencies. The useful electromagnetic frequency spectrum extends from very low frequencies (a few Hz) to several billions of Hz (10^{22} Hz). Fig. 1.1 depicts the electromagnetic frequency spectrum with various frequency bands.

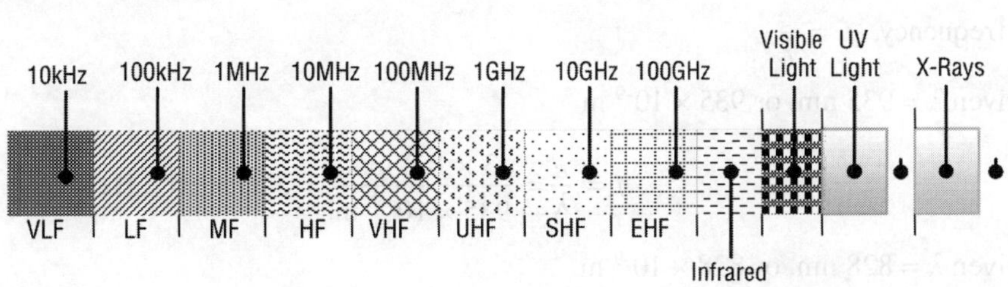

Fig. 1.1 The electromagnetic frequency spectrum

It is common to use the term wavelength rather than frequency when dealing with ultra-high frequency electromagnetic (em) waves such as light waves.

1.2.1 Relationship between Frequency and Wavelength

Definition of wavelength: The wavelength (λ in meter) is defined as the length occupied by one cycle of an electromagnetic wave in space. It is directly proportional to the velocity of propagation of light in free space ($c = 3 \times 10^8$ m/s) and inversely proportional to the frequency (f in Hz) of the em wave. Mathematically,

$$\lambda = \frac{c}{f} \tag{1.1}$$

Example 1.1 Calculation of Wavelength

Determine the wavelength for radio frequency of 100 MHz, cellular phone frequency of 1 GHz, and light wave frequency of 10^{15} Hz.

Solution:

We know that wavelength, $\lambda = \frac{c}{f}$

For f = 100 MHz, $\lambda = \frac{c}{f} = \frac{3 \times 10^8 \text{ m/s}}{100 \times 10^6 \text{ Hz}} = 3$ m **Ans.**

For f = 1 GHz, $\lambda = \frac{c}{f} = \frac{3 \times 10^8 \text{ m/s}}{1 \times 10^9 \text{ Hz}} = 0.3$ m or 30 cm **Ans.**

For $f = 10^{15}$ Hz, $\lambda = \frac{c}{f} = \frac{3 \times 10^8 \text{ m/s}}{10^{15} \text{ Hz}} = 3 \times 10^{-7}$m, or 300 nm **Ans.**

Thus, as frequency of the signal increases, the wavelength decreases.

Example 1.2 Calculation of Light Frequency

Determine the light wave frequency for the following wavelengths:

(a) 935 nm (b) 828 nm (c) 800 nm (d) 869 nm

Solution:

We know that the wavelength, $\lambda = \frac{c}{f}$

Or, the frequency, $f = \frac{c}{\lambda}$

(a) For given λ = 935 nm, or 935×10^{-9} m

$f = \frac{c}{\lambda} = \frac{3 \times 10^8 \text{ m/s}}{935 \times 10^{-9} \text{ m}} = 3.21 \times 10^{14}$ Hz **Ans.**

(b) For given λ = 828 nm, or 828×10^{-9} m

$f = \frac{c}{\lambda} = \frac{3 \times 10^8 \text{ m/s}}{828 \times 10^{-9} \text{ m}} = 3.62 \times 10^{14}$ Hz **Ans.**

(c) For given $\lambda = 800$ nm, or 800×10^{-9} m

$$f = \frac{c}{\lambda} = \frac{3 \times 10^8 \text{ m/s}}{800 \times 10^{-9} \text{ m}} = 3.75 \times 10^{14} \text{ Hz} \quad \textbf{Ans.}$$

(d) For given $\lambda = 869$ nm or 869×10^{-9} m

$$f = \frac{c}{\lambda} = \frac{3 \times 10^8 \text{ m/s}}{869 \times 10^{-9} \text{ m}} = 3.45 \times 10^{14} \text{ Hz} \quad \textbf{Ans.}$$

The velocity of electromagnetic waves differs in medium other than that in free space. Its value depends on the material and on the geometry of any waveguide structure such as optical fiber that may be present. Thus, the wavelength of a light beam (λ in meters) can be expressed as

$$\lambda = \frac{v}{f} \tag{1.2}$$

where, v = velocity of light beam in a guided medium (m/s)

f = frequency of the light beam (Hz)

It may be noted that the wavelengths of optical beams are of the order of 1 μm near the visible spectrum region. This corresponds to very short period of oscillation (the reciprocal of its frequency).

> ***Note:*** The frequency (or wavelength) of the optical signal is determined by the optical source. It does not change when the light beam (optical signal) travels from one type of material to another type of material. Instead, the velocity difference causes a corresponding change in wavelength so that the frequency remains the same.

Example 1.3 Electromagnetic Spectrum Band and Applications

Summarize various electromagnetic spectrum bands in tabular form and give typical applications for each band.

Solution:

The electromagnetic spectrum is divided into several narrower frequency ranges. Table 1.2 presents various spectrum bands along with their respective frequency range and typical application areas.

Table 1.2 Electromagnetic spectrum bands

Designation	*Frequency Range*	*Free-space Wavelength Range*	*Typical Applications*
ELF (Extremely Low Frequency)	30–300 Hz	10,000–1000 km	Power line communications
VF (Voice Frequency)	300–3000 Hz	1000–100 km	Telephone system for analog subscriber lines
VLF (Very Low Frequency)	3–30 kHz	100–10 km	Long-range navigation; submarine communications
LF (Low Frequency)	30–300 kHz	10–1 km	Long-range navigation; submarine communication radio beacons

Contd.

Contd.

Designation	Frequency Range	Free-space Wavelength Range	Typical Applications
MF (Medium Frequency)	300–3000 kHz	1000–100 m	AM broadcasting; Maritime radio; Direction finding radio
HF (High Frequency)	3–30 MHz	100–10 m	Long-distance aircraft and ship communication; Military communication; Amateur radio
VHF (Very High Frequency)	30–300 MHz	10–1 m	FM broadcasting; Two-way radio; VHF television; Aircraft navigational aids
UHF (Ultra High Frequency)	300–3000 MHz	100–10 cm	UHF television; Cellular mobile telephone; Microwave links; Radar: Personal communications systems (PCS)
SHF (Super High Frequency)	3–30 GHz	10–1 cm	Wireless local loop; Satellite communication; Radar: Terrestrial microwave links
EHF (Extremely High Frequency)	30–300 GHz	10–1 mm	Wireless local loop; specialized laboratory experiments
Infrared Light	300 GHz–300 THz	1 mm–1 nm	Infrared LANs; Consumer electronic applications; Astronomy
Visible Light	400–750 THz	0.75–0.40 nm	Optical fiber communications

Note: Electromagnetic signals higher than 300 GHz are not called radio waves; these are called rays (for example, X-rays, Gamma rays, Cosmic rays, etc.).

Note: Radio waves are invisible, whereas light waves are visible, and heat waves too can be seen as well as felt.

1.2.2 Light Frequency Spectrum

The light frequency spectrum can be divided into three general frequency bands:

1. *Infrared.* Infrared is the band of light frequencies which is quite high and cannot be seen by the human eye. Typical useful wavelengths range between 770 nm and 1600 nm. In the infrared spectrum, there are three regions (850 nm, 1300 nm, and 1550 nm) in which silica glass fibers are relatively efficient. Optical fiber systems generally operate in the infrared band.
2. *Visible.* It is the band of light frequencies (typically 390–770 nm wavelength range) which is visible to the human eye. Silica glass fibers are not very good transmitters of light in the visible

spectrum. They attenuate the light waves to such an extent that only short optical transmission links are useful.

3. *Ultraviolet.* It is the band of light frequencies which cannot be seen by the human eye. Typical wavelengths range between 10 nm and 390 nm. The fiber losses in the ultraviolet spectrum are even greater. This band is used in medical applications.

Facts to Know

Light waves and radio waves obey identical optical laws and have similar characteristics but the light waves occupy a much higher frequency range. All electromagnetic waves travel at a velocity of 3×10^8 m/s in free space and possess electric and magnetic fields associated with them.

Units of wavelengths with light frequencies

- It is customary to express the wavelength in *microns* (1 micron = 10^{-6} m, or 1μm), or in nanometers (1 nm = 10^{-9} m) with light frequencies.
- With optical spectrum, the unit *angstrom* is sometimes used to express the wavelength (1 angstrom = 10^{-10} m, or 0.0001 micron).

Example 1.4 **Expressing Wavelength in Angstroms**

Determine the wavelength in angstroms units for the light wave signal having frequency equal to 3.45×10^{14} Hz.

Solution:

We know that wavelength, $\lambda(m) = \dfrac{c(\text{m/s})}{f(\text{Hz})}$; and 1 angstrom (Å)= 10^{-10} m

For given $f = 3.45 \times 10^{14}$ Hz, $\lambda = \dfrac{c}{f} = \dfrac{3 \times 10^8 \text{ m/s}}{3.45 \times 10^{14} \text{ Hz}} \times 10^{10} = 8695$ Å **Ans.**

Example 1.5 **Converting Wavelength Å to nm**

Determine the light wave frequency corresponding to specified wavelength as 780 Å.

Solution:

We know that the wavelength, $\lambda(\text{m}) = \dfrac{c(\text{m/s})}{f(\text{Hz})}$, or, the frequency, $f(\text{Hz}) = \dfrac{c(\text{m/s})}{\lambda(\text{m})}$

First we have to convert the given wavelength in angstrom (Å) in meters.

We know that 1 Å = 10^{-10} m

Therefore, 780 Å = 780×10^{-10} m = 7.8×10^{-8} m

Hence, $f(\text{Hz}) = \dfrac{3 \times 10^8 (\text{m/s})}{7.8 \times 10^{-8} (\text{m})} = 3.85 \times 10^{15}$ Hz **Ans.**

1.2.3 Optical Bandwidth

Definition: The optical bandwidth of an optical fiber system is the range of frequencies of transmitted optical signal where the output optical power falls to 50% or 0.5 of its maximum value (corresponds to a loss of 3 dB, usually called half-power point).

In an optical fiber communication system, the power delivered by an optical source is directly proportional to the current supplied to it. Thus, the half-power point is equivalent to half-current point. In terms of currents, the optical bandwidth of an optical fiber system is the range of frequencies of transmitted optical signal where the output optical current falls to 50% or 0.5 of its maximum value.

1.2.4 Optical versus Electrical Bandwidth

We know that the electrical bandwidth of an electronic (all-electrical) communication system signifies the range of frequencies of transmitted electrical signal where the output power falls to 50% or 0.5 of its maximum value (half-power point). But in an electrical system, the power delivered is directly proportional to the square of the root-mean-square (RMS) value of the current. Thus the half-power point on a power scale corresponds to 0.707 (not equivalent as in optical fiber system) on a current scale.

From the above discussions, the relationship between optical bandwidth $(\Delta f)_{opt}$ and electrical bandwidth $(\Delta f)_{el}$ of an optical fiber communication system can be defined as

$$(\Delta f)_{opt} = \frac{(\Delta f)_{el}}{0.707} = \sqrt{2}\,(\Delta f)_{el} \tag{1.3}$$

Therefore, the optical bandwidth of an optical fiber system is more than its electrical bandwidth.

Facts to Know

Optical wavelengths are so small that most devices used in a fiber system have dimensions of many wavelengths. This is contrary to the situation at radio frequencies, where device sizes can be a wavelength or even less.

Section Practice Problems

1. Determine the wavelength in nanometers for the light wave signal having frequency equal to 3.21×10^{14} Hz. *[Ans.: 935 nm]*
2. Determine the wavelength in angstroms for the light wave signal having frequency equal to 3.62×10^{14} Hz. *[Ans.: 8287 Å]*
3. Determine the light wave frequency corresponding to the following specified wavelengths
   (a) 9350 Å *[Ans.: 3.21×10^{14} Hz]*
   (b) 8690 Å *[Ans.: 3.45×10^{14} Hz]*
   (c) 8280 Å *[Ans.: 3.62×10^{14} Hz]*
4. A commonly used wavelength for optical communication is 1550 nm.
   (a) What type of light spectrum is this?
   (b) Calculate the frequency corresponding to this wavelength, assuming free-space propagation.
   [Ans.: (a) Infrared; (b) 1.94×10^{14} Hz]

1.3 Optical Power Basics

Definition of optical power: It is described as the flow of light energy past a given point in the system in a specified time.

In fact, the optical power measures the rate at which electromagnetic waves transfer light energy. It is expressed in joules per second, or watts. Typical optical power values generated by light sources range from tens of microwatts to more than 100 milliwatts. Optical power is generally stated in decibels relative to a defined power level, such as 1 mW (dBm) or 1 μW (dBμ).

Note: dBm stands for an absolute power level with reference to fixed constant reference power level as 1 mW, and dBμ stands for an absolute power level with reference to fixed constant reference power level as 1 μW.

Mathematically,

$$dBm = 10\log\frac{P\ (mW)}{1\ mW} \tag{1.4}$$

$$dB\mu = 10\log\frac{P\ (\mu W)}{1\ \mu W} \tag{1.5}$$

Facts to Know

The design of an optical fiber communication link involves keeping a track of the optical power along the communication link from source to detector. The measurement of relative power levels in dB is convenient.

Example 1.6 Expressing Optical Power Levels

Express the optical power levels of 1 mW and 10 μW in dBm and dBμ units.

Solution:

(a) Expressing in dBm

We know that $dBm = 10\log\frac{P\ (mW)}{1\ mW}$

For P = 1 mW, $dBm = 10\log\frac{1\ mW}{1\ mW} = 0\ dBm$ **Ans.**

For P = 10 μW, $dBm = 10\log\frac{10\ \mu W}{1\ mW} = -20\ dBm$ **Ans.**

(b) Expressing in dBμ

We know that $dB\mu = 10\log\frac{P\ (\mu W)}{1\ \mu W}$

For P = 1 mW or 1000μW, $dB\mu = 10\log\frac{1000\mu W}{1\ \mu W} = 30\ dB\mu$ **Ans.**

For P = 10 μW, $dB\mu = 10\log\frac{10\mu W}{1\ \mu W} = 10\ dB\mu$ **Ans.**

It is important to take care while adding or subtracting different power levels and gains/losses (expressed in decibels) in a communication system. For example, the transmitted power level, $P_t(dBm)$, the system loss, $L(dB)$, and received power level, $P_r(dBm)$, are related by

$$P_r(dBm) = P_t(dBm) + L(dB) \tag{1.6}$$

It may be noted here that loss is always taken as negative value.

1.3.1 Transmission Efficiency

In general, efficiency of transmission between any two points in a communication link is given as the ratio of power at second point to that at the first point. Let us say the power at one point in the system is P_1 watts, and at some other point farther away along the communication link is P_2 watts ($P_2 < P_1$ because of transmission losses); then the ratio P_2 / P_1 (both expressed in the same unit) is termed as the fraction of the power transmitted between these two points, or power loss, or transmission efficiency. This ratio can be expressed in dB as

$$L(dB) = 10\log\frac{P_2}{P_1} \tag{1.7}$$

It may be noted that both power levels P_1 and P_2 must be expressed in the same unit (i.e., watt or milliwatt).

Example 1.7 Compute Received Power
An optical source radiates 2 mW power. Compute the power level (in mW) at the input of optical receiver if the system losses accumulate to 23 dB.

Solution:
We know that the received power level is given as

$$P_r(dBm) = P_t(dBm) + L(dB)$$

For given transmitter power of 2 mW, $P_t(dBm) = 10\log\frac{2\ mW}{1\ mW} = +3\ dBm$

The transmitted power is reduced by the system loss of 23 dB, that is, the received power is 23 dB less than the transmitted power. So, $L = -23$ dB.

Therefore, $P_r(\text{dBm}) = +3 + (-23\text{ dB}) = -20\text{ dBm}$

The corresponding power in mW can be computed by using the expression

$$P_r(dBm) = 10\log\frac{P_r(\text{mW})}{1\text{ mW}}$$

$$\Rightarrow \quad P_r(\text{mW}) = 10^{\frac{P_r(\text{dBm})}{10}} = 10^{\frac{-20}{10}} = 0.01\text{ mW}$$ **Ans.**

Example 1.8 Transmission Power Efficiency
A system has 23 dB of power loss. Compute its transmission power efficiency.

Solution:

We know that transmission power efficiency = $\frac{P_2}{P_1}$

Using the expression $L(dB) = 10\log\frac{P_2}{P_1}$; we have $\frac{P_2}{P_1} = 10^{\frac{L}{10}(dB)}$

For given L = –23 dB, $\frac{P_2}{P_1} = 10^{-\frac{23}{10}(dB)} = 0.005$, or 0.5% **Ans.**

1.3.2 Photon Energy

The performance of optical fibers can be analyzed completely by application of Maxwell's equations, which is quite complex. According to Maxwell theory, electromagnetic radiations contain a series of oscillating electric and magnetic fields in quadrature (at 90° angles). However, according to Einstein and Planck, when light is emitted or absorbed, it behaves not only like an electromagnetic wave but also as a tiny particle, known as a photon, which possesses energy proportional to its wavelength or frequency.

We know that an atom has several energy levels, also called states, the lowest of which is known as the ground state. Any other energy level above the ground state is known as an excited state. If an atom in a particular energy state falls (decays) to a lower energy state, there is a certain loss of energy (expressed in electron volts). This loss of energy is emitted as a photon of light. In other words, the photon energy is equal to the difference between the energy levels of two energy states. The process of decaying of an atom from one energy state to another energy state is called spontaneous emission.

It implies that atoms can also be illuminated by a light source whose energy is equal to the difference between an energy state and the ground state. This can cause an electron to change from one energy state to another energy state by absorbing light energy. The process of movement of electron from one energy level to another higher in energy is called absorption. When making the transition from one energy level to another, the atom absorbs a packet of energy (a photon). This process is similar to that of emission.

As stated earlier, the energy emitted (i.e., photon) or absorbed is exactly equal to the difference between higher and lower energy levels, or simply the two energy levels. Mathematically,

$$E_p = E_2 - E_1 \tag{1.8}$$

where, E_p is the photon energy in joules.

As per Planck's law, when high-frequency electromagnetic radiation in the visible light spectrum illuminates a metallic surface, electrons are generated (emitted). This phenomenon is known as the photoelectric effect. The emitted electrons produce an electric current. So, according to Planck's law, the photon energy in joules is,

$$E_p = hf \tag{1.9}$$

where, h represents Planck's constant (= 6.626×10^{-34} joules–sec), and f represents the frequency of photon generated (Hz).

Photon energy may also be expressed in terms of wavelength. Using the expression $\lambda = c/f$, or $f = c/\lambda$, we have

$$E_p = \frac{hc}{\lambda} \tag{1.10}$$

Note: A convenient unit of energy is the electron-volt (eV). The relationship between electron-volts and joules can be expressed as 1 eV = 1.6×10^{-19} J.

Example 1.9 Photon Energy in eV

Show that the energy in one photon at a wavelength of 1 μm is 1.24 eV.

Solution:

We know that the photon energy, $E_p = \frac{hc}{\lambda}$

where, $h = 6.626 \times 10^{-34}$ joule–sec; $c = 3 \times 10^8$ m/s.

For given $\lambda = 1 \mu m$, or $1 \times 10^{-6} m$, we get

$$E_p = \frac{6.626 \times 10^{-34} (J - s) \times 3 \times 10^8 (m/s)}{1 \times 10^{-6} m} = 1.99 \times 10^{-19} J$$

We know that 1 eV = 1.6×10^{-19} J. This simply means

$$1J = \frac{1eV}{1.6 \times 10^{-19}} = 6.25 \times 10^{18} eV$$

Therefore, $E_p = 1.99 \times 10^{-19} \times 6.25 \times 10^{18} eV = 1.24 \, eV$ Ans.

Example 1.10 Particle Nature of Light

Consider an optical source operating at wavelength of 0.8 μm. It generates 1 μW optical power. Determine the number of photons incident on a photo detector in

(a) 1 s (b) 1 ns

Solution:

We know that the photon energy, $E_p = \frac{hc}{\lambda}$

For given $\lambda = 0.8$ μm, we have

$$E_p = \frac{6.626 \times 10^{-34} (J \times s) \times 3 \times 10^8 (m/s)}{0.8 \times 10^{-6} (m)} = 2.48 \times 10^{-19} J$$

By definition, power signifies the rate at which photon energy is delivered.

Therefore, total energy radiated by optical source, $E_T = P_{opt} \times t$

(a) For the given $P_{opt} = 1 \mu W$, the equivalent energy in 1 s is $E_T = 1 \mu J$, or 1×10^{-6}J

Therefore, the number of photons required to make up $E_T = 1 \times 10^{-6}$J will be

$$N_p = \frac{E_T}{E_p} = \frac{1 \times 10^{-6} J}{2.48 \times 10^{-19} J/photon} \approx 4 \times 10^{12} \, photons$$ **Ans.**

(b) For the given $P_{opt} = 1\mu W$, the equivalent energy in 1ns is $E_T = 1 \times 10^{-6} \times 10^{-9}$J

Therefore, the number of photons required to make up $E_T = 1 \times 10^{-15}$J will be

$$N_p = \frac{E_T}{E_p} = \frac{1 \times 10^{-15} J}{2.48 \times 10^{-19} J / photon} \approx 4000\, photons$$ **Ans.**

Example 1.11 Optical Power versus Photons

Let there be 10^{10} number of photons per second that are incident on a photodetector at 800 nm wavelength. Determine the incident power on the photodetector.

Solution:

We know that the number of photons per second is given by the relationship $N_p = \frac{E_T}{E_p}$

where E_T is the total energy emitted by optical source or incident on the detector, and E_p is the photon energy given as $E_p = \frac{hc}{\lambda}$.

For given $\lambda = 800$ nm, we have $E_p = \frac{6.626 \times 10^{-34} (J \times s) \times 3 \times 10^8 (m / s)}{800 \times 10^{-9} (m)} = 2.48 \times 10^{-19} J$

For given $N_p = 10^{10}$ photons, $E_T = N_p \times E_p = 10^{10} \times 2.48 \times 10^{-19} = 2.48 \times 10^{-9} J$

By definition, power is the rate of change of energy at which it is delivered. That is,

$$P_{opt} = \frac{E_T}{t}$$

Therefore, the power incident on the detector in 1-sec, $P_{opt} = 2.48 \times 10^{-9} W$ **Ans.**

Facts to Know

Light is often interpreted in different ways to explain different observations and experiments. It has been established that at some time light behaves as an electromagnetic wave, and at some other time it behaves as a particle.

Section Practice Problems

1. Express the optical power levels of 0.1 mW and μW in dBm and dBμ units. *[Ans.: -10 dBm, -40 dBμ; -30 dBm, 0 dBμ]*
2. Convert -20 dBm and 10 dBμ into equivalent μW values. *[Ans.: 10μW, 10μW]*
3. Prove that 0 dBm = 30 dBμ.
4. Find the energy of a single photon in *eV* if the wavelength is specified as 0.8 μm. *[Ans.: 1.55 eV]*
5. Compute number of photons per second received at the input of a photodetector if the incident power is 1 nW at wavelength 1.3 μm. *[Ans.: 6.54 x 10^9 photons/sec]*

6. The power incident on a photodetector is 100 nW. Find the number of photons per second if the operating wavelength is
(a) 800 nm (b) 1550 nm
Which wavelength requires more number of photons per second to produce 100 nW of power?
[Ans.: (a) 4 x 10^{11} photons/s; (b) 7.8 x 10^{11} photons/s; 1550 nm]

1.4 Need of Optical Fiber Communications

The information-carrying capacity of any electronic communications system is directly proportional to the channel bandwidth. In an electronic communications system, the transmitter superimposes (modulates) low-frequency information signal on a radio frequency carrier signal. The modulated RF carrier signal is then transmitted through wireless or guided medium. The receiver retrieves the original information from the carrier signal. The carrier frequencies, in fact, restrict the information-carrying capacity as well as the rate of transfer of information. Increasing the carrier frequency, therefore, increases the transmission bandwidth which helps to eliminate these limitations.

The optical frequency is typically 10^{14} Hz, as compared to that of microwave frequency of 10^9 Hz. Thus, the optical carrier can offer 100,000 times more bandwidth. In addition, the optical region of the electromagnetic spectrum ranges from 50 nm (ultraviolet radiation) to about 100 μm; the visible portion lies in the 400–700-nm region. In optical fiber communication systems, the carrier frequencies are selected from the optical region (particularly the infrared part, ranging between 1700–800 nm).

Consequently, the only practical type of optical communications system is one that uses optical fibers as transmitting medium. For all practical purposes, optical fiber cables have an infinite bandwidth. In other words, they have the capacity to carry much more information than metallic cables or even the wireless communications.

Facts to Know

Transmission of light waves for any useful distance through the earth's atmosphere is impractical because water vapor, oxygen, and particles in the air absorb and attenuate the signals at light frequencies.

Example 1.12 Information Carrying Capacity

With the help of suitable data, show that the light frequencies used in optical fiber communication systems, have much larger bandwidth utilization ratio.

Solution:

Definition of bandwidth utilization ratio: Bandwidth utilization of an analog communications system is the ratio of system bandwidth to its carrier frequency, and is often expressed in percentage. For instance, a VHF radio communications system operating at a carrier frequency of 100 MHz with 10-MHz bandwidth has a bandwidth utilization ratio of 10%.

(i) For 10% bandwidth utilization ratio, a microwave radio communication system operating at a carrier frequency of 10 GHz would have 1 GHz of bandwidth available.

(ii) Light frequencies used in optical fiber communications systems are between 1×10^{14} Hz and 4 ×

10^{14} Hz (100,000–400,000 GHz). A bandwidth utilization ratio of 10% would have a bandwidth between 10,000–40,000 GHz.

Obviously, the higher the carrier frequency, the more the bandwidth available and greater the information-carrying capacity.

Optical frequencies, being of the order of 10^{14} Hz, can handle information signals requiring very high transmission bandwidths. Since optical communication through atmosphere requires optical line-of-sight (LOS), the applications were limited to short distance communication such as across a highway or hilly terrain or satellite-based deep space applications. Subsequently, the use of fibers as transmitting medium was considered a practical reality for long distance communication. The advantages of fiber optics can be linked to the wave nature of light which can offer a diverse range of communication possibilities such as:

- Due to wide bandwidth, many signals can be incorporated onto a single optical fiber channel, without much interference among them.
- The optical fiber systems can be made compact and efficient, mainly due to use of guided wave devices to control light transmitted via fibers.
- Many design and maintenance problems are eliminated due to non-existence of electromagnetic interferences.

> ***Note:*** Large information capacity, inherent immunity from electro-magnetic interference (EMI), security, increased repeater spacing, cost effectiveness, and convenience of operation are some of the factors that justify the need of optical fiber communication.

1.5 Light Wave System Components

Fig. 1.2 illustrates a simplified block diagram of a light wave communication system, also known as optical fiber communication system.

A brief description of each functional block is given below:

1. *Information Source.* The source information may be in the form of non-electrical, physical form such as voice or image/video. An input transducer is a device that converts physical information into an electrical signal. The input signal can either be an analog or digital (computer data).
2. *Voltage-to-current converter.* It serves as an electrical interface between the information source circuit and the optical (light) source. The amount of light emitted by the light source is generally proportional to the amount of its drive current. Thus, the voltage-to-current converter is necessary to convert an input signal voltage to a current that is used to drive the light source.
3. *Optical source.* In an optical transmitter, the optical carrier signal generated by optical source is modulated by an analog or a digital signal. The optical source is either a light-emitting diode (LED) or an injection laser diode (ILD), which generates an electromagnetic wave in the infrared region of the optical spectrum. In essence, the light intensity is modulated by the input signal. The optical sources are generally compact, lightweight, consume moderate amount of power, and are relatively easy to modulate.

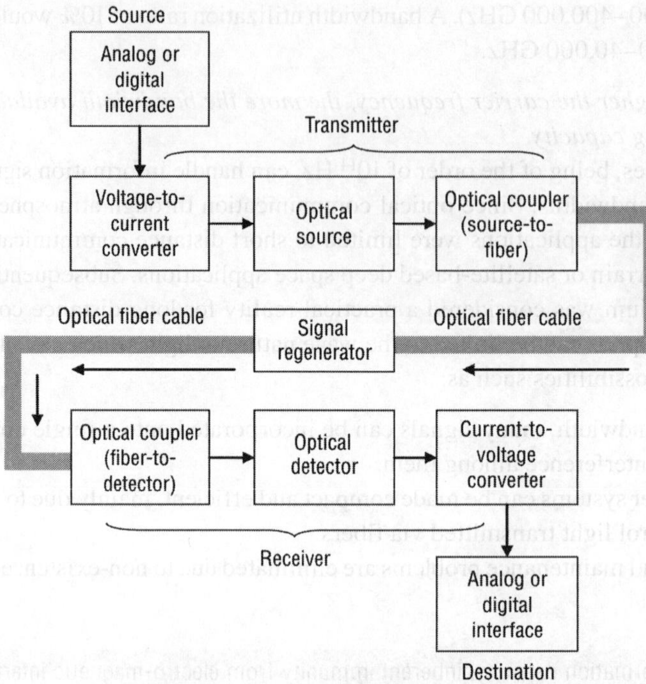

Fig. 1.2 Light wave system components

4. *Optical couplers.* The function of source-to-fiber coupler is to collect the light signals from the optical source and send it efficiently to the optical fiber cable. Similarly, the fiber-to-detector coupler is used at the other end of the fiber cable to direct the received light signals onto the photodetector.
5. *Optical fiber cable.* It is the guided transmission medium, which is either an ultrapure glass or a plastic cable. The optical fiber consists of a glass or plastic fiber core surrounded by a cladding and then encapsulated in a protective jacket. Techniques have been developed for the production of fibers with very low transmission losses (a few tenths of a dB/km at 1300 nm and 1550 nm optical wavelengths).
6. *Optical signal regenerators.* As the optical signals (in the form of intensity-modulated light pulses) propagate along the lengths of the optical fiber cable from the source to destination, they get attenuated (due to absorption, scattering, etc.) as well as broadened (due to dispersion). As a result, the signals may become weak and indistinguishable after a certain distance. Optical regenerators (or optical amplifiers, such as erbium-doped fiber amplifiers) are used at appropriate distances from the transmitter along the length of the fiber cables which help to restore the strength and shape of transmitted signal.
7. *Optical detector.* The optical detector is generally a *p–i–n* (*p*-type-intrinsic-*n*-type) diode, an avalanche photodiode (APD), or a phototransistor which converts an input optical signal into an equivalent electrical signal, usually in the form of electric current. The resultant output current is normally proportional to the incident optical signal level and hence to the input information signal. The optical detectors are generally compact, consume low power, and have flat spectral response, and long operating life.

8. *Current-to-voltage converter.* It transforms variations in photodetector current to corresponding variations in voltage. It produces an output voltage which is proportional to the original source information.
9. *Destination output.* Finally, the received information is presented in a form similar to that of input information source and suitable for destination device such as loud speaker, computer, or other machines.

> ***Note:*** The analog or digital interfaces at the input of transmitter and output of receiver are electrical interfaces that match impedances and signal levels between the information source and destination to the input and output circuitry of the optical fiber communication system.

1.6 Optical Fibers as a Communication Channel

An optical fiber is essentially a waveguide for light, usually in infrared spectrum. It consists of a core and a cladding that surrounds the core. Both are made of transparent material, either glass or plastic, but the main difference is in their index of refraction. The material used in cladding has lower refractive index than that used in the fiber core. This causes rays of light leaving the core to be refracted back into it which help the light beams to propagate automatically and continuously in the forward direction down the fiber. The actual fiber used in optical fiber communications is a very thin strand of material such as glass or plastic. This fiber has very little mechanical strength. So it is enclosed in a protective jacket that is usually made of plastic. Fig. 1.3 illustrates a typical cross-sectional view of the fiber.

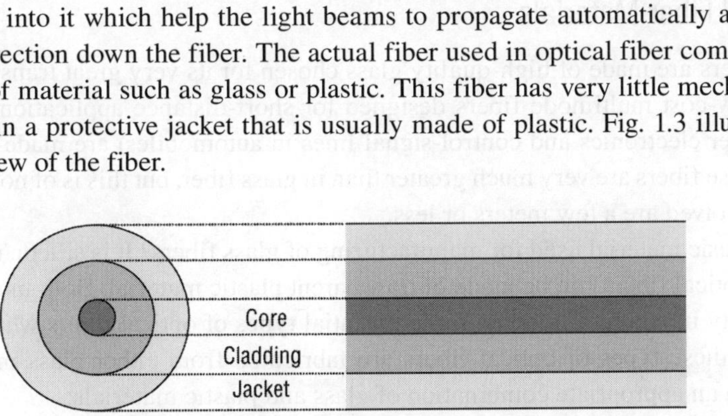

Fig. 1.3 Cross-sectional view of the optical fiber

1.6.1 Parts of Optical Fiber Cable

The fiber portion in an optical fiber cable is generally considered to include both the fiber core and its cladding. A special lacquer, silicone, or acrylate coating is generally applied to the outside of the cladding to seal and preserve the fiber's strength. Fig. 1.4 shows different parts of the optical fiber cable.

The coating also helps protect the fiber from moisture, which reduces the possibility of the occurrence of a detrimental phenomenon called stress corrosion (sometimes called static fatigue) caused by high humidity. Moisture causes silicon dioxide crystals to interact, causing bonds to break down and spontaneous fractures to form over a prolonged period of time. The protective coating is surrounded by a buffer jacket, which provides the cable additional protection against abrasion and shock. Materials commonly used for the buffer jacket include steel, fiberglass, plastic, flame-retardant polyvinyl chloride, Kevlar yarn, and paper. The buffer jacket is encapsulated in a strength member, which increases the tensile strength of the overall cable assembly. Finally, the entire cable assembly is contained in an outer polyurethane jacket.

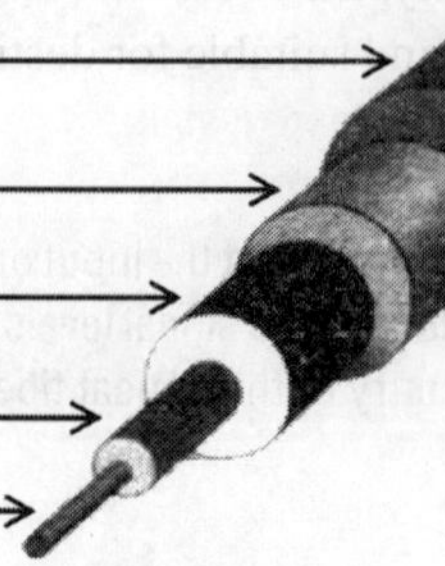

Fig. 1.4 Parts of optical fiber cable

> ***Note:*** Often, two or more fibers are included in one cable for increased bandwidth and redundancy in case one fiber breaks. It is also easier to build a full-duplex system using two fibers, one for transmission in each direction, than to send signals in both directions along the same fiber.

1.6.2 Optical Fiber Materials

Most optical fibers are made of high-quality glass chosen for its very great transparency to reduce losses. Some low-cost multimode fibers designed for short-distance applications (such as optical links in consumer electronics and control-signal lines in automobiles) are made of acrylic plastic. The losses in these fibers are very much greater than in glass fiber, but this is of no importance when the distances involved are a few meters or less.

What is the basic material used for manufacturing of glass fibers? It is silicon dioxide. However, some types of optical fibers can be made of transparent plastic material. Both materials are readily available in plenty in nature. There are three essential types of optical fibers which are commonly used today. All these types of optical fibers are fabricated from either glass material, or plastic material, or even an appropriate combination of glass and plastic materials:

- Optical fibers with plastic core and cladding, known as plastic fibers.
- Optical fibers with glass core and plastic cladding, known as plastic-clad silica (PCS) fibers.
- Optical fibers with glass core and cladding, known as silica-clad silica (SCS) fibers.

(a) *Optical fibers with plastic core and cladding.* Plastic fibers (fibers with plastic core and cladding) are more flexible and, consequently, more rugged than glass. Therefore, plastic cables are easier to install, can better withstand stress, are less expensive, and weigh approximately 60% less than glass. However, plastic fibers have higher attenuation characteristics and do not propagate light as efficiently as glass. Therefore, plastic fibers are limited to relatively short cable lengths, such as within a single building.

(b) *Optical fibers with glass core and plastic cladding (PCS fibers).* Generally, fibers with glass cores have less attenuation than plastic fibers. Therefore, plastic-clad silica fibers with glass core are also less affected by external electromagnetic radiations, thereby exhibiting greater immunity to interference.

(c) *Optical fibers with glass core and cladding (SCS fibers).* This type of optical fibers offers the best propagation characteristics. But they are the least rugged, and are more susceptible to external electromagnetic radiations and may lead to increase in signal attenuation.

Note: Photonic crystal fibers (PCFs), a new class of optical fibers, basically combine properties of classical fibers and 2D photonic crystals. They can guide light using photonic bandgap (PBG) mechanism, in addition to total internal reflection. They have a hexagonal lattice structure which help to propagate light along the fiber in defects (realized by removing central capillaries) of its crystal structure. PCFs have relatively high index of refraction. It is possible to design PCFs with zero or low dispersion at visible wavelengths, or with flattened dispersion over a very large optical range.

1.6.3 Protective Material

Generally, optical fiber cables have very low tensile (pulling) strength. For this reason, the fiber is reinforced with strengthening material so that it can withstand mechanical stresses it may typically undergo when being pulled and jerked through underground and overhead ducts and hung on poles during installation.

Materials commonly used to strengthen and protect fibers from abrasion and environmental stress are:

- Steel
- Fiberglass
- Plastic
- FR–PVC (flame-retardant polyvinyl chloride)
- Kevlar yarn
- Paper

Fig. 1.5 shows a pictorial view of optical fiber cable illustrating its parts.

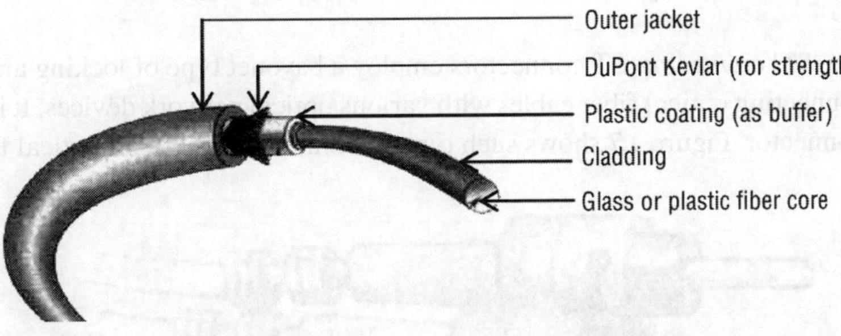

Fig. 1.5 A pictorial view of optical fiber cable

As it can be seen, the fiber part is at the center of the optical fiber cable. Basically, the fiber comprises of core and cladding which is responsible for propagation of light through it. The fiber is surrounded by plastic coating in order to provide sufficient cushion to the fiber, DuPont Kevlar (a strong material similar to that used in bulletproof vests) strands to provide strength, and finally an outer jacket made of either Teflon or PVC material. The type of cable construction used depends on the performance requirements of the system and both economic and environmental constraints.

> ***Note:*** For a given application, the selection of an optical fiber depends on specific requirements of the system. In addition to technical specifications, one has to consider overall economics as well as logistical aspects.

1.6.4 Connectors used with Optical Fiber

In an optical fiber communication system, connectors are needed for using optical fibers. Special connectors are required to couple the light from source to fiber cable at the transmitter end, and fiber cable to detector at the receiver end. In general, connectors used with optical fibers can be categorized in two different versions. These are: a fiber connector to provide flexible and detachable connections between the optical transmitter (or the optical receiver), and a fiber splice for providing permanent joint between two segments of optical fibers (usually 5–10 km long). Connectors are discussed in this section, whereas fiber splicing is covered in the next section. Depending on applications, three different types of connectors are used with optical fiber cables. These are:

1. *Subscriber channel (SC) connector.* SC connectors employ push–pull type of locking arrangement. Fig. 1.6 shows such type of connector used with optical fiber cables. SC connectors are mostly used in cable TV applications.

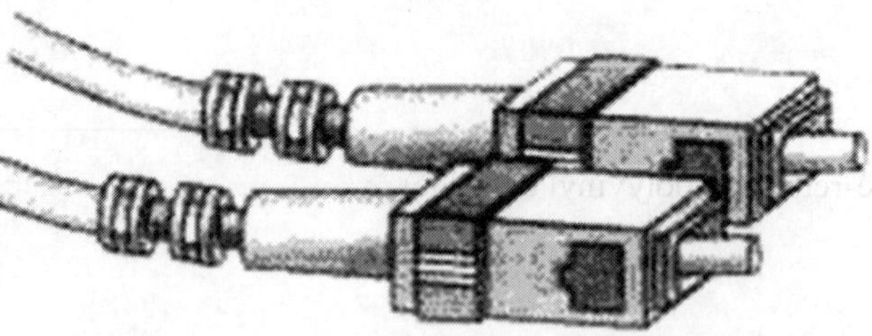

Fig. 1.6 Subscriber channel (SC) connector

2. *Straight-tip (ST) connector.* ST connectors employ a bayonet type of locking arrangement and used for connecting optical fiber cables with various optical network devices. It is more reliable than SC connector. Figure 1.7 shows such type of connector used with optical fiber cables.

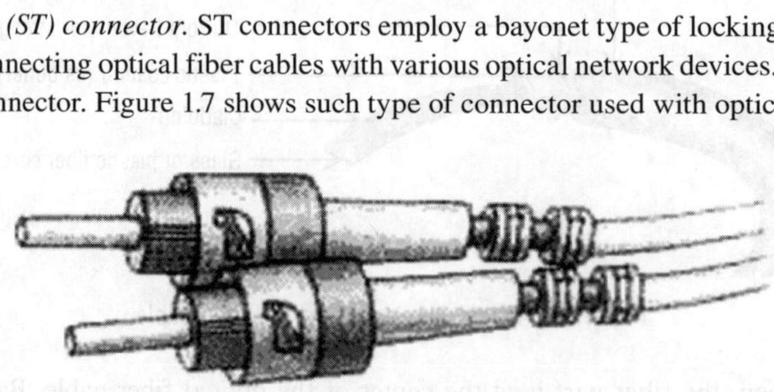

Fig. 1.7 Straight-tip (ST) connector

3. *MT–RJ connector.* It has a size similar to that of a standard RJ45 type connector used in telecommunications. Fig. 1.8 shows one such optical fiber cable which supports 1 Gbps data rate over 100 m distance.

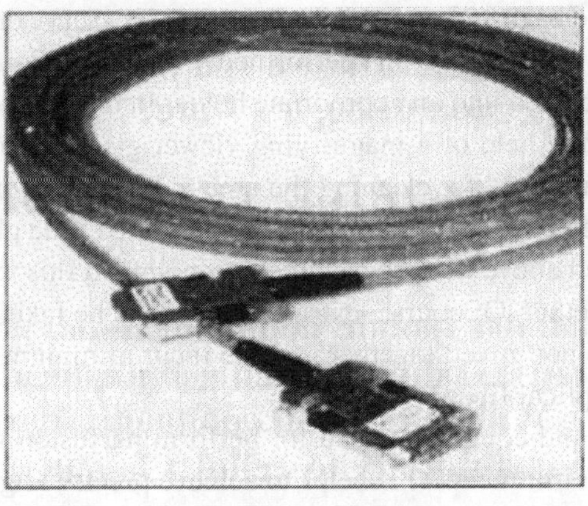

Fig. 1.8 An optical fiber cable with connector

It is worth mentioning here that any type of connector does have certain insertion loss. Typical insertion loss of connector may be on the order of about 0.3 dB.

1.6.5 Fiber Splicing

As stated earlier, fiber splicing is a technique to join two pieces or segments of optical fibers on permanent basis. Fiber splices are generally needed when the available optical fiber cable is not long enough for the intended application. This is due to the fact that usually one continuous length of optical fiber cable extends up to about 5 km only. So, in order to meet the requirement of 10 km length of optical fiber in a fiber–optic link, two different fiber lengths each having 5 km length is spliced together.

There are two techniques of fiber splicing:

- Mechanical splicing
- Fusion splicing

For an easy and quick process of fiber splicing, the *mechanical splicing* technique is normally used for joining two segments of optical fibers permanently. The first step in mechanical fiber splicing process is that the outer protective jacket of two optical fiber cable is stripped back, cleaned and then a precision cleave or cut (i.e., the cut on the fiber should be exactly at 90° to the fiber axis) is performed. In the second step, these ends of the fibers which are required to be spliced are placed together into a sleeve with accurate alignment. This helps to maximize the level of light transmission. Next, the two fiber segments are clamped in the right place. It is followed by the usage of a refractive-index matching gel which enhances the transmission of light across the fiber splicing. The major benefit of mechanical fiber splicing is that it takes very less time (e.g., may be about just 5 minutes) to make. Of course, it is accompanied with about 10% loss of the level of light. However, this loss is still less than that encountered while using a fiber connector for joining two segments of fiber cables.

The other type of fiber splice is known as the *fusion splicing.* In this technique, the two ends of the optical fibers are fused or melted together by using specialized equipment. For example, an electric arc may be used to weld two optical fiber cables together to perform fusion splicing. First, as in the mechanical splicing technique, the outer protective jacket from the ends of two optical fiber cables need to be spliced is stripped back and cleaned. Then a precision cleave or cut (i.e., the cut on the fiber should be exactly at 90° to the fiber axis) is performed with a precision cleaver tool. Next, the two optical fibers are properly placed into an appropriate holder in the optical fiber splicer. This enables automatic alignment with the help of a magnifying viewer glass for inspection. If needed, small electrical sparks may be employed for cleaning the area to be spliced of any dust. It is followed by applying a much intensive electric spark so that the temperature of the glass used for manufacturing the optical fiber is increased above its specified melting point. At this temperature, the two ends of optical fibers are fused together. Of course, extra care needs to be taken so that the molten form of fiber core and cladding are not mixed together so as to incur minimum light loss when propagated through two segments of optical fibers.

Nowadays all the tools and equipment used for performing fusion splicing are operated under the computer controlled environment. This helps to achieve precise alignment of the optical fibers, thereby exhibiting very low levels of loss, as low as 0.1 dB. However, the process of fusion splicing is quite expensive.

1.7 Advantages of Optical Fiber Cables

Optical communications through glass or plastic fibers offer several advantages over conventional metallic transmission media for both telecommunication and computer networking applications. Typical advantages of using optical fiber cables are briefly described below:

(i) *Larger bandwidth and greater information capacity.* Due to inherently available wider bandwidths at light frequencies, optical fibers have greater information-carrying capacity than that can be obtained with metallic cables as transmission medium. Typically, bandwidths up to several thousand GHz are available with optical fibers.

(ii) *Lower transmission loss.* Typical signal loss in modern sophisticated optical fibers is as small as a few-tenths-of-a-dB loss per km. As a result, optical amplifiers and regenerators can be spaced considerably farther apart as that can be offered by metallic transmission lines.

(iii) *Security.* Due to inherent property of optical fiber cable for propagation of light through it, it is almost impossible to tap the data flowing into an optical fiber cable without the knowledge of the user. Thus, we can say that optical fiber cables provide much higher data integrity and security than metallic cables. Moreover, it is not possible to detect the presence of optical fiber cables installed under the ground with metal detectors provided steel is not used alongwith fiber cables for reinforcement.

(iv) *Immunity to static noise.* Static noise usually occurs due to electromagnetic interference (EMI). It is mainly caused by various sources of man-made noise that include lightning, fluorescent lights, electric motors, relays, and other electrical appliances. Since fiber cables are nonconductors of electrical current, they do not radiate electromagnetic energy as well.

(v) *Immunity to crosstalk.* Optical fibers are made of glass and plastic materials which fibers are known to be nonconductors of electricity. Therefore, they are immune to crosstalk.

(vi) *Immunity to environmental variations.* Optical fiber cables tend to be more resistant to environmental and climatic conditions (including weather variations) than metallic cables. Optical fiber cables can also operate over a wider temperature range.

(vii) *Reliability.* Optical fiber cables are more reliable than metallic cables and last longer because they exhibit higher tolerance to changes in environmental conditions and are immune to corrosive materials including liquids and gases.

(viii) *Easier to install and maintain.* Optical fiber cables, in general, are quite easier to install as well as to maintain than metallic cables. Optical fibers are compact and much more lightweight than metallic cables. Consequently, they are more flexible, require less storage space, cheaper to transport, and easier to work with.

Facts to Know

Modern optical fiber communication systems can transmit thousands of Gbps data over hundreds of kilometers distance between source and destination. This allows millions of individual voice/data channels to be multiplexed together and propagated using one common optical fiber cable.

1.8 Disadvantages of Optical Fiber Cables

The difficulty to make connections to optical fiber cable is one of the major disadvantages of optical fiber cables versus metallic cables. Due to its small size and compactness, an optical fiber cable tends to be extremely susceptible to getting cut or bruised, thereby causing permanent damage during manufacturing or installation activities. Although the installation costs for optical fiber cables are drastically coming down by more than 60% a year, the installation cost of optical fiber cables is still relatively high. Special optical test equipment such as optical test probes and optical time domain reflectometer (OTDR) are required to be used for measurement of specified parameters as well as fault diagnostic purpose at most of the fiber endpoints. Certain disadvantages of using optical fiber cables are listed below.

(i) *Lower tensile strength.* Glass fiber is quite fragile as compared to copper wire, making it cumbersome to transport. As such optical fiber cables have considerably lower tensile strength than that exhibited by RF coaxial cables, which can be improved by Kevlar coating and a protective PVC jacket.

(ii) *Susceptible to bending losses.* Bending the optical fiber cable causes irregularities in the cable dimensions. Since electromagnetic waves propagate through it by total internal reflection, slight bending of cable results in a loss of signal power.

(iii) *Prone to manufacturing defects.* Excessive loss of optical signal power is experienced even with the minor manufacturing defect of the optical fiber cable. As a result, this may cause imperfect total internal reflection mechanism.

(iv) *Interfacing with electronic devices.* Optical fiber cables must be connected to standard electronic devices for communication purposes, which make the interfacing expensive.

(v) *Difficulty in locating faults.* Because of no electrical continuity, it is extremely difficult to locate physical or technical faults in optical fiber cables and maintain its proper functioning throughout the operating period.

(vi) *Need of specialized tools.* Special tools are needed to splice and repair optical fiber cables. In addition, special measuring test equipment is needed for making regular measurements by trained professionals.

(vii) *Reaction by chemicals.* The glass fiber is easily affected by number of chemicals such as hydrogen gas. This is really a serious concern while deploying optical fiber cables in underwater applications.

> ***Note:*** Occasionally, it is needed to use electrical power to remotely located optical equipments such as regenerators. This cannot be accomplished with the optical cables. So additional copper cables must be included alongwith optical fiber cables.

1.9 Applications

Optical Fiber Systems include processing of information before it is delivered to the communication channel and after it reaches the receiver, exactly the same way as Electronic Communications Systems. This allows incorporation of fibers into systems originally conceived for electrical signal transmission with only moderate modifications. For example, compatibility of optical fiber cables with existing structure of the telephone system. Similarly, optical fiber cables are used to transmit television signal, voice as well as data. The intensity of light wave is modulated that undergoes number of total internal reflections to reach destination where it is demodulated and the message is recovered.

1. Optical fiber cable has wide bandwidth and is widely used in backbone networks because it is capable of transferring data at a rate of 1600 Gbps. Moreover, it provides a cost-effective solution as transmission medium.
2. A hybrid CATV network is creating by using a combination of RF coaxial cable and optical fiber cable by some cable TV companies. RF coaxial cable is used to connect the end user directly. On the other hand, optical fiber cable is used as the backbone configuration. This type of arrangement offers an economical solution because the end user usually requires narrow bandwidth as compared to relatively very high bandwidth of an optical fiber cable.
3. The small size and large information-carrying capacity of optical fibers make them viable alternatives to traditional twisted-pair copper cables as trunk lines in modern telecommunication networks.
4. Optical fiber cables are also used in several types of local area networks (LANs). Examples of such LANs include 100Base- and 1000Base- Fast Ethernets.
5. Usually optical fiber cables have lower attenuation than that in a coaxial cable. This leads to greater repeater spacing in an optical fiber communication links. This is the reason that underwater optical fiber links are designed to span the oceans. More advanced systems use lower-loss fibers and optical amplifiers to reduce (or eliminate) the need for repeaters.
6. Because of the relative ease of transporting and laying the fibers due to low-weight as compared to coaxial cables, optical fiber cables have distinct edge for their use in submerged cable applications.
7. Due to availability of very large bandwidth, "fibered city" such as Hi-OVIS (Highly Interactive Optical Visual Information System) can provide reliable connectivity to home computers and video equipment provide live TV programs, recorded audio/video programs, etc by using optical fiber cables.
8. Optical fiber links are compatible with electrified railway tracks because they do not suffer from electromagnetic interference.

9. Optical fiber video transmission successfully competes with coaxial cable for surveillance and remote monitoring systems due to its EMI rejection and low susceptibility to lightning damage. Examples of such applications include surveillance of power-generating stations, parking areas, critical control points along railroad pathways, and the perimeter of military installations.
10. Fiber sensors have been used to measure temperature, pressure, linear and rotary positions, and liquid levels; for examples, the Optic Gyroscopes and Fiber Hydrophones.

Facts to Know

Optical fibers are safer to operate. Due to the non-conducting behaviour of glass and plastic fibers, no electrical voltages or currents can be associated with them. They can also be used around various types of gasses as well as volatile liquids without any risk of fires or explosions.

Besides telecommunications, fiber–optics enabled cable TV and local area networks are becoming very popular. A great deal of research has taken place in the area of synchronous optical networks (SONET) which synchronize optical and electrical interfaces, and fiber distributed data interface (FDDI). With the availability of ultra-narrow line-width single-mode lasers as optical source, coherent optical transmission and transmission of optical solitons through low-loss optical fibers (less than 0.01 dB/km) have now become a reality.

Points to Remember

- Fiber optics is a branch of optics that deals with the study of propagation of light (rays or beam) through optical fibers (transparent dielectric waveguides made up of glass or plastic).
- Optical fibers can be considered as light waveguides or photon conductors which can be constructed from transparent dielectrics such as plastic and glass materials.
- Optical fiber cables comprises of an inner core made of glass or plastic which is completely surrounded by cladding, and then properly encased in a jacket.
- Data signals are carried by optical fiber cables in the form of light using the principles of total internal reflection.
- Optical fiber has many advantages over copper cable for communications, including larger bandwidth, greater distance between repeaters, lower weight and smaller size, immunity from electrical interference, and even lower cost.
- Optical fiber communication is gaining widespread popularity due to its inherent advantages such as low attenuation of transmitted signals, resistance to noise, and high-bandwidth information-carrying capabilities.
- Nowadays optical fiber cable is widely deployed in Fast Ethernet networks, cable TV networks, and telecommunication backbone networks.

Important Equations

The wavelength, $\lambda = \frac{c}{f}$; *where c represents the velocity of light in free space (3×10^8 m/s) and f denotes the frequency in Hz.*

The energy of the photon, $E_p = hf$; *where h represents the Planck's constant (= 6.626×10^{-34} joules-sec) and f denotes the frequency of light (photon) emitted in Hz.*

Key Terms with Definitions

Electron–volt (eV) It is the energy given to or absorbed by an electron that moves through a potential difference of one volt (l eV = 1.6 × 10^{-19} joule).

Core In an optical fiber, the central part of the fiber in which the light propagates.

Cladding The material of lower refractive index that surrounds the fiber core in an optical fiber.

Photon A quantum of electromagnetic radiation.

Quantum The smallest amount in which energy can exist; the size of a quantum depends on the wavelength of the energy.

Short Answer Type Questions

1. What are typical advantages of using radio waves as means of transmitting wireless signals?

Radio waves provide the most common and effective means of transmitting wireless signals by using radio transmissions because no physical medium is required.

- Radio waves in the electromagnetic spectrum do not have distance limitations.
- Radio waves can penetrate non-metallic objects of any size, unlike light waves.
- Radio waves can travel much greater distances unlike light and heat waves.

2. Is light wave suitable for atmospheric propagation? Give example.

Unlike microwave, the light wave is greatly absorbed by atmosphere and is not suitable for atmospheric propagation. However, if LASER (Light Amplification by Simulated Emission of Radiation) is used as a light source, then it can penetrate atmosphere due to its high intensity and narrow beam width. But it necessitates that transmitter and reflector need to be perfectly aligned, which limits its use. Infrared remote control, infrared computer links are typical examples where light is used as communicating media through atmosphere.

3. In dealing with the emission and absorption of light, neither the particle theory nor the wave theory of light is appropriate. Justify that quantum theory explains all phenomena involving the transmission of light.

Electromagnetic energy can appear only in multiples of a discrete unit known as *quanta*. These quanta are called *photons* when the energy is radiated. A *photon* is an entity that is somewhere between a wave and a particle. It has a wavelength corresponding to the radiation and an amount of energy equal to one quantum, but no mass. The energy of the photon is given as $E_p = hf$; where h is Planck's constant (= 6.626 × 10^{-34} joules-sec) and f is frequency of light (photon) emitted in Hz. When light is incident on an atom, a photon can transfer its energy to an electron within this atom, thereby exciting it to higher energy level. Conversely, an electron in an excited state can drop to a lower energy state separated from it by an energy equivalent to that of a photon. Thus, the quantum theory indicates that optical radiation has particle as well as wave properties.

4. What do you understand by an optical fiber? Give its main parts in brief.

An optical fiber is an extremely thin strand of ultra-pure glass designed to transmit optical signals (in form of light pulses) from the optoelectronic source (LED or ILD) to the optoelectronic detector (semiconductor *p–i–n* or avalanche photodiode). In its simplest form, it consists of three main regions:

1. *Core* – a solid cylindrical region of diameter 8–100 μm
2. *Cladding* – a coaxial cylindrical region of diameter normally 125 μm
3. *Protective coating* – a primary or buffer coating of plastic to give strength

The refractive index of the cladding is necessarily kept lower than that of the fiber core made from the same type of material so that the light can propagate through it by following the basic principle of total internal reflection.

5. **What do you understand by an optical fiber communication link?**
   In an optical fiber communication link, the optical signal (carrying information) traverses along the cable consisting of a single fiber or a bundle of optical fibers. The key sections of system are a transmitter section comprising of an optical source (LED or Laser) and its associated drive circuitry, an optical fiber cable, and a receiver consisting of a photodetector along with amplification and signal-restoring circuitry. Additional components include couplers, regenerators, optical amplifiers, splices and connectors. The optical fiber cable is one of the most important elements in an optical fiber communication link. It can be installed in ducts, undersea, or buried directly in the ground.

6. **When two metallic conductors are placed physically near to each other, there is a possibility of crosstalk between them. What is its primary cause? What basically limits information-carrying capacity in metallic cables?**
   The primary cause of crosstalk between metallic conductors located physically close to each other is the changing magnetic field due to flow of current in them. The optical fiber cables do not carry electric currents. The non-electrical nature of the signals on optical fiber makes them immune to crosstalk between cables. The primary electrical constants (resistance, inductance, and capacitance) cause metallic cables to act like low-pass filters, which in turn limit their transmission frequencies, bandwidth, bit rate, and results in smaller information capacity.

7. **What are essential requirements in selecting materials for optical fibers so that fiber cable can function as reliable information channel?**
   The essential requirements in selecting materials for optical fibers are:
   1. It must be possible to make long, thin, flexible fibers from the material selected for optical fibers.
   2. It is desirable that the material to be used for fabrication of optical fibers should be transparent at a specified optical wavelength which can enable the fiber to guide light efficiently.
   3. It should offer low attenuation for the light frequencies being transmitted through it.
   4. The availability of physically compatible but having slightly different refractive indices materials must be ensured for the core and cladding in an optical fiber.
   5. It should offer a large light-gathering capacity.
   6. It should provide low dispersion in order to ensure minimum distortion of the propagating signal.

   Examples of materials that satisfy these requirements are glasses (consisting of either silica SiO_2 or a silicate), and plastics. High attenuation glass fibers with large cores are widely used for short-distance transmissions, whereas low-attenuation glass fibers are used for long-haul telecommunication applications. Plastic fibers are used in short-distance applications and in abusive environment.

8. **Optical fiber cable has several advantages over twisted pair or RF coaxial cable. List at least six unique advantages.**
   1. *Higher bandwidth.* Optical fiber cable is capable of supporting considerably higher transmission bandwidths (and thereby data rates) in comparison to that possible with other wireline transmission mediums such as twisted-pair cable, coaxial cable, etc. However, optimum utilization of bandwidth as well as requirement of very high transmission bit rates using optical fiber cable as medium of signal propagation are primarily constrained by the available optical signal generation and reception technologies.

2. *Low attenuation.* Due to low fiber attenuation, it is possible to achieve significantly greater transmission distances as compared to what can be obtained with any other guided medium. For example, an optical signal can be propagated for over 50 km transmission distance without the need of any signal regeneration. Whereas the transmission of electronic signal over twisted-pair cable or coaxial cable may require the use of repeaters after every 5 km distance.
3. *Immunity to electromagnetic interference (EMI).* Optical fiber cable as transmission medium is not affected by any type of external electromagnetic noise.
4. *Resistance to corrosion.* Glass material is mostly used as fiber material in optical fiber cables. It happens to be more resistant to corrosion as compared to that of copper material used in fabrication of twisted-pair or coaxial cable.
5. *Light weight.* Optical fiber cables are much lighter than copper cables.
6. *More immune to tapping.* Optical fiber cables are definitely more immune to tapping than copper cables which act as antennas that can easily be tapped.

9. There are certain limitations of using optical fiber cables. List at least three such limitations.

1. *Installation/maintenance.* Installation and maintenance of optical fiber cable need expertise.
2. *Unidirectional.* Propagation of light is unidirectional. Two optical fibers are needed for bidirectional communication.
3. *Cost.* The cable and the interfaces are relatively more expensive than those of other guided media. If the demand for bandwidth is not high, the use of optical fiber cannot be justified.

10. For long-haul communication links, optical fiber cables are economical to transport and much easier to lay (install) than metallic RF coaxial cables. Justify it with the help of example data.

A typical optical fiber cable has a fiber diameter of 125 μm enclosed in a plastic sheath with an outer diameter of 2.5 mm. The weight of this cable is 6 kg/km; the attenuation is 5 dB/km. The RG-19/U coaxial cable has an outer diameter of 28.4 mm. Its weight is 1110 kg/km; the attenuation is 22.6 dB/km at 100 MHz. Smaller and lighter coaxial cables have higher attenuation.

11. What do you understand by the terms RFI and EMI? Do fiber cables offer better RFI and EMI rejections?

Radio Frequency Interference (RFI) refers to interference caused by radio and television broadcast stations, radar, and other signals originating in electronic equipment. Electromagnetic Interference (EMI) refers to interference caused by natural phenomena such as lightening, or caused by man-made sources such as sparking. If RFI and EMI are not rejected, then these undesired signals could increase the system noise level beyond acceptable limits. Fiber cables offer excellent RFI and EMI rejection because of its ability to isolate itself from its environment.

12. Briefly describe some practical consequences because of insulating nature of a fiber. Can optical fiber cables operate near nuclear installation?

Optic fibers, glass or plastic, are insulators. There cannot be any flow of electric current through them. Moreover, fibers exhibit better rejection of radio-frequency interference (RFI), electro-magnetic interference (EMI) and electro-magnetic compatibility (EMP). In an environment in which high-voltage lines are present, there is no possibility of short-circuit or sparking which could damage a wire communication link severely. Moreover, optical coupling eliminates the need for a common ground between a fiber transmitter and receiver. It is also possible to repair the fiber while the system is on without causing any problem to the electronics at the transmitter and receiver. Since fibers are insulators, they will not propagate or pick up electromagnetic pulses (EMP) caused by nuclear explosions that can induce millions of volts in a conducting transmission line causing damage.

13. A fiber is well protected from interference and coupling with other electrical or optical communications channels. Also fibers offer a degree of security and privacy. Comment.

The light wave carrying information is trapped within the fiber. So, none leaks out during transmission to interfere with signals in other fibers. Conversely, light cannot couple into the fiber from its side. Fibers do not radiate the energy within them. So it is difficult for an intruder to detect the signal being transmitted. The fibers need to be physically broken, or a new fiber has to be fused to the transmitting fiber, to access the optical beam. In such activity, the signal power reaching the receiver would drop significantly. A sensitive receiver can measure this loss, and can provide sufficient knowledge about the location of occurrence of intrusion.

14. Why is loose-tube construction preferred when cables must be pulled through ducts?

In loose-tube cables, all the stress of cable pulling is taken up by the cable's strength members. The fiber is also free to expand and contract with temperature at a different rate from the rest of the cable. Loose-tube cables tend to be relegated to applications where their greater strength is important, such as telephone cables that have to be pulled for long distances through ducts.

15. List some examples where optical fiber is used as a substitute for a copper cable or a point-to-point microwave radio link.

Although the operating principle of optical fiber is that of a waveguide, it is used in practice as a substitute for a copper cable (either coaxial or twisted-pair) or a point-to-point microwave radio link. It is found in many applications; a few random examples include telephone cables, point-to-point transmission of television signals, and computer networks. In general, optical fiber cable has greater bandwidth than coaxial cable. Greater bandwidth provides the fiber cable the ability to handle greater data rates. Moreover, the increased bandwidth allows more signals to be multiplexed. Optical fibers can be built with lower loss than copper cables, increasing the allowable distance between repeaters. The fiber cable itself can be less expensive.

16. How is the use of optical fiber links comparable with that of radio links? How is fiber optics useful in medical applications?

Radio links such as point-to-point microwave links and geostationary satellite channels can be replaced with fiber optics. Radio links have the advantage of avoiding the laying of cables. This advantage is especially attractive in the case of satellites, which require no access to the terrain between source and destination (separated by thousands of kilometers). On the other hand, the delay of about one-half second between transmission and reception is a considerable nuisance in telephony via satellite. Optical fibers have greater bandwidth, and, of course, they are much more private. In medical electronics applications, fiber optics is often used to isolate the patient from circuits connected to the electrical power line and thus avoid shock hazards.

17. Give at least one application in which optical fibers cannot substitute for copper cable or microwave waveguide.

One such application is the transmission of signal power. For instance, optical fiber cables cannot be used to connect a transmitter to an antenna. Fiber optics are used with power levels in the milliwatts range and are strictly for the transmission of information, not energy.

18. Summarize the major application areas of fiber optics technology.

Fiber optics technology will have a major impact on number of application areas. Some are listed below for quick reference.

1. *Voice Communication:* Inter-office, Intercity, Intercontinental links
2. *Data Transfer:* Computers, LANs, Inter-office data links, Satellite Earth stations
3. Internet: Email, Access to webpages, Videoconferencing
4. *Video Communication and Entertainment:* TV broadcast, CATV, HDTV, Video phones, Video on demand, Video games, Wired city

5. *Industrial Applications:* Robotics, Dedicated and distributed sensors, Smart structures, Monitoring of power-generating stations and manufacturing plants
6. *Education:* Distance learning, Access to digital libraries, CCTV
7. *Healthcare:* Biomedical sensors, Endoscopes, Remote monitoring of patients, Minimal invasive diagnosis/surgery/therapy
8. *Transportation:* Traffic control in metro cities and high-speed electrified railways, Monitoring of aircrafts
9. *Business development:* Videoconferencing, Industrial CAD/CAM
10. *Military:* Tactical communication, Guided missiles, Sensors, Virtual wars

19. What is the type of optical source and type of fiber chosen for FDDI networks?

The FDDI (Fiber Distributed Data Interface) uses light-emitting diodes (LEDs) as source, operating at 1.3-μm wavelength optical transmitters. The FDDI network uses multimode fibers for fiber-optics LANs deployed in ring topology. The network can operate at typical data rate of 100 Mbps.

20. Where do we use mechanical splicing and fusion splicing to join two pieces of optical fiber cables?

The mechanical fiber splicing technique is used for those applications which need very quick splices. Some of the sleeves used for mechanical splicing allow easy connection and disconnection if the need arises. In other words, mechanical splicing may be deployed when less permanent splicing may be required. On the other hand, fusion splicing technique offers a low-loss solution with a high degree of permanence slicing. But due to high cost of fusion splicing equipment, they are mostly used for the long-haul higher data rate capacity telecommunication links based on optical fiber techniques.

Multiple Choice Questions

1. In optical fiber communications, the signal source is ____________ waves.

A. Light B. Infrared
C. Radio D. Very low-frequency

2. Which one of the following is not a guided medium of transmission?

A. Fiber–Optic cable B. Coaxial cable
C. Twisted-pair cable D. The atmosphere

3. An operating environment has many high-voltage devices. What would be the best medium of transmission?

A. The atmosphere B. Twisted-pair cable
C. Optical fiber D. Coaxial cable

4. Which of these converts the electrical signal to optical signals?

A. Optical photo detectors B. Demultiplexers
C. Multiplexers D. Optical modulators

5. Fiber optic system has three basic components, in the order. They are:

A. light guide, light source, light detector B. light source, light guide, light detector
C. light detector, light source, light guide D. light guide, light detector, light source

6. In optical fiber, the outer layer is ____________ and inner layer is ____________.

A. core, cladding B. cladding, core
C. transmit, reflect D. reflect, transmit

7. Optical fiber cables are highly immune to EMI because information is carried by:
A. light
B. electrical means
C. magnetic means
D. acoustic means

8. Which one of the following is based on laser beam technology?
A. Magnetic tape
B. Terminals
C. Optical disks
D. Keyboards

9. ________________ method allows a large number of selectable and independent user channels to co-exist on a single optical fiber link?
A. PCM
B. FDM
C. TDM
D. CDM

10. Usually various types of transmission media are categorized as:
A. Metallic or nonmetallic
B. Guided or unguided
C. Determinate or indeterminate
D. Fixed or unfixed

11. ________________ is a guided medium.
A. Microwave
B. Radio
C. Fiber-optic cable
D. Atmosphere

12. Which mechanism is used in Laser Technology for generation of light?
A. Dispersion
B. Absorption
C. Stimulated Emission
D. Spontaneous Emission

13. Optical splice provides a connection between
A. transmitter to fiber
B. receiver to fiber
C. fiber to fiber
D. fiber to repeater

14. Optical fibers are highly immune to EMI. Which one of the following four statements justifies it?
A. They transmit signals in as light rather than electric current.
B. They are readily shielded by outer conductors in cable.
C. They are too small for magnetic fields to introduce current in them.
D. Magnetic fields cannot penetrate the glass of the fiber.

15. In an optical fiber, the fiber core ________________ the cladding.
A. is denser than
B. has the same density as
C. is less dense than
D. is another name for

16. The material used for fabrication of inner core of an optical fiber is
A. glass or plastic
B. bimetallic
C. copper
D. liquid

17. Unlike wired media, optical fibers are highly resistant to
A. refraction
B. low-frequency transmission
C. electromagnetic interference
D. high-frequency transmission

Keys to Multiple Choice Questions

1. A	**2.** D	**3.** C	**4.** D	**5.** B	**6.** B	**7.** A	**8.** C	**9.** B	**10.** B
11. D	**12.** C	**13.** C	**14.** A	**15.** A	**16.** A	**17.** C			

Review Questions

1. The different generation of light wave system improves the performance of an optical communication system. Discuss.
2. In which form the signal is propagated in an optical fiber cable? How does this differ from the signal in twisted-pair cable and coaxial cable?
3. Give at least one advantage and one disadvantage of optical fiber communication link as compared with a geostationary satellite radio link.
4. The wave theory of light adequately accounts for all phenomena involving the transmission of light. Briefly describe the quantum nature of light which indicate that light radiation has particle as well as wave properties.
5. Outline the basic functional blocks of an optical fiber communication system. Which is the most vital component in the link and why?
6. Draw a functional block diagram of light-wave communication system and describe the function of various components.
7. Briefly describe the construction of an optical fiber cable with the help of suitable illustration, showing all parts of the cable.
8. What are the functions of the core and cladding in an optical fiber? Why should their refractive indices be different? What would happen if the light is propagated in the fiber core without cladding?
9. Contrast glass and plastic fiber cables. Describe the different application areas where glass and plastic fiber cables are used.
10. Why do optical fiber cables require strength members? What materials are most commonly used to add strength to fiber cables?
11. Discuss various types of materials that can be used for the manufacturing of optical fiber cables.
12. List four possible benefits of using optical fiber cables against twisted-pair or coaxial cables. Give reasons to support your answer.
13. What are the typical disadvantages of optical fiber as a transmission medium in short-distance and long-haul applications?
14. Contrast and compare the advantages and disadvantages of optical fiber cables and metallic cables.

Numerical Problems

1. Determine the wavelength in nanometers for the following light frequencies.
   (a) 3.21×10^{14} Hz *[Ans.: 935 nm]*
   (b) 3.62×10^{14} Hz *[Ans.: 828 nm]*
2. Determine the wavelength in angstroms for the light wave signal having frequency equal to 3.62×10^{14} Hz. *[Ans.: 8287 Å]*

3. Assuming free-space propagation, determine the frequency corresponding to each of the following given wavelengths of light.
(a) 400 nm (b) 670 nm
(c) 700 nm (d) 900 nm
[Ans.: (a) 750 THz; (b) 448 THz (c) 429 THz; (d) 333 THz]

4. Determine the wavelengths in nanometers and angstroms for the following light frequencies:
(a) 3.8×10^{14} Hz (b) 3.5×10^{14} Hz
(c) 3.2×10^{14} Hz
[Ans.: (a) 789 nm, 7890 Å; (b) 857 nm, 8570 Å; (c) 937 nm, 9370 Å]

5. Tabulate the wavelengths and respective region of the electromagnetic spectrum that include each one of these frequencies: 20 Hz, 50 Hz, 10^3 Hz, 2×10^4 Hz, 10^6 Hz, 10^9 Hz, 10^{10} Hz, and 10^{14} Hz.

6. Compute the bandwidth of the visible spectrum, that is, the difference between the highest and lowest visible frequencies. *[Ans.: 3.2×10^{14} Hz]*

7. The output power of an optical transmitter is 5 mW. What will be the input power at the optical receiver if the total system loss is 20 dB? *[Ans.: 0.05 mW]*

8. Prove that -20 dBm = 10 dBμ.

9. What is the difference (in watts) between -60 dBm and 60 dBm? *[Ans.: 1000W]*

10. An optical fiber receiver requires minimum -34 dBm power level. The total system losses add up to 31 dB from the optical source to receiver. How much power (in mW) is emitted by the optical source? *[Ans.: 0.5 mW]*

11. If the speed of light in fiber is 2×10^8 m/sec, what is the bandwidth of a fiber that passes light from 1000 nm to 1500 nm without significant loss in magnitude? *[Ans.: 6.67×10^{13} Hz]*

12. Two fiber cables are connected together, each having loss of 4-dB and the splice used between them has 2-dB loss. If the optical power at the input is 2 mW, compute the optical power output of combined fiber cables. *[Ans.: 0.2 mW]*

13. An optical receiver requires an input power of 1 nW. How much optical power must be transmitted by the source if the total system losses add up to 50 dB? *[Ans.: 10 mW]*

14. Consider an optical fiber system where the fiber losses are 25 dB, the light-source-to-fiber coupling loss is 15 dB, the connector losses are 5 dB. The system has a single optical amplifier having a gain of 20 dB. Compute the net loss (in dB) in the system. *[Ans.: 25 dB]*

15. The optical transmitter generates a power level of +4 dBm. If the optical fiber receiver has a specified sensitivity of -38 dBm (minimum power level at the input of receiver so as to produce acceptable information signal). The system losses are caused by inefficient coupling from transmitter into the fiber, connector and splice losses, and fiber losses. Calculate the allowable system losses. *[Ans.: 42 dB]*

16. Compute the energy of a photon at 1.3 μm. Planck's constant, $h = 6.626 \times 10^{-34}$ joules-sec. *[Ans.: 1.5×10^{-19} J]*

17. Compute the photon energy at 1.3 μm, 0.82 μm, and 0.6 μm wavelengths. Which one has more energy - a visible photon or an infrared photon?
[Ans.: 1.5×10^{-19} J, 2.4×10^{-19} J, 3.3×10^{-19} J; Visible photons]

18. There are 10^{10} number of photons that are incident on a photodetector in one second at given wavelength of 800 nm. If the conversion rate of photodetector from incident light to output electric current is 0.65 mA/mW, how much current is generated? *[Ans.: 1.6 nA]*

19. Calculate the energy contained in one photon of a light wave at a specified wavelength of 400 nm. Express the result in both joules and electron-volts. *[Ans.: 4.97×10^{-19} J; 3.11 eV]*

20. A particular digital communication system is operated at a data rate equal to 1% of the carrier frequency. Estimate the allowed data rate at operating optical wavelength of 1000 nm. How is it comparable with radio communication that operates at carrier frequency of 1 GHz? *[Ans.: 3000 Gbps; 100 Mbps]*

CHAPTER

Basics of Optical Fibers 2

Chapter Objectives

After studying this chapter, you should be able to

- ✓ describe the mechanism of propagation of light through an optical fiber cable;
- ✓ define modes of propagation and index profile;
- ✓ know optical fiber configuration types such as single-mode step index, and multimode step index and graded index;
- ✓ analyze key parameters such as critical angle, numerical aperture, and bandwidth–distance product for optical fiber;
- ✓ explain various types of losses that incur in optical fiber cables.

An optical fiber is the core component of an optical fiber communication link. Popularly known as optical fiber cables, they are the most promising type of guided transmission medium for virtually all forms of digital and data communications applications. With optical fibers, electromagnetic light waves propagate through the media composed of a transparent material without using electrical current flow. Optical fibers are mostly made of glass or plastic material having properties such that the phenomena of total internal reflection takes place that enables light waves to propagate within it in a properly guided manner similar to that of electromagnetic waves through a metallic transmission medium. This chapter begins with an easy-to-understand ray model of the propagation of light through optical fibers. It is followed by a discussion on the concept of modes and the modal analysis of step-index as well as the graded-index type of fibers. Finally, the type of losses and dispersions are explained to assess the limitation of optical fibers.

2.1 Review of Optical Ray Theory

In essence, an optical fiber communications system is one that uses light (optical signal) as the carrier of analog or digital information signal. Propagating light waves, carrying information, through the earth's atmosphere is difficult and often impractical. The optical energy in a light wave follows narrow paths, called light rays or beams. For most practical applications, the light rays are used to describe a number of optical phenomena geometrically. In fact, ray theory is known as geometric optics. It is these rays (geometrical paths traversed by light) which actually carry the optical energy.

2.1.1 Velocity of Propagation

Electromagnetic energy, such as light waves, travels at a velocity of $c = 3 \times 10^8$ m/sec approximately in free space (a vacuum). Moreover, the velocity of propagation is the same for all light frequencies in free space. However, it has been demonstrated that

- All light frequencies are not propagated with the same velocity.
- Since materials are denser (possess higher refractive index) than free space, electromagnetic waves travel slower in materials than in free space.
- When the velocity of an electromagnetic wave is reduced as it travels from one medium to another medium of denser material, the light ray refracts (i.e., bends or changes direction) toward the normal.
- Likewise, when an electromagnetic wave travels from a denser material into a lighter one, it gets refracted away from the normal.

It may be recalled that the normal line is simply an imaginary line which is drawn perpendicular to the intersection of two different materials at the point of incidence of the light rays.

Fig. 2.1 shows how a light ray is refracted (bent) as it passes from a less dense material into a more dense one (may be of the same type but having different refractive indices). (Actually, the light ray changes its direction of propagation at the interface of two different materials, not bent.)

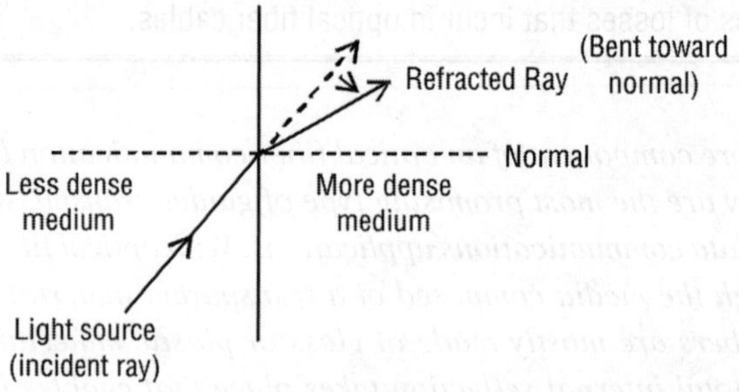

Fig. 2.1 Refraction of Light

For light-wave frequencies, electromagnetic waves travel through the earth's atmosphere (air) at approximately the same velocity as through vacuum (that is, the velocity of light).

2.1.2 Refractive Index

Definition of refractive index: It is the ratio of the velocity of propagation of a light ray in free space to that of in a specified material.

The extent of refraction that occurs at the intersection of two different materials having different values of index of refraction can be exactly predicted. Mathematically, refractive index, n is expressed as

$$n = \frac{c}{v} \tag{2.1}$$

where, c = velocity of propagation of a light ray travelling in a free space (3×10^8 m/sec)

v = velocity of propagation of a light ray travelling in a specified material (m/sec)

Note: Refractive index is dimensionless, that is, it does not have any unit. Although the refractive index also varies with the frequency of incident light, yet this variation is quite insignificant in most of the lightwave applications.

Table 2.1 depicts the value of index of refraction of several commonly used materials.

Table 2.1 Typical values of index of refraction

S. No.	*Type of Material*	*Index of Refraction (dimensionless)*
1.	Vacuum	1.0
2.	Air	1.0003 (approximately 1)
3.	Water	1.33
4.	Ethyl alcohol	1.36
5.	Fused silica	1.46
6.	Silica glass	Typical ≈1.5
7.	Diamond	2.0–2.42
8.	Indium phosphide (InP)	3.21
9.	Gallium arsenide (GaAs)	3.35
10.	Silicon (Si)	3.5
11.	Indium gallium arsenide phosphide (InGaAsP)	3.51
12.	Aluminum gallium arsenide (AlGaAs)	3.6
13.	Germanium (Ge)	4.0

Note: The refractive index of any material varies with a number of parameters including wavelength and temperature. The values given in Table 2.1 are not exact values under all operating conditions. However, these values are close enough to actual values to be used for all meaningful calculations.

Example 2.1 Wavelength of Light in Glass

Let the wavelength of light in free-space is 900 nm. Calculate the wavelength of the light that propagates in glass material having a refractive index of 1.5.

Solution:

First we have to determine the frequency of light corresponding to given wavelength of 900 nm using the relationship between frequency (f), velocity of light in free-space ($c = 3 \times 10^8$ m/s), and the wavelength (λ) as $f = \dfrac{c}{\lambda}$

$$f = \frac{c}{\lambda} = \frac{3 \times 10^8 \, m/s}{900 \times 10^{-9} \, m} = 3.33 \times 10^{14} \text{ Hz}$$

Next, the velocity of propagation of light in specified glass material of refractive index 1.5 can be determined using the expression, $n = \frac{c}{v}$. That is,

$$v = \frac{c}{n} = \frac{3 \times 10^8 \, m/s}{1.5} = 2 \times 10^8 \, m/s$$

Therefore, the wavelength of the light source in glass,

$$\lambda_g = \frac{v}{f} = \frac{2 \times 10^8 \, m/s}{3.33 \times 10^{14} \, Hz} = 6 \times 10^{-7} \text{ m} = 600 \text{ nm} \quad \textbf{Ans.}$$

Alternately,

$$\lambda_g = \frac{\lambda}{n} = \frac{900 \, nm}{1.5} = 600 \text{ nm} \quad \textbf{Ans.}$$

This implies that the wavelength of light in glass decreases as compared to the wavelength of light in free-space. This is due to decrease in the velocity of propagation of light in the material as compared to that of in free-space.

2.1.3 Snell's Law

Snell's law explains how a light ray reacts when it meets the intersection of two types of transparent media of uniform but different indices of refraction. Consider that a ray (or, a narrow beam) of light passes from a transparent medium of refractive index n_1 into another transparent medium of refractive index n_2. A refractive index model for Snell's law is shown in Fig. 2.2.

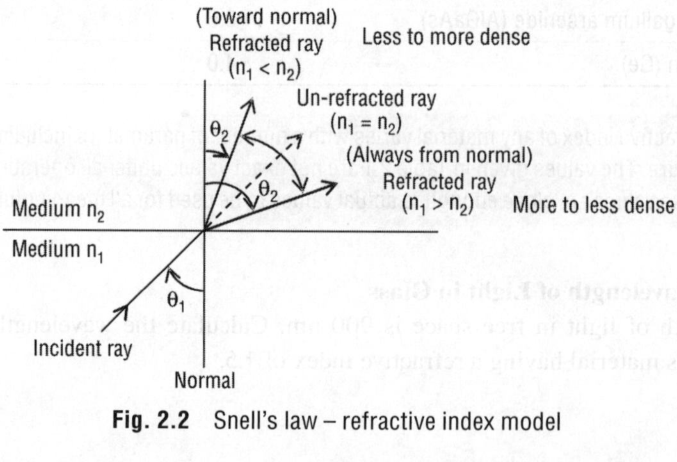

Fig. 2.2 Snell's law – refractive index model

Angle of Incidence: It is defined as the angle at which the light ray strikes the intersection of two different materials with respect to the normal in the first medium.

Angle of Refraction: It is defined as the angle formed between the refracted light ray and the normal in the second medium.

Normal: The normal is a straight line drawn perpendicular to the intersection of two different mediums at the point where the incident ray strikes it.

At the intersection of two different mediums—medium 1 and medium 2, the incident ray may be refracted toward the normal or away from it, depending on whether refractive index n_1 of the first medium is greater or less than refractive index n_2 of the second medium. Hence, the angle of refraction can be either smaller or larger than the angle of incidence, depending on the values of index of refraction of the two mediums under consideration.

Mathematically, according to the Snell's law,

$$n_1 \sin\theta_1 = n_2 \sin\theta_2 \tag{2.2}$$

where, n_1 and n_2 represents absolute value (dimensionless) of refractive index of material 1 and material 2, respectively, and θ_1 denotes the angle of incidence (degrees), and θ_2 denotes the resultant angle of refraction (degrees).

Example 2.2 Snell's Law – An Illustration

A light ray is refracted as it travels from a more dense (higher refractive index) material (Glass with $n_1 = 1.5$) into a less dense (lower refractive index) material (ethyl alcohol with $n_2 = 1.36$). If the angle of incidence made by the ray at the point of intersection of two materials is 30°, then determine the angle of refraction. Also interpret the results with the help of suitable illustration.

Solution:

According to Snell's law, we know that $n_1 \sin\theta_1 = n_2 \sin\theta_2$

For the given $n_1 = 1.5$, $\theta_1 = 30°$, and $n_2 = 1.36$, we have

$$1.5 \times \sin 30 = 1.36 \times \sin\theta_2$$

$$\Rightarrow \qquad \sin\theta_2 = \frac{1.5}{1.36} \times \sin 30 = 0.5515$$

$$\Rightarrow \qquad \theta_2 = \sin^{-1} 0.5515 = 33.47° \qquad \textbf{Ans.}$$

Since $\theta_2 > \theta_1$, it implies that the incident light ray gets refracted further away from the normal at the intersection surface of two materials. This is the expected result as the light was incident in a more dense material (having higher refractive index) and travelling into a less dense material (having lower refractive index). The same is illustrated in Fig. 2.3.

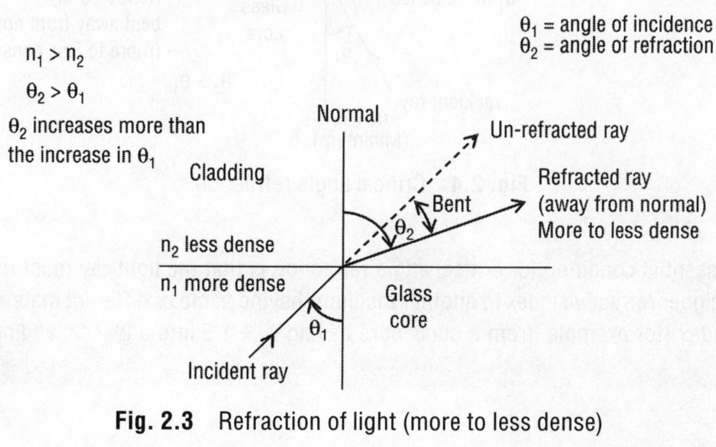

Fig. 2.3 Refraction of light (more to less dense)

It can be seen that the light ray changes direction at the intersection, and the angle of refraction (θ_2) is greater than the angle of incidence (θ_1). Therefore, we can say that when a light ray travelling in a more dense material and enters a less dense material (that is, $n_1 > n_2$), the refracted light ray bends away from the normal.

Note: When $n_1 > n_2$, then $\theta_2 > \theta_1$. This means when a ray of light enters from a medium having higher refractive index into another medium having lower refractive index, then the refracted ray in the second medium bends away from the normal in comparison with that of unrefracted ray.

2.1.4 Critical Angle

Definition of critical angle: It is the minimum possible angle of incidence at which if a light ray is incident at the intersection of two different mediums, then it gets refracted with an angle of refraction exactly equal to 90°. This phenomenon is usually called the critical angle refraction.

Let us consider the case $n_1 > n_2$. As the angle of incidence θ_1 in the first medium is increased, the angle of refraction θ_2 in the second medium will go on increasing until a critical situation is reached, when for a certain value of incidence angle $\theta_1 = \theta_c$, the angle of refraction θ_2 becomes exactly equal to 90°. In such situation, as it is evident, the refracted ray travels along the intersection surface of two mediums. It is, in fact, this angle of incidence which is known as the critical angle of incidence, θ_c. Fig. 2.4 shows such a condition in which an incident ray strikes the interface of two mediums (having desired refractive indexes) at an angle θ_1 such that the angle of refraction in the second medium θ_2 is exactly 90°. As a result, the refracted ray is along the interface.

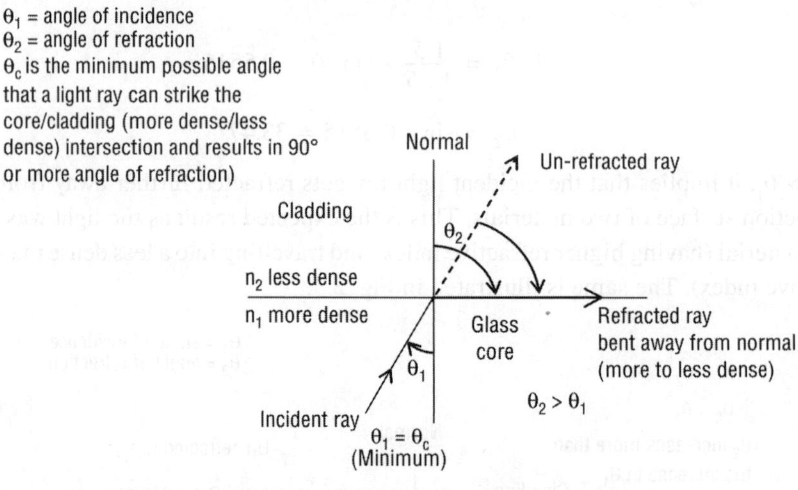

Fig. 2.4 Critical angle refraction

Note: The essential condition for critical angle refraction is that the light ray must travel from a medium of higher refractive index to another medium (having same or different material) of lower refractive index (for example, from a glass core having n_1 = 1.5 into a glass cladding having n_2 = 1.36).

Derivation for the critical incidence angle

According to Snell's law, we know that

$$n_1 \sin\theta_1 = n_2 \sin\theta_2 \tag{2.3}$$

Rearranging this expression, we can write,

$$\sin\theta_1 = \frac{n_2}{n_1}\sin\theta_2 \tag{2.4}$$

We know that for the angle of incidence θ_1 to be equal to the critical angle of incidence θ_c, the angle of refraction θ_2 must be equal to 90°. Therefore, the above expression can be re-written as

$$\Rightarrow \quad \sin\theta_c = \frac{n_2}{n_1}\sin 90° = \frac{n_2}{n_1} \qquad (\because \sin 90° = 1)$$

Hence,

$$\theta_c = \sin^{-1}\left(\frac{n_2}{n_1}\right) \tag{2.5}$$

> ***Note:*** The critical angle of incidence depends on the ratio of the index of refraction of less dense medium (the cladding n_2) to more dense medium (the core n_1), rather than their absolute values.

Example 2.3 Critical Angle

A light ray is refracted as it travels from a denser medium (with higher refractive index, n_1) into a less dense medium (with lower refractive index, n_2). Determine the critical angle of incidence if the ratio of two refractive indexes is

(a) $\frac{n_2}{n_1} = 0.77$

(b) $\frac{n_2}{n_1} = 0.63$

Solution:

We know that critical angle of incidence, $\theta_c = \sin^{-1}\left(\frac{n_2}{n_1}\right)$

(a) For the given $\frac{n_2}{n_1} = 0.77$, we get $\theta_c = \sin^{-1}(0.77) = 50.35°$ **Ans.**

(b) For the given $\frac{n_2}{n_1} = 0.63$, we get $\theta_c = \sin^{-1}(0.63) = 39°$ **Ans.**

What happens when the angle of incidence in higher refractive index medium is further increased beyond the critical angle of incidence (that is, $\theta_1 > \theta_c$)?

In the given situation, the light ray in lower refractive index medium is no longer refracted but is reflected back into the same medium. Fig. 2.5 shows the concept of the angle of refraction and the angle of reflection under three different situations of the angle of incidence to be (a) less than, (b) equal to, or (c) more than the critical angle of incidence.

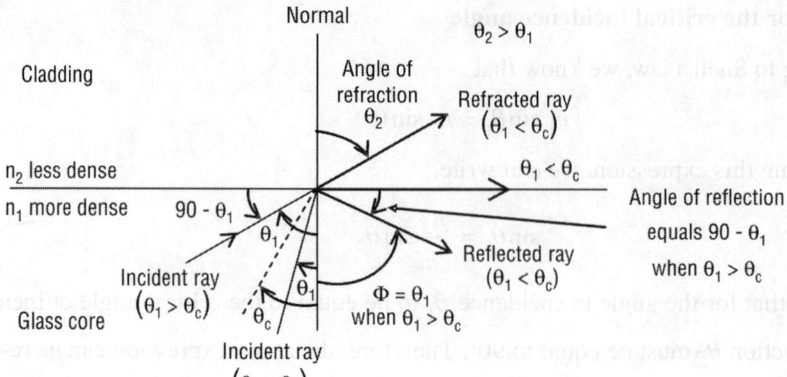

Fig. 2.5 Angle of refraction and reflection

It is concluded from the above discussion that

- If the angle of incidence made by the ray of light is less than the critical angle of incidence, then the phenomenon of refraction takes place.
- If the angle of incidence made by the ray of light is equal to the critical angle of incidence, then the angle of refraction is exactly 90°. It implies that the refracted ray is along the intersection of two mediums.
- If the angle of incidence made by the ray of light is more than the critical angle of incidence, then the phenomenon of reflection takes place. In this case, the angle of reflection will be exactly equal to the angle of incidence, as per laws of reflection.

Example 2.4 Critical Angle, Refraction and Reflection

A typical optical fiber cable has specification of refractive index of 1.6 and 1.4 for the core and the cladding, respectively. Determine the following:

(a) the critical angle of incidence (b) θ_2 for $\theta_1 = 30°$

(c) θ_2 for $\theta_1 = 75°$

Solution:

(a) We know that the critical angle of incidence, $\theta_c = \sin^{-1}\left(\frac{n_2}{n_1}\right)$

For the given $n_1 = 1.6$ and $n_2 = 1.4$, $\theta_c = \sin^{-1}\left(\frac{1.4}{1.6}\right) = 61°$ **Ans.**

(b) Since the given angle of incidence $\theta_1 = 30°$ is less than the calculated (in part a) critical angle, $\theta_c = 61°$, the phenomenon of refraction will take place. Therefore, the Snell's law is applicable, that is, $n_1 \sin\theta_1 = n_2 \sin\theta_2$.

For the given $n_1 = 1.6$ and $n_2 = 1.4$, and $\theta_1 = 30°$, we have

$$1.6 \times \sin 30° = 1.4 \times \sin\theta_2$$

$$\Rightarrow \quad \theta_2 = \sin^{-1}\left(\frac{1.6 \times \sin 30°}{1.4}\right) = 34.8°$$ **Ans.**

As expected, we get $\theta_2 > \theta_1$ since it is given that $n_1 > n_2$. The ray will be refracted away from the normal.

(c) Now the given angle of incidence $\theta_1 = 75°$ is greater than the calculated (in part a) critical angle, $\theta_c = 61°$, the phenomenon of refraction will not take place. Instead, the resultant ray will be reflected back in the same medium as per the laws of reflection and the angle of reflection in the first medium will be exactly equal to the angle of incidence.

Hence, $\theta_2 = 75°$ **Ans.**

Facts to Know

When the angle of incidence made by a ray of light is more than the specified critical angle of incidence, then the phenomenon, known as total internal reflection, will take place. This is the fundamental principle of propagation of light through optical fibers.

Section Practice Problems

1. Calculate the wavelengths of the optical signals produced by each of the light sources in glass material having a refractive index of 1.5 if the wavelengths of light propagating in air is specified as
   (a) 400 nm (b) 700 nm *[Ans.: (a) 266.7 nm; (b) 466.7 nm]*
2. The indices of refraction of the core and cladding material of an optical fiber are 1.5 and 1.45, respectively. Determine the speed of light in the core and the cladding. *[Ans.: 2×10^8 m/s; 2.07×10^8 m/s]*
3. Find the critical angle of incidence when a light beam travels from a glass material having a refractive index value of 1.5 to free space. *[Ans.: 41.8°]*
4. Determine the angle of refraction for an angle of incidence of 35° at a glass/quartz interface. Assume refractive index for glass as 1.5, and that of quartz as 1.38. *[Ans.: 38.6°]*
5. A plastic fiber of 1 mm diameter has $n_1 = 1.496$ and $n_2 = 1.40$. Calculate the critical angle of incidence. *[Ans.: 69.36°]*

2.2 Light Propagation in Optical Fibers

An optical fiber (looks like a waveguide) is a very thin long cylinder that consists of two circularly symmetric coaxial elements—the inner one (the core) is made of glass material having relatively higher refractive index and the outer one (cladding) is made of either glass or plastic material having relatively lower refractive index. The light rays (optical signals) are launched into the cylindrical fiber which can be automatically guided over long distance by an optical phenomenon known as total internal reflection.

2.2.1 Total Internal Reflection

Definition of total internal reflection. When a ray (or beam) of light travels from a medium with a higher refractive index (such as fiber core) to another medium with a lower refractive index (such as fiber cladding) and it happens to strike (incident) the core–cladding intersection at more than the known critical angle of incidence (at which the angle of refraction is 90°), then total light will be reflected back to the medium of incidence (i.e., the fiber core). This particular phenomenon is known as Total Internal Reflection.

Reiterating that when an incident light ray strikes the intersection of two different materials (or same materials but having different index of refraction) at an incidence angle which is exactly equal to the critical angle of incidence, then the angle of refraction in the second medium is exactly 90° (in other words, the refracted ray travels along the line of intersection of two materials). This implies that when the angle of refraction in the second medium happens to be 90° or more, the light ray is not allowed to penetrate it (provided the second medium is relatively less dense). The refracted ray, in fact, is reflected in the same material. In such circumstances we say that the phenomenon of total internal reflection has taken place at the intersection of two mediums with the resulting angle of reflection being exactly equal to the angle of incidence.

> ***Note:*** The whole concept of optical fiber communications is based on the fundamental principle of total internal reflection.

Fig. 2.6 illustrates the source end of a fiber cable and a light ray propagating into and then down the fiber length, depicting the basic principle of total internal reflection.

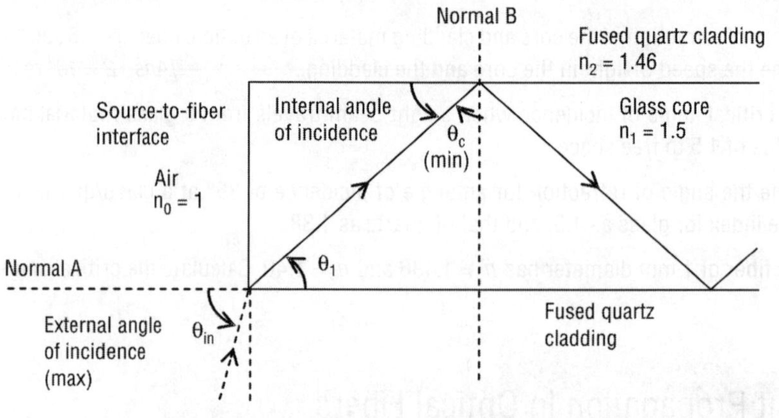

Fig. 2.6 Concept of total internal reflection

As seen in the figure, when a light ray enters the fiber core from the air medium, it strikes the air/glass interface at normal A. It may be noted here that the refractive index of air is approximately 1.0, and that of the glass core is 1.5. This means that the light ray enters the fiber cable traveling from a less dense to a denser medium, causing the light ray to refract toward the normal. This causes the light ray to change its direction and propagate diagonally down the fiber core at an angle of incidence which is less than the external angle of incidence, θ_{in}. That is, $\theta_1 < \theta_{in}$.

> ***Note:*** For a ray of light to propagate within the fiber cable, it must strike the internal core/cladding intersection at an incidence angle which must be greater than the critical incidence angle, θ_c.

Significance of Total Internal Reflection

When the light beam is incident at the core–cladding intersection of an optical fiber cable with angles of incidence greater than the critical incidence angle, θ_c, then all the light gets reflected back into the fiber core with high efficiency (as high as 99.9%). This is known as total internal reflection. This is the essential condition under which no part of the incident light will propagate in the cladding material. It means that the propagation of light takes place within the optical fiber core only (with very low loss of propagating optical signal). Thus, it is certainly the phenomenon of total internal reflection which keeps light propagating within an optical fiber. This is the essence of optical fiber communications.

> ***Note:*** Total internal reflection is a necessary condition to make optical fiber cable as a guided medium of propagation of light in optical fiber communication link.

Example 2.5 Total Internal Reflection

Consider a light ray traveling from a denser (i.e., higher refractive index, $n_1 = 1.5$) material into a less dense (lower refractive index, $n_2 = 1.47$) material. Show that the desired criterion of total internal reflection phenomenon is completely satisfied.

Solution:

As per Snell's Law, $n_1 \sin\theta_1 = n_2 \sin\theta_2$

For total internal reflection, the critical angle (corresponding to $\theta_2 = 90°$) $\theta_c = \sin^{-1}\left(\frac{n_2}{n_1}\right)$

For the given $\begin{matrix} n_1 = 1.5 \\ n_2 = 1.47 \end{matrix}$, we get $\theta_c = \sin^{-1}\left(\frac{1.47}{1.5}\right) = 78.5°$

When $\theta_c = 78.5°$, $\theta_2 = 90°$ which means refraction will not take place and the refracted ray will be along the intersection of two materials.

When $\theta_c > 78.5°$, $\theta_2 > 90°$ which means total internal reflection.

Thus the condition of total internal reflection is fully satisfied with the given situation that the ray of light happens to cross a less dense material to a denser material.

2.2.2 Acceptance Angle

Definition of acceptance angle: It is defined as the maximum (not minimum) external angle of incidence at which the external light rays must strike the air/glass (fiber core) intersection and enters the fiber core and propagate within it.

Derivation of acceptance angle:

Fig. 2.7 represents the related geometry.

From the geometry, it can be seen that the maximum angle that external light rays that strikes the air/glass intersection and subsequently enter the core and propagate down the fiber is

$$\theta_{in(\max)} = \sin^{-1}\left(\frac{\sqrt{n_1^2 - n_2^2}}{n_0}\right) \tag{2.6}$$

where, $\theta_{in(\max)}$ = acceptance angle (in degrees) = θ_a

n_1 = refractive index of glass fiber core (dimensionless, typical value = 1.5)

n_2 = refractive index of quartz fiber cladding (dimensionless, typical value = 1.46)

n_0 = refractive index of air (dimensionless, typical value = 1)

Substituting n_0 =1, we get

$$\theta_a = \sin^{-1}\left(\sqrt{n_1{}^2 - n_2{}^2}\right) \tag{2.7}$$

It may be noted that the acceptance angle is dependent upon the value of the refractive index of fiber core as well as cladding, but certainly not on the fiber core diameter.

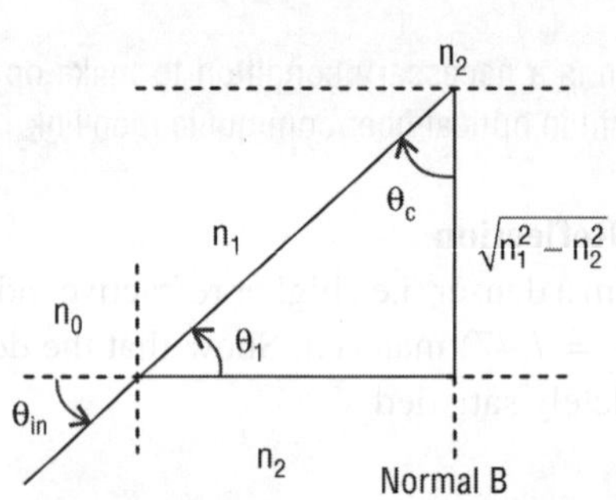

Fig. 2.7 Geometrical relationship

> ***Note:*** The acceptance angle is the maximum value of incidence angle made by a light ray that enters the interface of open air and fiber core (of glass material) that will not ensure total internal reflection. In case it exceeds the acceptance angle, then the light ray is likely to enter the cladding material, resulting in signal loss, because then it will not travel within the fiber core at all.

Rotating the acceptance angle around the fiber core axis describes the cone of acceptance at the input of the fiber. Fig. 2.8 depicts the acceptance cone.

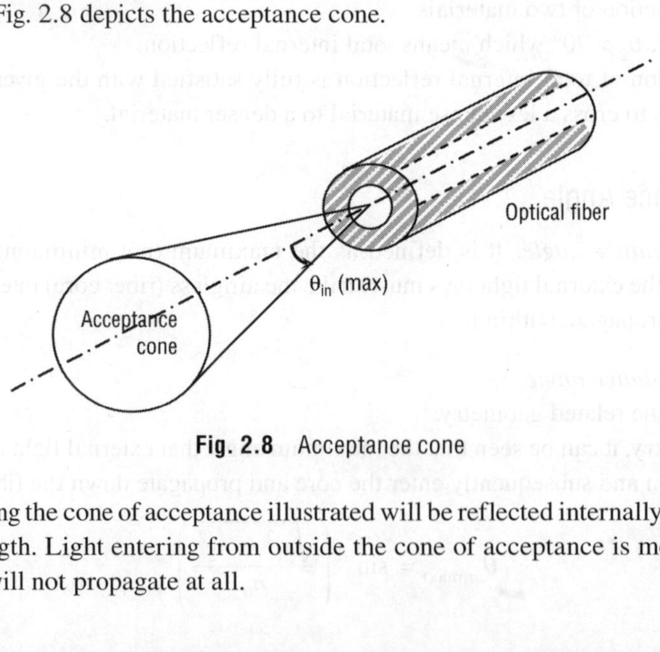

Fig. 2.8 Acceptance cone

Any light entering the cone of acceptance illustrated will be reflected internally and may propagate along the fiber length. Light entering from outside the cone of acceptance is merely refracted into the cladding and will not propagate at all.

Note: It is worth mentioning here that the critical incidence angle is the minimum value, whereas acceptance angle is the maximum one.

Fig. 2.9 depicts the relationship between the critical incidence angle and the acceptance angle for total internal reflection to take place in optical fiber cable.

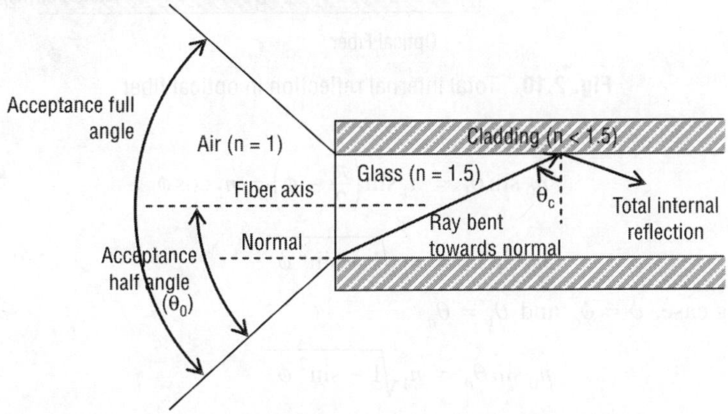

Fig. 2.9 Relationship between acceptance angle and critical angle

Note: It is evident that light rays launched into the optical fiber core at angles less than a certain value (that is, acceptance angle) are guided, whereas light rays launched at angles greater than the acceptance angle are refracted (not reflected back into the fiber core) towards the cladding and thereby lost.

2.2.3 Numerical Aperture

Definition: The numerical aperture (NA) is expressed as the sine (sinusoidal) of the maximum angle which a light ray (being launched into the optical fiber) can make with the central axis of the fiber core and can travel through the fiber using the principle of total internal reflection.

Numerical aperture is the figure of merit which is used to describe the capability of an optical fiber to gather the light efficiently. This parameter is closely associated with the acceptance angle. In fact, the value of numerical aperture can be used to measure the magnitude of the acceptance angle.

Derivation for numerical aperture

Consider a light ray that enters from one end of the fiber from air and is incident on the axis of the fiber core at a particular angle, as shown in the Fig. 2.10.

Assuming, $\theta_1 < \theta_a$, at the intersection of air and fiber core, as per the Snell's law

$$n_0 \sin\theta_1 = n_1 \sin\theta_2$$

From figure, we get $\phi = \frac{\pi}{2} - \theta_2$; and $\theta_2 = \frac{\pi}{2} - \phi$

where, $\phi > \phi_c$ at the intersection of fiber core and cladding of the optical fiber cable for total internal reflection to take place which is necessary for propagation of light through optical fiber cable.

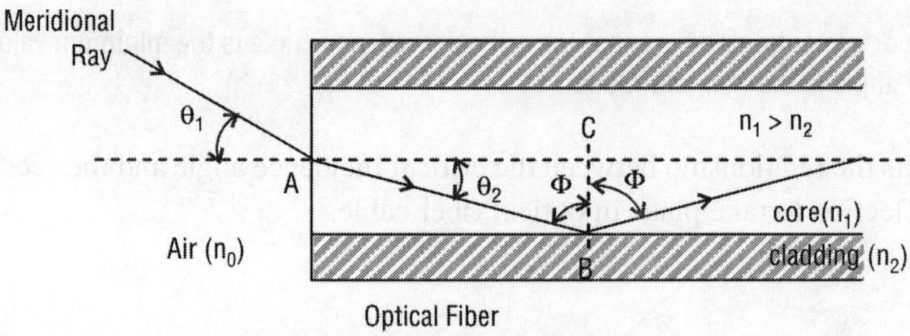

Fig. 2.10 Total internal reflection in optical fiber

$$\therefore \quad n_0 \sin\theta_1 = n_1 \sin\left(\frac{\pi}{2} - \phi\right) = n_1 \cos\phi \tag{2.8}$$

$$\Rightarrow \quad n_0 \sin\theta_1 = n_1\sqrt{1 - \sin^2\phi} \tag{2.9}$$

In the limiting case, $\phi = \phi_c$ and $\theta_1 = \theta_a$

$$\therefore \quad n_0 \sin\theta_a = n_1\sqrt{1 - \sin^2\phi_c} \tag{2.10}$$

But

$$\sin\phi_c = \frac{n_2}{n_1} \qquad \left(\begin{aligned} \because n_1 \sin\phi_c &= n_2 \sin 90^\circ \text{ Snell's Law} \\ &= n_2 \end{aligned}\right) \tag{2.11}$$

$$\therefore \quad n_0 \sin\theta_a = n_1\sqrt{1 - \left(\frac{n_2}{n_1}\right)^2} \tag{2.12}$$

$$n_0 \sin\theta_a = \sqrt{n_1^2 - n_2^2} \tag{2.13}$$

$$\text{NA} = \sin\theta_a = \sqrt{n_1^2 - n_2^2} \qquad (\because n_0 \approx 1 \text{ for air}) \tag{2.14}$$

Hence, the numerical aperture (NA) for light rays that enter the glass fiber core from an external air can be described as

$$\text{NA} = \sin\theta_a \tag{2.15}$$

where, θ_a represents the acceptance angle, given as $\theta_a = \sin^{-1}\left(\sqrt{n_1^2 - n_2^2}\right)$ (2.16)

$$\Rightarrow \quad \sin\theta_a = \sqrt{n_1^2 - n_2^2} \tag{2.17}$$

Therefore,

$$\text{NA} = \sqrt{n_1^2 - n_2^2} \tag{2.18}$$

Note that it is assumed here that light enters the fiber from free space, that is, air. Thus, numerical aperture depends solely on the values of refractive index of the fiber core and the cladding materials used for optical fibers. The numerical aperture is a dimensionless quantity. Typically, the value of NA is less than unity (usually between 0.1 and 0.5).

It implies that acceptance angle, $\theta_a = \sin^{-1}(\text{NA})$ (2.19)

Thus, the incident light rays entering the optical fiber with incidence angles ranging from $0 \leq \theta_1 \leq \theta_a$ will be definitely propagated and confined within the optical fiber.

> ***Note:*** In essence, the term numerical aperture signifies the ability to couple the light rays (optical signals) into the optical fiber cable from an optical source. It implies that if the value of numerical aperture of a particular optical fiber cable is greater, then larger amount of external light will be accepted by the fiber for propagation.

Numerical aperture can also be related with the relative difference or fractional change in the refractive index of the fiber core and the cladding of the optical fiber cable. Let us define another term known as **relative refractive index difference**, Δ as

$$\Delta = \frac{n_1^2 - n_2^2}{2n_1^2} \tag{2.20}$$

where, n_1 and n_2 are the refractive index of the fiber core and the cladding materials, respectively.

For $\Delta << 1$,
$$\Delta \approx \frac{n_1 - n_2}{n_1} \tag{2.21}$$

$$\Rightarrow \quad n_1\Delta = n_1 - n_2; \text{ for } \Delta << 1$$

$$\therefore \quad \text{NA} = \sqrt{n_1^2 - n_2^2} = \sqrt{(n_1 + n_2)(n_1 - n_2)} \tag{2.22}$$

Using $n_1 \approx n_2$ in the first term and $(n_1 - n_2) = n_1\Delta$ in second term,

$$\text{NA} = \sqrt{(2n_1)(n_1\Delta)} = \sqrt{n_1^2 2\Delta} \tag{2.23}$$

$$\text{NA} = n_1\sqrt{2\Delta} \tag{2.24}$$

Hence,
$$\text{NA} = \sqrt{n_1^2 - n_2^2} = n_1\sqrt{2\Delta} \tag{2.25}$$

Thus, numerical aperture can be calculated from the given values of refractive index of the fiber core and cladding used in the optical fiber cable. It may be noted that it is independent of the diameter of the fiber core. In other words, a larger diameter core does not necessarily produce a larger numerical aperture. However, in practice, larger core fibers tend to have larger numerical apertures. Numerical aperture (a property concerned at the input of the fiber) is generally measured by looking at the output of the fiber because of the simple reason that the light guiding properties of an optical fiber cable are symmetrical. It means that the light leaving a fiber cable spreads out (i.e., scattered) over an angle which is almost same as that of the acceptance angle.

Example 2.6 Numerical Aperture

An optical fiber cable has values of refractive index of 1.50 for the fiber core and 1.47 for the cladding. Determine the numerical aperture.

Solution:

We know that $\text{NA} = \sqrt{n_1^2 - n_2^2}$

For the given $n_1 = 1.50$ and $n_2 = 1.47$, we have

$$NA = \sqrt{1.50^2 - 1.44^2} = 0.30 \qquad \textbf{Ans.}$$

Example 2.7 Numerical Aperture

An optical fiber cable has values of refractive index of 1.46 for the fiber core. If the fractional change in the index of refraction of the optical fiber is specified as 0.01, then find its numerical aperture.

Solution:

We know that $NA = \sqrt{n_1^2 - n_2^2}$

First we have to determine the refractive index of cladding to use this formula.

By definition, the fractional change in the refractive index $\Delta = \dfrac{n_1^2 - n_2^2}{2n_1^2}$

For the given $n_1 = 1.46$ and $\Delta = 0.01$, we have $0.01 = \dfrac{1.46^2 - n_2^2}{2 \times 1.46^2}$

$$\Rightarrow \qquad n_2^2 = 1.46^2 - 0.01 \times 2 \times 1.46^2$$

$$\Rightarrow \qquad n_2 = 1.445$$

Therefore, $NA = \sqrt{1.46^2 - 1.445^2} = 0.2$ **Ans.**

Alternatively, using the relationship, $NA = n_1\sqrt{2\Delta}$, we obtain

$$NA = 1.46\sqrt{2 \times 0.01} = 0.2 \qquad \textbf{Ans.}$$

Table 2.2 gives the typical ranges of numerical apertures for preferred sizes of multimode glass optical fibers.

Table 2.2 Typical values of numerical apertures

S. No.	*Preferred sizes of multimode glass optical fibers*		*Numerical aperture (range of values)*
	Core diameter (μm)	*Cladding diameter (μm)*	
1.	50	125	0.19–0.25
2.	62.5	125	0.27–0.31
3.	85	125	0.25–0.30
4.	100	140	0.25–0.30

Example 2.8 NA and Acceptance Cone

An optical fiber cable has specified refractive index values of 1.6 and 1.4 for the fiber core and cladding, respectively. Determine the numerical aperture and the total width of the acceptance cone.

Solution:

We know that $NA = \sqrt{n_1^2 - n_2^2}$

For the given $n_1 = 1.6$ and $n_2 = 1.4$, we have $NA = \sqrt{1.6^2 - 1.4^2} = 0.775$ **Ans.**

We also know that the acceptance angle, $\theta_a = \sin^{-1}(NA)$

Therefore, $\theta_a = \sin^{-1}(0.775) = 50.8°$

Total width of the cone of acceptance $= 2\theta_a = 2 \times 50.8° = 101.6°$ **Ans.**

Example 2.9 Critical Angle and Solid Acceptance Angle

An optical fiber cable having a core refractive index value of 1.46 and a typical relative refractive difference of 1%, is used for long distance data transmission. Calculate

(a) The critical angle of incidence of the optical fiber cable at the intersection of the fiber core and cladding.

(b) Solid acceptance angle in air.

Solution:

(a) To calculate the critical angle

We know that the critical angle of incidence, $\theta_c = \sin^{-1}\left(\frac{n_2}{n_1}\right)$

Given $n_1 = 1.46$, firstly we have to calculate n_2.

We know that relative refractive difference, $\Delta = \frac{n_1 - n_2}{n_1}$ for $\Delta << 1$

For the given $\Delta = 1\%$, or 0.01, we have

$$0.01 = \frac{1.46 - n_2}{1.46}$$

$$\Rightarrow \quad n_2 = 1.46 - 0.01 \times 1.46 = 1.4454$$

Therefore, $\theta_c = \sin^{-1}\left(\frac{1.4454}{1.46}\right) = 81.9°$ **Ans.**

(b) To calculate solid acceptance angle in air

We know that the solid acceptance angle, $\theta_a = \sin^{-1}(NA)$

where, $NA = n_1\sqrt{2\Delta}$

Using $n_1 = 1.46$ and $\Delta = 1\%$, or 0.01, we have

$$\text{NA} = 1.46 \times \sqrt{2 \times 0.01} = 0.2064$$

Therefore, $\theta_a = \sin^{-1}(0.2064) = 11.91°$ **Ans.**

Example 2.10 Numerical Aperture and Acceptance Angle

The velocity of light in the fiber core of a step–index fiber is specified as 2.01×10^8 m/s. The critical angle of incidence at the intersection of the fiber core and the cladding is 80°. Estimate the values of the numerical aperture as well as the acceptance angle of incident light rays for the optical fiber placed in air. Assume that the fiber has a core diameter which is suitable to be considered by optical ray analysis.

Solution:

We know that $NA = \sqrt{n_1^2 - n_2^2}$

First we have to find the refractive index n_1 and n_2 of the fiber core and cladding, respectively. Since the velocity of light in the core,

$$v = \frac{c}{n_1}$$

$$\Rightarrow \quad n_1 = \frac{c}{v}$$

For the given $v = 2.01 \times 10^8$ m/s, $c = 3 \times 10^8$ m/s (in air), we have

$$\Rightarrow \quad n_1 = \frac{3 \times 10^8 \text{ m/s}}{2.01 \times 10^8 \text{ m/s}} = 1.49$$

To find cladding refractive index n_2, we use the relationship $\sin\theta_c = \left(\frac{n_2}{n_1}\right)$

For the given critical angle, $\theta_c = 80°$, we have $\sin 80° = \left(\frac{n_2}{1.49}\right)$

$$\Rightarrow \quad n_1 = 1.467$$

Therefore, NA = $\sqrt{n_1^2 - n_2^2} = \sqrt{(1.49)^2 - (1.467)^2} = 0.259$ **Ans.**

We know that the acceptance angle, $\theta_a = \sin^{-1}(NA)$

Therefore, $\theta_a = \sin^{-1}(0.259) = 15.01°$ **Ans.**

Example 2.11 Determination of the Velocity of Light in Fiber Core

In a given optical fiber, the relative refractive difference between the core and cladding is 0.9 % and a solid acceptance angle in air is 0.15 radians. Determine the velocity of light in the fiber core, assuming the fiber has a core diameter such that ray theory can be applied for analysis.

Solution:

We know that velocity of light in fiber core, $v = \frac{c}{n_1}$

Where the core refractive index n_1 can be calculated from the given values of the acceptance angle, $\theta_a = 0.15$ radians $= 8.594°$, and $\Delta = 0.9\% = 0.009$

Using the relationship $\sin\theta_a = \sqrt{n_1^2 - n_2^2}$, and $\Delta = \frac{n_1^2 - n_2^2}{2n_2^2}$; we have

$$\sin(8.594) = \sqrt{n_1^2 - n_2^2} = 0.1494$$

$$\Rightarrow \quad n_1^2 - n_2^2 = (0.1494)^2 = 0.022$$

And $$0.009 = \frac{0.022}{2n_2^2}$$

$$\Rightarrow \qquad n_2 = \sqrt{\frac{0.022}{2 \times 0.009}} = 1.10554$$

Substituting it in $n_1^2 - n_2^2 = 0.022$; we get

$$n_1^2 - (1.10554)^2 = 0.022$$

$$\Rightarrow \qquad n_1^2 = \sqrt{\left(0.022 + (1.10554)^2\right)} = 1.115$$

Hence, the velocity of light in the fiber core, $v = \dfrac{c}{n_1} = \dfrac{3 \times 10^8 \text{ m/s}}{1.115} = 2.69 \times 10^8 \text{ m/s}$ **Ans.**

> ***Note:*** The greater the value of the acceptance angle, the greater is the proportion of light that can be gathered and guided within the fiber core. The sine (sinusoidal) function of this maximum acceptance angle is popularly known as the numerical aperture (NA) of the fiber, and gives its measure of the light gathering efficiency.

Section Practice Problems

1. For the optical fiber cable having glass core ($n_1 = 1.5$) and quartz cladding ($n_2 = 1.41$), determine the values of the following parameters:
   (a) the critical angle of incidence
   (b) numerical aperture of the fiber
   (c) the acceptance angle
   (d) the acceptance cone *[Ans.: (a) 70°; (b) 0.512; (c) 30.8°; (d) 61.6°]*

2. The refractive index of the fiber core is specified as 1.46 for a typical optical fiber cable. If the fractional change in the index of refraction of the fiber is specified as 0.01, then show that the fiber would accept light that is incident over a cone with semi-angle of about *11.5°*.

3. The refractive index of the fiber core and the cladding of a given optical fiber cable is specified as 1.5 and 1.48, respectively. Find its numerical aperture, acceptance angle, and critical angle. *[Ans.: 0.244; 14.1°; 80.6°]*

4. The speed of light in a step-index fiber core is specified as 2 x 10^8 m/sec. When the fiber is placed in air, the critical angle at the intersection of the fiber core and the cladding is *75°*. Calculate the Numerical Aperture of the fiber. *[Ans.: 0.388]*

5. A step-index fiber has a refractive index of 1.46 for the cladding and numerical aperture of 0.17. Determine the followings:
   (a) the acceptance angle of the fiber when it is placed in water (refractive index of 1.33).
   (b) the critical angle at the core-cladding interface. *[Ans.: (a) 7.34°; (b) 83.35°]*

2.3 Classification of Optical Fibers

There are various ways to classify optical fibers. These can be based on different characteristics related to the nature (or types) of refractive index profiles of the fiber core, modes of propagation of light in the fiber, and the improvement in the propagation properties.

1. Based on the refractive index profiles of the fiber core, there can be two types of optical fibers, namely,
   - Step-index optical fiber
   - Graded-index optical fiber
2. Based on the number of modes (rays or light wave patterns) propagated within the optical fiber, there can be two types of optical fibers. These are
   - Single-mode optical fiber
   - Multi-mode optical fiber
3. Based on the improvement in propagation properties, there can be three different types of optical fibers, such as
   - Polarization-maintaining optical fiber
   - Dispersion-shifted optical fiber
   - Dispersion-flattened optical fiber

> ***Note:*** The mechanism of propagation of light within an optical fiber cable mainly depends on two aspects: the index profile (step-index, or graded-index) and the mode of propagation (single-mode, or multi-mode).

2.3.1 Step-Index Optical Fiber

A step-index optical fiber cable has a central core having a uniform index of refraction (i.e., constant density throughout the core) which is surrounded by a cladding that too has a uniform index of refraction but of lesser value as that of the core. Thus, there is an abrupt change in the refractive index at intersection of the fiber core and the cladding of the optical fiber cable. A typical index profile of a step-index optical fiber is illustrated in Fig. 2.11.

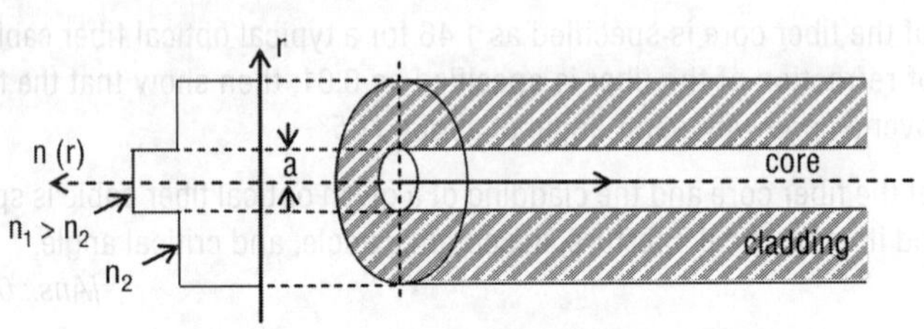

Fig. 2.11 Index profile of step-index fiber

Mathematically, the reference index profile of a step-index fiber may be expressed as:

$$n(r) = \begin{cases} n_1; & r < a \text{ (core)} \\ n_2; & r \geq a \text{ (cladding)} \end{cases} \tag{2.26}$$

where, *n(r)* represents the refractive index profile of a step-index fiber as a function of its radius (*r*), and *a* denotes the radius of the fiber core.

> ***Note:*** The word 'index' in step-index fiber basically refers to refractive index or the index of refraction. It is related to the density of the medium or material.

2.3.2 Graded-Index Optical Fiber

Graded-index optical fiber, as the name suggests, is different from step-index optical fiber with respect to the nature of their index profile. It has a central core with a non-uniform index of refraction. This simply means that the refractive index of the fiber core is maximum in the middle and then it gradually decreases towards its outer edge up to core–cladding interface.

Thus, graded-index optical fibers do not possess a constant refractive index even within the fiber core. Instead, the fiber core has a decreasing refractive index *n(r)* with respect to the radial distance. It has a maximum value of refractive index n_1 at the central axis to another constant value n_2 which is the refractive index of the cladding beyond the core radius *a*. Fig. 2.12 shows index profile in a typical graded-index optical fiber.

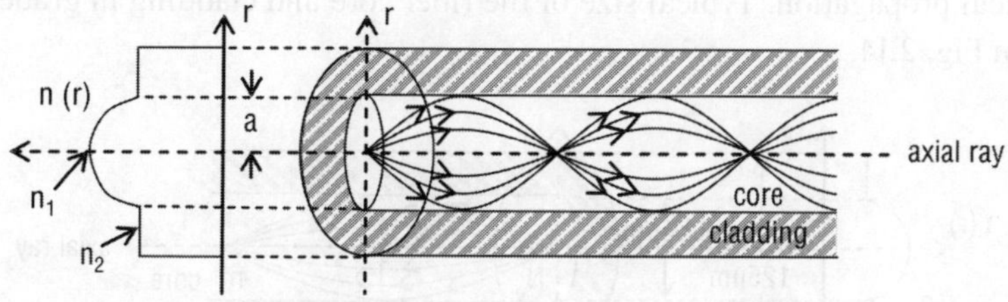

Fig. 2.12 Index profile of a graded-index fiber

Mathematically, the index profile can be expressed as

$$n(r) = \begin{cases} n_1\left[1 - (2\Delta)\left(\dfrac{r}{a}\right)^{\alpha}\right]^{1/2} & r < a \quad (core) \\ n_1(1 - 2\Delta)^{1/2} = n_2 & r \geq a \quad (cladding) \end{cases} \tag{2.27}$$

where, n_1 represents the refractive index at the axis of the fiber core; n_2 represents the refractive index of the cladding; *r* denotes the radial distance from the fiber core axis towards the cladding; *a* represents the actual radius of the fiber core, and Δ signifies the relative refractive index difference for the given graded-index optical fiber which can be expressed in terms of n_1 and n_2 as

$$\Delta = \frac{n_1^2 - n_2^2}{2n_1^2} \approx \frac{n_1 - n_2}{n_1} \tag{2.28}$$

Let us introduce another important parameter, α, known as the index profile parameter of the fiber core which specifies the characteristic refractive index profile. It is a dimensionless quantity. For example,

$\alpha = 1$ for triangular profile of graded-index fiber

$\alpha = 2$ for parabolic profile of graded-index fiber

$\alpha = \infty$ (infinity) for a step-index fiber, where $n(r) = n_1$

Fig. 2.13 shows the α-profile parameter of the optical fiber for different refractive index profiles.

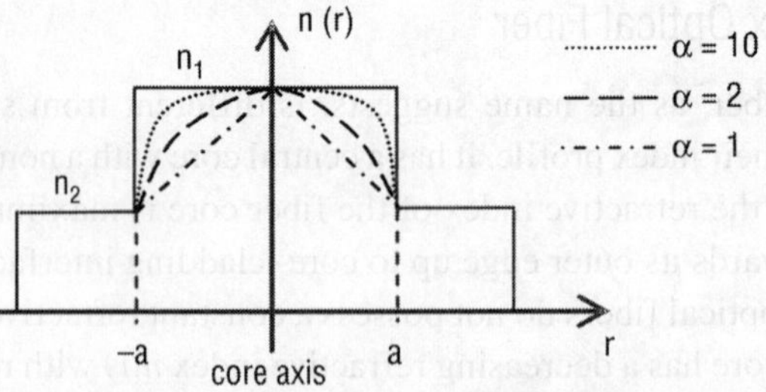

Fig. 2.13 Refractive index profile *(α)* of graded-index fiber

It is worth mentioning here that graded-index profile optical fiber produces the best results for multimedia optical propagation. Typical size of the fiber core and cladding in graded-index profile fiber is shown in Fig. 2.14.

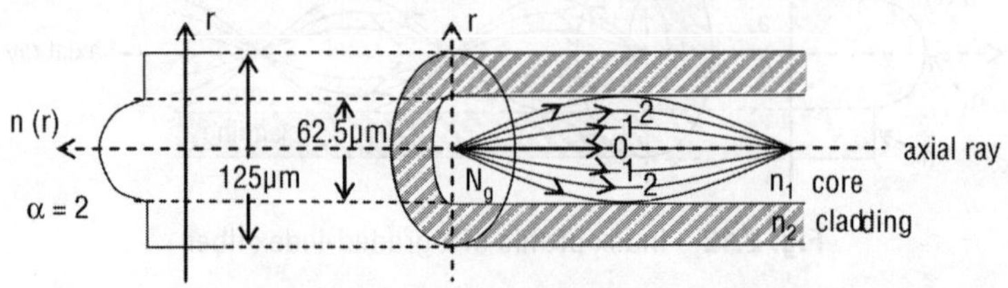

Fig. 2.14 A typical graded-index fiber

Fig. 2.15 depicts the propagation of light rays through a multimode graded-index optical fiber.

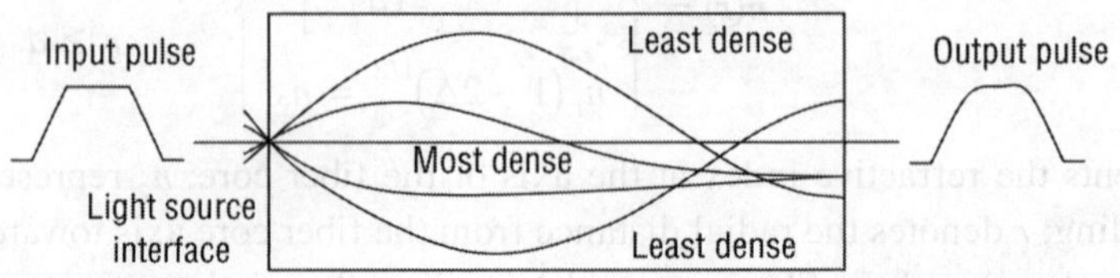

Fig. 2.15 Propagation of light rays through graded-index fiber

It may be noted here that light propagate through graded-index fiber through the phenomena of refraction rather than total internal reflection. Let us try to understand it. As a light ray propagates diagonally across the width of the fiber core toward its center, it is continuously traversing from a lesser dense to more dense medium. As a consequence, the light rays experience constant refraction (i.e., continuous bending of the light rays). It is quite evident that practically light beam enters the optical fiber at several different angles.

As the light rays traverse through the optical fiber in the forward direction progressively, the light rays propagating in the outermost area of the fiber core (near the cladding) travel a much larger distance than the light rays propagating close to the fiber axis. As stated earlier that in the graded-index profile fiber, the value of index of refraction decreases with distance from the core axis and the velocity being inversely proportional to refractive index, the light rays traveling far away from the core axis travel at a relatively higher velocity of propagation of light. As a result, they take almost identical amount of time to traverse the given length of the optical fiber cable.

The numerical aperture for graded-index fibers is a function of the radial distance from the axis of the fiber, expressed as

$$\text{NA}(r) = \begin{Bmatrix} \left[n^2(r) - n_2^2\right]^{1/2} & ; \text{ for } \; r \le a \\ 0 & ; \text{ for } \; r > a \end{Bmatrix} \tag{2.29}$$

So, numerical aperture for graded-index fiber is quite complex as compared to that of step-index fibers. However, it can also be easily deduced that the axial numerical aperture is same as that of step-index fiber, that is,

$$\text{NA}(0) = \left[n_1^2 - n_2^2\right]^{1/2} \tag{2.29a}$$

where, *NA(0)* represents the value of numerical aperture at fiber axis. How does NA of a graded-index fiber vary as *r* moves from the fiber axis to the intersection of the fiber core and the cladding? It is obvious that it decreases gradually from *NA(0)* to zero.

2.4 Propagation Modes

Definition of mode: In fiber–optics terminology, the word `mode' simply means `path', the path that is traversed by the propagation of light within the fiber. This path can be best described by a well-defined set of guided electromagnetic (em) waves. These are known as the propagation modes, or simply modes, of the optical fiber.

Modes are distinguished by their propagating angles (incidence or reflected). Modes are designated by the order of a specific mode. The extent of propagating angle is directly related with the order of the mode of an optical fiber. It implies that the smaller the propagating angle the lower will be the order of the propagating mode. Thus, we can say that:

- The fundamental mode, also called the zero-order mode, is the propagating mode of the light that travels precisely along the central axis of the optical fiber.
- The highest-order mode is the propagating mode of the light that travels at critical angle, θ_c, of incidence.

As stated previously, in an optical fiber cable, the light rays can be propagated using the law of total internal reflection. In reality, an optical fiber can be considered as a waveguide. Therefore, the light can propagate in a number of specific modes only. For instance,

- if the diameter of a fiber is sufficiently narrow, then it may support only one mode of propagation of light;
- if the diameter of a fiber is relatively large, light entering at different angles will excite different modes of propagation of light.

Current fiber–optic technology supports mainly two types of propagation modes of light along the optical fiber cable, namely, single mode and multimode. Of course, each one of them does require a specific type of fiber having unique physical characteristics. Further, multimode type of light propagation through the optical fiber can be supported in two different categories of optical fibers (depending on nature of index profiles of fiber core): step-index optical fiber, and graded-index optical fiber. Various types of modes are shown in Fig. 2.16.

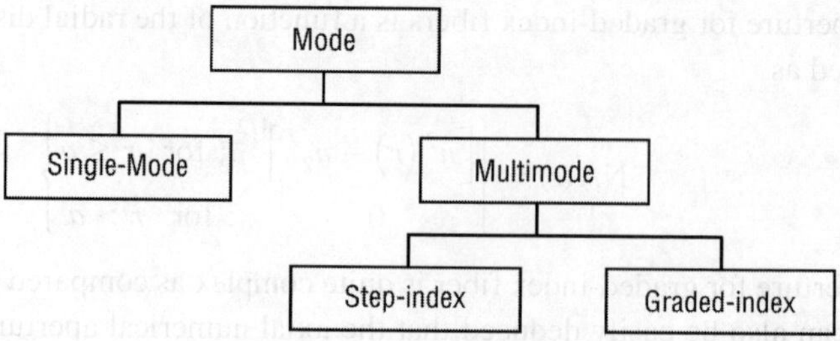

Fig. 2.16 Propagation modes

2.4.1 Single Mode Propagation

Definition: As the name suggests, if there is only one possible path for light rays to travel through an optical fiber cable, it is called single mode propagation.

Fig. 2.17 shows single mode propagation of light rays through an optical fiber.

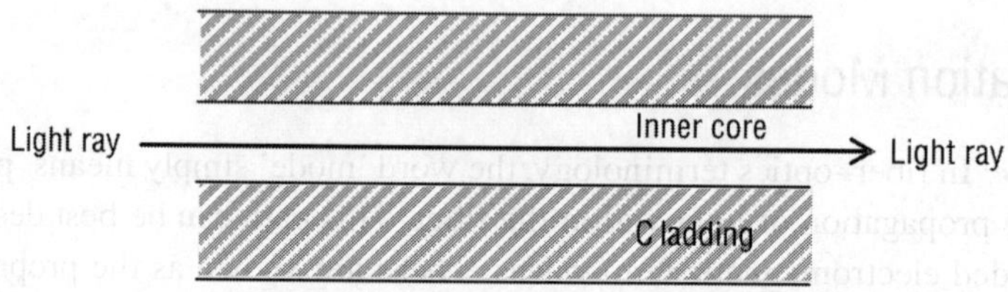

Fig. 2.17 Single mode propagation

As it is evident from the figure that with single-mode propagation of light, there is one and only one path for light rays to travel, which is directly along the axis of the fiber cable. Single-mode propagation uses step-index configuration of optical fiber cable. It requires a highly focused optical source that emits light beam with a very narrow beamwidth. Moreover, the single-mode optical fiber cable itself is manufactured having considerably lower index of refraction and much smaller diameter as compared to that of multimode optical fiber cable. The lower refractive index results in a critical angle of incidence that is approximately equal to 90° so as to make the propagation of light rays almost horizontal to the central axis of the fiber core. This also ensures that different light rays propagate through the optical fiber with negligible propagation delays.

Now, we should know the key parameters on which the maximum allowable diameter for a single-mode fiber depends. Obviously, it varies directly with the wavelength of the propagating light and inversely with the numerical aperture of the given optical fiber cable. The relationship among these parameters is given as

$$d_{max} = \frac{0.766\lambda}{NA} \tag{2.30}$$

where, d_{max} = the maximum diameter of the core

λ = wavelength

NA = numerical aperture

Note that d_{max} and λ should have same units.

Example 2.12 Maximum Core Diameter of Single-mode Fiber

A single-mode fiber cable has specified value 0.15 for numerical aperture. What is the maximum core diameter it could have for use with infrared light with a wavelength of 850 nm?

Solution:

The maximum allowable diameter for a single-mode fiber is given as:

$$d_{max}\,(m) = \frac{0.766\lambda(m)}{NA}$$

For the given λ = 850 nm, or 850 × 10^{-9} m, and NA = 0.15, we have

$$d_{max}(m) = \frac{0.766 \times 850 \times 10^{-9}(m)}{0.15} = 4.34 \times 10^{-6} m \text{, or } 4.34\ \mu m$$ **Ans.**

To summarize, a single-mode step-index optical fiber cable has a central core whose diameter is significantly smaller as compared to that of multimode fiber cables. Actually, the core diameter of single-mode step-index optical fiber cable is kept sufficiently small (typically of the order of 7–10 μm) so as to ensure only one transverse electromagnetic mode of propagation path that light may travel through it.

Facts to Know

Why are single-mode step-index fibers the most prominent type of optical fiber cables employed in today's data networking and telecommunications applications? Due to their single mode of propagation, they exhibit low intermodal dispersion (i.e., broadening of transmitted light pulse) as they propagate down the fiber.

Example 2.13 Single-Mode Step-Index Fiber with Air Cladding

In the crudest form of a single-mode step-index optical fiber, the cladding may be just external air. In this case, the index of refraction of the glass core (n_1) is approximately 1.5, and that of the air cladding (n_2) is 1. With the help of suitable geometrical representation, determine (a) critical angle; (b) acceptance angle. Comment on the results obtained.

Solution:

Fig. 2.18 shows the geometrical diagram of the given fiber.

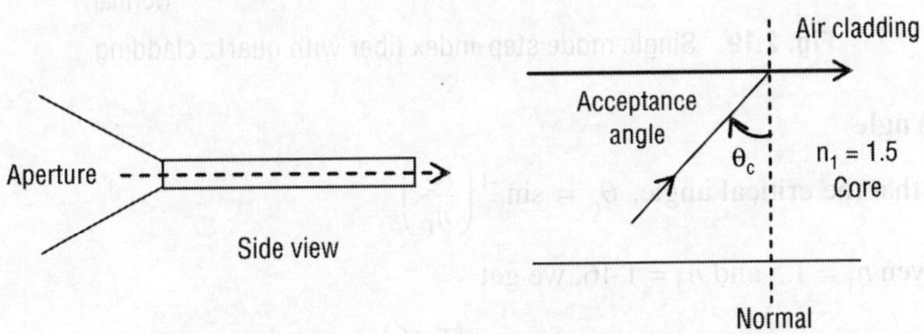

Fig. 2.18 Single mode step-index optical fiber (with air cladding)

(a) **Critical Angle**

We know that the critical angle of incidence, $\theta_c = \sin^{-1}\left(\frac{n_2}{n_1}\right)$

For the given $n_1 = 1.5$ and $n_2 = 1$ (for air), we get

$$\theta_c = \sin^{-1}\left(\frac{1}{1.5}\right) = 41.8° \quad \textbf{Ans.}$$

(b) **Acceptance Angle**

From the figure, it is seen that

Acceptance angle, $\theta_a = 90 - \theta_c$

For $\theta_c = 41.8°$, we have $\theta_a = 90 - 41.8 = 48.2°$ **Ans.**

Comment on the results obtained: The large difference in the refractive indexes results in a small critical angle (41.8°) at the glass/air interface. This, in turn, results in a wider acceptance angle (48.2°). This implies that we can easily couple light from an external optical source into such type of optical fiber cable. But due to air cladding, it is quite fragile, and practically almost impossible to terminate or splice.

Example 2.14 Single-Mode Step–Index Fiber Cable (Glass Core and Quartz Cladding)

A practical single-mode step-index type of optical fiber cable has the core made of glass (n_1=1.5) and the cladding made of quartz (n_2=1.46). With the help of suitable geometrical representation, determine (a) critical angle; (b) acceptance angle. Comment on the results obtained.

Solution:

Fig. 2.19 shows the geometrical diagram of the given fiber.

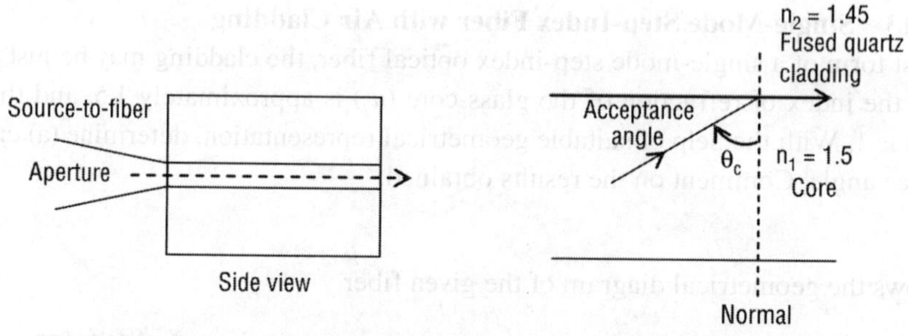

Fig. 2.19 Single mode step-index fiber with quartz cladding

(a) **Critical Angle**

We know that the critical angle, $\theta_c = \sin^{-1}\left(\frac{n_2}{n_1}\right)$

For the given $n_1 = 1.5$ and $n_2 = 1.46$, we get

$$\theta_c = \sin^{-1}\left(\frac{1.46}{1.5}\right) = 76.7° \quad \textbf{Ans.}$$

(b) **Acceptance Angle**

From the figure, it is seen that

Acceptance angle, $\theta_a = 90 - \theta_c$

For $\theta_c = 76.7°$, we have $\theta_a = 90 - 76.7 = 13.3°$ **Ans.**

Comment on the results obtained: The given index of refraction for the cladding material (n_2=1.46) is slightly smaller than the given refractive index of the fiber core (n_1=1.5). Moreover, the refractive index is uniform. So, obviously this type of cable is physically stronger than the air-clad fiber (described in the previous example). On the other hand, it exhibits much higher critical angle (76.7°) which, in turn, results in a small acceptance angle (13.3°). This implies that the fiber cable would have a narrow source-to-fiber aperture. In such situation, it will be quite difficult to couple (or launch) light from the optical source into the fiber.

2.4.2 Multimode Propagation

Definition: If there is more than one path for light rays to travel through an optical fiber cable, it is called multimode propagation.

As the name suggests, the 'multimode' refers to 'multiple light rays' from an optical source that propagate through the fiber core following different paths. The movement of light rays within the optical fiber cable depends on the core structure. As stated previously, there are many higher-order modes possible with multimode propagation. In fact, the light rays, following several paths, propagate within the optical fiber cable in a zigzag manner. Fig. 2.20 shows multimode propagation of light rays through an optical fiber.

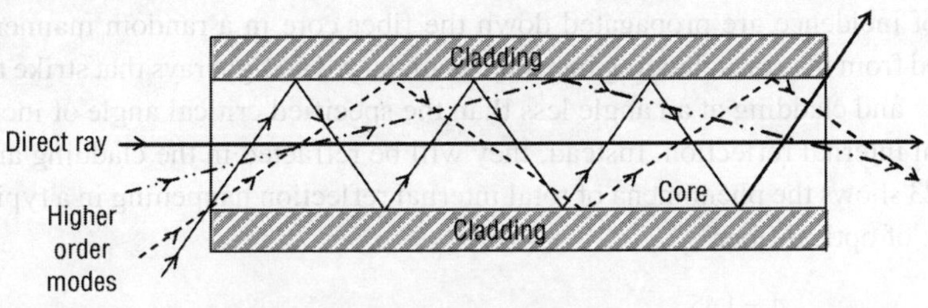

Fig. 2.20 Multimode propagation

In terms of propagation of light through fiber, modes can be categorized in the following three types:

1. *Guided mode*: which experiences total internal reflection
2. *Radiation mode*: which experiences total refraction in the cladding and propagate outside the fiber core
3. *Leaky mode*: in which some light rays are partially reflected within the fiber core and some are partially refracted toward the cladding

2.4.3 Multimode Step-index Fiber

Multimode step-index fiber has a larger core diameter (typically of the order of 50 μm or more) than that of single-mode step-index fiber. It allows the propagation of several modes (different possible paths) of light rays within the fiber core.

Fig. 2.21 shows the core index profile of a multimode step index fiber.

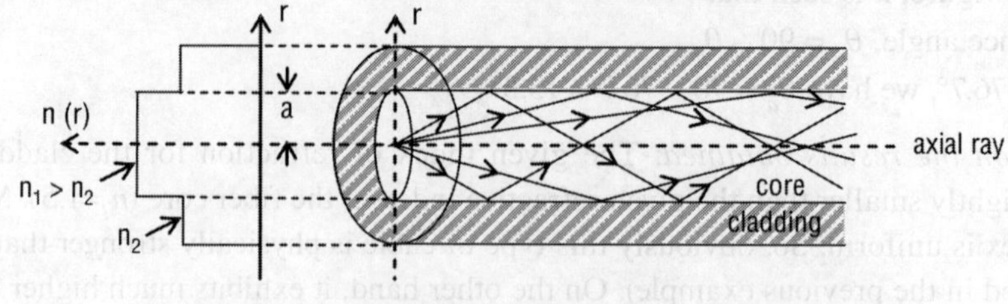

Fig. 2.21 Core index profile in multimode step-index fiber

It is quite evident that multimode step-index fiber has a greater source-to-fiber aperture. As a consequence, more external light is allowed to be coupled with the fiber cable. This is shown in Fig. 2.22.

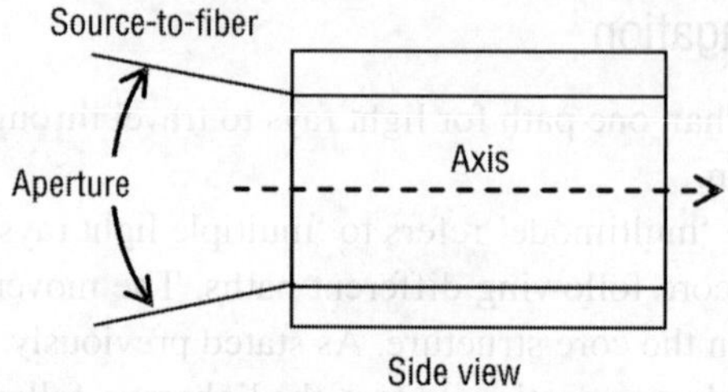

Fig. 2.22 Multimode step-index fiber

The light rays that enter the intersection of the core and cladding at an angle more than the specified critical angle of incidence are propagated down the fiber core in a random manner, continuously getting reflected from the core-cladding interface boundary. The light rays that strike the intersection of the fiber core and cladding at an angle less than the specified critical angle of incidence will not experience total internal reflection. Instead, they will be refracted in the cladding and are likely to be lost. Fig. 2.23 shows the phenomena of total internal reflection happening in a typical multimode step-index type of optical fiber.

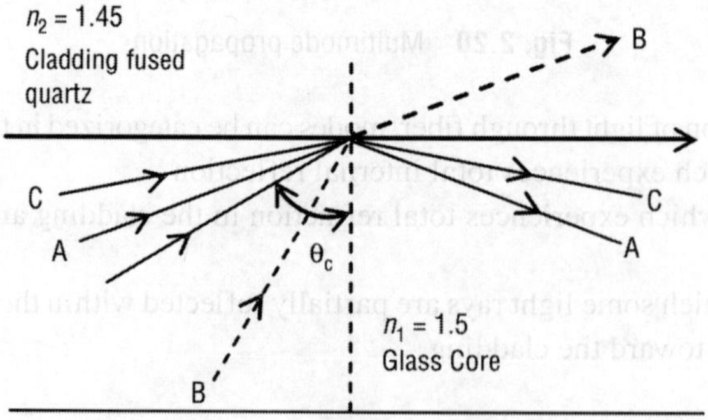

Fig. 2.23 Total internal reflection in multimode step-index fiber

It can be seen that there are several propagation paths followed by a light ray as it propagates down the multimode step-index fiber. It implies that all light rays may not follow identical paths—some paths may be shorter and some may be longer. Therefore, different light rays may take different amount of time to travel the given length of the optical fiber cable.

> ***Note:*** The above discussion leads to considerable dispersion in multimode step index fiber, mainly because of different group velocities of the propagating modes. This, in turn, restricts the maximum transmission bandwidth that can be attained.

2.4.4 V-Parameter

Now we are in a position to define an important parameter that determines the cut-off condition for the number of propagation modes significant for multimode optical fibers. This parameter is called the normalized frequency parameter *V*, or simply the *V*-number, or *V*-parameter. It depends on the core radius, operating wavelength, and refractive index of core and cladding. Mathematically, it can be expressed as

$$\text{V} \approx \frac{2\pi a}{\lambda}\sqrt{n_1^2 - n_2^2} \approx \frac{\pi d}{\lambda}\sqrt{n_1^2 - n_2^2} \tag{2.31}$$

where, a = core radius (meters)

d = core diameter (meters) = $2a$

λ = wavelength of incident light rays (meters)

n_1 = refractive index of fiber core (dimensionless)

n_2 = refractive index of fiber cladding (dimensionless)

Using the expression for numerical aperture NA = $\sqrt{n_1^2 - n_2^2}$, we have

$$\Rightarrow \qquad \text{V} \approx \frac{2\pi a}{\lambda}(\text{NA}) \approx \frac{\pi d}{\lambda}(\text{NA}) \tag{2.32}$$

Thus, by a proper choice of core diameter and the numerical aperture, single mode fiber at any wavelength can be realized.

Substituting, NA = $n_1\sqrt{2\Delta}$, we can re-write

$$\text{V} \approx \frac{2\pi a}{\lambda} n_1\sqrt{2\Delta} \tag{2.33}$$

<u>Example 2.15</u> Normalized Frequency Parameter

A light of wavelength 850 nm is transmitted through a commercial step-index fiber whose numerical aperture is specified as 0.17 and core diameter as 100 μm. Determine the normalized frequency parameter V.

Solution:

We know that $V \approx \frac{\pi d}{\lambda}(NA)$

For the given d = 100 μm, λ = 850 nm, *NA* = 0.17, we get

$$\text{V} \approx \frac{\pi \times 100 \times 10^{-6}}{850 \times 10^{-9}} \times 0.17 = 62.83$$ **Ans.**

Example 2.16 Normalized Frequency Parameter

A light of wavelength 1.55 μm is used to excite the modes in a step-index fiber having the following parameters:

- Fiber core diameter = 7.2 μm
- Fiber core refractive index = 1.46
- Relative refractive index difference = 1%

Determine the normalized frequency parameter V.

Solution:

We know that $V \approx \frac{2\pi a}{\lambda} n_1 \sqrt{2\Delta}$

For the given 2a = 7.2 μm, λ = 1.55 μm, n_1 = 1.46, and Δ = 1% or 0.01, we get

$$V \approx \frac{\pi \times 7.2 \times 10^{-6}}{1.55 \times 10^{-6}} \times 1.46 \times \sqrt{2 \times 0.01} = 3 \quad \textbf{Ans.}$$

2.4.5 V-number versus Propagation Constant

A mode of propagation is normally described by its propagation constant. The propagation constant in vacuum, or the propagation vector, is defined as

$$k = \frac{2\pi}{\lambda} \tag{2.34}$$

where, λ represents the wavelength of light in air or vacuum.

Let the index of refraction within the fiber core is designated as n_1. The wavelength of light in the fiber core is reduced to $\lambda_m = \frac{\lambda}{n_1}$, and the propagation vector is increased to

$$\beta_1 = \frac{2\pi}{\lambda_m} = \frac{2\pi n_1}{\lambda} = kn_1 \tag{2.35}$$

If the propagation vector β_1 makes an angle θ with the fiber core axis, then the effective propagation vector of the guided wave will be

$$\beta = \beta_1 \cos\theta \tag{2.36}$$

The maximum value that β can have is β_1, which corresponds to $\theta = 0$, that is, the waves travel parallel to the axis of the fiber core.

The limiting value of θ, that is, the acceptance angle, θ_a is related to the critical angle of incidence θ_c at the intersection of the fiber core and the cladding. It is given by

$$\sin\theta_c = \cos\theta_a = \frac{n_2}{n_1} \tag{2.37}$$

Thus the minimum value of β will be determined by the acceptance angle, θ_a, or

$$\beta_{\min} = \beta_1 \cos\theta_a = \beta_1 \frac{n_2}{n_1} = \beta_2 \tag{2.38}$$

Substituting, $\beta_1 = kn_1$, we get

$$\beta_2 = kn_1 \frac{n_2}{n_1} = kn_2 \tag{2.39}$$

So it is expected that β will lie between β_1 and β_2, or $\beta_2 (= kn_2) < \beta < \beta_1 (= kn_1)$.

The normalized propagation constant b can be defined as

$$\text{Normalized propagation constant,} \quad b = \frac{\beta^2 - \beta_2{}^2}{\beta_1{}^2 - \beta_2{}^2} = \frac{\beta^2 - k^2 n_2{}^2}{k^2 n_1{}^2 - k^2 n_2{}^2} \tag{2.40}$$

Since β lies between β_1 (= kn_1) and β_2 (= kn_2) for the guided modes, the value of b will lie between 0 (for $\beta = \beta_2$) and 1 (for $\beta = \beta_1$).

$$\Rightarrow \quad b = \frac{(\beta/k)^2 - n_2{}^2}{n_1{}^2 - n_2{}^2} \tag{2.41}$$

In this expression, the term β/k is called the mode index, $\tilde{n}$. Therefore,

$$\Rightarrow \quad b = \frac{\tilde{n}^2 - n_2{}^2}{n_1{}^2 - n_2{}^2} \tag{2.42}$$

$$\Rightarrow \quad b = \frac{(\tilde{n} - n_2)(\tilde{n} + n_2)}{(n_1 - n_2)(n_1 + n_2)} \tag{2.43}$$

For most types of fibers, $n_1 \approx n_2$. Thus, we can re-write

$$\Rightarrow \quad b = \frac{(\tilde{n} - n_2)}{(n_1 - n_2)} \tag{2.44}$$

$$\text{Therefore, the mode index,} \quad \tilde{n} \approx n_2 + b(n_1 - n_2) \tag{2.45}$$

Fig. 2.24 illustrates a plot between normalized frequency parameter (or, V-number) and normalized propagation constant b for various modes of propagation.

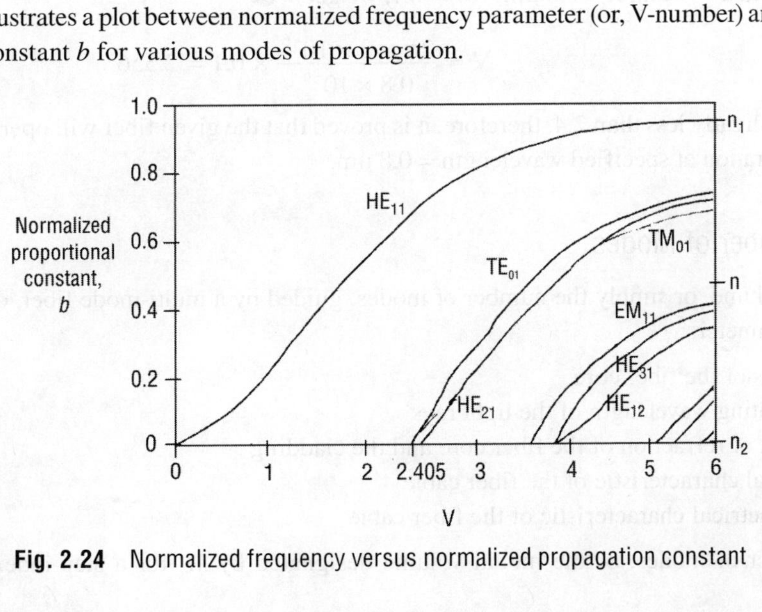

Fig. 2.24 Normalized frequency versus normalized propagation constant

A mode ceases to be guided, when $\beta^2 < \beta_2^2$. Such modes are known as radiation modes, in which the rays undergo refraction (that is, leaked into the cladding away from the core) rather than total internal reflection at the intersection of the fiber core and the cladding of the optical fiber cable. The condition $\beta < \beta_2$ corresponds to what is called the cut-off condition of a propagation mode. Thus, at $\beta = \beta_2$; $b = 0$; and $V = V_c$, where V_c is the normalized cut-off frequency.

It is seen that each mode can exist only above a certain value of *V*-number. It may be noted that for V ≤ 2.405, only the lowest order mode exists and other modes are cut-off.

> ***Note:*** The essential condition to be fulfilled for an optical fiber to be single-mode is that the normalized frequency V ≤ 2.4.

From the expression $V \approx \frac{2\pi a}{\lambda} n_1 \sqrt{2\Delta}$ with $V = 2.4$, it is observed that the single-mode propagation is possible for fairly large variations in core size and core-cladding index difference. However, practically, the single-mode fibers can be designed with variations in the core-cladding index difference to be limited within 0.2–1%, and the core diameter should be chosen to be slightly less than the cutoff value of the first higher-order mode, that is, for V to be just less than or closer to 2.4.

Example 2.17 Normalized Frequency for Single-Mode Operation

A typical fiber is specified to have a numerical aperture of 0.1 and a fiber core radius of 3 μm. Prove that this optical fiber will operate in a single-mode configuration for a given wavelength = 0.8 μm.

Solution:

We know that for a fiber to operate in a single-mode configuration, the value of normalized frequency parameter, or *V*-number must be slightly less than or equal to 2.4.

We know that $V \approx \frac{2\pi a}{\lambda}(\mathrm{NA})$

For the given $a = 3$ μm, $\lambda = 0.8$ μm, $NA = 0.1$, we get

$$V \approx \frac{2\pi \times 3 \times 10^{-6}}{0.8 \times 10^{-6}} \times 0.1 = 2.356 \quad \textbf{Ans.}$$

Since *V* is slightly less than 2.4, therefore, it is proved that the given fiber will operate in a single-mode configuration at specified wavelength = 0.8 μm.

2.4.6 Number of Modes

The modal volume, or simply the number of modes, guided by a multi-mode fiber, depends on the following parameters:

- the radius of the fiber core
- the operating wavelength of the light rays
- the index of refraction of the fiber core and the cladding
- the optical characteristic of the fiber cable
- the geometrical characteristic of the fiber cable

The number of modes, or the modal volume designated by M_s for a step-index fiber can be expressed as

$$M_S \approx \frac{1}{2}\left[\frac{2\pi a}{\lambda}\sqrt{n_1^2 - n_2^2}\right]^2 \tag{2.46}$$

But we know that $V \approx \frac{2\pi a}{\lambda}\sqrt{n_1^2 - n_2^2}$

$$\therefore \quad M_S \approx \frac{V^2}{2} \tag{2.47}$$

The number of modes, or the modal volume designated by M_g for a graded-index fiber can be expressed as

$$M_g = \frac{1}{2}\left(\frac{\alpha}{\alpha + 2}\right)\left[\frac{2\pi a}{\lambda}\sqrt{n_1^2 - n_2^2}\right]^2 \tag{2.48}$$

where, α is the profile parameter of the graded-index fiber. It may be noted here that n_1 represents the index of refraction of the fiber core at its axis (which decreases gradually towards the intersection of the fiber core and the cladding, as defined by the profile parameter).

Using $V \approx \frac{2\pi a}{\lambda}\sqrt{n_1^2 - n_2^2}$; we have

$$\therefore \quad M_g \approx \left(\frac{\alpha}{\alpha + 2}\right)\frac{V^2}{2} \tag{2.49}$$

For a parabolic profile, $\alpha = 2$. Therefore,

$$M_g \approx \frac{V^2}{4} \tag{2.50}$$

In order to support a single mode in a graded-index fiber, the cut-off value of the normalized frequency parameter, designated by V_c, is given by the expression

$$V_c = 2.405\sqrt{\left(1 + \frac{2}{\alpha}\right)} \tag{2.51}$$

Note: In practical optical fibers, $(n_1 - n_2) << 1$. Natural modes will combine (degenerate) into linear–polarized (LP) modes.

Example 2.18 Number of Propagating Modes

A multimode step-index fiber cable has the following specifications:

- the fiber core diameter = 50 μm
- the index of refraction of a fiber core = 1.6
- the index of refraction of cladding = 1.584.

If the wavelength of light propagating through it is 1300 nm, determine the approximately number of propagating modes.

Solution:

We know that the number of propagating modes is given as $M_s \approx \frac{V^2}{2}$ in a multimode step-index fiber.

Where $V \approx \frac{\pi d}{\lambda}\sqrt{n_1^{\ 2} - n_2^{\ 2}}$

For the given $d = 50$ μm, $\lambda = 1300$ nm, $n_1 = 1.6$, and $n_2 = 1.584$, we have

$$V \approx \frac{\pi \times 50 \times 10^{-6}}{1300 \times 10^{-9}}\sqrt{1.6^2 - 1.584^2} = 27.27$$

Therefore, $M_s \approx \frac{27.27^2}{2} = 372$ **Ans.**

Example 2.19 Propagating Modes versus Fiber Size

The core refractive index and a relative refractive index difference of a multimode step-index fiber are specified as 1.5 and 2%, respectively. At operating wavelength of 1300 nm, the approximate number of propagating modes is 1000. Determine the diameter of the fiber core.

Solution:

We know that the number of propagating modes is given as $M_s \approx \frac{V^2}{2}$ in a multimode step-index fiber.

$$\Rightarrow \qquad V = \sqrt{2 \times M_s}$$

For the given $M_s = 1000$, we have $V = \sqrt{2 \times 1000} = 44.72$

We know that $$V = \frac{2\pi a}{\lambda} n_1 \sqrt{2\Delta}$$

$$\Rightarrow \qquad 2a \approx \frac{V \times \lambda}{\pi \times n_1 \sqrt{2\Delta}}$$

For the given $\lambda = 1300$ nm, $n_1 = 1.5$, and $\Delta = 2\%$ or 0.02, we get

$2a \approx \frac{44.72 \times 1300 \times 10^{-9}}{\pi \times 1.5\sqrt{2 \times 0.02}} = 6.2 \times 10^{-5} m$, or 62 μm **Ans.**

2.4.7 Mode Field Diameter

As stated earlier, a single-mode optical fiber cable is designed in such a way that it supports only the fundamental mode of propagation of light rays, and all higher-order propagation modes are cut-off at the specified wavelength of operation. It is also mentioned that in a step-index fiber the single-mode operation is possible over a normalized frequency parameter range specified as $0 < V < 2.405$. However, for a particular value of λ, the V-parameter can be made lower than its cut-off value of $V_c = 2.405$, by reducing either the core diameter ($2a$), or the relative refractive index difference (Δ). Practically it is not recommended because of the following reasons:

- With small core diameter, it becomes difficult not only to launch the light into the optical fiber but also to join two pieces of fibers in the field (splicing).
- It is extremely difficult to manufacture fibers with very low values of Δ.

A higher value of V_c (and hence larger core diameter) is possible with graded-index fiber in which,

$V_c = 2.405\left(\sqrt{1+\frac{2}{\alpha}}\right)$; where, α is its profile parameter.

- For a triangular index profile ($\alpha = 1$), V_c increases by a factor of $\sqrt{3}$
- For a parabolic index profile ($\alpha = 2$), V_c increases by a factor of $\sqrt{2}$

It is observed that with low values of *V*-parameter and Δ for the single-mode operation, the modal field distribution of the fundamental mode of propagation (due to the radial distribution of the optical power) is approximately Gaussian for step-index optical fibers as well as for parabolic-profile graded-index optical fibers that operate near the specified cut-off wavelength, λ_c. Therefore, the *mode field diameter* (MFD), or the *mode spot size*, of the propagating mode constitutes a key parameter characteristic of a single-mode fiber. It may be noted here that the mode field diameter for single-mode fibers is analogous to the fiber core diameter in multimode fibers, except that all the light propagated through the fiber is not carried within the fiber core in single-mode fibers.

For a step-index fiber, the mode field diameter (MFD), designated by $2w$, is given by

$$\text{MFD} = 2w \approx 2a\left(0.65 + \frac{1.619}{\sqrt{V^3}} + \frac{2.879}{V^6}\right) \tag{2.52}$$

where, w represents the mode field radius, and a represents the fiber core radius.

This expression gives the value of MFD to within about 1% of accurate value for *V*-parameter in the range of 0.8–2.5. It is obvious that for a fiber of given diameter, the normalized mode spot size increases as *V* becomes smaller (or as λ becomes longer).

> ***Note:*** At longer operating wavelength, the modal field is less confined within the fiber core. Therefore, the design of a single-mode fiber should be such that the cut-off wavelength is not too far away from the operating wavelength.

Example 2.20 Mode Field Diameter

The core diameter of a typical step-index fiber is specified as 8 μm. The values of core refractive index and relative refractive index difference are specified as 1.46 and 0.3%, respectively. Calculate the mode field diameter (MFD) at operating wavelength of 1550 nm.

Solution:

We know that $\text{MFD} \approx 2a\left(0.65 + \frac{1.619}{\sqrt{V^3}} + \frac{2.879}{V^6}\right)$

$$V = \frac{2\pi a}{\lambda} n_1 \sqrt{2\Delta}$$

For the given $2a = 8$ μm, $\lambda = 1550$ nm, $n_1 = 1.46$, and $\Delta = 0.3\%$, or 0.003, we have

$$V = \frac{\pi \times 8 \times 10^{-6}}{1550 \times 10^{-9}} \times 1.46\sqrt{2 \times 0.003} = 1.834$$

$$\text{Therefore, } MFD \approx 8 \times 10^{-6}\left(0.65 + \frac{1.619}{\sqrt{(1.834)^3}} + \frac{2.879}{(1.834)^6}\right) = 11.0196\ \mu\text{m} \qquad \textbf{Ans.}$$

Section Practice Problems

1. For an optical fiber cable with indices of refraction as 1.46 and 1.41 for the core and cladding, respectively, calculate the maximum diameter which the fiber core could have so as to operate for single-mode propagation at a given wavelength of 1.5 μm. *[Ans.: 3 μm]*
2. A single-mode fiber is required to be designed having n_1 = 1.505 and n_2 = 1.502 as refractive index of core and cladding, respectively. If it is to be operated at wavelength of 1300 nm, then find the dimension of the fiber core. [Ans.: *a* = 5.23 μm]
3. A light of wavelength 0.85 μm is transmitted through a multimode step-index fiber whose core diameter and numerical aperture are specified as 100 μm and 0.17, respectively. Compute the approximate number of guided propagating modes in this fiber. [Ans.: 1974]
4. A graded-index fiber has a well-defined parabolic index profile. It has the following specifications:
   - Fiber core diameter = 75 μm
   - Fiber core refractive index = 1.45
   - Relative refractive index difference = 2%
   - Number of guided modes supported = 700.

   (a) Determine the normalized frequency parameter *V*.
   (b) Compute the wavelength of light travelling through the given fiber.
   (c) Calculate the maximum diameter of the fiber core so that the fiber can function in the single-mode configuration. [Hint: Use the cut-off value of the normalized frequency, $V_c = 2.405\sqrt{2}$]

   [Ans.: (a) 52.9; (b) 1300 nm; (c) 4.85 μm*]*
5. A step-index fiber has the following specifications:
   - Diameter of the fiber core = 8 μm
   - Refractive index of the fiber core = 1.46
   - Relative refractive index difference = 0.3%.

   Calculate the mode field diameter at operating wavelength of 1300 nm. *[Ans.: 9.4116 μm]*

2.5 Dispersion in Optical Fibers

Definition: Dispersion, also known as pulse spreading, means spreading out of an optical pulse of light energy in time as it propagates down a fiber. Dispersion occurs because of the difference in the propagation time taken by the light rays that traverse different propagation paths within the fiber. If the pulse spreading is sufficiently severe, one pulse may interfere with another.

2.5.1 General Concept of Dispersion

1. *Dispersion in a single-mode step-index fiber*: The light rays propagating down a single-mode step-index fiber (assuming radial dimension of the fiber being sufficiently small) is depicted in Fig. 2.25.

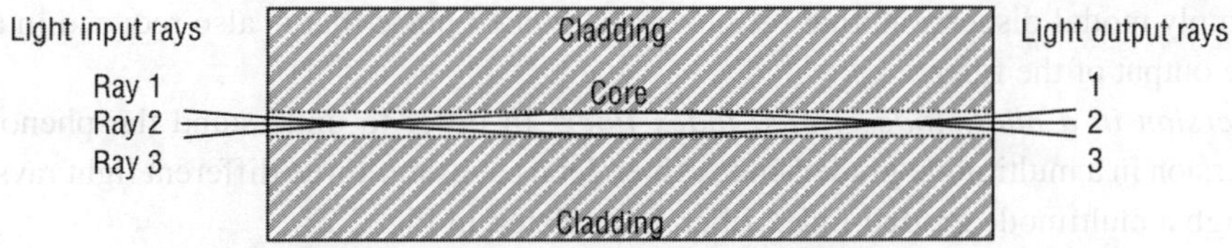

Fig. 2.25 Light propagation down a single-mode step-index fiber

The following is observed:

- There is only a single transmission path that all rays must follow as they propagate through the fiber length.
- Each ray of light travels the same distance in a given period of time, and there is virtually no intermodal dispersion.
- However, a light pulse, although propagating as a fundamental mode, has a number of spectral components, and the group velocity of the fundamental mode varies with frequency, that results in broadening of the transmitted optical pulse.
- This phenomenon is known as intramodal, or group velocity dispersion (GVD).

2. *Dispersion in a multimode step-index fiber:* Let us try to understand the phenomenon of dispersion that occurs in a multimode step-index fiber. Consider three different light rays that propagate down a multimode step-index optical fiber as illustrated in Fig. 2.26.

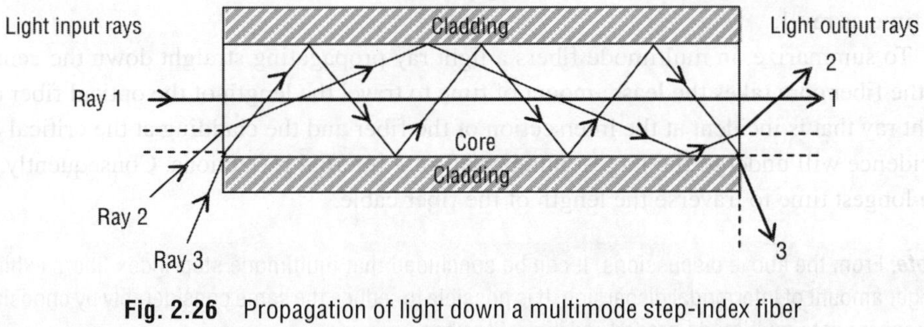

Fig. 2.26 Propagation of light down a multimode step-index fiber

From the figure, the following is observed.

- The lowest-order propagation mode (i.e., light ray 1) travels in a path parallel to the central axis of the optical fiber.
- The middle-order propagation mode (i.e., light ray 2) bounces several times at the interface before traveling the length of the fiber.
- The highest-order propagation mode (i.e., light ray 3) makes many trips back and forth across the fiber as it propagates the entire length. It is quite evident that light ray 3 travels a considerably longer distance than ray 1 over the length of the cable.

Thus, if the three rays of light were emitted into the fiber at the same time, each ray would reach the far end at a different time. It implies that the propagated light energy would spread out with respect to time. As mentioned earlier, this is what is known as intermodal dispersion, or simply modal dispersion, that results in a stretched pulse that is also reduced in amplitude at the output of the fiber.

3. *Dispersion in a multimode graded-index fiber:* In order to understand the phenomenon of dispersion in a multimode graded-index fiber, let us consider three different light rays traveling through a multimode graded-index fiber, as depicted in Fig. 2.27.

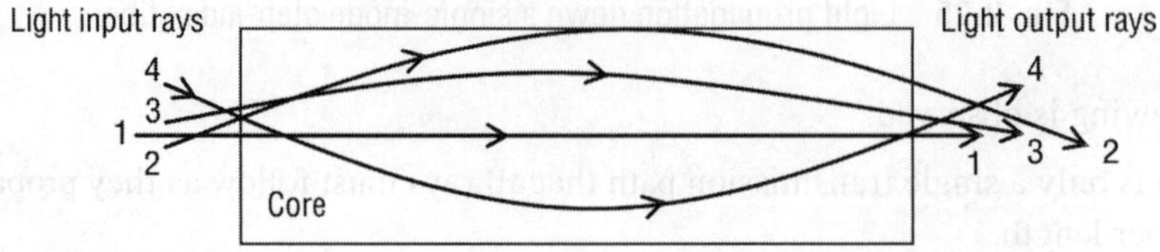

Fig. 2.27 Light propagation down a multimode graded-index fiber

From the figure, the following is observed.

- Three rays are shown traveling in three different modes.
- Although the three rays travel different paths, they all take almost identical amount of time for propagating through the length of the optical fiber.
- This is due to the reason that the index of refraction decreases with radial distance from the center axis of the fiber core toward the cladding. We know that the velocity at which a ray travels is inversely proportional to the refractive index. So, the rays 2 and 3 travel farther away from the center of the cable, they propagate faster and reach the end almost at the same time as ray 1.

To summarize, in multimode fibers a light ray propagating straight down the central axis of the fiber core takes the least amount of time to travel the length of the optical fiber cable. A light ray that is incident at the intersection of the fiber and the cladding at the critical angle of incidence will undergo the largest number of total internal reflections. Consequently, it takes the longest time to traverse the length of the fiber cable.

Note: From the above discussions, it can be concluded that multimode step-index fibers exhibit larger amount of intermodal dispersion. It is possible to reduce the same considerably by choosing an appropriate multimode graded-index profile fiber.

2.5.2 Bandwidth × Length Product

For multimode propagation, dispersion is often expressed as a bandwidth-length product (B × L). Bandwidth-length product indicates what signal frequencies can be propagated through a given distance of fiber cable. Mathematically, (B × L) is expressed as the product of distance and bandwidth (sometimes called linewidth). Bandwidth-length products are often expressed in MHz-km units. As the length of an optical fiber cable increases, the transmission bandwidth (and thus the data rate) decreases in the same proportion.

The effects of modal dispersion increase with the length of the optical fiber cable because the difference in the velocity of two optical signals propagating down the fiber length produces a difference in the time required to reach the other end of the fiber. In fact, this time difference is directly proportional to the length of the fiber. The greater the bandwidth of the transmitted optical signal, the greater will be the effect of a given amount of modal dispersion. Practically, multimode step-index or graded-index fibers are rated in accordance with the product of transmission bandwidth and fiber length.

> ***Note:*** It is the modal dispersion that occurs in multimode step-index fibers. However, it can be considerably reduced by using graded-index fibers. It can be almost entirely eliminated if single-mode step-index fibers can be used.

Example 2.21 Bandwidth of an Optical Fiber Cable

Determine the transmission bandwidth of a 300-meter optical fiber cable having specified Bandwidth-Length Product (BLP) of 600 MHz-km.

Solution:

Given Bandwidth × Length = 600 MHz-km

For 300-meter length of optical fiber cable, we have

$$\text{Bandwidth of optical fiber} = \frac{600 \text{ MHz-km}}{300 \times 10^{-3} \text{ km}} = 2 \times 10^3 \text{ MHz, or 2 GHz}$$ **Ans.**

Example 2.22 Maximum Distance between Repeaters

An optical fiber cable has a specified bandwidth-length product of 500 MHz-km. If it is required to obtain a transmission bandwidth of 85 MHz for a particular mode of propagation, determine the maximum distance that can be used between repeaters.

Solution:

Given Bandwidth × Length = 500 MHz-km

$$\text{That is, the maximum distance between repeaters} = \frac{500 \text{ MHz-km}}{\text{Bandwidth}}$$

For the given bandwidth of 85 MHz for a particular mode of transmission, we have

$$\text{Maximum distance between repeaters} = \frac{500 \text{ MHz-km}}{85 \text{ MHz}} = 5.9 \text{ km}$$ **Ans.**

Facts to Know

Dispersion increases with the bandwidth of the light source. For LEDs and lasers, the term linewidth is generally used rather than bandwidth. Where minimum dispersion is important, the laser diode is greatly preferred over the LED because of its much narrower linewidth.

Table 2.3 gives typical values of bandwidth-length product of different types of fibers.

Table 2.3 Typical values of bandwidth-length product

S. No.	*Fiber Type*	*Bandwidth × Length Value (Typical)*
1.	Step-index	20 MHz-km
2.	Graded-index	1 GHz-km
3.	Single mode	100 GHz-km

2.5.3 Pulse Spread and Maximum Data Rate

Definition of pulse-spreading constant: It is defined as the absolute time difference between the fastest and slowest light rays propagating through an optical fiber cable of unit length. It is also known as pulse-stretching constant. The unit of pulse-spreading constant is nanoseconds per kilometer (ns/km).

We are now interested to determine total pulse spread (ΔT). It can be easily derived from the basic definition of pulse-spreading constant. It is given as the product of pulse-spreading constant and the total fiber length (L). That is,

$$\Delta T_{(ns)} = \Delta t_{(ns/km)} \times \Delta L_{(km)} \tag{2.53}$$

Example 2.23 Maximum Permissible Pulse-Spreading Constant

A 45-km length of the optical fiber cable is required not to lengthen transmitted optical pulses by more than 100 ns. Find the maximum permissible value for the pulse-spreading constant.

Solution:

By definition, the total pulse spread (ΔT) is given as the product of the pulse-spreading constant (Δt) and the length (L) of the optical fiber cable. That is,

$$\Delta T_{(ns)} = \Delta t_{(ns/km)} \times \Delta L_{(km)}$$

Given length of the optical fiber, $L = 45$ km; and maximum pulse spread, $\Delta T = 100$ ns.

Therefore, maximum pulse-spreading constant can be determined as

$$\Delta t_{(ns/km)} = \frac{\Delta T_{(ns)}}{L_{(km)}} = \frac{100}{45} = 2.22 \text{ ns/km}$$ **Ans.**

1. *Maximum data rate for UP–NRZ transmissions:*
   We know that for unipolar non-return-to-zero (UP–NRZ) type of digital transmission signalling format, the maximum transmission data rate is given as

$$f_{b(Gbps)} = \frac{1}{2 \times \Delta T_{(ns)}} \tag{2.54}$$

$$\Rightarrow \quad f_{b(Gbps)} = \frac{1}{2 \times \Delta t_{(ns/km)} \times L_{(km)}} \tag{2.55}$$

Example 2.24 Transmission Rate in UP–NRZ

Determine the maximum digital transmission rates for Unipolar non-return-to-zero data transmissions over an optical fiber 10 km long with specified pulse-spreading constant of 5 ns/km.

Solution:

We know that for unipolar non-return-to-zero (UP–NRZ) type of digital transmission signalling format, the maximum transmission data rate can be expressed as

$$f_{b(Gbps)} = \frac{1}{2 \times \Delta t_{(ns/km)} \times L_{(km)}}$$

For the given pulse-spreading constant, $\Delta t = 5$ ns/km; and optical fiber length, L = 10 km,

$$f_{b(Gbps)} = \frac{1}{2 \times 5_{(ns/km)} \times 10_{(km)}} = 0.01\ Gbps,\ or\ 10\ \text{Mbps} \quad \textbf{Ans.}$$

2. *Maximum data rate for UP–RZ transmissions:*
   For unipolar return-to-zero (UP–RZ) digital transmission, the maximum data transmission rate is given as

$$f_{b(Gbps)} = \frac{1}{\Delta T_{(ns)}} \tag{2.56}$$

$$\Rightarrow \quad f_{b(Gbps)} = \frac{1}{\Delta t_{(ns/km)} \times L_{(km)}} \tag{2.57}$$

Example 2.25 Transmission Rate in UP–RZ

Determine the maximum digital transmission rates for unipolar return-to-zero data transmissions over an optical fiber 10-km long with specified pulse-spreading constant of 5 ns/km.

Solution:

We know that for unipolar return-to-zero (UP–RZ) digital transmission, the maximum data transmission rate is given as

$$f_{b(Gbps)} = \frac{1}{\Delta t_{(ns/km)} \times L_{(km)}}$$

For the given pulse-spreading constant, $\Delta t = 5$ ns/km; and optical fiber length, L = 10 km,

$$f_{b(Gbps)} = \frac{1}{5_{(ns/km)} \times 10_{(km)}} = 0.02\ Gbps,\ or\ 20\ \text{Mbps} \quad \textbf{Ans.}$$

> ***Note:*** UP–RZ transmissions can tolerate twice as much delay or spread as UP–NRZ transmissions. Accordingly, digital transmission rate possible for same optical fiber is twice as high for UP–RZ as for UP–NRZ transmission.

2.6 Types of Dispersions

Signal dispersion in optical fibers can be broadly classified in three main categories:

1. *Chromatic or Intramodal Dispersion:* This type of dispersion results from the finite spectral linewidth of the optical source. Propagation delay differences between the different spectral components of the transmitted light signal causes broadening of each transmitted mode and hence collectively known as intramodal dispersion. It may occur in all types of optical fibers (single-mode as well as multimode, step-index as well as graded-index). It is caused by the dispersive properties of the fiber material (material dispersion) and also guidance effects within the fiber structure (waveguide dispersion).
2. *Intermodal Dispersion or Mode Dispersion:* The main reason for intermodal dispersion, also known as mode dispersion, is that the difference in propagation delay between various propagation modes within a multimode fiber (hence, it is not applicable for single mode fiber). Different modes of a transmitted light pulse travel at different group velocities. Different transmission times between the fastest and slowest modes of propagation yields in broadening of transmitted optical pulse at the output of the fiber cable.
3. *Polarization Mode Dispersion:* Due to variations in the fiber core diameter in a practical single-mode fiber, two orthogonal, linearly polarized modes are supported that are degenerate. The variations in the fiber core diameter may be due to the presence of non-uniform stress, bends, stress, etc. along the length of the fiber. As discussed earlier, propagation times are different because of difference in the group velocities among light rays, which give rise to pulse broadening or pulse dispersion.

2.6.1 Intramodal Dispersion

A light source such as LED emits light which contains several wavelengths in a beam. We have earlier seen that refractive index is a function of operating wavelength. So, we can express refractive index n as $n(\lambda)$. We know that the velocity of light in any medium or material through which it propagates is expressed as $v = \dfrac{c}{n} = \dfrac{c}{n(\lambda)}$. This clearly shows that 'Light of different wavelengths travels along the length of the optical fiber cable at different velocities'. Even if all the rays of an incident light beam traverse the same propagation path within the optical fiber, they will arrive at the other end of the fiber at different times. As a result, the output light (optical) pulse will appear to be spread (stretched) in time. This phenomenon is called Intramodal dispersion, popularly known as chromatic dispersion (sometimes called wavelength dispersion also). It limits the transmission bandwidth of the optical fiber.

Facts to Know

Chromatic dispersion results from the presence of different wavelengths of light. The wavelength of visible light corresponds to its color. In optical communications, infrared light is commonly used than visible light, but even then the term 'chromatic dispersion' is more popular.

Chromatic dispersion is expressed directly in units of time rather than bandwidth. The chromatic dispersion is then the duration of the output pulse when an infinitesimally short pulse of light is applied to the input. The linewidth of the optical source is very important in determining the chromatic dispersion for single-mode fiber. Fig. 2.28 depicts the phenomenon of broadening of light pulse transmitted through a step-index single mode optical fiber.

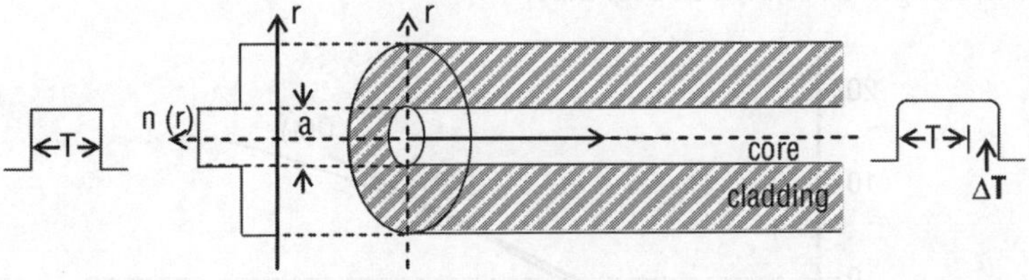

Fig. 2.28 Broadening of transmitted light pulse – single-mode fiber

Fig. 2.29 depicts the phenomenon of broadening of light pulse transmitted through a step-index multi-mode optical fiber.

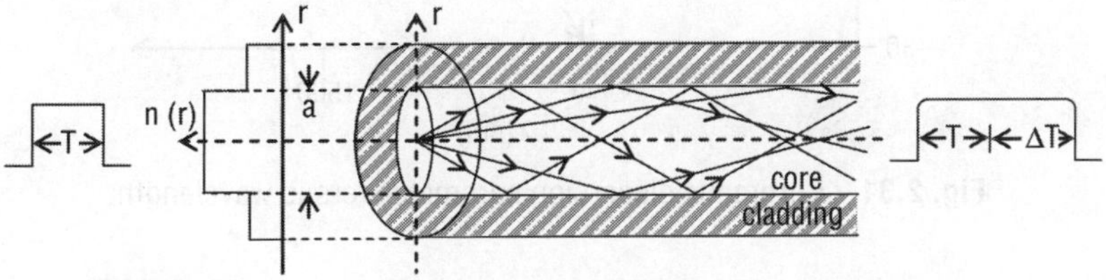

Fig. 2.29 Broadening of transmitted light pulse – multimode fiber

Fig. 2.30 depicts the phenomenon of broadening of light pulse transmitted through a graded-index multi-mode optical fiber.

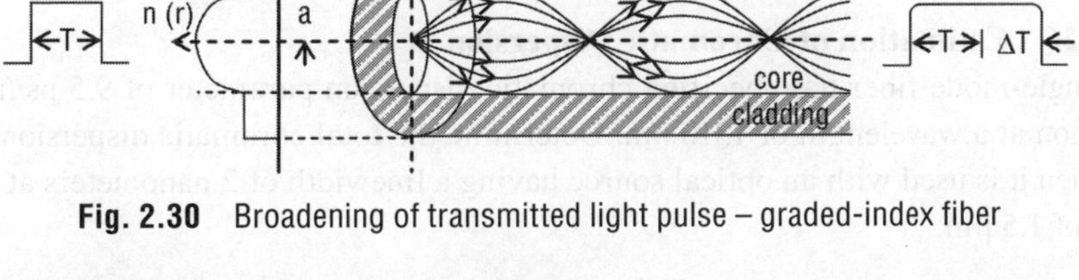

Fig. 2.30 Broadening of transmitted light pulse – graded-index fiber

It can be easily seen that chromatic dispersion is proportional to source linewidth as well as to the length of the optical fiber. In a single-mode fiber, the *chromatic dispersion parameter* is given in picoseconds (amount of pulse broadening) per nanometer of source linewidth for each unit of kilometer of the length of the fiber (abbreviated as ps/(nm–km). The expression for chromatic dispersion can be written as

$$(\Delta t_c) = D_c L(\Delta\lambda) \tag{2.58}$$

where, Δt_c = chromatic dispersion in nanoseconds (ns)

D_c = chromatic dispersion parameter in ps/(nm–km), generally specified by the manufacturer for a particular type of single-mode fiber

$\Delta\lambda$ = linewidth of the optical source in nanometers (nm)

L = the fiber length in kilometers (km)

It may be noted that the chromatic dispersion parameter D_c is zero at the specific wavelength, known as the *zero-dispersion wavelength*, λ_{ZD}. Obviously, the chromatic dispersion parameter varies with operating wavelength as shown in Fig. 2.31.

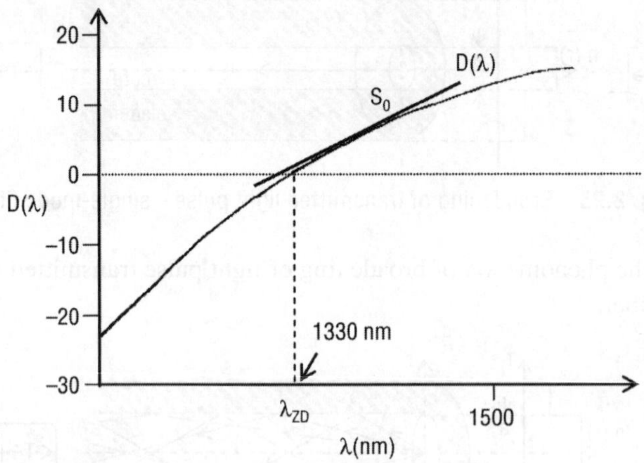

Fig. 2.31 Chromatic dispersion parameter versus wavelength

Facts to Know

The chromatic dispersion at a given wavelength also depends on its proximity to the zero-dispersion wavelength, which can also vary to some extent. There is indeed one wavelength (approximately only) where the dispersion is zero. Similarly, only the maximum value of the rate of variation of dispersion with wavelength is specified.

Example 2.26 Estimation of Chromatic Dispersion

A typical single-mode fiber has specified chromatic dispersion parameter of 9.5 ps/(nm–km) and zero dispersion at a wavelength of 1310 nm. Determine the total chromatic dispersion of 50 km of this fiber when it is used with an optical source having a linewidth of 2 nanometers at an operating wavelength of 1.5 μm.

Solution:

We know that total chromatic dispersion, $\Delta t_c = D_c(\Delta\lambda)L$

For the given $D_c = 9.5$ ps/(nm–km), $\Delta\lambda = 2$ nm, and fiber length $L = 50$ km, we have

$$\Delta t_c = 9.5(ps / nm - km) \times 2(nm) \times 50(km) = 950 \text{ ps}$$ **Ans.**

Note: Chromatic dispersion occurs only in single mode fibers (SMFs). It can be minimized to almost zero by using a monochromatic optical source. Example of such light source is an injection laser diode (ILD).

Effect of Chromatic Dispersion

Let us assume that a square wave signal corresponding to NRZ data with alternating 1s and 0s is applied to the fiber. When the dispersion is large enough that a pulse merges into the next bit period, intersymbol interference takes place. This limits the fiber's capacity for most applications. This intersymbol interference takes place when the dispersion is one-half the period of the square wave. The dispersion can be said to be equal to one-half of the period of an optical pulse at the maximum transmission bandwidth of the optical fiber, that is,

$$\Delta t_c = \frac{T_{\min}}{2}$$

where, $T_{\min}$ is the period of a signal at the maximum frequency transmitted by the fiber, related by $T_{\min} = \dfrac{1}{f_{\max}}$.

Therefore, $\Delta t_c = \dfrac{1}{2f_{\max}} \approx \dfrac{1}{2B_{\max}}$

where, $B_{\max}$ is the approximate maximum bandwidth of the fiber (same as f_{max}).

$$\Rightarrow \qquad B_{\max} \approx \frac{1}{2\Delta t_c} \approx \frac{1}{2D_c(\Delta\lambda)L} \tag{2.59}$$

Example 2.27 Bandwidth of Single-Mode Fiber (SMF)

A typical single-mode fiber has chromatic dispersion specification as 950 ps. Calculate its maximum possible transmission bandwidth.

Solution:

We know that maximum bandwidth, $B_{\max} \approx \dfrac{1}{2\Delta t_c}$

For the given $\Delta t_c = 950$ ps, we have

$$B_{\max} \approx \frac{1}{2\times 950\,ps} \approx 526.3 \text{ MHz}$$ **Ans.**

2.6.2 Dispersion Parameter

Consider that an optical pulse is launched at the source-end of a given single-mode fiber (SMF). Then, a spectral component of a given frequency ω of the light pulse would travel with a group velocity v_g given by

$$v_g = \frac{1}{d\beta/d\omega} \tag{2.60}$$

If the total length of the fiber is L, then this spectral component of the optical pulse would arrive at its other end with the group delay time, Δt_g given by

$$\Delta t_g = \frac{L}{v_g} = L\frac{d\beta}{d\omega}$$

For the spectral components that are $\Delta\omega$ apart, total delay difference Δt_g over fiber length L may be given as

$$\Delta T_g = \frac{d\Delta t_g}{d\omega}\Delta\omega = \frac{d}{d\omega}\left[L\frac{d\beta}{d\omega}\right]\Delta\omega = L\left(\frac{d^2\beta}{d\omega^2}\right)\Delta\omega \qquad (2.61)$$

The factor $\left(\frac{d^2\beta}{d\omega^2}\right)$ is defined as the group velocity dispersion parameter, or simply GVD parameter. Basically, it signifies the amount of pulse broadening as the light pulse travels down the fiber.

Using $\omega = 2\pi f = 2\pi\frac{c}{\lambda}$, and $\Delta\omega = \left(-\frac{2\pi c}{\lambda^2}\right)\Delta\lambda$.

Therefore,
$$\Delta T_g = \frac{dt_g}{d\lambda}\Delta\lambda = \frac{d}{d\lambda}\left[\frac{L}{v_g}\right]\Delta\lambda = L\frac{d}{d\lambda}\left[\frac{1}{v_g}\right]\Delta\lambda \qquad (2.62)$$

The intramodal dispersion parameter, D, is expressed as the optical pulse spread per unit fiber length per unit spectral width of the optical source. Mathematically,

$$D = \frac{1}{L}\frac{d\Delta t_g}{d\lambda} = \frac{d}{d\lambda}\left(\frac{1}{v_g}\right) \qquad (2.63)$$

The unit of intramodal dispersion parameter is ps/(nm–km), same as that of chromatic dispersion parameter.

> **Note:** Intramodal dispersion is the combination of dispersion effects due to material dispersion as well as waveguide dispersion present together in an optical fiber.

2.6.3 Material Dispersion

Material dispersion can be considered as the spreading of the transmitted optical pulse through the optical fiber due to dispersive properties of the fiber core material. It mainly occurs as a result of the variation of the refractive index of the fiber core material with respect to the operating wavelength of the light propagating through the fiber. A beam of light may consist of many different wavelengths which propagate at different velocities in the glass material of the fiber core. We know that the group velocity of a propagation mode of the light through the fiber is a function of the core refractive index. It implies that the various spectral components of any mode may propagate with different group velocities. Therefore, if the spectral width (also called the linewidth) of an optical source is quite large, then the broadening of the transmitted optical pulse due to this effect may be significant. This is termed as material dispersion.

Now let us determine the material dispersion parameter in case of a single-mode fiber. It is expressed as

$$D_m = \frac{\lambda}{c}\left|\frac{d^2n}{d\lambda^2}\right| \qquad (2.64)$$

The units of D_m is ps/(nm–km). It may be noted that D_m changes sign at the wavelength of zero material dispersion λ_{ZD}. For pure silica, λ_{ZD} = 1276 nm.

The broadening of the optical pulse due to material dispersion can then be expressed as

$$\Delta t_m = D_m L(\Delta\lambda) \tag{2.65}$$

$$\Rightarrow \quad \Delta t_m = \left|\frac{\lambda}{c}\frac{d^2 n}{d\lambda^2}\right| L(\Delta\lambda) \tag{2.66}$$

$$\Rightarrow \quad \Delta t_m = \left|\frac{\Delta\lambda}{\lambda}\right| \times \left|\frac{\lambda^2}{c}\frac{d^2 n}{d\lambda^2}\right| L \tag{2.67}$$

Thus, the pulse broadening per unit fiber length due to material dispersion is

$$\frac{\Delta t_m}{L} = \left|\frac{\Delta\lambda}{\lambda}\right| \times \left|\frac{\lambda^2}{c}\frac{d^2 n}{d\lambda^2}\right| \tag{2.68}$$

The term $\left|\frac{\Delta\lambda}{\lambda}\right|$ on the right-hand-side of this expression represents the relative spectral width of the optical source, and is denoted as γ. Then,

$$\frac{\Delta t_m}{L} = \frac{\gamma}{c}\left|\lambda^2 \frac{d^2 n}{d\lambda^2}\right| \tag{2.69}$$

> ***Note:*** The pulse broadening due to material dispersion can be substantially minimized by using appropriate optical source having a very narrow spectral width around zero dispersion wavelength, λ_{ZD}.

Example 2.28 Effect of Material Dispersion

Let the spectral width of the optical source is specified as 30 nm. Calculate the total amount of pulse broadening due to material dispersion in the following cases:

(a) For an optical fiber of 80-km length when the optical source emits at 850-nm wavelength. The material dispersion parameter 850 nm is specified as -105.5 ps/(nm–km).

(b) For an optical fiber of 80-km length when the optical source emits at 1300-nm wavelength. The material dispersion parameter at 1300 nm is specified as -2.8 ps/(nm–km).

Solution:

We know that $\Delta t_m = \left|D_m L(\Delta\lambda)\right|$

(a) For the given $|D_m|$ = 105.5 ps/(nm–km), L = 80 km, and $\Delta\lambda$ = 30 nm, we get

$$\Delta t_m = 105.5 \times 80 \times 30 = 253{,}200\, ps = 253.2\, ns$$ **Ans.**

(b) For the given $|D_m|$ = 2.8 ps/(nm–km), L = 80 km, and $\Delta\lambda$ = 30 nm, we get

$$\Delta t_m = 105.5 \times 80 \times 30 = 6720\, ps = 6.72\, ns$$ **Ans.**

It can be seen that 1300-nm wavelength yields much less amount of pulse broadening due to material dispersion for given parameters.

2.6.4 Waveguide Dispersion

Another source of dispersion (i.e., optical pulse broadening) in a single-mode fiber is called waveguide dispersion. It may be noted that it is significant in single-mode fiber only, and is almost negligible in multimode fiber. The waveguide dispersion occurs because some part of the optical energy may propagate in the cladding rather than in the fiber core. Because the fiber core and cladding do have different refractive indexes, the light propagates at different velocities in the core and the cladding. Assuming no material dispersion (that is, n_1 and n_2 do not vary with λ), the time taken by a propagation mode to travel the fiber length L (due to waveguide dispersion only) is represented by

$$\Delta t_w = \frac{L}{v_g} = \frac{Ln_2}{c}\left[1 + \Delta\frac{d}{dV}(bV)\right] \tag{2.70}$$

where, Δ = the relative refractive index difference which is equal to $\frac{n_1 - n_2}{n_2}$, as specified earlier;

V = normalized frequency parameter, or V-parameter given by $V = \frac{2\pi a n_1\sqrt{2\Delta}}{\lambda}$

b = normalized propagation constant within the fiber, related to V, and ω

For an optical source of spectral width $\Delta\lambda$, the corresponding pulse broadening due to waveguide dispersion can also be determined using $\frac{d\Delta t_w}{d\lambda}\Delta\lambda$. Therefore,

$$\Delta t_w = \Delta t_w = D_w L(\Delta\lambda) \tag{2.71}$$

where, $D_w = -\frac{n_2\Delta}{c\lambda}\left[V\frac{d^2}{dV^2}(bV)\right]$ is called the waveguide dispersion parameter, the factor inside the bracket is given by an empirical formula

$$V\frac{d^2}{dV^2}(bV) = 0.08 + 0.549(2.834 - V)^2 \tag{2.72}$$

Therefore,

$$D_w = -\frac{n_2\Delta}{c\lambda}\left[0.08 + 0.549(2.834 - V)^2\right] \tag{2.73}$$

It may be noted that the value of D_w is negative for λ in the 1000–1700 nm range.

Example 2.29 Intramodal Dispersion

A single-mode step-index fiber has the following specifications:

- Fiber core refractive index = 1.45
- Relative refractive index difference = 0.5%
- Material dispersion = 6 ps/(nm–km) at wavelength = 1550 nm.

Determine the diameter of the fiber core needed to make total dispersion of the fiber zero at specified wavelength.

Solution:

It is required that the total dispersion, $D = D_m + D_w = 0$.

For the given $D_m = 6 \text{ ps}/(nm - km)$, we obtain the value of $D_w = -6 \text{ ps}/(nm - km)$

Using
$$\Delta = \frac{n_1 - n_2}{n_2}$$

$$\Rightarrow \quad 0.005 = \frac{1.45 - n_2}{n_2}$$

$$\Rightarrow \quad n_2 = \frac{1.45}{1.005} = 1.4427$$

We know that $D_w = -\frac{n_2 \Delta}{c\lambda}\left[0.08 + 0.549(2.834 - V)^2\right]$

$$\Rightarrow \quad -\frac{6 \times 10^{-12} s}{1 \times 10^{-9} m \times 1 \times 10^3 m} = -\frac{1.4427 \times 0.005}{3 \times 10^8 (m/s) \times 1550 \times 10^{-9} m}\left[0.08 + 0.549(2.834 - V)^2\right]$$

On solving it, we get $V = 2.0863$

We know that
$$V = \frac{2\pi a n_1 \sqrt{2\Delta}}{\lambda}$$

$$\Rightarrow \quad 2a = \frac{V\lambda}{\pi n_1 \sqrt{2\Delta}}$$

$$\Rightarrow \quad 2a = \frac{2.0863 \times 1550 \times 10^{-9}}{\pi \times 1.45\sqrt{2 \times 0.005}} = 7.1 \times 10^{-6} \text{m} = 7.1 \,\mu\text{m}$$ **Ans.**

> ***Note:*** Material dispersion and waveguide dispersion are not necessarily additive. In fact they can and do cancel each other out at some wavelength. For a normal glass fiber, this wavelength is about 1300 nm in the infrared region.

2.6.5 Intermodal Dispersion

From the basics of digital optical transmission, we represent a logic 1 with the presence of an optical pulse, and a logic 0 with the absence of an optical pulse (i.e., no optical pulse). Such type of optical pulses is generated by an optical source such as an LED or an injection laser diode. When these optical pulses are coupled into an optical fiber, then each optical pulse is divided into a number of small optical pulses, each one is carried by an individual mode of propagation. At the output of the fiber, individual optical pulses may recombine due to overlapping. In such situation, the optical receiver sees one long optical pulse having peculiar properties—its rising edge has the fundamental mode, whereas its falling edge has the critical mode of light propagation.

Intermodal Dispersion in Step-index Fiber

Fig. 2.32 depicts a typical phenomenon of pulse broadening created by four different modes of an incident light beam propagating through the optical fiber. It is known as the *intermodal dispersion.*

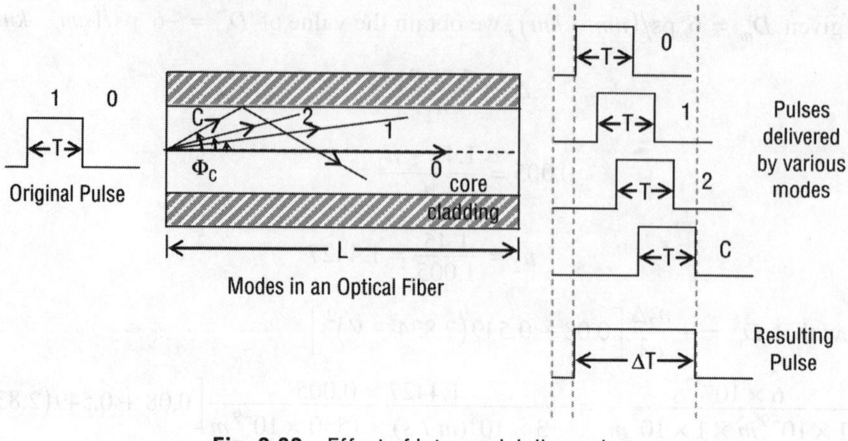

Fig. 2.32 Effect of intermodal dispersion

Intermodal dispersion basically occurs because of differences in propagation delays among various modes through a multimode fiber. As a result, the pulse broadening depends upon the following factors:

- transmission time of the fastest mode of propagation
- transmission time of the slowest mode of propagation

The difference in time delay between the axial ray (the fastest mode propagation) and the extreme meridional ray (the input optical ray incident at critical angle) depends on their respective path lengths travelled within the multimode step-index fiber. That is,

$$\therefore \quad \Delta t_{SI} = t_c - t_o = \frac{L}{v\cos\theta_c} - \frac{L}{v} = \frac{L}{v}\left[\frac{1}{\cos\theta_c} - 1\right] \tag{2.74}$$

where, $v = \dfrac{c}{n_1}$ represents the velocity of light within the fiber core.

It is assumed here that the refractive index (n_1) of the fiber core is constant throughout its length and all the modes travel at the same velocity.

Substituting $\cos\theta_c = \dfrac{n_2}{n_1}$ (Using Snell's Law)

$$\Delta t_{SI} = \frac{Ln_1}{c}\left[\frac{n_1}{n_2} - 1\right] \tag{2.75}$$

$$\Rightarrow \quad \Delta t_{SI} = \frac{Ln_1}{c}\frac{(n_1 - n_2)}{n_2} \times \frac{n_1}{n_1} \tag{2.76}$$

$$\Rightarrow \quad \Delta t_{SI} = \frac{Ln_1^2}{cn_2} \times \left(\frac{n_1 - n_2}{n_1}\right) = \frac{Ln_1^2}{cn_2}\Delta$$

where $\Delta \ll 1$,

$$\Delta = \frac{n_1 - n_2}{n_1} \tag{2.77}$$

$\because$ $$\text{NA} = n_1\sqrt{2\Delta}$$

$\Rightarrow$ $$\Delta = \frac{1}{2}\left(\frac{NA}{n_1}\right)^2 \tag{2.78}$$

$$\Delta t_{SI} = \frac{Ln_1^2}{cn_2}\frac{(NA)^2}{2n_1^2} \approx \frac{L(NA)^2}{2n_1c} \tag{2.79}$$

$\because$ $$n_1 \approx n_2$$

This is also referred to as multipath time dispersion of the multimode step-index fiber.

Example 2.30 Intermodal Dispersion in Step–Index Multimode Fiber

A step-index multimode fiber has the following specifications:

- Fiber core diameter = 100μm
- Core refractive index = 1.5
- Cladding refractive index = 1.48.

Calculate the pulse broadening per unit length of the above-mentioned optical fiber cable due to multipath dispersion.

Solution:

We know that $$\Delta t_{SI} = \frac{Ln_1}{n_2} \times \left(\frac{n_1 - n_2}{c}\right)$$

$\Rightarrow$ $$\frac{\Delta t_{SI}}{L} = \frac{n_1}{n_2} \times \left(\frac{n_1 - n_2}{c}\right)$$

For the given $n_1 = 1.5$, and $n_2 = 1.48$, we have

$$\frac{\Delta t_{SI}}{L} = \frac{1.5}{1.48} \times \left(\frac{1.5 - 1.48}{3 \times 10^8 m/s}\right) = 6.75 \times 10^{-11} s/m = 67.5 \text{ ns/km}$$ **Ans.**

Example 2.31 Maximum Bit Rate due to Modal Dispersion

A step-index multimode fiber has specified core refractive index = 1.486 and numerical aperture = 0.2. Assuming modal dispersion, calculate the maximum transmission bit rate (bps) for 1 km fiber length.

Solution:

We know that $\Delta t_{SI} \approx \dfrac{L(NA)^2}{2n_1c}$

For the given L = 1 km, NA = 0.2, and $n_1 = 1.486$, we have

$$\Delta t_{SI} = \frac{1 \times 10^3 \text{ m} \times (0.2)^2}{2 \times 1.486 \times 3 \times 10^8 \text{ m/s}} = 4.48 \times 10^{-8} \text{ s} = 44.8 \text{ ns}$$

Number of bits per second can be determined as

$$\frac{1}{\Delta t_{SI}} = \frac{1}{44.8 \times 10^{-9} s} = 0.2229 \times 10^8 \text{ bps} = 22.29 \text{ Mbps} \quad \textbf{Ans.}$$

Intermodal Dispersion in Graded-index Fiber

We know that in graded-index optical fiber, the index profile (with profile parameter $\alpha = 2$) can be expressed as:

$$n(r) = n_1\left[1 - 2\Delta\left(\frac{r}{a}\right)^2\right]^{1/2} \qquad r < a \text{ (core)} \quad (2.80)$$

$$= n_1\left[1 - 2\Delta\right]^{1/2} = n_2 \qquad r \geq a \text{ (cladding)}$$

Since the core refractive index n_1 is generally not uniform within the fiber core, the meridional rays (apart from axial ray) follow the sinusoidal trajectories having different path lengths. This, in turn, may result in index grading, as shown in Fig. 2.33.

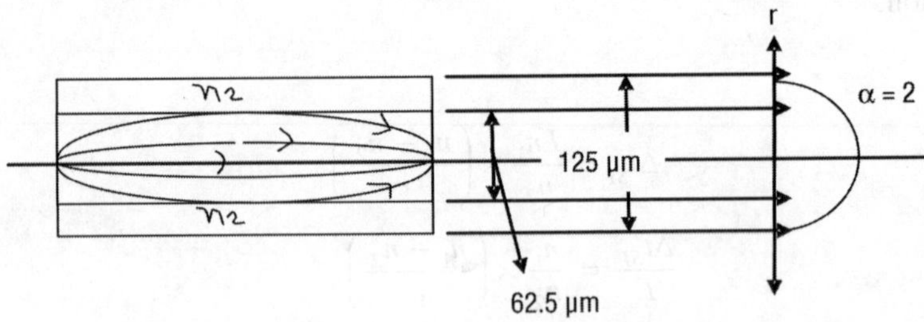

Fig. 2.33 Light propagation in a graded-index fiber

However, the group velocity $\quad v_g = \dfrac{c}{n_g} \qquad (2.81)$

where, n_g represents the group index of the multimode graded-index fiber core.

i.e., $\quad v_g \propto \dfrac{1}{n_g}$

This means that longer the sinusoidal paths traversed by light rays, higher will be their velocity of propagation in the lower refractive index medium which is away from the axis of the fiber. Hence, there is an equalization of the transmission times of various path trajectories vis-à-vis propagation time of the axial ray that encounters the highest refractive index region, thereby having the lowest velocity of propagation.

Thus, the fractions of an input pulse delivered by the different modes arrive at the receiver and more or less simultaneously. Therefore, intermodal dispersion will be reduced and the bit rate will be increased. The pulse broadening in graded-index fiber for minimum dispersion corresponding to $\alpha = 2(a - \Delta)$is given by

$$\Delta t_{GI} = \Delta t_{GI} = \frac{Ln_g\Delta^2}{8c} \tag{2.82}$$

$\because \quad ng \approx n_1;$ $\qquad NA = n_1\sqrt{2\Delta}$

$$\Rightarrow \qquad \Delta = \frac{1}{2}\frac{NA^2}{n_1^2}$$

$$\Rightarrow \qquad \Delta t_{GI} = \frac{Ln_1}{8c}\frac{1}{4}\frac{NA^4}{n_1^4} \tag{2.83}$$

$$\Delta t_{GI} = \frac{L \times NA^4}{32cn_1^3} \tag{2.84}$$

It may be observed that the multipath time dispersion, or intermodal dispersion in grade-index fibers will be lesser than in step-index fibers. This is due to the reason that the rays near the core axis have to travel shorter paths compared to those near the intersection of the fiber core and the cladding. The velocity of the light rays near the axis of the fiber core will be lesser than that of the meridional rays because they have to travel through a region of high refractive index ($v = c/n$). Thus, both the light rays will arrive at the other extreme of the fiber almost simultaneously. As a result, there is reduction in multipath time dispersion.

> ***Note:*** If the refractive index profile of a graded-index fiber is chosen in such a way so that the time taken for the axial light ray and the most oblique light ray is nearly same, then the multipath dispersion will be zero. In practice, a parabolic profile ($\alpha = 2$) reduces this type of dispersion considerably.

Example 2.32 Intermodal Dispersion

A typical fiber has core refractive index = 1.50, and relative refractive index difference = 2%. Show that the graded-index profile fiber exhibits much less modal dispersion as compared to that of step-index fiber for a given fiber length of 5 km.

Solution:

(a) Intermodal dispersion in step-index fiber

Using the relationship, $\qquad \Delta t_{SI} = \dfrac{Ln_1\Delta}{c}$

For the given L = 5 km, n_1 = 1.5, and Δ = 2%, or 0.02, we have

$$\Delta t_{SI} = \frac{Ln_1\Delta}{c} = \frac{5 \times 10^3 \times 1.50 \times 0.02}{3 \times 10^8} = 500\ ns \qquad \textbf{Ans.}$$

(b) Intermodal dispersion in graded-index fiber

We know that $\qquad \Delta t_{GI} = \dfrac{Ln_1\Delta^2}{8c}$

For the given L = 5 km, n_1 = 1.5, and Δ = 2%, or 0.02, we have

$$\Delta t_{GI} = \frac{Ln_1\Delta^2}{8c} = \frac{5 \times 10^3 \times 1.50 \times 0.02^2}{8 \times 3 \times 10^8} = 1.25\ ns \qquad \textbf{Ans.}$$

Hence it is seen that graded-index profile fiber exhibit much less amount of intermodal dispersion vis-à-vis step-index profile fiber for identical parameters.

Example 2.33 Total Dispersion and Bit Rate

An optical source having line width of 40 nm excites a 2-km long multimode fiber cable which has the following specifications:

- Modal dispersion = 1 ns/km
- Dispersion parameter due to chromatic dispersion = 100 ps/(nm–km).

Find the total system dispersion and the achievable transmission data rate.

Solution:

We know that total dispersion, $\Delta t_{Total} = \sqrt{(\Delta t_{Modal})^2 + (\Delta t_{Chromatic})^2}$

where, $\Delta t_{Modal} = D_{\text{modal}} \times L$; and $\Delta t_{Chromatic} = D_{chromatic} \times \Delta\lambda \times L$

For the given D_{modal} = 1 ns/km and L = 2 km, we have

$$\Rightarrow \qquad \Delta t_{Modal} = 1\ (ns\ /\ km) \times 2\ km = 2\ \text{ns}$$

Similarly, for the given $D_{chromatic}$ =100 ps/(nm–km); $\Delta\lambda$ = 40 nm; and L = 2 km, we get

$$\Rightarrow \qquad \Delta t_{Chromatic} = 100\ \text{ps/(nm-km)} \times 40\ \text{nm} \times 2\ \text{km} = 8000\ \text{ps} = 8\ \text{ns}$$

Hence, $\Delta t_{Total} = \sqrt{(2\ \text{ns})^2 + (8\ \text{ns})^2} = 8.24\ \text{ns}$ **Ans.**

We know that achievable data rate = $\dfrac{1}{\Delta t_{Total}} = \dfrac{1}{8.24\ \text{ns}} = 121\ \text{Mbps}$ **Ans.**

2.6.6 Polarization Mode Dispersion

Definition: Polarization mode dispersion (PMD) is defined as the difference in the propagation times (that is, pulse broadening) between two orthogonally polarized components of the light pulse in a single-mode fiber.

We know that the fundamental mode of the fiber has two orthogonally polarized components. When the input light pulse excites both of these components, then they travel with difference group velocities. As a result, both components do not arrive at the other end of the fiber having length L km at the same time. This difference in their propagation times gives rise to pulse broadening. This is known as the polarization mode dispersion.

> ***Note:*** Polarization mode dispersion (PMD) can be a major concern to achieve high transmission data rates in long-haul fiber-optic communication networks.

Modal Birefringence

As mentioned earlier, a single-mode fiber supports two orthogonally linearly-polarized propagation modes that are degenerate. Assume that a single-mode fiber is manufactured with a perfectly cylindrical fiber core having constant diameter throughout the length of the fiber. Ideally, these two modes travel with the same velocity because they have the same mode index, designated by $\tilde{n}$, as well as the propagation constant, designated by β. But in a practical fiber, the core is not perfectly cylindrical and it may have non-uniform diameter due to the presence of bends, stress, twists, etc. along the length of the fiber. Hence the propagation constants of these two orthogonal linearly polarized components become different.

Definition: Modal birefringence represents the actual difference between the mode indices of two orthogonally linearly-polarized components of degenerative propagation modes in a single-mode fiber in the x- and y-direction.

Mathematically,

Modal birefringence,
$$\delta n = |n_x - n_y| \tag{2.85}$$

where, n_x and n_y represent the respective mode indices of the polarized components in the x- and y-direction.

The corresponding change in the propagation constants may be given by

$$\delta\beta = \frac{2\pi}{\lambda}\delta n = \frac{2\pi}{\lambda}|n_x - n_y| \tag{2.86}$$

where, λ represents the wavelength of light in vacuum.

$$\Rightarrow \qquad \delta n = \frac{\lambda}{2\pi}\delta\beta \tag{2.87}$$

Assume that initially both components of orthogonally linearly-polarized light possess the same amplitude levels and there is no phase difference as these modes are excited into a single-mode fiber. As the light travels through the fiber length in the forward direction, the two components produce elliptically polarized light due to different phase propagation constants. At a phase difference of $\pi/2$, circularly polarized light will be produced. Thus, the polarization changes from a linear shape to an elliptical one, and then to a circular shape to an elliptical one, and back to linear shape as the light travels along the fiber length.

Let us now introduce another relevant term, `beat length' here.

Definition of beat length: It is defined as the length of the fiber over which the phase difference between two orthogonally polarized modes becomes exactly equal to 2π radians.

Beat length,
$$L_p = \frac{2\pi}{\delta\beta} = \frac{2\pi}{\frac{2\pi}{\lambda}\delta n} = \frac{\lambda}{\delta n} = \frac{\lambda}{|n_x - n_y|} \tag{2.88}$$

Assume that the fiber has constant modal birefringence. If the group velocities of the two polarized components are v_{gx} and v_{gy} in the x- and y-direction respectively, then both components will arrive at the other end of the fiber after taking times L/v_{gx} and L/v_{gy} respectively, where L is the length of the fiber. The time delay between two orthogonally polarized components results into pulse broadening due to polarization mode dispersion. It is given by the following expression:

$$\Delta t_{PMD} = \left| \frac{L}{v_{gx}} - \frac{L}{v_{gy}} \right| \tag{2.89}$$

However, in practical single-mode fibers, the modal birefringence varies randomly in both magnitude as well as direction along the length of the fiber. Therefore, the above expression for polarization mode dispersion is not valid. In fact, linearly polarized light launched in such a fiber quickly attains an arbitrary state of polarization. An approximate estimate of PMD-induced pulse broadening can be made using the following expression:

$$\Delta t_{PMD} = D_{PMD}\sqrt{L} \tag{2.90}$$

where, D_{PMD} is the average PMD parameter which normally varies between 0.01 $ps/\sqrt{km}$ and 10 $ps/\sqrt{km}$, and L is the fiber length in km.

Example 2.34 Modal Birefringence

A beat length of 10 cm is observed in a single-mode fiber when light rays of wavelength 1000 nm is propagated through it. Calculate the amount of modal birefringence for the two orthogonally polarized modes.

Solution:

We know that the beat length, $L_P = \dfrac{2\pi}{\delta\beta}$

$\Rightarrow \quad \delta\beta = \dfrac{2\pi}{L_p}$

For the given $L_p = 10$ cm, $\delta\beta = \dfrac{2\pi}{10 \times 10^{-2}\, m}$

We know that modal birefringence, $\delta n = \dfrac{\lambda}{2\pi}\delta\beta$.

Therefore, $\delta n = \dfrac{1000 \times 10^{-9}\, m}{2\pi} \times \dfrac{2\pi}{10 \times 10^{-2}\, m} = 10 \times 10^{-6}\, m = 10 \mu m$ **Ans.**

Example 2.35 PMD-induced Pulse Broadening

An average PMD parameter $D_{PMD} = 10\ ps/\sqrt{km}$ is specified for a typical single-mode fiber. If the fiber length is 100 km, calculate the PMD-induced pulse broadening.

Solution:

We know that PMD-induced pulse broadening may be estimated by

$$\Delta t_{PMD} = \Delta t_{PMD} = D_{PMD}\sqrt{L}$$

For the given $D_{PMD} = 10\ ps/\sqrt{km}$ and $L = 100$ km, we get

$$\Delta t_{PMD} = 10\sqrt{100}\ \text{ps} = 100\ \text{ps}$$ **Ans.**

Facts to Know

By changing the refractive index profile in the core as well as in the cladding, single-mode fibers can be designed that exhibit different dispersion properties. Examples of such types of single-mode fibers include dispersion-shifted fibers, dispersion-compensated fibers, dispersion-flattened fibers, and polarization-maintaining fibers.

Section Practice Problems

1. A light pulse is launched into source end of a 30-km long optical fiber with specified total dispersion value 20 ns/km. What will be the pulse width at the other end of the fiber? *[Ans.: 600 ns]*
2. An optical source emits a mean wavelength of $\lambda = 0.90$ μm having spectral width $\Delta\lambda = 18$ nm. What is its relative spectral width? *[Ans.: 0.02]*
3. Calculate the total pulse broadening that occurs in an optical fiber of 80 km length due to material dispersion when an optical source such as laser diode having spectral width as 2 nm emits at
   (a) 850 nm wavelength. The material dispersion parameter is specified as -105.5 ps/(nm–km).
   (b) 1300 nm wavelength. The material dispersion parameter is specified as -2.8 ps/(nm–km).
   [Ans.: (a) 16.88 ns; (b) 0.448 ns]
4. A given optical source with the relative spectral width of 0.035 at $\lambda = 850$ nm is coupled to a pure silica fiber with specified parameter $\left|\lambda^2 \frac{d^2 n}{d\lambda^2}\right| = 0.021$ at 850 nm. Calculate the pulse broadening per kilometer fiber length if only material dispersion occurs. *[Ans.: 2.45 ns/km]*
5. A fiber has core refractive index = 1.50 and numerical aperture = 0.3. Show that the graded-index profile fiber exhibit less amount of modal dispersion as compared to that exhibited by step-index profile fiber for a total fiber length of 5 km.
6. In a conventional single-mode fiber, it is specified that the difference in effective refractive indices for two orthogonally linearly-polarized modes varies from 9.3×10^{-7} to 10.1×10^{-5}. Compute the corresponding variations in the beat lengths of the fibers at $\lambda = 1300$ nm. *[Ans.: 0.12–1.4 m]*

2.7 Attenuation in Optical Fibers

As the optical signal propagates down the optical fiber cable, it gets attenuated. The term 'attenuation' simply means the reduction in the signal strength. Attenuation is probably the most important parameter of an optical fiber cable. It limits the performance of an optical fiber communication system, as the optical power reaching the receiver is reduced. Every receiver needs a certain minimum input power for optimum signal detection. Signal attenuation has had several adverse effects on the performance of optical fiber communication system.

Mathematically, the total power loss in an optical fiber cable is given by

$$A_{(dB)} = 10\log\left(\frac{P_{out}}{P_{in}}\right) \tag{2.91}$$

where, $A_{(dB)}$ represents attenuation (i.e., total reduction in power level) in dB, P_{out} is the cable output power in watts, and P_{in} is the cable input power in watts.

It can be easily seen that a 1-dB attenuation decreases the output power to about 79% of the input power, and 3-dB attenuation decreases the output power to 50% of the input power.

Example 2.36 Power Loss in dB versus % Output Power

Tabulate the typical values of power loss in dB as percent output power of input power.

Solution:

Table 2.4 gives the output power as a percentage of input power for an optical fiber cable with several typical values of power loss specified in decibels.

Table 2.4 Power loss in dB versus percentage output power

S. No.	Power Loss, A(dB)	Output Power (%)
1.	1	79
2.	2	63
3.	3	50
4.	6	25
5.	9	12.5
6.	10	10
7.	13	5
8.	20	1

Note: Although total power loss expressed in decibel (dB) is of primary importance in an optical fiber cable, attenuation is generally expressed in dB per unit length of the fiber cable. Attenuation is expressed as a positive dB value because by definition it is a loss.

Mathematically, the measured optical power in watts at a given distance from a power source can be determined as

$$P = P_t \times 10^{-(A\times L)/10} \tag{2.92}$$

where, P = measured optical power level (watts)

P_t = transmitted optical power level (watts)

A = cable attenuation (dB/km)

L = cable length (km)

Expressing in dB, we have $P(dBm) = P_t(dBm) - (A \times L)(dB)$ (2.93)

The product $(A \times L)$ is called cable power loss, or attenuation expressed in dB.

Facts to Know

If we compare the attenuation performance of optical fiber cables with that of twisted-pair or RF coaxial cables, the attenuation is quite flat with respect to fiber cable length. This means that less number of (may be as less as 10 times) signal regenerators are needed with the use of optical fiber cables.

Example 2.37 Determination of Optical Power

For a single-mode fiber cable having specified attenuation as 0.25 dB/km, determine the measured optical power at a distance of 100 km from a 0.1-mW optical source.

Solution:

We know that measured optical power, $P = P_t \times 10^{-(A \times L)/10}$

Substituting given values of P_t = 0.1 mW (1 x 10^{-4} W), A = 0.25 dB/km, L = 100 km,

$$P = 1 \times 10^{-4} \times 10^{-(0.25 \times 100/10)} = 0.316\ \mu\text{W} \qquad \textbf{Ans.}$$

Converting it in dBm, we have

$$P = 10 \log \frac{0.316 \times 10^{-3}\ \text{mW}}{1\ \text{mW}} = -35\ \text{dBm} \qquad \textbf{Ans.}$$

Alternatively,

Using the expression $$P(dBm) = P_t(dBm) - (A \times L)(dB)$$

where $$P_t(dBm) = 10 \log \left(\frac{0.1\ mW}{1\ \text{mW}} \right) = -10\ \text{dBm}$$

Therefore, $P(dBm) = -10\ \text{dBm} - 0.25\ (dBm\ /\ km) \times 100\ \text{km} = \text{-35 dBm}$ **Ans.**

An ideal fiber would have no signal attenuation, and $P_{out} = P_{in}$. This corresponds to 0-dB/km attenuation. But, in practice, it is impossible. For example, a practical low-loss fiber may have a 3-dB/km average loss at 900 nm. This means that the optical signal power would decrease by 50% (one-half) over a 1-km fiber length and would decrease by 75% (a 6-dB loss) over a 2-km fiber length. Note that the loss contributions expressed in dBs are additive.

2.7.1 Fiber Attenuation versus Wavelength

Attenuation of light propagating through fiber made of glass material depends on the operating wavelength. Typical wavelength regions commonly used for optical fiber communications systems are centered around 1550 nm, 1300 nm, and 850 nm.

Table 2.5 lists attenuation in dB/km for several types of standard optical fiber cables.

Table 2.5 Attenuation (dB/km) in standard optical fiber cables

S. No.	Type of Optical Fiber Cable	Fiber Core Diameter (μm)	Fiber Cladding Diameter (μm)	Fiber Cable Attenuation (dB/km)
1.	Single-mode step-index fiber	5	125	0.4 at 1300 nm
2.		8	125	0.5 at 1300 nm

S. No.	Type of Optical Fiber Cable	Fiber Core Diameter (μm)	Fiber Cladding Diameter (μm)	Fiber Cable Attenuation (dB/km)
3.	Multimode step-index fiber	200	380	6 at 850 nm
4.		300	440	6 at 850 nm
5.	Multimode graded-index fiber	50	125	4 at 850 nm
6.		100	140	5 at 850 nm

Note: **In general, multimode fibers tend to have more attenuation than single-mode fibers, primarily because of the increased scattering of the light wave produced from the dopants in the glass.**

For the kind of glass typically used for optical communications systems, the 1300-nm and 1550-nm bands have less than 5% loss per kilometer. The 850-nm band experiences almost 20% loss per kilometer. Fig. 2.34 shows the typical measurement of attenuation (loss in dB/km) versus wavelength in optical fiber cable.

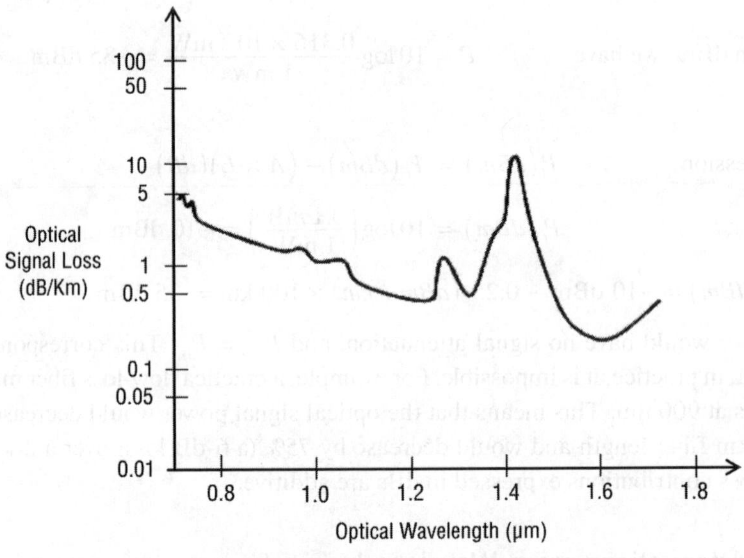

Fig. 2.34 Attenuation versus wavelength in optical fiber cable

From the above figure, the following observations can be made:

- Loss in glass fibers is lowest at wavelengths around 1550 nm, in the infrared region of the spectrum.
- Loss in plastic fibers performs best at about 660 nm, which corresponds to visible red light.
- The sharp peak at about 1400 nm is due to the presence of water ions, and operation near this wavelength should be avoided.
- Optical fibers are not always used at the wavelength exhibiting lowest loss.
- Many short-range systems use a wavelength of about 820 nm, as LEDs for this wavelength are less costly than those designed for longer wavelengths.

- On the other hand, long-distance high-data-rate links may be limited more by dispersion than by power loss. In this case, they can operate at the lowest-dispersion wavelength of 1300 nm.

Another possibility is to modify the fiber by changing its composition or the way in which the refractive index varies to bring its zero-dispersion wavelength closer to the minimum-loss wavelength.

> ***Note:*** Fiber losses are an important factor that limits the distance between repeaters, because even with laser diodes as optical sources, the optical power output of the order of a few mW only can be coupled into the optical fiber cable.

2.8 Transmission Losses in Optical Fiber Cable

Various types of optical signal transmission losses in optical fiber cables result in less output power than required, thereby reducing the overall system capacity, net transmission bandwidth, transmission data rate, and operating efficiency. Losses in optical fibers result from attenuation in the material itself and from scattering, which causes some of the light to strike the intersection of the fiber and the cladding at incidence angle less than the desired critical angle of incidence so as to produce total internal reflection, so vital for propagation of optical signals within the optical fiber cable. Some part of the incident light may get refracted from the core-cladding interface and enter into the cladding, which does not contribute to the transmission of optical signals through the optical fiber cable.

Let us turn our attention to various factors that may cause signal attenuation in optical fiber cables. It is emphasized here that many of these factors are influenced by the waveguide structure, and the composition of the basic material (including preparation and purification) used for the fiber core and cladding. Some of the major transmission losses in optical fiber cables may be categorized as follows:

1. Absorption loss
2. Scattering (Linear and non-linear) losses
3. Bending losses
4. Coupling losses

2.8.1 Absorption Losses

Due to the presence of impurities in the fiber material, some part of propagating light can be absorbed and converted into heat. Such type of transmission loss in optical fiber cable is termed absorption loss. We can say that absorption loss is directly related with the actual composition of the material and the manufacturing process used. It is analogous to power dissipation in copper cables when electric current flows through it.

Optical fibers are normally made of silica-based glass material. As light energy passes through the fiber, it may be partially absorbed by this material itself. This is called *intrinsic absorption*. There may be some impurities within the fiber material which may absorb some part of light energy. This is called *extrinsic absorption*.

> ***Note:*** Although, 99.9999% ultrapure glass is normally used to fabricate optical fibers nowadays, yet the absorption loss of the order of 1 dB/km is quite normal.

1. *Intrinsic Absorption.* As the name suggests, it is mainly caused by the interaction with major components of the fiber material. Two factors cause intrinsic losses - Material resonances in UV and IR regions, and Rayleigh scattering. Silica displays heavy absorption in the UV and IR regions and is wavelength dependent. Rayleigh scattering occurs because the EM field excites the irregularities of the molecules of SiO_2. Essentially, there are three factors that contribute to the intrinsic absorption in optical fibers: ultraviolet absorption, infrared absorption, and ion resonance absorption.
   - *Ultraviolet absorption.* Ultraviolet absorption is caused by valence electrons in the silica material from which fibers are manufactured. Light ionizes the valence electrons into conduction. The ionization contributes to the transmission losses of the fiber.
   - *Infrared absorption.* Infrared absorption occurs due to absorption of photons by the atoms of the glass core molecules. The absorbed photons are converted to random mechanical vibrations typical of heating.
   - *Ion resonance absorption.* During the manufacturing of optical fiber cables, water molecules may get trapped in the glass process. This results in OH^- ions in the material, which cause ion resonance absorption. Iron, copper, and chromium molecules also cause ion resonance absorption.

   Fig. 2.35 shows typical absorption losses in optical fiber cables due to ultraviolet, infrared, and ion resonance absorption.

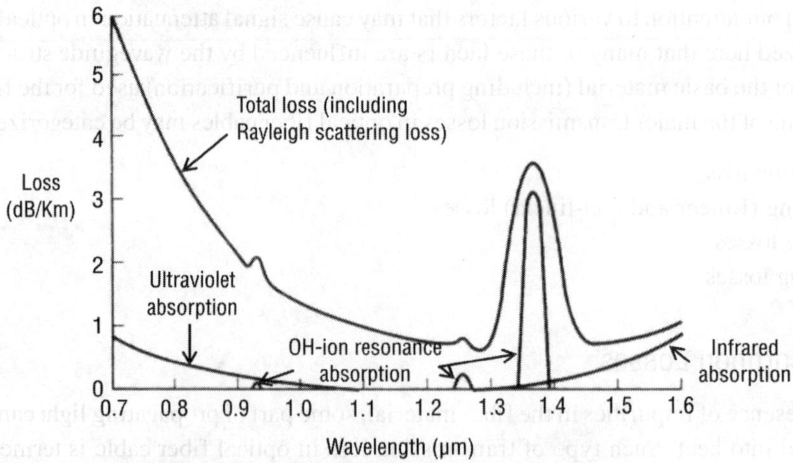

Fig. 2.35 Absorption losses in optical fiber cables

2. *Extrinsic Absorption.* Contrary to intrinsic absorption, extrinsic absorption is caused by different types of impurities usually contained within the fiber material (i.e. the glass). Common metallic ion impurities in glasses are Cr^{3+}, Cu^{2+}, Fe^{2+}, Fe^{3+}, Ni^{2+}, Mn^{3+}, which absorb light energy strongly in the wavelength range of 600–1600 nm. Another source of extrinsic absorption is the presence of hydroxyl ions, OH^-. These molecules enter the silica in the form of water vapor during the manufacturing process.

   Fig. 2.36 shows attenuation (dB/km) versus photon energy (eV) as well as wavelength.

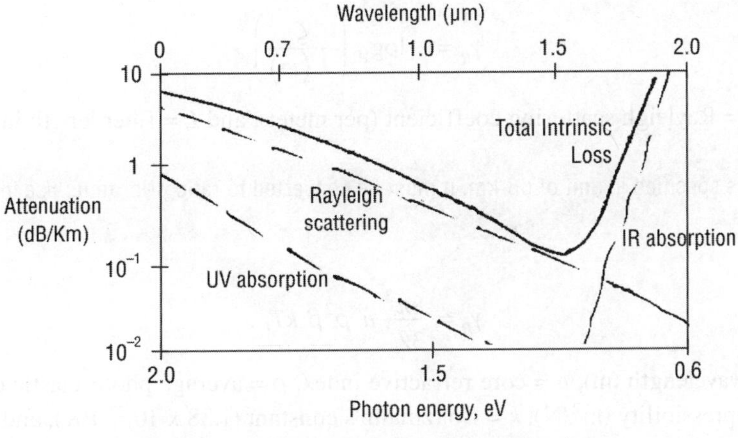

Fig. 2.36 Intrinsic and extrinsic absorption losses

> ***Note:*** It is essential to reduce the impurity content of metal ions below 1 part per billion so as to have fiber loss less than 1 dB/km. Similarly, the OH^- ion concentration should be less than 10 parts per billion to have fiber loss less than 10 dB/km at 1380 μm.

2.8.2 Scattering Losses

Scattering loss occurs due to transfer of the optical power available within one guided propagation mode into a different mode of propagation. The scattering may happen when a light beam propagating within the fiber strikes at an imperfection in a core material and then changes its direction. This scattering effect prevents attainment of desired total internal reflection at the intersection of the fiber core and the cladding. This may result in a power loss. Even very minute variation in the specified value of the core refractive index can be considered as an optical obstacle by a propagating light beam.

Linear Scattering Losses

Rayleigh scattering and Mie scattering are examples of linear scattering mechanisms.

1. *Rayleigh scattering.* It is the prominent intrinsic loss mechanism which occurs because of core refractive index fluctuations due to density and compositional variations in the glass lattice on cooling. During manufacturing of optical fibers, glass is drawn into long fibers of very small diameter. During this process, the glass is in a plastic state (not liquid and not solid). The tension applied to the glass causes the cooling glass to develop permanent submicroscopic irregularities. When light rays propagating down a fiber strike one of these impurities, they are diffracted. Diffraction causes the light to disperse or spread out in many directions. Some of the diffracted light continues down the fiber, and some of it escapes through the cladding. The light rays that escape represent a loss in light power. Mathematically,

   Transmissivity or transmission loss factor, $\zeta = 10^{(-\gamma_R L)}$ (2.94)

$$\Rightarrow \qquad \gamma_R = -\log_{10}\left(\frac{\zeta}{L(m)}\right) \tag{2.95}$$

where, γ_R = Rayleigh scattering coefficient (per meter), and L = fiber length in m

> Note: If ζ is specified in unit of dB/km, it must be converted to ratio prior to its use in the above expression.

$$\gamma_R = \frac{8\pi^3}{3\lambda^4} n^8 \rho^2 \beta_c k T_F \tag{2.96}$$

where, λ = wavelength (m), n = core refractive index, ρ = average photo-elastic coefficient, β_c= isothermal compressibility (m^2/N), k = Boltzmann's constant (1.38×10^{-23} J/K), and T_F = estimated fictive temperature (K).

Fig. 2.37 graphically shows the Rayleigh scattering loss (dB/km) as a function of operating wavelength (μm).

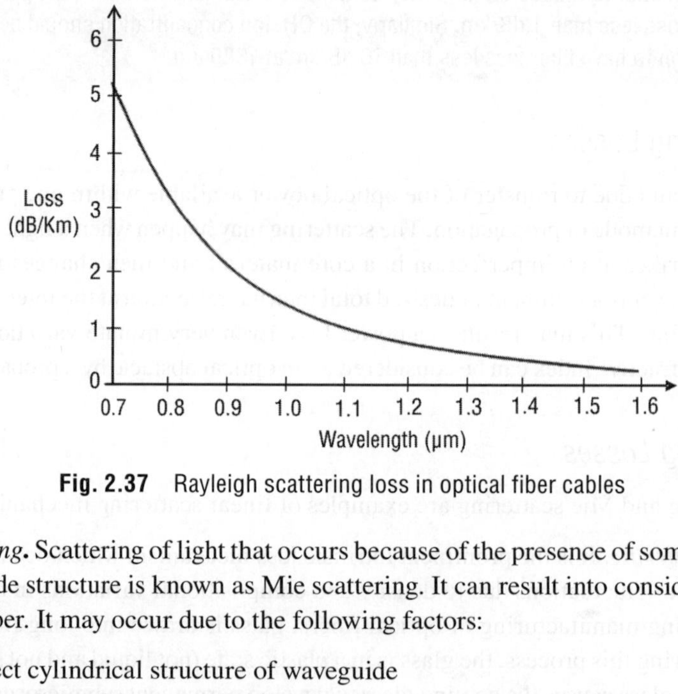

Fig. 2.37 Rayleigh scattering loss in optical fiber cables

2. *Mie Scattering*. Scattering of light that occurs because of the presence of some imperfections in the waveguide structure is known as Mie scattering. It can result into considerable attenuation in optical fiber. It may occur due to the following factors:
   - non-perfect cylindrical structure of waveguide
   - irregular surface of the core-cladding interface
   - difference in refractive indexes of the fiber core and cladding along the length of the fiber
   - fluctuations in the fiber core diameter
   - presence of strains and bubbles in the fiber

When the dimension of the scattering inhomogeneity exceeds λ/10, then the effect of Mie scattering may be very large, mainly in the forward direction. Mie scattering can cause significant losses in fiber, depending upon the fiber material, design and manufacturing process.

Example 2.38 Rayleigh Scattering Loss

A typical glass core optical fiber has a specified transmission loss factor of 0.046 dB/km due to Rayleigh scattering. At wavelength of 1000 nm, the glass material possesses an approximate fictive temperature value of 758K, Isothermally compressibility parameter of 8.4×10^{-11} m²/N, and photo-elastic coefficient of 0.245. Determine the core refractive index from the view of theoretical considerations.

Solution:

The core refractive index (n) can be determined from the expression of Rayleigh scattering coefficient given as $\gamma_R = \dfrac{8\pi^3}{3\lambda^4} n^8 \rho^2 \beta_c k T_F$

where $$\gamma_R = -\log_{10}\left(\frac{\zeta}{L(m)}\right)$$

Converting given transmission loss factor ζ of 0.046 dB/km into ratio, we have

$$-0.046 = 10\log\zeta$$

$$\Rightarrow \quad \zeta = 10^{-\frac{0.046}{10}} = 0.9895$$

For $L = 1$ km, we have $\gamma_R = -\log_{10}\left(\dfrac{0.9895}{1000 \text{ m}}\right) = 0.0046 \times 10^{-3}$ per meter

Substituting the given values of various parameters, we have

$$0.0046 \times 10^{-3} = \frac{8\pi^3}{3 \times \left(1000 \times 10^{-9}\right)^4} n^8 \times (0.245)^2 \times 8.4 \times 10^{-11} \times 1.38 \times 10^{-23} \times 758$$

$\Rightarrow \quad n \approx 1.54$ **Ans.**

Non-Linear Scattering Losses

In general, a non-linear optical effect is one when its parameters are functions of optical signal intensity. Non-linear scattering is also called *stimulated scattering* when light power essentially increases. Stimulated scattering means transfer of light energy from the incident wave to scattered wave at longer wavelength, with the small energy difference being released in the form of phonons. A *phonon* is an elementary particle analogous to a photon but differs from a photon in its quantum properties. Scattered optical waves are also known as Stokes' optical waves. Stimulated scattering increases fiber losses at a high level of transmitted optical power (negligible at low power levels). Stimulated scattering is characterized by three major parameters, namely,

- *Threshold power.* It is the power of incident light at which loss due to stimulated scattering is 3 dB over fiber length L. The intensity of scattered light goes exponentially when the power of incident light exceeds threshold power.
- *Gain coefficient.* It refers to peak gain of stimulated scattering at given λ.
- *Spectral Bandwidth.* Range of frequencies within which scattering is effective.

It is quite evident that several nonlinear effects occur at relatively higher optical power levels. Due to occurrence of non-linear scattering, the optical power from one propagation mode is transferred to the same mode, or some other modes, at a different wavelength in either the forward or backward direction. Non-linear effects mainly depend upon the optical power density available within the optical fiber cable. Obviously, these effects become dominant only when optical power density exceeds the specified threshold power levels. There are two major types of nonlinear or stimulated scattering:

1. *Stimulated Raman Scattering (SRS).* The scattered light moves mostly in the forward direction (with some light in the backward direction), and the phonons associated with this process are optical ones. The threshold optical power in single-mode fiber for stimulated Raman scattering may be given as

$$P_{SRS}\text{ (watts)} = 5.9 \times 10^{-2} d^2 \lambda \alpha_{dB} \tag{2.97}$$

   where, d = the core diameter (m), λ = wavelength (m), α_{dB} = attenuation in dB/km.

2. *Stimulated Brillouin Scattering (SBS).* The scattered light moves only in the backward direction, and the phonons associated with this process are acoustic ones. The threshold optical power in single-mode fiber for stimulated Brillouin scattering may be given as

$$P_{SBS}\text{ (watts)} = 4.4 \times 10^{-3} d^2 \lambda^2 \alpha_{dB} B \tag{2.98}$$

   where, d = the core diameter (m), λ = wavelength (m), α_{dB} = attenuation in dB/km, and B = bandwidth of optical source (GHz)

Example 2.39 SRS Threshold Power

Find the threshold optical power level for stimulated Raman scattering in single-mode fiber which has fiber core diameter of 5 μm and operating at wavelength of 900 nm. The fiber attenuation parameter is specified as 1.8 dB/km at 900 nm.

Solution:

We know that the threshold optical power level for stimulated Raman scattering is

$$P_{SRS}\text{ (watts)} = 5.9 \times 10^{-2} d^2 \lambda \alpha_{dB}$$

For d = 5 μm or 5×10^{-6} m, λ = 900 nm or 900×10^{-9} m, α_{dB} = 1.8 dB/km, we get

$$P_{SRS}\text{ (watts)} = 5.9 \times 10^{-2} \times \left(5 \times 10^{-6}\right)^2 \times \left(900 \times 10^{-9}\right) \times 1.8 = 2.388\text{ W} \qquad \textbf{Ans.}$$

Example 2.40 SBS Threshold Power

In a typical single-mode fiber that uses 800-MHz bandwidth optical source at 850 nm wavelength, the threshold optical power level for stimulated Brillouin scattering is given as 127 mW. The fiber attenuation is typically 2 dB/km at 850 nm. Find the fiber core diameter.

Solution:

We know that in a single-mode fiber, the threshold optical power level for stimulated Brillouin scattering is given as

$$P_{SBS}\,(watts) = 4.4 \times 10^{-3} d^2 \lambda^2 \alpha_{dB} B$$

$$\Rightarrow \qquad d = \sqrt{\frac{P_{SBS}\,(watts)}{4.4 \times 10^{-3} \lambda^2 \alpha_{dB} B(GHz)}}$$

For the given P_{SBS} = 127 mW or 127 × 10^{-3} W, λ = 850 nm, α_{dB} = 2 dB/km, and B = 800 MHz or 0.8 GHz, we get

$$\Rightarrow \quad d = \sqrt{\frac{127 \times 10^{-3}}{4.4 \times 10^{-3} \times \left(850 \times 10^{-9}\right)^2 \times 2 \times 0.8}} = 5 \times 10^{-6}\text{m or } 5\ \mu\text{m} \qquad \textbf{Ans.}$$

Note: Stimulated Brillouin scattering limits the optical power level to be launched into a single-mode fiber cable to be of the order of a few mW only mainly because of its extremely low threshold power level.

Table 2.6 shows a comparative study of some of the major aspects of SRS and SBS.

Table 2.6 Comparison between SRS and SBS

S. No.	*Stimulated Raman Scattering (SRS)*	*Stimulated Brillouin Scattering (SBS)*
1.	The scattered light moves mostly in forward direction.	The scattered light moves only in backward direction.
2.	Optical phonons participate in SRS mechanism.	Acoustic phonons participate in SBS mechanism.
3.	There is a frequency shift by about 13 THz in the SRS scattered light, that is, Raman shift ≈ 13 THz.	There is a frequency shift by about 10 GHz only in the SBS scattered light, that is, Stokes shift ≈ 10 THz.
4.	The SRS gain spectrum extends over 20-30 THz frequency range.	The SBS gain spectrum is less than 100 MHz, i.e., extremely narrow.
5.	SRS has a negligible impact on single-channel system but becomes an issue in WDM system.	SBS does not produce inter-channel crosstalk in WDM system because 10 GHz shift is much less than channel spacing usually employed in WDM.

Note: Contrary to Rayleigh scattering, stimulated scattering (SRS or SBS) lowers the frequency of the scattered signal. Fiber loss due to stimulated Brillouin scattering can be reduced for depressed-cladding and dispersion-shifted fibers.

Non-Linear Optical Effects

Non-linear optical effects, also known as *Kerr nonlinearities*, are usually described by the light-intensity dependent characteristics of refractive index of the fiber. When the intensity of the optical signal propagating through the optical fiber varies, the core refractive index varies. As a result, there will be a possible phase shift in the optical pulse within the fiber. Moreover, the phase shift would be different at the peak of the optical pulse as compared to the near the leading or trailing edges of the pulse. It is quite evident that the spectrum of the optical pulse will be modified. This phenomenon is called *self-phase modulation* (SPM). Due to variation in the phase with time, a time-varying frequency will be created. Therefore, we can conclude that the impact of SPM will be broadening of the frequency spectrum of the optical pulse.

Kerr nonlinearities also allow a pulse to be modified by another pulse which can be at different polarization or alternatively just a different mode of the fiber. In fact, self-phase modulation imposes a *frequency chirp* i.e., a positive frequency sweep on the propagating optical pulse.

2.8.3 Bending Losses

Bending an optical fiber too sharply can increase fiber losses by causing some of the light to meet the intersection to the fiber core and the cladding at less than the critical angle of incidence, thereby preventing the desired phenomenon of total internal reflection. Fig. 2.38 shows how this can happen.

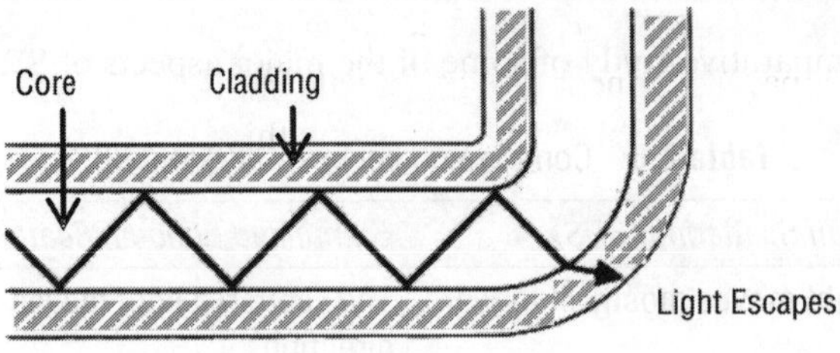

Fig. 2.38 Bending an optical fiber

Losses vary greatly with the type of fiber. For example,

- Plastic fiber may have losses of several hundred decibels per kilometer.
- Graded-index multimode glass fiber has attenuation figure of about 2–4 dB/km, while single-mode glass fiber has attenuation figure of about 0.4 dB/km or less.

Definition of microbend: It is a minor geometric imperfection present along the fiber axis that actually represents a discontinuity at which Rayleigh scattering can occur.

- Microbends may occur due to differences in the thermal contraction rates for the materials used for fabrication of the fiber core and the cladding.
- Microbends, sometimes called constant-radius bends, result in radiation losses in the fiber.

Microbends may also refer to some imperfections in the geometry of the interface boundary between the fiber core and the cladding, as shown in Fig. 2.39.

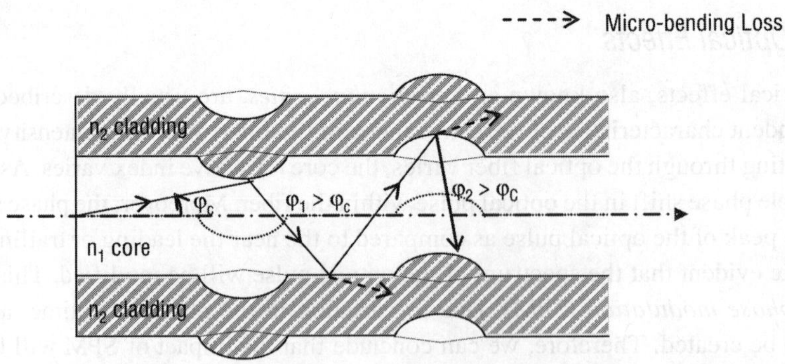

Fig. 2.39 Microbending loss

The beam, which initially travels at the critical propagation angle, after being reflected at these imperfection points, will change the angle of propagation. This results in partial refraction and will leak out of the core, which is called microbending loss. The condition of total internal reflections will not be met at microbends.

> ***Note:*** Microbending loss occurs mainly due to by micro deformations of the fiber axis. Fiber-axis is determined by the core-cladding geometry. So, micro deformations of fiber axis are micro deformations of core-cladding boundary. Fiber losses due to microbends typically contribute < 20% of the total transmission losses in an optical fiber.

Now the question arises as to whether we can reduce microbending loss in a single-mode fiber. One way is to choose the *V*-parameter near the cut-off value (that is, $V_c = 2.405$ for step-index profile fiber), such that the mode energy is mainly constrained within the fiber core.

2.8.4 Coupling Losses

In optical fiber cables, coupling losses can occur at any of the following three types of optical junctions:

- connections between light source and the fiber at the transmitter end (source-to-fiber power launching)
- fiber-to-fiber interconnections (to extend the length of the optical fibers)
- connections between the fiber and the photodetector at the receiver end

Coupling the fiber to sources and detectors creates losses, especially when it involves mismatches in numerical aperture or in the size of optical fibers.

In an optical fiber communication system, the losses in splices and connectors can easily be more than in the fiber cable itself. Coupling losses are normally caused by the imperfect physical connections in optical fiber communication system. Junction losses are most often caused by one of the following alignment problems:

- lateral or axial misalignment
- gap misalignment
- angular misalignment
- imperfect surface finishes

1. *Lateral or axial misalignment.* Lateral or axial misalignment or displacement is the lateral or axial displacement between two pieces of adjoining fiber cables. Fig. 2.40 shows lateral misalignment in optical fibers.

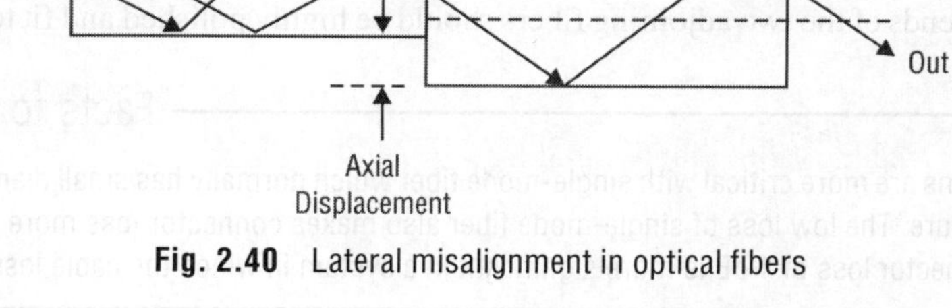

Fig. 2.40 Lateral misalignment in optical fibers

The amount of coupling loss may vary from a tenths of a dB to several dBs. This loss is generally negligible if the fiber axes are aligned to within 5% of the smaller fiber's diameter.

2. *Gap misalignment.* Gap misalignment or displacement occurs when splices are made in optical fibers, and the optical fibers should actually touch. If the fibers are kept far apart, there will be more loss of optical signal. Fig. 2.41 shows gap misalignment in optical fibers.

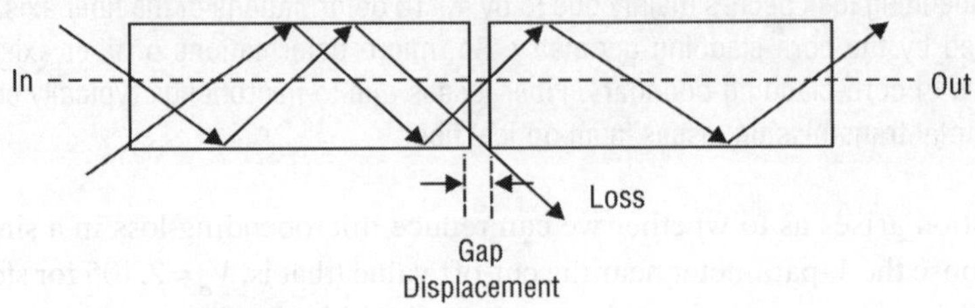

Fig. 2.41 Gap misalignment in optical fibers

If two fibers are joined with a connector, the ends should not touch because the two ends rubbing against each other in the connector could cause damage to either or both fibers.

3. *Angular misalignment.* Angular misalignment or displacement is shown in Fig. 2.42.

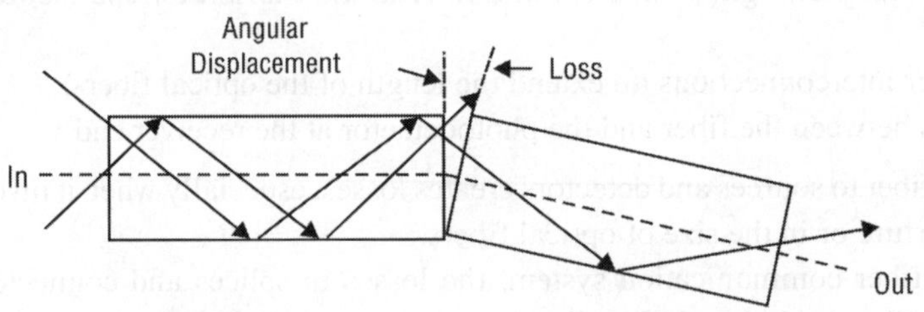

Fig. 2.42 Angular misalignment in optical fibers

If the angular displacement is less than 2°, the loss will typically be less than 0.5 dB.

4. *Imperfect surface finish.* Imperfect surface finish is shown in Fig. 2.43.

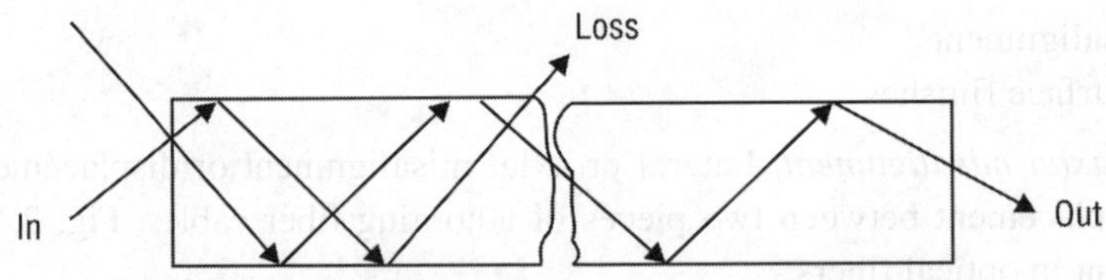

Fig. 2.43 Imperfect surface finish in optical fibers

If the fiber ends are less than 3° off from perpendicular, the losses will typically be less than 0.5 dB. The ends of the two adjoining fibers should be highly polished and fit together squarely.

Facts to Know

Good connections are more critical with single-mode fiber which normally has small diameter and small numerical aperture. The low loss of single-mode fiber also makes connector loss more significant. For example, a connector loss of 1 dB is more significant in a system in which the cable loss is 1 dB/km.

2.9 Comparison of Optical Fibers

As outlined earlier, optical fibers can be categorized either based on propagation modes as single-mode or multimode, or based on their refractive index profiles as step-index and graded-index. Although there are a wide variety of combinations of propagation modes and index profiles, there are only three practical configurations of optical fibers as follows.

1. Single-mode step-index optical fiber
2. Multimode step-index optical fiber
3. Multimode graded-index optical fiber

The index of refraction in step-index fibers (single-mode or multimode) changes sharply between core and cladding. Graded-index fiber is a multimode fiber, but the refractive index gradually decreases as we move away from the axis of the fiber core, so that light can travel faster near the outer edge of the core.

Note: It is an established fact that graded-index profile fiber exhibits much less dispersion than exhibited by a multimode step-index fiber. But single-mode fibers are still preferred for the most demanding optical fiber communication applications.

There is another aspect for the proper selection of the material to be used for manufacturing of optical fiber cables. Single-mode and graded-index profile multimode fibers are always manufactured using glass material, whereas multimode step-index fibers that have a relatively large core diameter of the order of about 1 mm are fabricated using plastic material. Typically, core diameters for the single-mode and multimode step-index profile fiber are chosen as 8.2 μm and 1000 μm, respectively, whereas for multimode graded-index profile fiber, it is 50 μm.

Table 2.7 International standards for optical fibers

ITU/Equivalent Standards	*About*	*Description, if any*
ITU-T-G.651	Multimode Fiber	--
ITU-T-G.652, TIA/ EIA-492CAAB, IEC 60793-2	Single-mode Fiber	Four different categories (A, B, C,D) differ in the peak attenuation around the 1383-nm optical window
ITU-T-G.653	Zero Dispersion Shifted Fiber	Zero dispersion around the 1550-nm optical window
ITU-T-G.654	Low attenuation and Cutoff shifted Fiber	Designed mainly for submarine applications
ITU-T-G.655	Non-zero Dispersion Shifted Fiber	Low dispersion around the 1550-nm and 1625-nm optical windows. Suited for backbone and longhaul applications
ITU-T-G.656	Medium Dispersion Fiber	Designed for local access as well as longhaul applications
ITU-T-G.657	Designed to bend at small radius (7.5–10 mm)	Latest standard for Fiber to the Home (FTTH) application

2.9.1 Advantages of Single-mode Step-index Fibers

1. *Minimum dispersion*: In a single-mode step-index fiber, an optical pulse carrying information propagating down the optical fiber cable can be reproduced quite accurately at the receiver. This is because of the fact that all light rays travelling within the single-mode step-index fiber traverse almost the same path, thereby, arrive at the same time.
2. *Higher information transmission rates*: Since transmitted optical pulses can be reproduced with high accuracy at the receiving end in single-mode step-index fiber, it is possible to obtain higher information transmission rates (bps) due to available wider bandwidths.

2.9.2 Disadvantages of Single-mode Step-index Fibers

1. In a single-mode step-index fiber, the size of the fiber core is extremely small. It implies that the optical source-to-fiber aperture is very less. As a result, the coupling of light from source to fiber at the transmitting end as well as from fiber to photodetector at the receiving end becomes quite cumbersome.
2. Due to extremely small size of the fiber core, it is mandatory to use a highly focused optical source such as laser, for coupling light into a single-mode step-index fiber.
3. The manufacturing process of single-mode step-index fibers is quite complex.
4. Single-mode step-index fibers are relatively expensive.

2.9.3 Advantages of Multimode Step-index Fibers

1. Due to relatively large source-to-fiber aperture of a multimode step-index fiber, it becomes quite easy to couple light into and out of it.
2. The manufacturing process of multimode step-index fibers is relatively simple.
3. Multimode step-index fibers are inexpensive.

2.9.4 Disadvantages of Multimode Step-index Fibers

1. In multimode step-index fibers, light rays traverse several paths, different from each other within the optical fiber cable. Consequently, there is large difference in their propagation times which results in spreading out of the transmitted optical pulse by the time it arrive at the receiving end. This amounts to higher distortion as compared to other types of optical fibers.
2. The transmission bandwidth and achievable rate of information are quite less than that are possible with other types of optical fiber cables.

2.9.5 Pros and Cons of Multimode Graded-index Fibers

1. It is relatively much easier to couple light into and out of a multimode graded-index fiber in comparison with that of a single-mode step-index fiber. However, the same is more difficult than that is possible with a multimode step-index fiber.
2. There is greater possibility of distortion in the received optical pulses because of dispersion in multimode graded-index fibers, which occurs due to multiple propagation as compared to that in single-mode step-index fibers but relatively lesser than in multimode step-index fibers.

Note: As per the foregoing discussions, one can say that the multimode graded-index profile fibers are considered to be an intermediate fibers as compared to single-mode as well as multimode step-index fibers.

Points to Remember

- Optical fibers are waveguides for light.
- Signal propagation in optical fibers can be single-mode (i.e., essentially a single light beam from an optical source), or multimode (multiple light beams from an optical source).
- A single-mode fiber is an optical fiber that supports only the fundamental mode of propagation of light through it using the principle of total internal reflection.
- In single-mode step-index fibers, single-mode operation can be achieved over a normalized frequency parameter range of $0 < V < 2.405$. In α-profile graded-index fibers, the cut-off normalized frequency parameter is governed by $V_c = 2.405\sqrt{(1 + 2/\alpha)}$.
- When light propagates through a multimode step-index fiber, the fiber core density remains constant. Therefore, the light rays change direction (due to total internal reflection) suddenly at the intersection between the fiber core and the cladding surfaces.
- When light propagates through a multimode graded-index profile fiber, the fiber core density decreases with distance from the fiber-axis towards the core-cladding interface. This results bending of the light beams within the fiber core.
- Single-mode fibers have greater bandwidth and lower losses, while multimode fibers are cheaper and easier to use.
- In single-mode fibers, dispersion (i.e., pulse broadening) occurs due to group velocity dispersion (GVD)—a combination of material and waveguide dispersion. Typically, total dispersion is zero near 1320 nm wavelength.
- Signal attenuation in optical fibers occurs due to several factors such as material absorption, scattering and bending. The minimum loss in silica fiber is around 1550 nm wavelength.
- Splices and connectors are important in fiber systems. Even a well-made splice can have as much loss as equivalent to that of 1 km of fiber.

Important Equations

The refractive index, $n = \frac{c}{v}$; *where c represents the velocity of propagation of light in free space (3×10^8 m/sec) and v is the velocity of propagation of light in a particular material (m/sec).*

The Snell's law states that $n_1 \sin\theta_1 = n_2 \sin\theta_2$; *where n_1 represents the index of refraction of material 1, θ_1 denotes the angle of incidence (degrees), n_2 represents the index of refraction of material 2, and θ_2 denotes the angle of refraction (degrees).*

The critical angle, $\theta_c = \sin^{-1}\left(\frac{n_2}{n_1}\right)$; *where n_1 and n_2 represents the index of refractions of the fiber core and the cladding, respectively.*

The numerical aperture, $NA = \sqrt{n_1^2 - n_2^2}$; *where n_1 and n_2 represents the index of refractions of the fiber core and the cladding, respectively.*

The acceptance angle, $\theta_a = \sin^{-1}\left(\sqrt{n_1^2 - n_2^2}\right)$; *where n_1 and n_2 represents the index of refractions of the fiber core and the cladding, respectively.*

The V-parameter, $V \approx \frac{2\pi a}{\lambda}\sqrt{n_1^2 - n_2^2}$; *where a represents the core radius (meters), λ is the wavelength of light rays (meters), n_1 and n_2 are the index of refractions of the fiber core and the cladding, respectively.*

The V-parameter, $V \approx \frac{2\pi a}{\lambda}(NA) \approx \frac{\pi d}{\lambda}(NA)$; *where NA is the numerical aperture and d is the core diameter.*

For a multimode fiber, the number of possible modes, $M_s \cong \frac{V^2}{2}$ *for a step-index profile ($V > 20$); and* $M_g \cong \frac{V^2}{4}$ *for graded index profile.*

The chromatic dispersion, $\Delta t_c = D_c(\Delta\lambda)L$; *where Δt_c is chromatic dispersion expressed in nanoseconds, D_c represents the chromatic dispersion constant in ps/(nm–km), $\Delta\lambda$ represents the linewidth of the optical source in nm, and L is the fiber length in km.*

Maximum bandwidth of optical fiber, $B_{\max} \approx \frac{1}{2\Delta t_c}$; *where Δt_c is chromatic dispersion.*

Total power loss in an optical fiber cable, $A_{(dB)} = 10\log\left(\frac{P_{out}}{P_{in}}\right)$; *where $A_{(dB)}$ represents total reduction in optical power level (i.e., attenuation) in dB, P_{out} is cable output power (watts), and P_{in} is cable input power (watts).*

Key Terms with Definitions

Angle of incidence The angle made by an incident light ray with reference to the normal to a reflecting or refracting surface.

Angle of reflection The angle made by a reflected ray with reference to the normal to a reflecting surface.

Angle of refraction The angle made by a refracted ray with reference to the normal to a refracting surface.

Angle of acceptance The maximum angle made by a light ray entering the optical fiber cable with reference to the fiber core axis.

Critical angle The maximum incidence angle made by a light ray for which refraction takes place.

Dispersion Broadening of transmitted optical pulse due to variations in propagation velocity with varying wavelengths.

Graded-index fiber An optical fiber in which the core refractive index decreases gradually with increasing distance from the core axis.

Multimode fiber A type of an optical fiber in which light travels through it in multiple (more than one) waveguide mode, at different velocities.

Normal A line perpendicular to a reflecting or refracting surface.

Numerical aperture	The sine (sinusoidal) of the acceptance angle for an optical fiber.
Single-mode fiber	A type of optical fiber which has extremely small fiber core diameter such that only one waveguide mode (i.e., fundamental mode) of incident light can propagate within it.
Step-index fiber	An optical fiber that has one constant value of refractive index for the fiber core and another constant lower value of refractive index for the cladding, with a sharp transition between them.
Total internal reflection	The phenomenon that signifies reflection (not refraction) at the interface between two different media (differing in their refractive indexes) provided the incidence angle is greater than that of the critical angle of incidence.

Short Answer Type Questions

1. List two methods for understanding and analyzing the propagation aspects of light along a fiber.

(1) By an application of electromagnetic theory and evolving solution the Maxwell's wave equations with appropriate boundary conditions

(2) By an application of the geometrical optics or ray theory approach

Since the core size of multimode fibers is much larger the wavelength of the light (around 1 μm), an intuitive picture of the propagation mechanism in an ideal multimode step-index optical fiber is most easily visualized by a simple ray (geometrical) optics representation. However, to attain a more detailed understanding of the optical power propagation mechanism in a fiber, it is necessary to solve Maxwell's equations subject to the cylindrical boundary conditions at the core-cladding interface of the fiber.

2. State the criteria to choose the refractive index of cladding for a practical step-index fiber so that the electromagnetic energy at optical frequencies is made to propagate along the fiber waveguide through total internal reflection at the core-cladding interface.

In a practical step-index fiber, the core having radius *a*, has an index of refraction n_1 which is typically equal to 1.48. This is surrounded by a cladding having slightly lower value of index of refraction n_2 such that $n_2 = n_1 (1 - \Delta)$, where the parameter Δ is known as the *core–cladding index difference.* The values of n_2 are chosen such that Δ is nominally 0.01. For single-mode fibers, typical values of Δ range from 0.2% to 1.0% and for multimode fibers the range is from 1% to 3%.

3. What are two different types of light rays that can propagate in an optical fiber? Distinguish between them.

The two types of rays that can propagate in an optical fiber are meridional rays and skew rays. *Meridional rays* are confined to the meridian planes of the fiber which contain the core axis. Since it lies in a single plane, its path is easy to track as it travels along the fiber. *Skew rays* are not confined to the single plane, but instead tend to follow a helical-type path along the fiber. These rays are more difficult to track as they travel along the fiber. Although skew rays constitute a major portion of the total number of guided rays, the examination of meridional rays will be sufficient to obtain a general picture of rays propagating in a fiber.

4. Define critical angle with suitable illustration.

The critical angle is the angle of incidence at which a light ray strikes at the intersection of the fiber core and cladding surfaces where the angle of refraction is 90° and the refracted ray emerges parallel to the core-cladding interface. Fig. 2.44 shows a pictorial view of an incident ray entering the interface at critical angle.

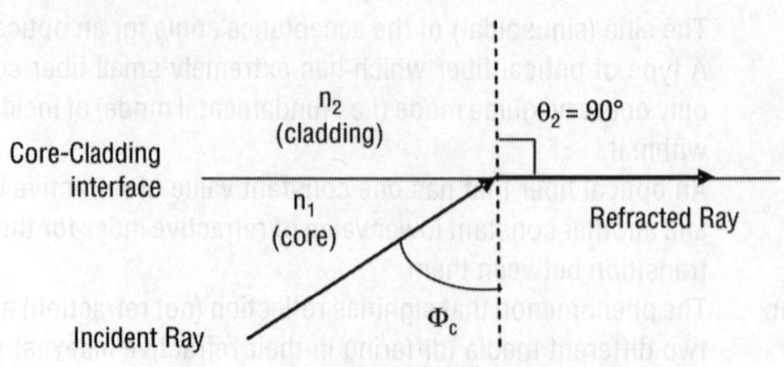

Fig. 2.44 A pictorial view of critical angle

The critical angle of incidence is the limiting case of the refraction, that is, the ray is not refracted in the cladding. This is significant because it is required to confine the light to travel within the fiber core, not the cladding. When the angle of incidence is greater than the critical angle, the condition of total internal reflection is satisfied, that is, the light is totally reflected back into the glass core.

5. Draw suitable diagrams to illustrate the following conditions in an optical fiber:

(a) Actual angle of incidence is less than the specified critical angle of incidence

(b) Actual angle of incidence is equal to the specified critical angle of incidence

(c) Angle of incidence is greater than the critical angle of incidence

For an optical fiber, the fiber core with refractive index n_1 is the first medium, and the cladding with refractive index n_2 is the second medium, and n_1 is always greater than n_2. As per definition, the critical angle of incidence is an incidence angle made by a light ray at the intersection of the fiber core and the cladding for which the refraction angle is exactly 90°.

(a) When the incidence angle is less than the critical angle

In this case, the light ray will be refracted with $\theta_2 > \theta_1$, as shown in Fig. 2.45.

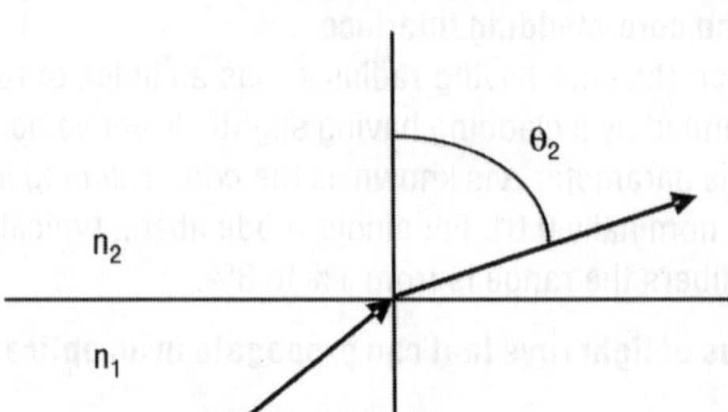

Fig. 2.45 The incidence angle is less than critical angle

As per Snell's Law, we know that $n_1 \sin\theta_1 = n_2 \sin\theta_2$

(b) When the incidence angle is equal to critical angle

In this case, the light ray will be refracted along the interface between the two media having $\theta_2 = 90°$, as shown in Fig. 2.46.

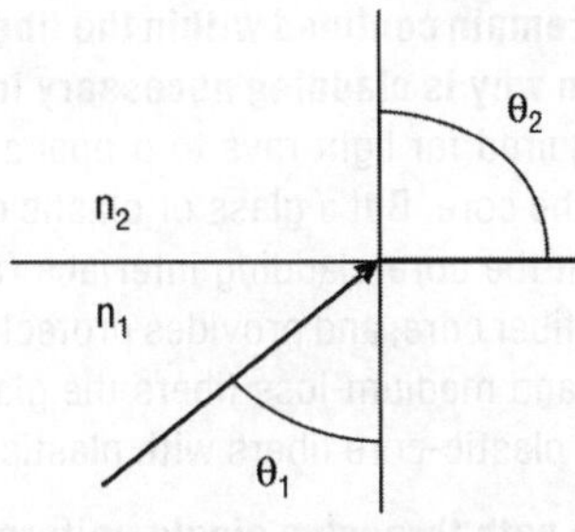

Fig. 2.46 The incidence angle is equal to critical angle

As per Snell's Law, we know that $n_1 \sin\theta_1 = n_2 \sin 90° = n_2$.

In this case, $\theta_1 = \theta_c$, the critical angle.

(c) When the incidence angle is greater than the critical angle
In this case, the light ray will be reflected instead of refracted, as shown in Fig. 2.47.

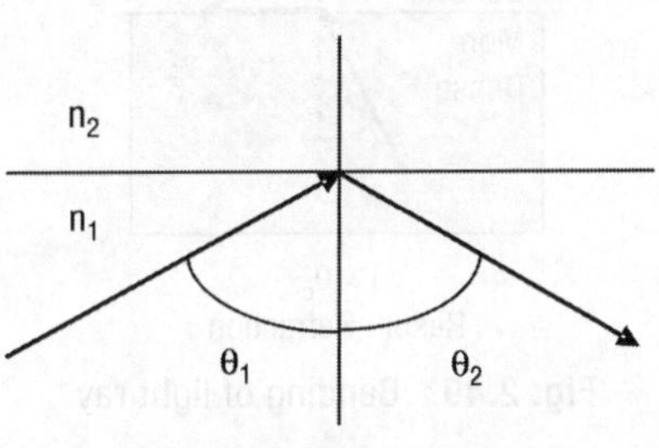

Fig. 2.47 The incidence angle is greater than critical angle

In fact, the Snell's Law is not applicable in such situation. As per the laws of reflection, the angle of reflection in the same medium is equal to the angle of incidence, that is, $\theta_1 = \theta_2$.

6. How is light propagated through optical fiber cable?

An optical fiber is made up of a central fiber core of glass material surrounded by the cladding of glass material with lower refractive index than that of the fiber core. It is then covered by a protective layer. It propagates signals in the form of light (i.e., optical signals). The light rays undergo total internal reflection when it moves forward from fiber core having higher refractive index to cladding having relatively lower refractive index and goes back and forth as it travels through the core along the fiber length. The relative difference in refractive indexes of two materials (may be same type as glass) must be such that a light propagating within the fiber core is reflected off the core-cladding interface surface instead of getting refracted into the cladding, as shown in Fig. 2.48.

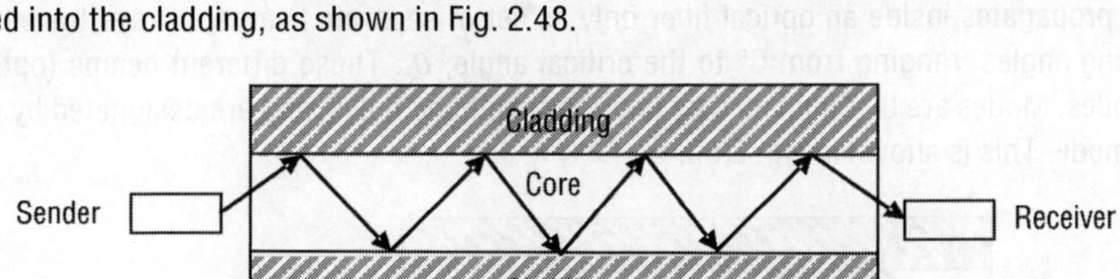

Fig. 2.48 Light propagation through optical fiber cable

Number of fibers can be bundled together as rod or sheet and this can be flexible enough for bending and twisting for installation purpose.

7. **As it is essential that the light should remain confined within the fiber core and no part of light should escape from it towards cladding. Then why is cladding necessary in a practical optical fiber?**
In principle, a cladding may not be required for light rays to propagate within the fiber core, as the air itself can be the medium surrounding the core. But a glass or plastic cladding serves several purposes. The cladding reduces light scattering at the core-cladding interface (thereby reducing scattering loss), adds mechanical strength to the fragile fiber core, and provides protection against surface contamination to which the cable is exposed. In low- and medium-loss fibers the glass core is surrounded by either a glass or a plastic cladding. Higher-loss plastic-core fibers with plastic claddings are also widely in use.

8. **Light usually traverses a straight line path through a single uniform medium. Imagine a ray of light traveling through a denser (higher refractive index) medium to a less dense (slightly lower refractive index) medium. Draw suitable diagrams to demonstrate this phenomenon.**
Fig. 2.49 shows three different cases when a light ray changes direction when it moves from a relatively denser medium to a less dense medium.

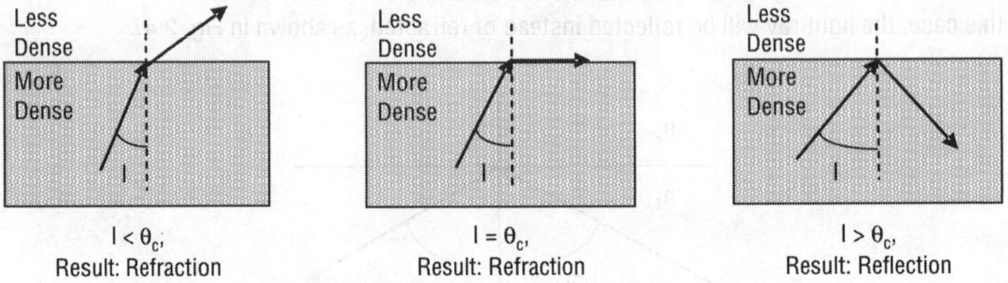

Fig. 2.49 Bending of light ray

The following is observed:

- If the **incidence angle (I)** in the first medium is less than the **critical angle of incidence (θ_c)**, then the light ray **refracts** in the second medium and moves closer to the interface (away from the normal).
- If the **incidence angle (I)** in the first medium is exactly equal to the **critical angle of incidence (θ_c)**, then the light ray bends along the intersecting horizontal line of two mediums.
- If the **incidence angle (I)** in the first medium is greater than the **critical angle of incidence (θ_c)**, then the light ray **reflects** back in the first medium.

It may be added here that the critical angle of incidence is a property of the material used in the two mediums under consideration, and its value is different from one type of material to another type of material.

9. **Define propagation modes. How are different modes distinguished?**
The light propagates inside an optical fiber only as set of separate beams, or rays traveling at distinct propagating angles, ranging from 0° to the critical angle, θ_c. These different beams (optical rays) are called modes. Modes are distinguished by their propagating angles, and are designated by the order of a specific mode. This is shown in Fig. 2.50.

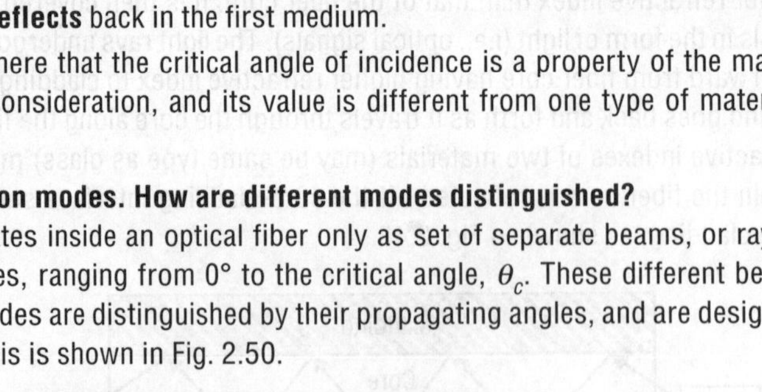

Fig. 2.50 Order of propagation modes

For low-order modes of light propagation, the electromagnetic fields are tightly concentrated quite close to the central axis of an optical fiber core. There is almost negligible chance of penetration into the cladding region. Contrary to this, the electromagnetic fields for higher-order modes are distributed towards the outer edges of the fiber core. It is quite likely that the fields can penetrate further into the cladding region.

10. Categorize propagation modes in optical fibers.

Propagation modes can be broadly categorized as either single-mode or multimode. Although a finite number of guided modes are supported, the optical fiber waveguide exhibits almost an infinite range of *radiation modes* that are not trapped in the fiber core and guided by it. Because of the finite radius of the cladding, some of this radiation gets trapped in the cladding, thereby causing radiation modes (also called refracted modes) to appear. As the guided and radiation modes propagate along the fiber, there is mode coupling between the higher-order modes and the radiation modes. In addition to guided and refracted modes, leaky modes are also present in optical fibers. These leaky modes are only partially confined to the core region and attenuate by continuously radiating their power out of the core as they propagate along the fiber.

11. How does the light ray propagate in single-mode fiber?

Single-mode fibers are fabricated with very small dimensions of the core diameter (usually 8–12 λ) and by having small index differences between the core and the cladding. All the light rays arrive at the destination together and can be recombined with little distortion to the signal, as shown in Fig. 2.51.

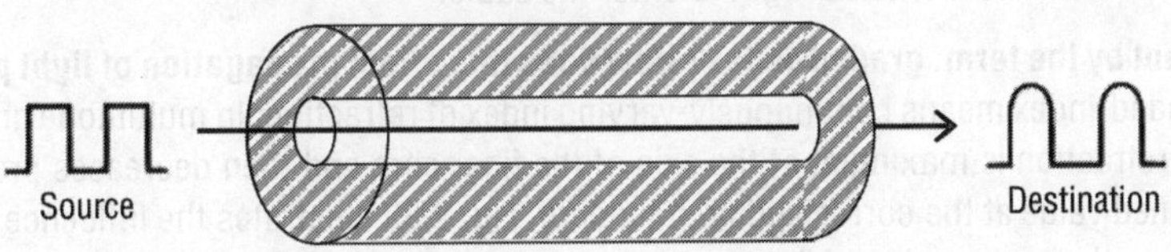

Fig. 2.51 Light ray propagation in single-mode

A mode remains guided as long as the propagation factor β satisfies the condition $n_2k < \beta < n_1k$; where n_1 and n_2 denote the respective refractive indexes of the fiber core and cladding, and $k = \frac{2\pi}{\lambda}$. The boundary between truly guided modes and leaky modes is defined by the *cut-off condition* $\beta = n_2k$. This occurs when $V \leq 2.405$. This is the basis of operation of the single-mode fibers. With V = 2.4, single-mode propagation is possible for fairly large variations in values of the physical core size *a* (i.e., radius) as well as the relative difference between refractive indexes between the fiber core and the cladding, Δ.

12. Since the electromagnetic field of a guided mode in an optical fiber cable extends partly into the cladding, the fractional power flows in the core and cladding for a given mode. Estimate the fraction of the average optical power residing in the cladding in case of step-index fiber.

As the *V*-parameter for a particular mode of propagation of light approaches the specified cutoff condition, most of the part of its optical power contained in this mode goes in the cladding region. Therefore, the mode becomes radiative with complete power residing in the cladding region. For large value of *V*-parameter, the fractional average power available in the cladding can be estimated by the expression, $\frac{P_{clad}}{P_{total}} \approx \frac{1.33}{\sqrt{M_s}}$; where $M_s \cong \frac{V^2}{2}$ is the number of modes possible for a multimode step-index fiber for V > 20. Thus, the power flow in the cladding decreases as V increases. However, increasing the number of modes in the fiber is not desirable for a high-bandwidth requirement.

13. What is meant by the term: step-index? Describe it by taking the example of multimode step-index fiber with appropriate illustration.

The term **step-index** refers to the sudden variation in the density or the index of refraction at the core-cladding interface in the optical fiber cable. In case of **multimode step-index fiber,** the core refractive index remains same from the axis to the boundary of core-cladding interface. A ray of light travels through this uniform refractive-index fiber core in a straight line before it arrives at the core-cladding interface boundary. At this junction, the angle of the direction of the movement of the light ray changes due to a sudden change to a lower refractive index value. Fig. 2.52 shows propagation of light rays through multimode step-index fiber.

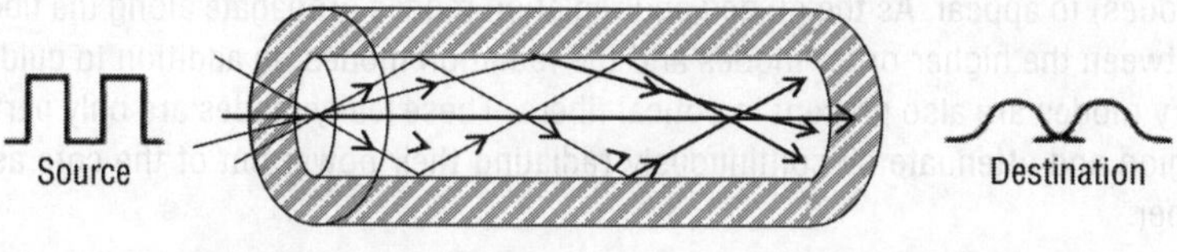

Fig. 2.52 Propagation through multimode step-index fiber

Multimode step-index fibers are similar to the single-mode step-index fibers, except the center core is much larger with the multimode configuration. This type of fiber has a large light-to-fiber aperture and, consequently, allows more external light to enter the cable.

14. What is meant by the term: graded-index? How does this effect propagation of light pulse through it?

The term graded-index means continuously-varying index of refraction. In multimode graded-index fiber, the index of refraction is maximum at the axis of the fiber core and then decreases progressively to its lowest specified value at the core-cladding boundary. Fig. 2.53 illustrates the influence of variable index of refraction on the propagation of light rays.

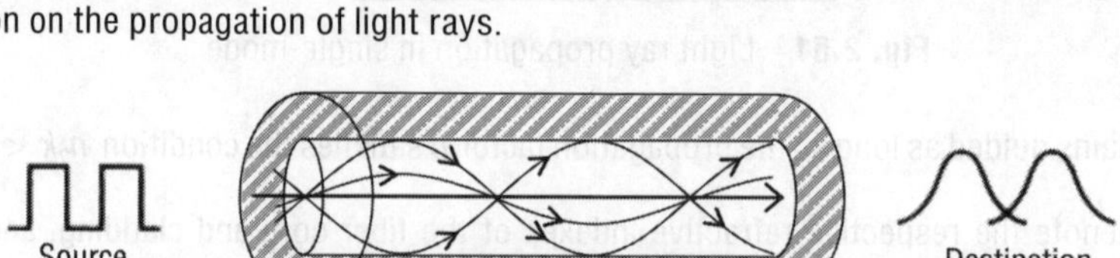

Fig. 2.53 Propagation through multimode graded-index fiber

One of the major advantages of multimode graded-index fiber is reduction in the optical signal distortion through it.

15. How is graded-index profile optical fiber cable different from step-index profile optical fiber cable?

In graded-index profile optical fiber cable, the core refractive index is non-uniform. It is maximum at the core-axis (i.e., at the center of the fiber core) and decreases progressively as we move across its radius toward the outer edge. The index profile shows a core refractive index that is maximum at the center and decreases symmetrically with distance from the center across its diameter. In other words, the index of refraction between the fiber core and cladding is changed gradually that is responsible for bending of the light rays slowly and make it follow a more sinusoidal path.

On the other hand, in step-index profile optical fiber cable, there is an abrupt change in the index of refraction at the intersection of the fiber core and the cladding. This is true for both single-mode and multimode step-index profile optical fiber cables. Although step-index fibers are more economical but spreads an optical pulse more as compared to graded-index fibers.

16. What is understood by index profile of an optical fiber? What are the two basic types of index profiles?

In an optical fiber, the index profile is a graphical representation of the absolute value of the magnitude of the refractive index across its diameter. It implies that the index-profile plot is between the core refractive index (taken on the horizontal axis) and the radial distance from the center of the core (taken on the vertical axis) across its diameter. There are two basic types of index profiles, one is called step-index profile fiber and the other one is known as the graded-index profile fiber.

- A step-index profile is one which has a uniform refractive index of the core.
- A graded-index profile fiber is one which has a non-uniform refractive index of the core, i.e., maximum in the center of the core and then decreases gradually with radial distance from the center toward the outer edge.

17. Why do most of the modern long-haul optical fiber communications networks use wavelengths which are around 1550 nm?

Most of the modern long-haul optical fiber communication networks employ glass fiber. Fig. 2.54 shows approximately how the fiber loss varies with wavelength for typical glass fiber.

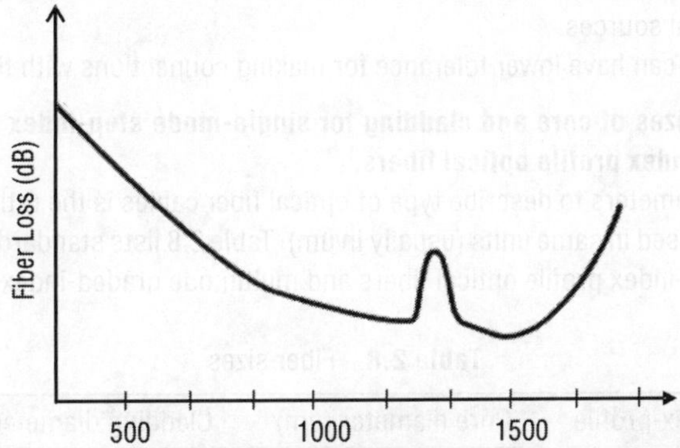

Fig. 2.54 Fiber loss versus wavelength (nm) in glass fiber

It is observed that the fiber loss in glass fiber is the lowest at a wavelength of approximately 1550 nm.

18. The simplest form of a single-mode step-index profile optical fiber is the one which has air as the outside cladding layer. Why is it not useful for fiber-optic communications?

The large difference in the refractive indexes in this type of optical fiber results in a small critical angle at the glass/air interface. As a result, it exhibits relatively wider acceptance angle. Consequently, it is fairly convenient to launch light into this type of optical fiber cable from an external optical source. The biggest disadvantage is that it is quite fragile and extremely difficult to terminate with connector or splice with another segment of similar type of optical fiber. Therefore, it is not useful for fiber-optic communications.

19. What is the main advantage of single-mode step-index profile optical fibers?

With single-mode step-index profile optical fibers, the light is propagated within the fiber using the principle of total internal reflection. Light rays that enter the fiber either propagate straight down the core or, perhaps, are reflected only a few times. Consequently, all light rays follow approximately the same path down the cable and take approximately the same amount of time to travel the length of the cable. This is one overwhelming advantage of single-mode step-index fibers.

20. What are typical benefits of graded-index profile optical fiber cables over step-index profile optical fiber cables?

(a) Graded-index profile optical fiber cables exhibit far less intermodal dispersion than exhibited by the multimode step index profile optical fiber cables mainly due to the nature of their refractive index profiles.
(b) Different group velocities of the propagating modes get normalized with respect to the index grading.
(c) It has large core diameter (>30 μm).
(d) Its bandwidth is greater than multimode step-index fiber bandwidth, but less than single-mode step index fiber bandwidth.
(e) Graded index fibers accept less light.

21. What are typical advantages of multimode fibers?

Multimode fibers offer several advantages for lower bandwidth applications. Some of them are given below:

(a) Spectrally incoherent optical sources such as LEDs can be efficiently coupled to non-metallic fibers.
(b) For large values of fiber core diameter and numerical aperture, it is relatively easier to couple them to external optical sources.
(c) Fiber connectors can have lower tolerance for making connections with the fibers.

22. Tabulate standard sizes of core and cladding for single-mode step-index profile optical fibers and multimode graded-index profile optical fibers.

One of the major parameters to describe type of optical fiber cables is the ratio of diameters of core and cladding, both expressed in same units (usually in μm). Table 2.8 lists standard sizes of core and cladding for single-mode step-index profile optical fibers and multimode graded-index profile optical fibers.

Table 2.8 Fiber sizes

Mode type	Index-profile	Core diameter (μm)	Cladding diameter (μm)	Fiber Type
Single-mode	Step-index	7	125	7/125
Multi-mode	Graded-index	50	125	50/125
		62.5	125	62.5/125
		100	125	100/125

23. What is meant by pulse-width dispersion?

As the pulse propagates down the optical fiber cable, the light rays that make up the pulse spread out in time, causing a corresponding reduction in the pulse amplitude and stretching of the pulse width. This is called pulse-width dispersion, or pulse spreading. This causes errors in digital transmission because as light energy from one pulse falls back in time, it will interfere with the next pulse, causing intersymbol interference.

24. Contrast the extent of dispersion occurring in single-mode and multimode fibers.

Multimode propagation causes more dispersion, just as it does in waveguides. Dispersion results in the spreading of pulses and limits the usable bandwidth of the fiber. Single-mode fiber has much less dispersion, but it is more expensive to manufacture. Moreover its small diameter, coupled with the fact that its numerical aperture is less than that of multimode fiber, makes it more difficult to couple light into and out of the fiber. Multimode fiber is easier to work with than single-mode fiber because it has a much larger core diameter, and also a larger numerical aperture, than single-mode fiber.

25. Name two problems which are encountered when light travels through a fiber cable.

Attenuation and dispersion are two problems to contend with when light travels through a fiber cable. *Attenuation* means reduction in optical power as the light travels distance in the cable. The attenuation is due to various imperfections along the fiber and is measured in dB/km. The level of attenuation decides when the repeater is to be connected or what type of receiver is to be used. *Dispersion* gives rise to pulse spread and restricts bandwidth when signal is sent as pulses of light energy. The signal distortion due to dispersion limits the information-carrying capacity of an optical fiber.

26. List various factors that are responsible for reduction in dispersion in graded-index multimode fibers.

Graded-index multimode fibers reduce dispersion by taking advantage of the fact that

- Signals propagating in higher-order modes occupy more time near the outer edge of the core than do the low-order modes.
- The reduction of the index of refraction toward the outside of the fiber core results in increased velocity in this region.
- Higher-order-mode components of the signal, which have farther to travel, propagate more quickly.
- This results in significant reduction of dispersion, although not fully eliminated.

27. Illustrate pulse-width dispersion in an optical fiber cable.

Fig. 2.55 shows the relative time/energy relationship of a pulse of light as it propagates down an optical fiber cable, depicting pulse-width dispersion.

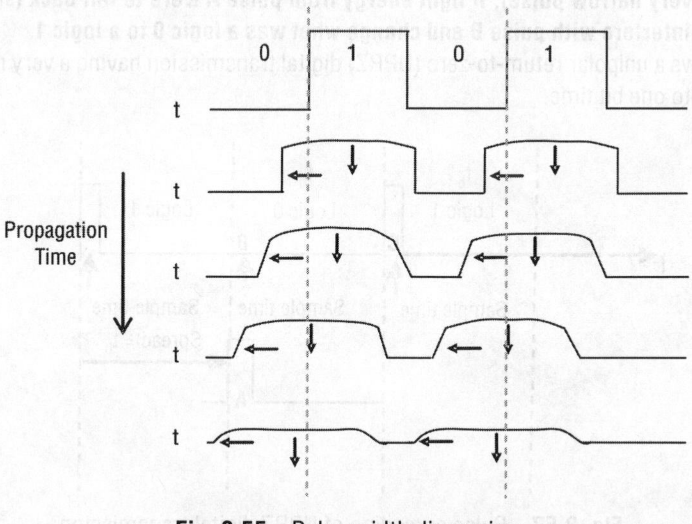

Fig. 2.55 Pulse-width dispersion

As an optical pulse propagates through the fiber length for a longer propagation time, pulse-width dispersion results in higher bit errors at the receiver.

28. Show with the help of suitable diagram that with a unipolar non-return-to-zero (UPNRZ) digital transmission, if energy from pulse A were to fall back one-half of a bit time, it would interfere with pulse B.

Fig. 2.56 shows a unipolar non-return-to-zero (UPNRZ) digital transmission where each pulse is equal to the bit time.

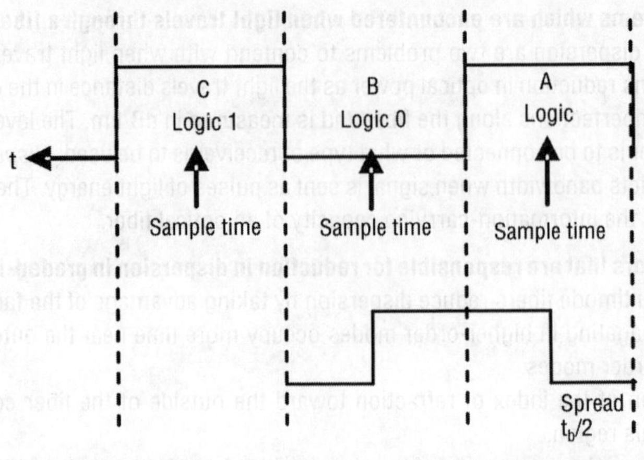

Fig. 2.56 Pulse spreading of UPNRZ digital transmission

It is observed that with UPNRZ transmission, if energy from pulse A were to fall back one-half of a bit time, it would interfere with pulse B.

29. Show with the help of suitable diagram that with a unipolar return-to-zero (UPRZ) digital transmission (assuming a very narrow pulse), if light energy from pulse A were to fall back (spread) one bit time (t_b), it would interfere with pulse B and change what was a logic 0 to a logic 1.

Fig. 2.57 shows a unipolar retum-to-zero (UPRZ) digital transmission having a very narrow pulse in time as compared to one bit time.

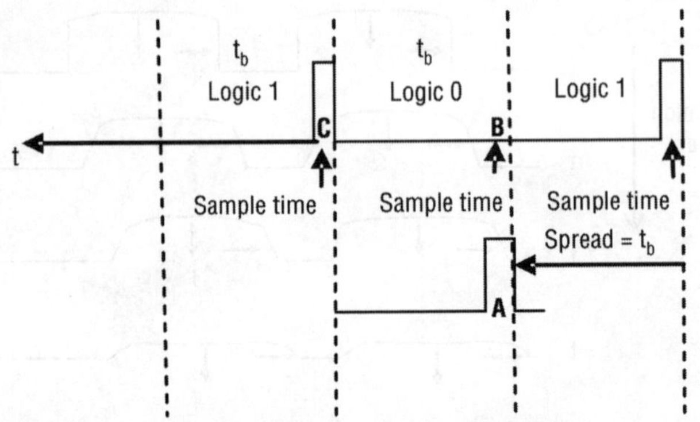

Fig. 2.57 Pulse spreading of UPRZ digital transmission

It is observed that with UPRZ transmission, if light energy from pulse A were to fall back (spread) one bit time (t_b), it would interfere with pulse B and change what was a logic 0 to a logic 1. However, UPRZ transmissions can tolerate twice as much pulse spread as compared to that of UPNRZ transmissions.

30. What do you understand by spectral width, or linewidth of a light source?

We know that a practical light source, such as LED or laser, radiates a band of wavelengths. These wavelengths concentrate near the peak wavelength (where the relative light power is maximum). The more a wavelength deviates from the peak wavelength, the less is its amplitude. Hence, spectral width is the width of wavelengths in nanometers at half of maximum light power. The greater the spectral width, more the wavelengths emitted by the light source, more the material dispersion, and thus increased pulse spreading. Typically, the spectral width of LED is tens of nanometers; whereas that of a laser diode is about one nanometer and even less.

31. Fiber losses result from axial misalignment of the fibers. Illustrate this problem showing the core of the fiber only for simplicity.

Fig. 2.58 illustrates the problem from axial misalignment of the fibers which allow light to escape at various angles.

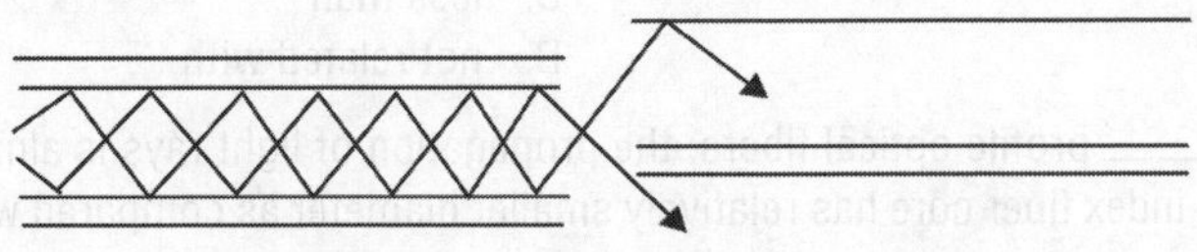

Fig. 2.58 Axial misalignment of the fibers

32. Distinguish between Splice and Connector. What are the main reasons for less loss in a properly made splice less than that in a connector?

The terms splice and connector are related but not equivalent.

- Generally, a splice is a permanent connection, while connectors are removable.
- Splices are necessary where sections of cable are joined. For practical reasons, the length of a spool of cable is limited to about 10 km, so longer spans between repeaters require splices to be made in the field.
- Connectors are needed between sources and detectors and the fiber cable.
- Generally, the loss in a properly-made splice is less than that in a connector. Splices can have losses of 0.02 dB or less, while connector losses are often about 2 dB.

One of the reasons for loss in a properly made splice being less than that in a connector is that the ends of the fibers touch in a well-made splice (there is no air gap), while in a connector there is a small air gap (that ensure that the polished fiber surfaces will not be damaged during the process of connecting or disconnecting).

Multiple Choice Questions

1. The light is propagated within the fiber core by the phenomenon
   A. total internal reflection at core-cladding intersection
   B. refraction at core-cladding intersection
   C. total internal reflection at the outer surface of the cladding
   D. change in the velocity of light within the fiber core

2. A step-index fiber has specified parameters for refractive index of fiber core and cladding as 1.50 and 1.46, respectively. Its numerical aperture is
   A. 0.344 B. 0.156
   C. 0.486 D. 0.244

3. A step-index fiber has specified parameters for refractive index of fiber core and cladding as 1.50 and 1.33, respectively. Its acceptance angle will be approximately
   A. 25° B. 20°
   C. 15° D. 10°

4. Consider a ray of light propagating from one medium to another medium having different indexes of refraction. If the incidence angle is greater than the specified critical angle, then ______ occurs.

A. reflection B. refraction
C. diffraction D. scattering

5. When the incidence angle is ______ the specified critical angle, the light rays bend along the intersection line of two different mediums of propagation.

A. more than B. less than
C. equal to D. not related with

6. In ______ profile optical fibers, the propagation of light rays is almost horizontal provided the low-refractive index fiber core has relatively smaller diameter as compared with those of other types of optical fibers.

A. multimode step-index B. multimode graded-index
C. multimode single-index D. single-mode

7. Dispersion (i.e., distortion in the transmitted optical pulse) is maximum in ______ type of optical fibers.

A. Multimode step-index B. Multimode graded-index
C. Multimode single-index D. Single-mode

8. In ______ type of optical fiber cables, the density of the fiber core varies.

A. multimode step-index B. multimode graded-index
C. multimode single-index D. single-mode

9. In optical fibers, the index of refraction in the fiber core is always

A. greater than that of cladding B. less than that of cladding
C. equal to that of cladding D. not at all related with that of cladding

10. For single-mode step index fibers, V-number should be less than

A. 2.4 B. 2.8
C. 4.2 D. 8

11. Which one of the following types does not exist in optical fibers?

A. single-mode step-index B. single-mode graded-index
C. multimode step-index D. multimode graded-index

12. The essential condition for total internal reflection to take place within the optical fiber is when the incidence angle exceeds the specified value of

A. critical angle B. refraction angle
C. reflection angle D. acceptance angle

13. The rays which do not intersect the core axis are called

A. meridional rays B. radial rays
C. helical rays D. skew rays

14. Intramodal dispersion is associated with

A. single-mode fibers B. multi-mode fibers
C. coaxial cables D. copper wires

15. Modal Birefringence is the main cause of

A. chromatic dispersion B. polarisation mode dispersion
C. modal dispersion D. profile dispersion

16. A typical optical fiber cable has specified values of n_1 = 1.82 and n_2 = 1.73. The critical angle of incidence is calculated as

A. 71.90°
B. 0.95°
C. 18.1°
D. 1.81°

17. Which out of the following options is not a non-linear effect in optical fiber communication?

A. Four Wave Mixing
B. Cross Phase Modulation
C. Rayleigh Scattering
D. Raman Scattering

18. In optical fibers, the term `dispersion' also signifies

A. scattering
B. multiple reflections
C. distortion
D. broadening of pulse width

19. V-number signifies to

A. relative refractive index
B. relative velocity
C. normalized frequency
D. real function

20. If the wavelength of the light in optical fiber is 1.3 micrometers, then the material dispersion is

A. infinite
B. Zero
C. 40
D. –100

21. Which of the following is not true for optical fiber?

A. high security
B. immunity to electrical noise
C. reduced size and weight
D. low performance

22. Which of the following is not a part of chromatic dispersion?

A. Waveguide dispersion
B. Material Dispersion
C. Intramodal Dispersion
D. Intermodal dispersion

23. $(B \times L)$ stand for

A. bandwidth distance product
B. bit distance product
C. bandwidth dispersion parameter
D. bandwidth distance parameter

24. For single-mode fibers, the preferred optical source is

A. LED
B. Laser
C. Maser
D. Transistor

25. The V-number for a multimode step-index profile optical fiber is given as 8. How many modes can propagate through this fiber?

A. 64
B. 16
C. 8
D. 32

26. The scattering of light signifies the loss of optical energy, mainly because of one of the following reasons.

A. imperfections in fiber core
B. group velocity
C. time delay
D. atmospheric stress

27. Typical optical wavelength, for which chromatic dispersion and fiber losses are almost zero, is

A. 1560.606 nm
B. 1552.52 nm
C. 1539.766 nm
D. 1332.45 nm

28. Light remains confined within the core of a simple optical fiber by one of the following phenomena.

A. total internal reflection
B. diffraction
C. Refraction
D. all of these

29. For long distance fiber-optic communication networks, which one of the following types of optical fibers is preferred?
A. Single-mode step-index profile optical fiber
B. Multimode step-index profile optical fiber
C. Single-mode graded-index profile optical fiber
D. Multimode graded-index profile optical fiber

30. For a triangular refractive index profile, the value of profile parameter is typically
A. zero B. unity
C. two D. infinity

31. Broadening of an optical pulse broadening which occurs due to intermodal dispersion can be minimized using
A. Single-mode step-index profile optical fiber
B. Multimode step-index profile optical fiber
C. Single-mode graded-index profile optical fiber
D. co-axial cables

32. Optical fiber communications is based on the principle of
A. total internal reflectance B. laser technology
C. photo-electric effect D. Tyndall effect

33. When V-number is less than 2.405, then how many modes can an optical fiber support?
A. 1 B. 2
C. 3 D. infinity

34. _______________ is used to cancel the effect of Chromatic Dispersion in an installed optical network
A. Dispersion Shifted Fiber B. Polarization Maintaining Fiber
C. Dispersion Compensating Fiber D. Dispersion Flattened Fiber

35. Microbending losses in an optical fiber can be reduced
A. if radius of bending is controlled
B. by removing waveguide imperfections
C. by controlling the bend radius and removing waveguide imperfections
D. neither by controlling bend radius nor by removing waveguide imperfections

36. At zero-dispersion wavelength, which one of these is zero?
A. Polarization mode dispersion B. Higher order dispersion
C. Intermodal dispersion D. Chromatic dispersion

37. The number of modes that are supported by parabolic refractive index profile optical fiber having normalized frequency parameter of 2.2 will be
A. 1 B. 2
C. 3 D. 4

38. The axial refractive index of the graded-index profile fiber core is specified as 1.50. If the maximum relative refractive index difference is 1%, then the index of refraction of the cladding is
A. 1.485 B. 1.50
C. It will depend on the profile parameter. D. It will depend on the radius of the core.

Keys to Multiple Choice Questions

1.A	2.A	3.C	4.A	5.C	6.D	7.A	8.B	9.A	10.A
11.B	12.A	13.D	14.A	15.B	16.A	17.C	18.D	19.C	20.B
21.D	22.D	23.A	24.B	25.D	26.A	27.B	28.A	29.A	30.B
31.C	32.A	33.A	34.C	35.A	36.D	37.A	38.A		

Review Questions

1. Define the terms: velocity of propagation in a medium other than air, refraction, and refractive index.
2. Using ray theory, explain the basic mechanism for the propagation of light in an optical fiber cable. In optical fibers, light travels faster in cladding as compared to core. Justify it with the help of suitable example data.
3. State Snell's law for refraction of light. Highlight its significance in context with optical fiber cables.
4. What is total internal reflection? Under what conditions does it occur? State the reasons as why is it necessary to meet the condition of total internal reflection at the intersection of the fiber core and the cladding of an optical fiber cable.
5. What is an acceptance angle? What is the significance of the acceptance angle and the acceptance cone for a fiber cable? How are they related to each other?
6. What do you understand by the term `numerical aperture' of an optical fiber cable? What happens if light moves from one fiber to another with a lower numerical aperture?
7. Derive an expression for acceptance angle and show its relationship with numerical aperture.
8. The numerical aperture of a fiber is generally used to describe the light acceptance or light gathering capability from an external optical source. What are the factors on which numerical aperture depends so as to determine the optical source-to-fiber power coupling efficiency?
9. Two most commonly used fiber types depends on the variations in the material composition of the fiber core. Name them. Specify the major difference in their construction.
10. What is meant by index profile of a fiber and mode of operation of light rays through the optical fibers?
11. Differentiate between single-mode and multimode step-index profile fibers. Explain how does the light propagate in these fibers with a suitable ray diagram?
12. Using ray theory, outline the basic mechanism for propagation of light in a multimode graded-index optical fiber.
13. Compare and contrast the chief benefits as well as drawbacks of single-mode and multimode propagation in step-index and graded-index profile optical fibers.
14. Single-mode propagation is not possible with graded-index profile optical fibers. Justify your answer giving sufficient reasons.
15. Discuss the concept of modes for propagating light along optical fiber cables.
16. Explain briefly the difference between single-mode and multimode fiber. Which one gives better performance and why?
17. What is a single-mode fiber? What is meant by the cut-off condition?

18. Distinguish between multimode step-index and graded-index profile optical fibers. Bring out clearly the difference between single-mode and multimode fibers.

19. What is typical bandwidth length product for single-mode fibers having step-index and graded-index profiles?

20. Define mode field diameter. How is it related to the *V*-parameter?

21. What are the impairments faced by single-mode fibers for long distance optical transmission?

22. What is pulse spreading? Define pulse spreading constant. Give its mathematical representation.

23. List and briefly describe the losses associated with fiber cables. What are coupling losses?

24. List the three types of optical fiber, and order them in terms of dispersion and fiber loss.

25. Describe the mechanisms by which dispersion takes place in optical fibers. Which of these mechanisms apply to single-mode fiber?

26. Explain why the maximum bit rate that can be transmitted using an optical fiber decreases as the fiber length increases. Which type of fiber has the highest bandwidth-distance product? Why?

27. Discuss various dispersion induced limitations in optical fibers. How does dispersion limit the maximum data rate that an optical fiber can carry?

28. Describe the various design issues for the fabrication of optical fibers.

29. There are different types of transmission losses that an optical signal incur when it propagates through the optical fiber cable. Which one is responsible that affects the power and shape of the transmitted optical pulse?

30. Explain multipath time dispersion and material dispersion. How can these be minimized?

31. Differentiate between Intramodal and intermodal types of dispersion. What are the components of Intramodal dispersion in a single-mode optical fiber?

32. Define mode birefringence and beat length of a single-mode optical fiber. Explain the effect of modal birefringence on propagation of pulse broadening in single-mode fibers.

33. How non-linear effects degrade the performance of an optical communication system? When and why do the non-linear scattering losses occur in optical fiber communication? Classify them.

34. What are the causes of attenuation in optical fibers? Why could bending loss in single-mode fibers be severe? What can be done to minimize this loss?

Numerical Problems

1. For a glass core (n_1 = 1.5) and quartz cladding (n_2 = 1.41) interface, find the angle of refraction if the incidence angle is 38°. *[Ans.: 42°]*

2. In an optical fiber cable, the core has a refractive index of 1.5 and the cladding has a refractive index of 1.45. Determine the critical angle for a ray moving from the core to the cladding. *[Ans.: 75°]*

3. For a glass core (n_1 = 1.5) and quartz cladding (n_2 = 1.38) optical fiber cable, determine the critical angle of incidence for a light ray moving from the fiber core to the cladding. *[Ans.: 66.9°]*

4. In an optical fiber cable, the core has a refractive index of 1.5 and the cladding has a refractive index of 1.45. Show that its numerical aperture is 0.384.

5. In an optical fiber cable, the core has a refractive index of 1.5 and the cladding has a refractive index of 1.38. Determine its numerical aperture. *[Ans.: 0.588]*

6. In an optical fiber cable, the core has a refractive index of 1.5 and the cladding has a refractive index of 1.45. What is the maximum angle (from the axis of the fiber) at which the light will be accepted? *[Ans.: 22.6°]*

7. In an optical fiber cable, the core has a refractive index of 1.5 and the cladding has a refractive index of 1.38. Determine the acceptance angle. *[Ans.: 56°]*

8. A step-index profile optical fiber has an acceptable angle of 20° in air medium. It has a relative refractive index difference value of 3%. Determine the critical angle at the core-cladding interface and numerical aperture of the fiber. *[Ans.: 0.34; 76°]*

9. A typical step-index profile optical fiber has a fiber core having radius = 4 μm and refractive index = 1.46. The relative refractive index difference is specified as 0.3%. Determine the normalized frequency parameter V at the following operating wavelength
   (a) 1300 nm (b) 1550 nm *[Ans.: (a) 2.186; (b) 1.834]*

10. A single-mode step-index profile optical fiber cable has a fiber core whose radius = 8 μm and refractive index = 1.46. The relative refractive index difference is specified as 0.52%. Calculate the cut-off wavelength for the fiber. *[Ans.: 1556 nm]*

11. A multi-mode step-index profile fiber cable has a fiber core whose diameter = 50 μm and refractive index = 1.46. The relative refractive index difference is specified as 0.3%.fiber has a core diameter of 50 μm, a core refractive index of 1.46, and a relative refractive index difference of 1%. At an operating wavelength of 1300 nm, determine
   (a) The cladding refractive index.
   (b) The normalized frequency parameter *V*.
   (c) Total number of guided modes. [Ans.: (a) 1.445; (b) 25; (c) 312]

12. A multi-mode step-index profile optical fiber cable has a refractive index = 1.5. The relative refractive index difference is specified as 1%. There are approximately 1100 modes that propagate at a given wavelength of 1300 nm. Determine the diameter of the fiber core. *[Ans.: 91.54 μm]*

13. A step-index profile optical fiber has the specification of a normalized V-parameter as 26.6 at a given wavelength of 1300 nm. If the diameter of the fiber core is 50 μm, then determine the numerical aperture. *[Ans.: 0.22]*

14. A triangular graded-index profile single-mode optical fiber cable has a core axis refractive index = 1.5 and relative refractive index difference = 1.3%. At operating wavelength of 1300 nm, determine the core diameter of the fiber. *[Ans.: 7.1 μm]*

15. A graded-index cubic profile optical fiber that supports the propagation of 600 guided modes has the following specifications:
   - Core diameter = 75 μm
   - Core axis refractive index = 1.45
   - Relative refractive index difference = 2%

   Calculate the wavelength of light propagating in this fiber. *[Ans.: 1400 nm]*

16. A graded-index cubic profile optical fiber has a core axis refractive index = 1.45, a relative refractive index difference = 2%, and number of guided modes supported = 600. Determine the diameter of the fiber core needed that can ensure single-mode operation at λ = 1400 nm. *[Ans.: 6.85 μm]*

17. A graded-index triangular profile single-mode optical fiber has a core diameter = 8.5 μm, a core axis refractive index = 1.5, a relative refractive index difference = 1.3%. Calculate the operating wavelength for single-mode operation. *[Ans.: 1550 nm]*

18. A graded-index triangular profile optical fiber has the following specifications:

- Core diameter = 75 μm
- Core axis refractive index = 1.46
- Number of propagation guided modes supported = 900

If the operating wavelength propagating through the fiber is 1300 nm, then determine the followings:
(a) The relative refractive index difference
(b) The maximum radius of the fiber core that would give single-mode operation at the same wavelength. *[Ans.: (a) 0.02; (b) 2.88 μm]*

19. A graded-index parabolic profile optical fiber cable has the following specifications:

- Diameter of the fiber core = 70 μm
- The core axis refractive index = 1.47
- The cladding refractive index = 1.45.

If the operating wavelength propagating through the fiber is 1300 nm, determine
(a) The normalized frequency parameter V
(b) The number of propagation modes supported by this fiber. *[Ans.: (a) 40.9; (b) 418]*

20. Consider an optical fiber cablewhose fiber core diameter = 50 μm, core refractive index = 1.48, and relative refractive index difference = 1%. If the operating wavelength is 840 nm, then determine the following:
(a) The normalized frequency parameter, V (i.e., V-number)
(b) The number of propagation modes in the fiber
(c) The fractional average power in the cladding
(d) The number of propagation modes within the fiber and the fractional average power in the cladding if the relative refractive index difference is reduced to 0.3% in order to decrease signal dispersion. *[Ans.: (a) 39; (b) 760; (c) 5%; (d) 242, 9%]*

21. Determine the maximum core diameter if a fiber is required to be operated in single-mode at λ = 1550 nm. Given NA = 0.12. *[Ans.: 9.9 μm]*

22. A fiber-optic cable has a bandwidth-distance product of 600 MHz-km. What bandwidth can be used with a cable that runs 30 km between repeaters? *[Ans.: 20 MHz]*

23. A fiber is installed over a distance of 15 km, it is found experimentally that the maximum operating bandwidth is 700 MHz. Determine the bandwidth-distance product for this fiber. *[Ans.: 10.5 GHz-km]*

24. A single-mode optical fiber cable has a specified chromatic dispersion parameter of 10 ps/(nm–km). Calculate the total dispersion over a distance of 10 km if the linewidth of an optical source is 40 nm. If the linewidth of the optical source is 5 nm, then how much is the chromatic dispersion? *[Ans.: 4 ns; 500 ps]*

25. The fiber has zero dispersion at λ = 1315 nm. It has a specified zero-dispersion slope of 0.075 ps/(nm^2 - km). Calculate the total dispersion of 120 km of this fiber when it is used with a source having a linewidth of 1.5 nm at λ = 1560 nm.

Hint: Use $D_c(\lambda) = \frac{S_0}{4}\left[\lambda - \frac{\lambda_0^{\,4}}{\lambda^3}\right]$ =14.5 ps/(nm - km). *[Ans.: 2610 ps]*

26. The fiber has zero dispersion at a wavelength of 1310 nm. It has the specification of zero-dispersion slope of 0.05 ps/(nm^2 - km). Calculate the total dispersion of 50 km of this fiber when an optical source having a linewidth of 2 nm at a wavelength of 1550 nm is used.

Hint: Use $D_c(\lambda) = \frac{S_0}{4}\left[\lambda - \frac{\lambda_0^4}{\lambda^3}\right]$ =9.49 ps/(nm - km). *[Ans.: 949 ps]*

27. A source with a power level of -20 dBm is connected to one end of a piece of fiber. The fiber length is 120 m. If the measured optical power level at the far end of the cable is measured as -22.5 dBm, then what is the fiber loss? *[Ans.: 20.8 dB/km]*

28. An optical fiber cable has specified fiber attenuation as 0.35 dB/km. If an optical source with a power output of 25 μW is connected to one end of a 20 km length of this fiber, how much power is available at the other end of the cable? *[Ans.: 5 μW]*

29. Estimate the maximum transmission data rate for NRZ encoding signaling format for the following given values of pulse-spreading constant and cable length:

(a) Δt = 10 ns/m; and L = 100 m

(b) Δt = 20 ns/m; and L = 1000 m

(c) Δt = 2000 ns/km; and L = 2 km [Ans.: (a) 500 kbps; (b) 25 kbps; (c) 125 kbps]

30. Calculate the maximum transmission data rate for RZ encoding signaling format for the following given values of pulse-spreading constants and cable lengths:

(a) Δt = 10 ns/m; and L = 100 m

(b) Δt = 20 ns/m; and L = 1000 m

(c) Δt = 2000 ns/km; and L = 2 km *[Ans.: (a) 1000 kbps; (b) 50 kbps; (c) 250 kbps]*

31. Compute the maximum transmission data rate for NRZ as well as RZ encoding signaling formats for the following given values of pulse-spreading constants and cable lengths:

(a) Δt = 10 ns/m; and L = 50 m

(b) Δt = 14 ns/m; and L = 200 m

(c) Δt = 20 ns/km; and L = 200 m

[Ans.: (a) NRZ - 1 Mbps, RZ - 2 Mbps; (b) NRZ - 179 kbps, RZ 357 kbps; (c) NRZ - 125 kbps, RZ - 250 kbps]

32. A laser diode has a relative spectral width of 0.002 and is emitting a mean wavelength of 1000 nm. What is its spectral half-width? *[Ans.: 2 nm]*

33. A given optical source with the relative spectral width of 0.03 at λ = 850 nm is coupled to a pure silica fiber whose specified parameter $\left|\lambda^2 \frac{d^2n}{d\lambda^2}\right|$ = 0.02 at 850 nm. Calculate the pulse broadening per kilometer fiber length that occurs because of material dispersion. *[Ans.: 2 ns/km]*

34. A single-mode step-index profile fiber cable has the following specifications:

- Core diameter = 8.2 μm
- Core refractive index = 1.45
- Relative refractive index difference = 0.3%

Determine the V-parameter as well as the waveguide dispersion parameter for this fiber at λ = 1300 nm. *[Ans.: 2.2256; -3.149 ps/(nm–km)]*

35. Determine the waveguide dispersion parameter for a single-mode step-index optical fiber cable at λ = 1550 nm. The fiber has the following specifications:

- Core diameter = 8.2 μm
- Core refractive index = 1.45
- Relative refractive index difference = 0.3%. *[Ans.: -5.536 ps/(nm–km)]*

36. Determine the waveguide dispersion parameter for a single-mode step-index optical fiber at λ = 1320 nm. The fiber has the following specifications:

- Core radius = 4.4 μm
- Core refractive index = 1.48
- Relative refractive index difference = 0.27% *[Ans.: -2.51 ps/(nm–km)]*

37. A single-mode fiber has specified parameter of waveguide dispersion as -4 ps/(nm–km). It is excited by an optical source operating at λ = 1550 nm and has spectral width = 1 nm. Calculate the pulse broadening mainly caused by the waveguide dispersion. *[Ans.: 400 ps]*

38. A step-index single-mode optical fiber cable has the following specifications:

- Core refractive index = 1.48
- Relative refractive index difference = 1%

If the material dispersion at 1550 nm is 7 ps/(nm–km), then what should be the value of the core diameter for total dispersion at this wavelength to be zero. *[Ans.: 5.48 μm]*

39. A single-mode optical fiber cable has the specification of beat length = 8 cm at λ = 1300 nm. Calculate the modal birefringence. *[Ans.: 1.63 x 10^{-5}]*

40. The modal birefringence of a typical single-mode fiber operating at λ = 1300 nm varies from 10^{-6} to 10^{-5}. Determine the following:

(a) The range of the differences of propagation constants for two orthogonally polarized modes of propagation

(b) The range of the beat length *[Ans.: (a) 4.833 m^{-1} – 48.33 m^{-1}; (b) 13 cm – 1.3 m]*

41. A typical single-mode step-index optical fiber cable has the following specifications:

- Core radius = 4.1 μm
- Effective core refractive index = 1.4677 at λ = 1310 nm
- Relative refractive index difference = 0.36%

Determine the value of mode field diameter. *[Ans.: 8.904 μm]*

42. A typical single-mode step-index optical fiber cable has a core diameter = 8.2 μm, effective core refractive index = 1.4682 at operating wavelength of 1550 nm, and relative refractive index difference = 0.36%. Determine the mode field radius. *[Ans.: 5.0427 μm]*

43. For a 30-km long optical fiber cable that has specification of fiber attenuation of 0.8-dB/km at λ = 1300 nm. Estimate the measured optical output power (in dBm and μW) at the other end of the cable if 0.2 mW of optical power is launched by an optical source into the fiber. *[-31 dBm; 0.8 μW]*

Optical Sources and Transmitters

CHAPTER

3

Chapter Objectives

After studying this chapter, you should be able to

- ✓ know the essential requirements for optical sources
- ✓ describe the characteristics and advantages of light emitting diode (LED) as an optical source
- ✓ understand the operating principle of stimulating emission (lasing)
- ✓ get familiarized with the characteristics of injection laser diode (ILD) as an optical source
- ✓ outline advantages of ILDs over LEDs

An optical source provides the electrical–optical signal conversion efficiently that enables the optical output to be effectively coupled and launched into the optical fiber. It is an active device that requires external power supply for operation in optical fiber communications. This chapter begins with the major requirements or desirable properties for the optical source in general. Light emitting diodes (LEDs) and injection laser diodes (ILDs) {which are simply known as laser diodes (LDs)} are the most popular semiconductor optical sources. LED is an incoherent optical source. It can support many propagation modes of light within its structure. This is the reason why is it employed as a multimode optical source. Whereas, ILD is a highly coherent optical source that has a very narrow spectrum and fast response time. Thus, it is mostly used as a single-mode optical source in single-mode propagation requirements. In this chapter the basic principle of operation of both these optical sources along with their major structures and configurations are described. For producing the light signal, the emission can be a spontaneous emission, as in the case of LED, or a stimulated emission as in the case of ILD. The spontaneous emission takes place when electrons are brought to a very high energy level, and an unstable state. The electrons will return spontaneously (within few picoseconds) to a stable state, and will consequently emit photons.. The optical wavelength is determined by the amount of energy the electron releases. A laser diode operates with stimulated emission in which the electrons enters and stays in a high-energy state for a few microseconds. Then it changes its state spontaneously. During this state, the photon stimulates so as to emit the energy in the form of another photon. Thus, the laser produces the light signal. Finally, a typical functional block schematic including features and operations of optical transmitter is covered to assess the utility of optical sources.

3.1 Requirements for an Optical Source

We know that electrical signals (in the form of voltage or current) have to be converted into optical signals (light pulses) at the transmitter end in an optical fiber communication system. So, the fundamental function of an optical source is to transform electrical energy (i.e., the current) into an equivalent optical energy (i.e., light pulse) as efficiently as possible. In essence, there are mainly two types of optical sources: monochromatic incoherent sources such as LEDs, and monochromatic coherent sources such as laser diodes. In general, the major requirements for an optical source are:

- Ideally, the light output must be focused (i.e., highly directional) so that it can be launched into an optical fiber efficiently.
- The size and configuration of an optical source should be compatible so as to couple light efficiently into an optical fiber.
- The optical source must have linear output. This means that the output optical signal should be in direct proportion to the input electrical signal so as ensure minimum noise and distortion.
- Optical sources should be able to generate optical signals at wavelengths where the fiber attenuation is minimum. Moreover, optical pulse dispersion should be low so that optical detectors can operate efficiently.
- Optical sources must emit sufficient optical power so as to compensate for transmission losses due to fiber cable as well as the connectors used in the link. This is required in order to ensure adequate optical power level necessary to operate the optical detector at the receiver end.
- Optical sources should be capable of providing signal modulation over a wide bandwidth (ranging from audio frequencies to several GHz). However, they should have narrow linewidth so as to produce negligible fiber dispersion.
- Optical sources must generate a stable output optical signal which should not vary with operating temperature and other environmental conditions.
- Optical sources should be highly reliable and inexpensive.

Laser diodes are commonly used in high-capacity optical communication networks for sending optical signals to the optical fiber. In fact, they are very small semiconductor devices which are specially designed to transmit very specific and precise wavelengths. These devices operate on the principle of an *excited state*, which means that the electrons in certain parts of the semiconductor material have more energy than the electrons in other parts of the semiconductor material. When the electrons of an excited state loses some of its energy and falls to a ground state, then the energy is released in the form of photons, known as light energy. When we apply external electrical current to a laser diode, it produces many electrons in an excited state. Of course, as more and more electrons fall from an excited state to the ground state, they emit higher levels of light energy in the form of many photons.

As mentioned earlier, heterojunction LEDs and injection laser diodes are generally used as optical sources. The term 'Heterojunction' simply means that a *p*–*n* semiconductor junction is formed using a single crystal of semiconductor material in such a way that the materials used on either sides of the junction are quite different. For example,

- A heterojunction is formed between two compound semiconductor materials—GaAs and GaAlAs, in GaAs diode lasers. These types of optical sources are normally used at 800 nm optical wavelength region.
- InP–InGaAsP heterojunction type of laser diodes can be used at longer wavelengths.

In general, heterojunction optical sources (LEDs or lasers) exhibit much better performance than traditional homojunction optical sources. They provide threshold current density (minimum 10 Amp/mm^2), relatively higher output optical power of the order of 10 mW at low input electric current (as low as less than 500 mA), a high degree of coherence and monochromaticity, stable operation for a much longer life, etc.

We can also use a double heterojunction structure stripe laser in which the width of active junction region is few microns only. This translates to considerable reduction in threshold current density. The stripe geometry offers many advantages such as higher stability associated with a longer lifetime for the optical sources, higher output optical power, higher degree of coherence and directionality, improved efficiency, and continuous wave operation. By forming heterojunction using two different materials, the carriers and the optical fields remain confined in the central active region. The bandgap differences of adjacent regions restrict the charge carriers, whereas the step change in the refractive indices of adjacent regions restricts the optical field to the central active region. This type of dual confinement leads to an efficient waveguide structure which can provide higher output optical power.

The optical source LEDs are extensively used as major multi-mode optical transmission source, for giving acceptable coupling efficiencies into multimode fibers, for increasingly wider bandwidth, and for long haul applications. On the other hand, the optical source LDs are extensively used as a single-mode optical transmission source and as coherent optical source.

In order to understand the principle of operation, efficiency, and structural designs of an LED or ILD, it is essential to be familiarized with the basic properties of semiconductors including the p–n junction (homojunctions and heterojunctions), and light-emission processes.

Generally, the classification of semiconductor materials is done as direct and indirect bandgap semiconductor materials.

- When the charge carriers in a semiconductor material can make a transition from conduction band to the valence band without any change in the momentum, it is classified as direct bandgap semiconductor material. For example, GaAs.
- When the charge carriers in a semiconductor material need phonon-assisted transitions from conduction band to the valence band so as to conserve the momentum, it is classified as indirect band semiconductor material. For example, Si.

Therefore, GaAs semiconductor material is the most appropriate kind for generation of light, and hence used for manufacturing of optical sources.

An intrinsic semiconductor has no charge carriers at absolute zero temperature, but it develops an equal number of two types of charge carriers—negative charge carriers, known as electrons, and positive charge carriers, known as holes. The intrinsic carrier density, n_I, can be related to temperature T by the relation

$$n_1 = 2 \times \left(\frac{2\pi kT}{h^2}\right)^{3/2} \times (m_e m_h)^{3/4}\, e^{\left(-\frac{E_g}{2kT}\right)} \tag{3.1}$$

where, k represents the Boltzmann's constant (= 1.38×10^{-23} J/K), h represents the Planck's constant (= 6.626×10^{-34} J-s), m_e and m_h denote the effective mass of an electron and a hole, respectively, and E_g represents the energy gap (also called the forbidden gap, or a band gap in which no energy levels exist) between the conduction band and the valence band, usually expressed in eV (1 eV = 1.6×10^{-19} J). It may be noted that the mass of an electron in free space is given as $m = 9.11 \times 10^{-31}$ kg.

In extrinsic semiconductors, the doping concentration is the main factor (not the temperature) which determines the number of free charge carriers available for conduction. There are two types of extrinsic semiconductor: *n*-type and *p*-type.

- In *n*-type extrinsic semiconductor, donor impurities (also known as *n*-type impurities such as antimony, phosphorous, and arsenic) are added to an intrinsic semiconductor, and they donate excess electron carriers. This process of doping not only increases the number of electrons but also decreases the number of holes originally present in the intrinsic semiconductor due to recombination of electrons and holes.
- In *p*-type extrinsic semiconductor, acceptor impurities (also known as *p*-type impurities such as boron, gallium, and indium) are added to an intrinsic semiconductor, and they create excess holes which can accept electrons.

Facts to Know

For moderate doping concentrations, the product of electron and hole densities is almost independent of the dopant concentrations at normal temperature.

A p–n semiconductor junction is a transition region between p-type and n-type doped semiconductor materials of the same single-crystal. Therefore, in a p–n junction, holes from the p-region will tend to diffuse into the n-region as the holes concentration is much higher in the p-region as compared to that of in the n-region. The diffusion of holes creates a negative space charge near the junction. Similarly, electrons from the n-region semiconductor material will tend to diffuse into the p-region as the electron concentration is higher in the n-region. The process of diffusion of electrons creates a positive space charge near the junction. This double space charge on either side of the junction sets up an internal field in a narrow region (called the depletion region) at equilibrium (that is, with no applied voltages or thermal gradients). This transition region establishes a diffusion potential V_D between the two sides given as

$$V_D = \left(\frac{kT}{e}\right)\ln\left(\frac{N_a N_d}{n_i^2}\right) \tag{3.2}$$

where, k represents the Boltzmann's constant ($= 1.38 \times 10^{-23}$ J/K), T denotes the room temperature, $e = 1.6 \times 10^{-19}$ J, N_a and N_d represent the acceptor and donor concentrations per unit volume, respectively, and n_I is the intrinsic carrier density.

If a p–n junction is formed by doping the same semiconductor with p-type as well as n-type impurities, then it is known as p–n homojunction. Example of p–n homojunction is GaAs semiconductor material with p-type and n-type impurities. If a p–n semiconductor junction is formed between two semiconductors (grown together as a single crystal) that have the same lattice parameters but different energy band gaps by doping with p- and n-type impurity atoms, then it is known as p–n heterojunction. For example, a p–n heterojunction may be formed between GaAs and its ternary alloy $Ga_{1-x}Al_xAs$. In general, the semiconductor materials used for fabrication of optical sources should comply with the following criteria:

(a) *Formation of p–n junction* – The semiconductor materials must possess desired characteristics for the purpose of carrier injections.

(b) *Efficient electroluminescence* – The semiconductor materials used may be of either direct or indirect bandgap semiconductor materials having suitable impurities. This will result in a relatively higher probability of radiative transitions. Consequently, they exhibit quite high internal quantum efficiency.

(c) *Useful emission wavelength* – It is desirable that the semiconductor materials produce light in a desired wavelength region that is applicable for available optical fibers and photodetectors (usually in the region 800–1700 nm). Moreover, the variation in the energy bandgap may be allowed with suitable doping concentrations and fabrication techniques.

Table 3.1 shows some commonly used semiconductor materials for manufacturing optical sources.

Table 3.1 Semiconductor materials used in optical sources

S. No.	*Substrate Material*	*Material for Optical Source*	*Useful Wavelength Range*
1.	GaAs	$GaAs/Al_xGa_{1-x}As$	0.8–0.9 μm
2.	GaAs	$GaAs/In_xGa_{1-x}P$	0.9 μm
3.	GaAs	$Al_yGa_{1-y}As/Al_xGa_{1-x}As$	0.65–0.9 μm
4.	InP	$In_{1-x}Ga_xAs_yP_{1-y}/InP$	0.92–1.7 μm
5.	GaSb	$Ga_{1-y}Al_yAs_{1-x}Sb_x/GaSb$	1.0–1.7 μm

Facts to Know

From the point of view of application in fiber–optic communications, LEDs formed with a p–n heterojunction possess higher efficiencies than p–n homojunction.

Recall that in a semiconductor material, the radiative recombination of electrons and holes in a bandgap material generates light efficiently. For a semiconductor material having bandgap energy level of E_g (eV), the emission wavelength λ (in microns) can be given as

$$\lambda(\mu m) = \frac{1.24}{E_g(eV)} \tag{3.3}$$

Fig. 3.1 illustrates the bandgap structure (electron energy versus momentum plot) of direct (only recombination) type semiconductors, whereas Fig. 3.2 illustrates the bandgap structure (electron energy versus momentum plot) of indirect (recombination plus momentum change creating photons) type semiconductors.

Table 3.2 depicts energy bandgap values for some types of direct and indirect bandgap semiconductor materials.

It is quite evident that the emission wavelength must coincide with a low-attenuation region in the loss–spectrum of fibers (for example, 1300 nm) so as to use it in optical fiber communications. The generation of light in electro–luminescent type of diodes is based on radiative recombination and minority carrier injection phenomena.

- The efficiency of radiative recombination phenomena solely depends on the band structure of the semiconductor material.
- The minority carrier injection occurs in p–n semiconductor junction under forward bias condition.

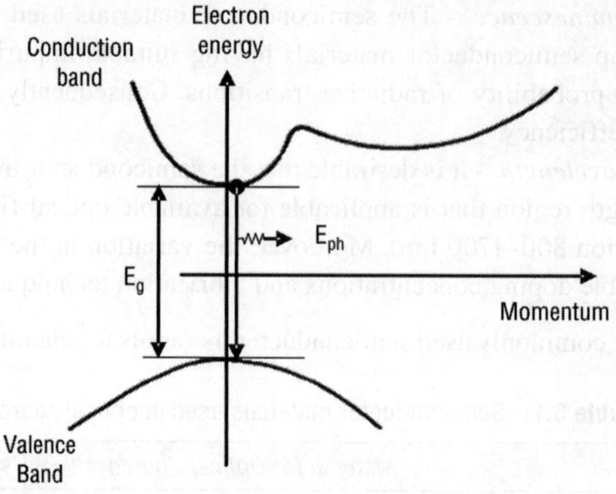

Fig. 3.1 Bandgap structure of a direct semiconductor

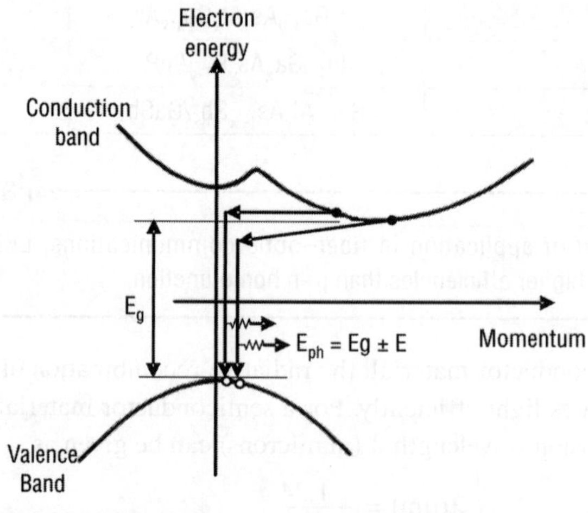

Fig. 3.2 Bandgap structure of an indirect semiconductor

Table 3.2 Energy bandgap values

S. No.	*Semiconductor Material*	*Energy Bandgap Value*	*Direct/Indirect*
1.	GaAs	1.43 eV	Direct
2.	GaSb	0.73 eV	Direct
3.	InAs	0.35 eV	Direct
4.	InSb	0.18 eV	Direct
5.	Si	1.12 eV	Indirect
6.	Ge	0.67 eV	Indirect
7.	GaP	2.26 eV	Indirect

Example 3.1 Emission Wavelength

The bandgap of a semiconductor material used for fabrication of a laser diode is specified as 1.3 eV. Find the emission wavelength.

Solution:

We know that the emission wavelength is given as

$$\lambda(\mu m) = \frac{1.24}{E_g(eV)}$$

For the given E_g = 1.3 eV, we get

$$\lambda(\mu m) = \frac{1.24}{1.3\ eV} = 954 \text{ nm}$$ **Ans.**

3.2 Light Emitting Diodes (LEDs)

Basically light emitting diodes (LEDs) are semiconductor p–n junction devices which are made to operate under forward bias condition. LEDs are mostly made with heterojunctions—single or double, in order to achieve high efficiency. The light produced by LEDs generally consists of many propagation modes. Thus, an LED can be termed as an optical source having a broad spectrum. Moreover, they can be directly intensity modulated with requirement of very less complex drive circuitry. These optical sources are widely used in optical fiber communications applications which operate at lower transmission data bit rates as low as about 50 Mbps. However, they perform moderately in terms of radiation power, efficiency and response time. Table 3.3 depicts typical values of important parameters of some commercially available LEDs.

Table 3.3 Properties of commercially available LEDs

Parameter	*Siemens IRED*	*Fujitsu FED130k4TF*	*Lasertron QLD3M504*	*UDT IR-1550*
Peak wavelength (nm)	900	1300	1300 (TE cooler)	1550
Spectral width (nm)	40	140	90	210
Typical frequency response (MHz)	100	240	200	100
Forward voltage (V)	1.3	1.5	1.8	1.5
Typical forward current (mA)	120	100	150	100
Optical output power (mW)	0.02 (fiber SI 200/280)	0.02 (50/125 fiber pigtail)	0.05 (50/125 fiber pigtail)	0.009 (GI fiber pigtail)

LED as an optical source, has certain advantages. Some of these are given below:

1. *Simpler Fabrication* – Due to no requirement of having mirror facets or striped geometry in some LED structures, it is fairly simple to fabricate them.

2. *Simple Drive Circuitry* –LEDs require lower value of drive currents which too is less dependent on operating temperature. So, drive circuitry is quite simple.
3. *Linearity* – Ideally, an LED exhibits quite linear optical output power characteristics with respect to input electric current. This has an advantage in analog communications.
4. *Less Temperature Dependence* – Since an LED is not a threshold operating device, so any increase in operating temperature may not cause an increase in threshold current requirement above the specified operating point.
5. *Reliability* – An LED is insensitive to regular degradation in its performance. Moreover, it is immune to modal noise and self-pulsation phenomena.
6. *Cost* – LED is less costly due to simpler construction.

Facts to Know

LEDs are suitable primarily for LAN applications for data transmission at the rate of 10–100 Mbps over few kilometers distance.

Two of the most often used LED structures are

- Edge Emitting LED structure
- Burrus Type Surface Emitting LED structure

As illustrated in Fig. 3.3, in an edge emitting LED structure (ELED), the light is taken out through an edge of the structure which is directly coupled to the optical fiber. The edge is a guiding layer with low refractive index and located on both sides of the active layers together with stripe contact type geometrical structure. The fiber is axially positioned so that the output optical power can be effectively coupled.

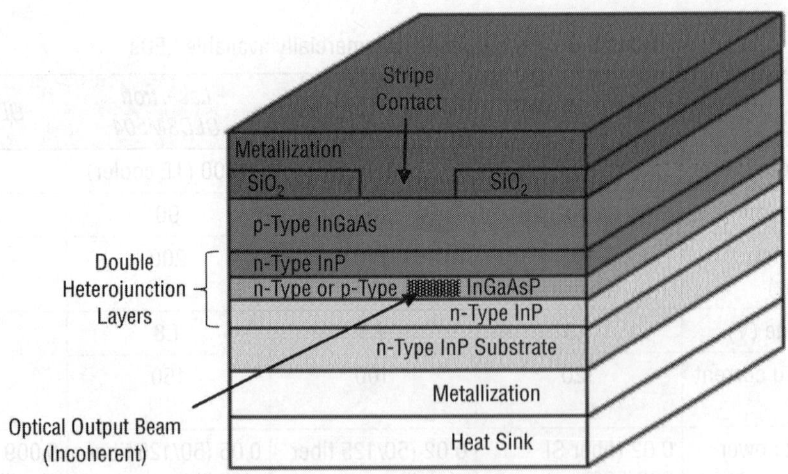

Fig. 3.3 Edge emitting LED structure

As seen, edge-emitting LED is a double heterostructure (structure that has junctions between different band gap materials). It is used to achieve carrier confinement and recombination in an active layer but additional layers of relatively low refractive index are included to produce optical guide.

A large fraction of the photons is therefore confined between two plates of material and emerge at the edge of the device as a highly incoherent beam. There is a narrow strip which is just below the semiconductor substrate. It acts as a primary active region. An appropriate part of the semiconductor substrate is cut and polished in such a way that the actual emission strip layer appears across the front end and back side. The rear edge of the semiconductor substrate is also polished in such a way that it becomes highly reflective, whereas the front side is coated with anti-reflective material. In this way, the light will only emit from the front edge and will get reflected from the rear side.

The dimensions of the active regions and strips are carefully designed so that they match with the specified diameters of fiber core (usually 50–100 μm). Typically, the length of the active regions is chosen as 100–150 μm and the width of the strips is kept as 50–70 μm. This enables it to emit light at a relatively narrower angle that is preferred, so as to achieve higher coupling efficiency as compared to that which can be obtained with surface emitting LEDs.

On the other hand, in the Burrus type surface emitting LED (SLED) structure, as shown in Fig. 3.4, a well is etched across the surface of the semiconductor substrate. An optical fiber is kept quite close to the emitting surface of SLED structure for efficient transfer of optical power.

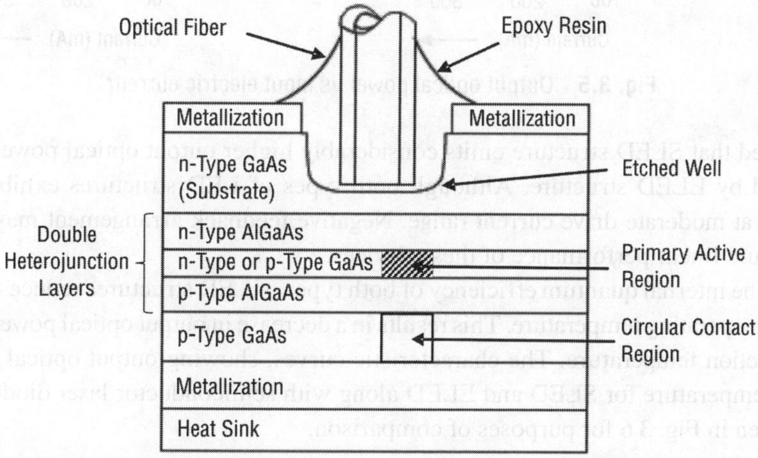

Fig. 3.4 Surface emitting LED (SLED) structure

In Burrus type surface emitting LED (SLED) structure, the emitting area is defined by oxide isolation, with the metal contact area (a circle having diameter of about 10–15 μm). The surface layer is kept as thin as possible (10–15 μm) to minimize re-absorption. A planar GaAs/AlGaAs double heterojunction LED exhibits a lifetime of 9×10^7 hours at 25°C. The edge emitting LED (ELED) shows more temperature dependence as compared to that exhibited by surface emitting LED (SLED). The surface emitting LED provides comparatively higher optical power output. However, both edge-emitting as well as surface-emitting LED structures have linear output power characteristics at medium drive current levels.

As mentioned earlier, the primary active region in SLEDs is a very small circular area (typically 20–50 μm diameter, 2.5 μm thick) placed just beneath the surface of the semiconductor substrate. A well is directly etched in the semiconductor substrate so as to enable efficient coupling of emitted light to the optical fiber located closer to it. The type of light radiated is isotropic in nature and occurs in Lambertian pattern. The light emission surface area of semiconductor substrate is normal to the central axis of the optical fiber. In order to optimize the coupling efficiency, the optical fiber is cemented to the surface of the semiconductor substrate by using epoxy resin which has matching refractive index.

An ideal characteristics curve between light output power against current is reasonably linear. But practically, LEDs have a considerable amount of non-linear characteristics, as depicted in Fig. 3.5, for ELED and SLED.

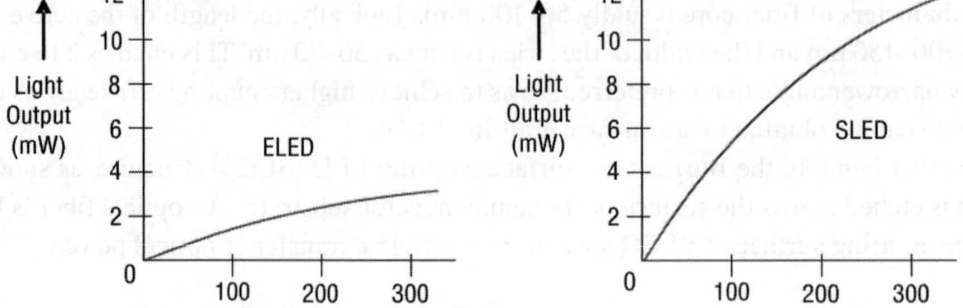

Fig. 3.5 Output optical power vs input electric current

It is observed that SLED structure emits considerably higher output optical power as compared to that emitted by ELED structure. Although both types of LED structures exhibit quite linear characteristics at moderate drive current range. Negative feedback arrangement may be employed in order to ensure linear performance of these devices.

In general, the internal quantum efficiency of both types of LED structures reduces exponentially with increase in operating temperature. This results in a decrease in output optical power with increase in the p–n junction temperature. The characteristic curves, showing output optical power against variations in temperature for SLED and ELED along with semiconductor laser diode (SLD) at λ = 1.3 μm are given in Fig. 3.6 for purposes of comparison.

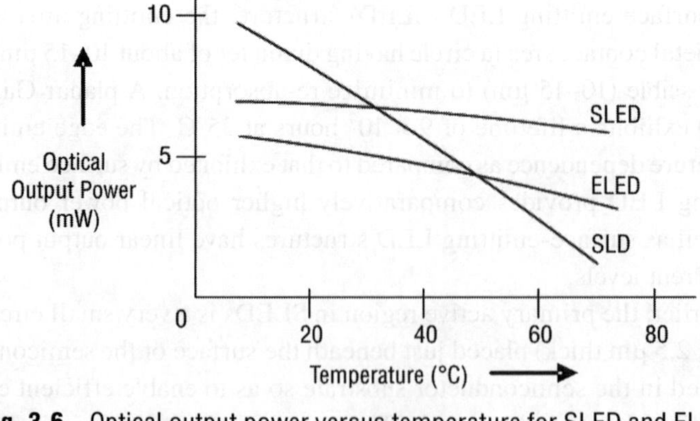

Fig. 3.6 Optical output power versus temperature for SLED and ELED

It can be clearly seen that the edge-emitting LED (ELED) has more temperature dependence than that of surface-emitting LED (SLED).

Fig. 3.7 shows the deviation in optical power output at a specified drive current for operating temperature range of 0°C to 40°C for ELED.

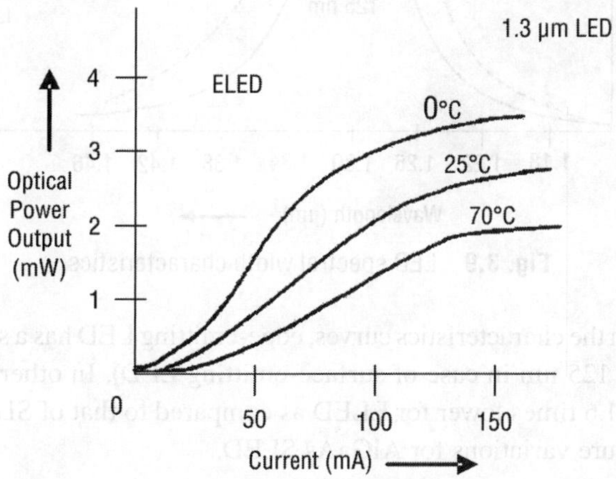

Fig. 3.7 Optical output power vs temperature for ELED

The spectral linewidth of an LED operating at room temperature is usually 25–40 nm in 800–900 nm optical band. However, at 1100–1700 nm optical band, the linewidth increases to about 50–160 nm. Fig. 3.8 shows relative intensity versus wavelength characteristics for AlGaAs SLED and InGaAsP SLED.

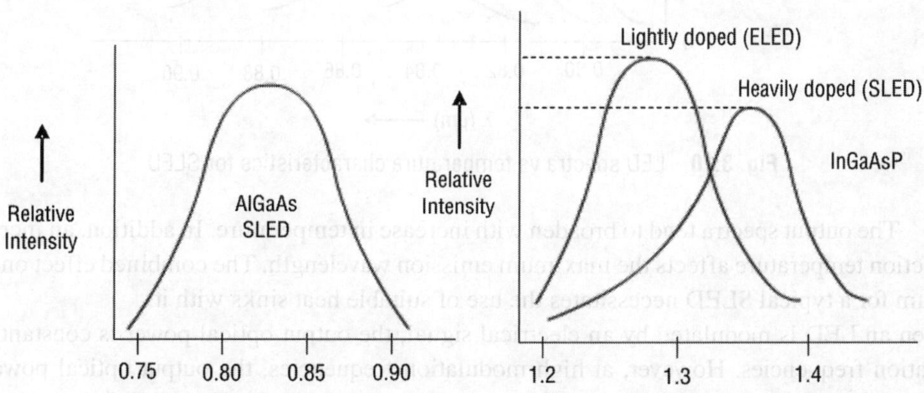

Fig. 3.8 LED output spectrum

It can be seen that there is an increase in the wavelength and a shift in the peak intensity levels due to increased doping levels by comparing characteristic curves of lightly doped and heavily doped InGaAsP SLEDs. The difference in the output spectra between SLEDs and ELEDs (both using InGaAsP semiconductor material) caused by self-absorption is shown in Fig. 3.9.

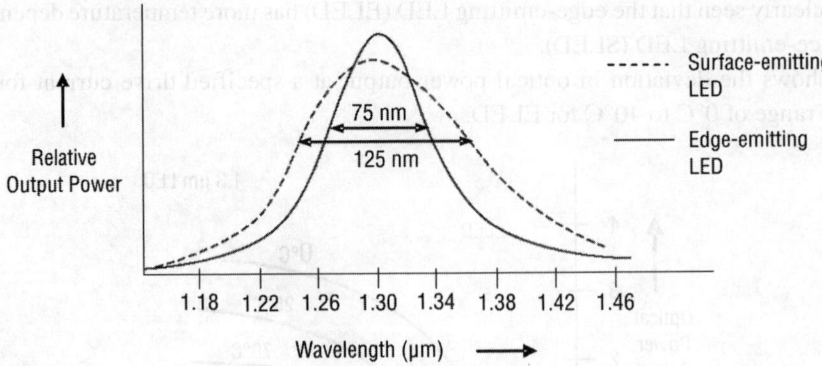

Fig. 3.9 LED spectral width characteristics

As it can be seen from the characteristics curves, edge-emitting LED has a slightly narrow linewidth (75 nm as compared to 125 nm in case of surface-emitting LED). In other words, full width half-power points are about 1.6 times lower for ELED as compared to that of SLED. Fig. 3.10 shows the spectra versus temperature variations for AlGaAs SLED.

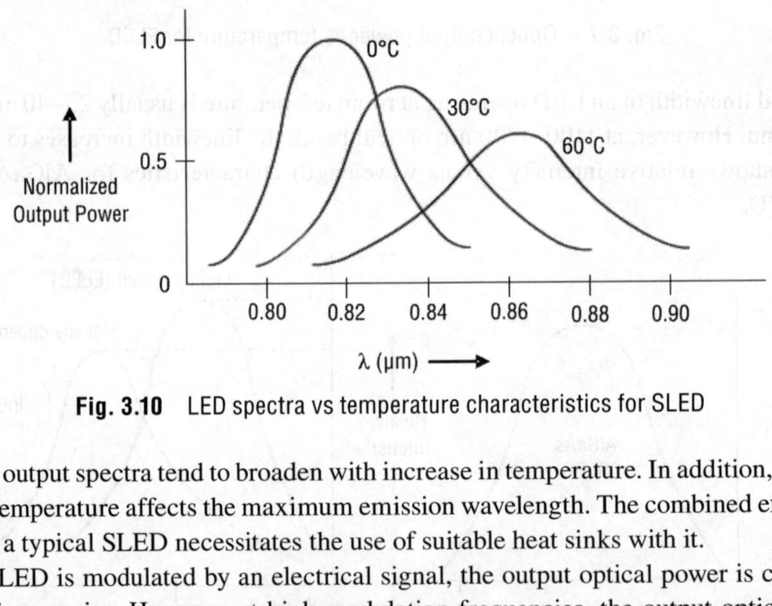

Fig. 3.10 LED spectra vs temperature characteristics for SLED

The output spectra tend to broaden with increase in temperature. In addition, an increase in the junction temperature affects the maximum emission wavelength. The combined effect on output spectrum for a typical SLED necessitates the use of suitable heat sinks with it.

When an LED is modulated by an electrical signal, the output optical power is constant at low modulation frequencies. However, at high modulation frequencies, the output optical power falls due to the delay in the recombination process of electrons and holes. The modulation response is described by the relationship

$$P(f) = \frac{P_o}{\sqrt{1+(2\pi f\tau)^2}} \tag{3.4}$$

where, $P(f)$ denotes the optical power output as a function of modulation frequency, P_o is the output power at dc current, f is the modulation frequency, and τ is the carrier lifetime.

If the modulation frequency has a 3 dB limit at which we may take the output optical power to have reduced to one reduced to one half of the output power at dc current. That is, f is termed as the 3-dB modulation bandwidth at $P(f)/P_o = 0.5$. Fig. 3.11 depicts the comparison between optical bandwidth and electrical bandwidth.

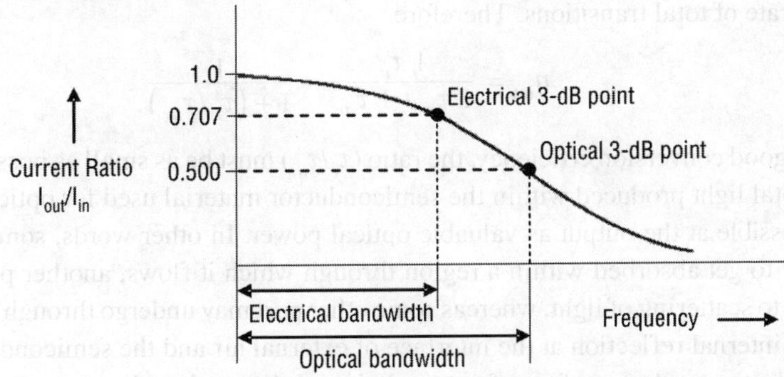

Fig. 3.11 3-dB optical vs electrical bandwidth

It is obvious that LED's modulation bandwidth is generally determined by the amount of doping concentration in the active region, the decline in radiative lifetime, and its parasitic capacitance. The modulation bandwidth is inversely proportional to the output power, as shown in Fig. 3.12.

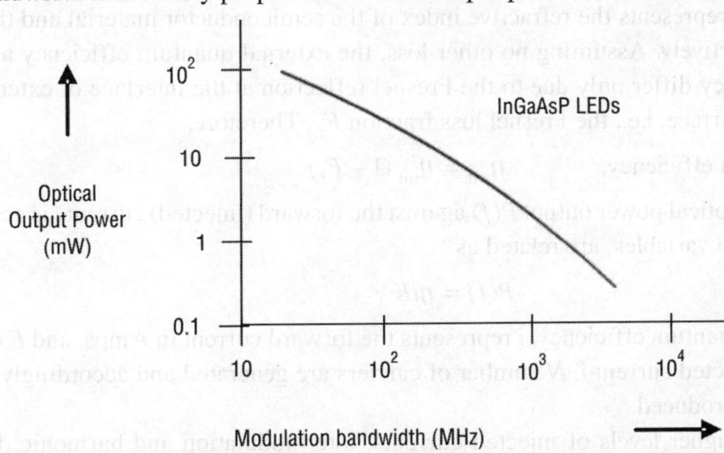

Fig. 3.12 Modulation response of an LED

Facts to Know

There is a rapid degradation in the performance of LEDs due to development of dislocation and precipitate-type deformations in their active regions. This results in various defects such as dark line and dark spot with ageing device. A planar GaAs/AlGaAs DH LED exhibit a lifetime of 9×10^7 hours at 25°C.

In case of heterojunction LED structure, the carrier lifetime is given by

$$\frac{1}{\tau} = \frac{1}{\tau_r} + \frac{1}{\tau_{nr}} + \frac{2V}{d} \tag{3.5}$$

where, τ_r represents the radiative lifetimes (i.e., recombination times), τ_{nr} represents the non-radiative lifetimes, V denotes the recombination velocity, and d represents thickness of active region.

By definition, the internal quantum efficiency, η_{int} (also known as conversion efficiency) represents the fraction of charges that recombine radiatively. In other words, it is the ratio of the rate of radiative transitions to the rate of total transitions. Therefore,

$$\eta_{\text{int}} = \frac{1/\tau_r}{1/\tau_r + 1/\tau_{nr}} = \frac{1}{1 + (\tau_r/\tau_{nr})} \tag{3.6}$$

Obviously, for good conversion efficiency, the ratio (τ_r/τ_{nr}) must be as small as possible. This also implies that the total light produced within the semiconductor material used for optical source may not be easily accessible at the output as valuable optical power. In other words, some part of light produced is likely to get absorbed within a region through which it flows, another part of it might have been lost due to scattering of light, whereas some other part may undergo through wither normal reflection or total internal reflection at the interface of external air and the semiconductor surface.

For normal reflection at the interface of external air and the semiconductor surface, the Fresnel loss fraction can be expressed as

$$F_L = \frac{(n_1 - n_2)^2}{(n_1 + n_2)^2}$$

where, n_1 and n_2 represents the refractive index of the semiconductor material and the air (which is usually 1), respectively. Assuming no other loss, the external quantum efficiency and the internal quantum efficiency differ only due to the Fresnel reflection at the interface of external air and the semiconductor surface, i.e., the Fresnel loss fraction F_L. Therefore,

external quantum efficiency, $\eta_{\text{ext}} = \eta_{\text{int}}\,(1 - F_L)$

A plot of the optical power output $P(f)$ against the forward (injected) current i is generally a linear curve and the two variables are related as

$$P(f) = \eta i E \tag{3.7}$$

where, η is the quantum efficiency, i represents the forward current in Amps, and E denotes energy in eV. For an injected current i, N number of carriers are generated and accordingly ηN number of photons will be produced.

However, at higher levels of injected currents, intermodulation and harmonic distortions may occur. Therefore, it is important to design electronic modulation circuits which can produce the linear optical power output and thereby keep the intermodulation and harmonic distortion to a minimum possible. At room temperature, the spectral linewidth of an LED is usually 25–40 nm in the wavelength region of 800–900 nm optical band. At 1100–1700 nm optical band, the linewidth increases to about 50–160 nm.

Facts to Know

Due to the linear curve between the optical power output and the forward current, LEDs are suitable for analog amplitude modulation applications such as those needed in the case of analog video transmission.

Example 3.2 Carrier Lifetime

Find the carrier lifetime of a double heterojunction LED structure in which the recombination velocity for both electrons and holes is 8 m/sec, the radiative recombination time is 12 ns, and the non-radiative recombination time is 35 ns. The thickness of the active layer is 0.4 μm.

Solution:

We know that in case of heterojunction LED structure, the carrier lifetime τ is given by

$$\frac{1}{\tau} = \frac{1}{\tau_r} + \frac{1}{\tau_{nr}} + \frac{2V}{d}$$

where, τ_r represents the radiative lifetimes (i.e., recombination times), τ_{nr} represents the non-radiative lifetimes, V denotes the recombination velocity, and d represents thickness of active region.

For the given values of $\tau_r = 12$ ns and $\tau_{nr} = 35$ ns, we get

$$\frac{1}{\tau} = \frac{1}{12 \times 10^{-9}} + \frac{1}{35 \times 10^{-9}} + \frac{2 \times 8}{0.4 \times 10^{-6}} \approx 152 \times 10^{6} s^{-1}$$

Therefore, $\tau = 6.6$ ns **Ans.**

Example 3.3 Internal Quantum Efficiency

If the radiative and non-radiative recombination times of a double heterodyne LED are specified as 12 ns and 35 ns, respectively, then find the internal quantum efficiency.

Solution:

We know that the internal quantum efficiency

$$\eta_{\text{int}} = \frac{1}{1 + (\tau_r/\tau_{nr})}$$

Using given values of $\tau_r = 12$ ns and $\tau_{nr} = 35$ ns, we get

$$\eta_{\text{int}} = \frac{1}{1 + (12/35)} = 0.75, \text{ or } 75\%$$ **Ans.**

Example 3.4 External Quantum Efficiency

Find the external quantum efficiency assuming normal incidence at the semiconductor having refractive index of 3.5 and air interface for specified internal quantum efficiency of a double heterojunction LED of 0.75.

Solution:

We know that the external quantum efficiency

$$\eta_{\text{ext}} = \eta_{\text{int}}(1 - F_L)$$

where, the Fresnel loss fraction, $F_L = \dfrac{(n_1 - n_2)^2}{(n_1 + n_2)^2}$

Using $n_1 = 3.5$ and $n_2 = 1$, we get

$$F_L = \frac{(3.5 - 1)^2}{(3.5 + 1)^2} = 0.3$$

Using given values of $\eta_{int} = 0.75$, we get

$$\eta_{ext} = 0.75 \times (1 - 0.3) = 0.52, \text{ or } 52\%$$ **Ans.**

Example 3.5 3-dB Bandwidth versus carrier lifetime

Find the 3-dB modulation bandwidth of a double heterojunction LED structure for a specified carrier lifetime of 6.6 ns.

Solution:

The modulation response is determined by the variation of the optical power output versus electrical input as a function of frequency, and is given by the relationship

$$P_{opt}(f) = \frac{P_o}{\sqrt{1 + (2\pi f \tau)^2}}$$

$$\Rightarrow \quad \frac{P_{opt}(f)}{P_o} = \frac{1}{\sqrt{1 + (2\pi f \tau)^2}}$$

where, $P_{opt}(f)$ denotes the optical power output as a function of modulation frequency, P_o is the output power at dc current, f is the modulation frequency, and τ is the carrier lifetime.

If the modulation frequency has a 3 dB limit at which we may take the output optical power to have reduced to one half of the output power at dc current. That is, f is termed as the 3-dB modulation bandwidth at $P_{opt}(f) / P_o = 0.5$.

Therefore, $$0.5 = \frac{1}{\sqrt{1 + (2\pi f_{3dB} \tau)^2}}$$

On squaring, we have $$0.25 = \frac{1}{1 + (2\pi f_{3dB} \tau)^2}$$

$$\Rightarrow \quad 1 + (2\pi f_{3dB} \tau)^2 = \frac{1}{0.25} = 4$$

$$\Rightarrow \quad (2\pi f_{3dB} \tau)^2 = 4 - 1 = 3$$

$$\Rightarrow \quad 2\pi f_{3dB} \tau = \sqrt{3}$$

$$\Rightarrow \quad f_{3dB} = \frac{\sqrt{3}}{2\pi\tau}$$

By using given value of $\tau = 6.6$ ns, we get

$$f_{3dB} = \frac{\sqrt{3}}{2\pi \times 6.6 \times 10^{-9}} = 42 \text{ MHz}$$ **Ans.**

Section Practice Problems

1. An LED has quantum efficiency of 3%. Find the power radiated by it at the operating wavelength of 670 nm. Assume typical current as 50 mA. *[Ans.: 0.93 mW]*

2. An LED emits light having a peak wavelength of 890 nm and has a radiative recombination time of 100 ns. If the bulk recombination life time is 130 ns and drive current is 14 mA, determine the non-radiative recombination time. *[Ans.: 433 ns]*
3. A planar LED structure has been manufactured using GaAs semiconductor material having a refractive index = 3.6. Assuming the transmission factor at the intersection of the semiconductor crystal and air to be 0.68, calculate the optical power emitted by the planar LED structure into air and specify it as a percentage of the internal optical power. *[Ans.: 1.3%]*
4. A double hetero-structure type of surface emitter LED (SLED) device has a light emitting area of 50-µm diameter. It is coupled with a step-index profile optical fiber having 80 µm diameter and given numerical aperture = 0.15. The radiance parameter of SLED is specified as 30 W-Sr^{-1}cm^{-2} at 40 mA input drive current. Compute the output optical power coupled by SLED into this fiber if the Fresnel reflection coefficient is 0.01 at the index-matched fiber surface. *[Ans.: 41.1 µm]*

3.3 Laser Diodes

Recall the acronym for LASER as Light Amplification by Stimulated Emission of Radiation. As compared to incoherent and wide spectral emission characteristics of LED optical source, a laser diode (LD) exhibits much better performance in terms of high transmission data rate for long-haul optical fiber communication networks. However, it has the disadvantage of having complex drive circuits, temperature-dependent optical power output, lower reliability and being expensive.

Some of the distinct advantages of laser diodes are given below:

1. Higher optical power output (of the order of mW), mainly due to amplifying effect of stimulated emissions.
2. Narrow bandwidth (less than or equal to 1 nm) which helps to minimize the impact of material dispersion (for example, group velocity dispersion).
3. Modulation capabilities up to GHz range.
4. Coherent output which allows heterojunction coherent detection in high capacity system.
5. Efficient coupling of optical power output into the low numerical aperture fibers by focusing the light into a tiny spot by using an external lens.
6. Size of injection laser diode is compatible with optical fiber.
7. Better error performance.

Table 3.4 depicts typical values of important parameters of some commercially available laser diodes.

Now the question arises how to realize the operation of the laser diode practically? In fact, a laser diode can be considered as an optical oscillator (or, an optical resonator) because an electromagnetic wave is formed within a cavity that provides highly coherent monochromatic light radiations at its output. This means stimulated emission process (sometimes known as lasing) with properties such as identical energy (as well as frequency), same polarization and in phase to that of incident photon. In order to achieve only stimulated emission rather than spontaneous emission or absorption, it is essential to increase the radiation density as well as the population density of the upper energy states as compared to that of the lower energy states. This condition is known as population inversion. For this to happen, it becomes quite essential that atoms having upper energy states be excited. Thus, a non-equilibrium distribution is obtained. The population inversion is made possible by employing an external energy source. This process is generally referred to as pumping. The lasing operation is depicted in Fig. 3.13.

Table 3.4 Properties of commercially available LDs

Parameter	*LDT-60008*	*Fujitsu FLD130F1CJ*	*Siemens SFH4423*	*Lasertron QLM3S860*	*Lasertron QLM5S990*
Peak wavelength (nm)	1300	1300	1305±25	1300	1550 (DFB laser)
Spectral width (nm)	5	1	3.5	4	0.1
Typical frequency response (MHz)	500	500	500	400	5000
Forward voltage (V)	2.0	1.2	1.3	1.5	1.8
Typical forward current (mA)	20	20	25	20	10
Optical output power (mW)	0.03 (MM fiber)	0.03 (SM fiber)	0.004 (SM fiber)	0.02 (SM fiber)	0.05 (SM fiber)

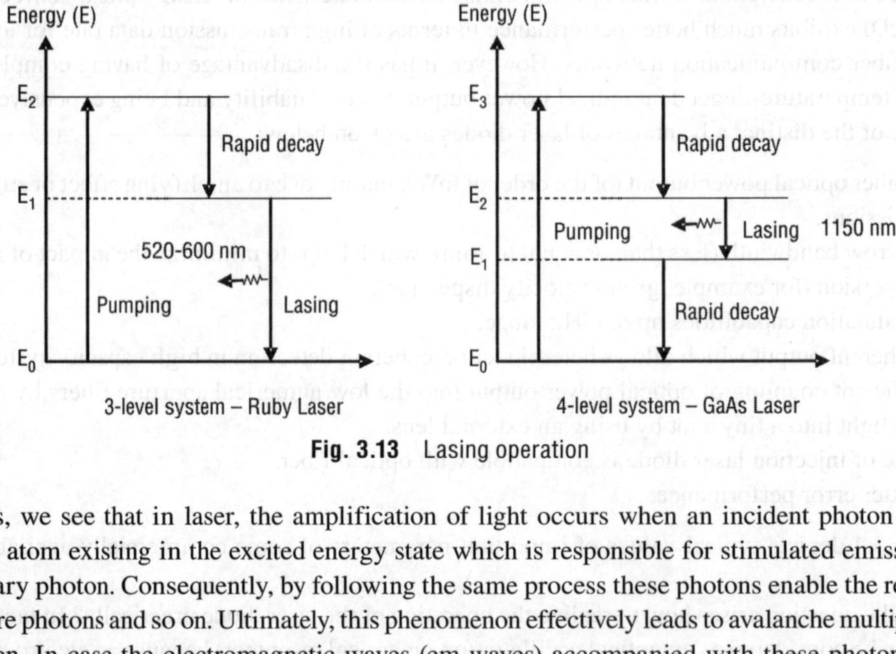

Fig. 3.13 Lasing operation

Thus, we see that in laser, the amplification of light occurs when an incident photon collides with an atom existing in the excited energy state which is responsible for stimulated emission of a secondary photon. Consequently, by following the same process these photons enable the release of two more photons and so on. Ultimately, this phenomenon effectively leads to avalanche multiplication condition. In case the electromagnetic waves (em waves) accompanied with these photons are in phase with one another, then the necessary condition of amplified coherent light emission is met.

There are some other aspects which must be considered in order to achieve the laser action. The first aspect is that all the photons must be confined within the laser medium. The second aspect is that all the photons must remain in phase with each other so as to obtain coherent emission of light. To accomplish these aspects, two mirrors are normally placed for reflection of photons on both sides of the amplifying medium. The optical cavity thus formed effectively provides positive feedback. Hence, the optical signal once generated is fed back several times, which results in amplification each time it crosses the medium. This type of laser diode structure functions similar to that of a Fabry–Perot resonator, as shown in Fig. 3.14.

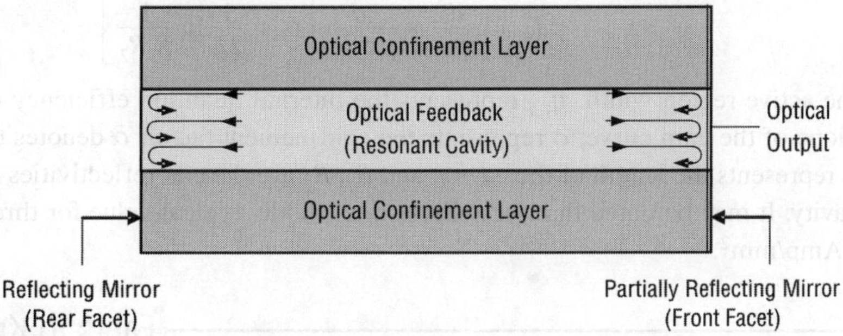

Fig. 3.14 Fabry–Perot resonator

As can be seen, there is one reflecting mirror at one end (rear facet) of the cavity while the other end has a partially reflecting mirror (front facet) that enables partial emission of light. The remaining emission gets reflected through this cavity for possible amplification of light having specified wavelengths only. This process is also known as optical feedback. The basic construction of Fabry–Perot resonator type of laser is quite identical to that of the edge-emitting LEDs (ELEDs). A stable optical power output is attained at saturation level at which the net optical gain is almost balanced by losses (due to absorption and scattering) in the amplifying medium. There may be certain losses due to scattering, absorption, or diffraction at mirrors.

If R_1 and R_2 represent reflectivities of the feedback, then the threshold condition for laser action is expressed as

$$R_1 R_2 e^{[2L(g-\alpha)]} \geq 1 \tag{3.8}$$

where, L is the length of the laser cavity (or, the physical spacing between two mirrors located at the ends of the cavity), g is called the gain coefficient, and α represents the absorption or loss coefficient.

The threshold gain, g_{th} is defined as the minimum gain for which laser emission is possible and below this gain, laser emission cannot occur at all. Using the above expression, the threshold gain is obtained as

$$g_{th} = \alpha + \frac{1}{2L}\ln\left(\frac{1}{R_1 R_2}\right) \tag{3.9}$$

At threshold gain, the amplification due to stimulated emission is just enough to compensate for transmission losses and absorption. We know that in case of semiconductors, the estimation of the electron-hole pairs and the possession of energy states is in accordance with the Fermi–Dirac distribution. In addition to this, the formation of additional electron–hole pairs (EHPs) due to population inversion is accomplished by injecting external current in a p–n junction. In fact, threshold gain is directly proportional to the injected current density which is given as

$$J_{th} = \frac{1}{\beta}\left[\alpha + \frac{1}{2L}\ln\frac{1}{R_1 R_2}\right] \tag{3.10}$$

where, β is a proportionality factor, known as slope of the gain curve, which can be determined experimentally.

For a laser diode, threshold current density can be determined accurately by using the following expression:

$$J_{th} = \frac{d}{\eta_{\text{int}}} + \frac{d}{\beta\sigma\eta_{\text{int}}}\left[\alpha + \frac{1}{2L}\ln\frac{1}{R_1R_2}\right] \tag{3.11}$$

where, d is the active region width, η_{int} represents the internal quantum efficiency of the laser, β denotes the slope of the gain curve, σ represents the confinement factor, α denotes the absorption coefficient, L represents the length of the cavity, and R_1, R_2 are the end reflectivities of the mirrors used in the cavity. It may be noted that for a GaAs laser diode, typical value for threshold current density is 25 Amp/mm^2.

Facts to Know

At room temperature, threshold current density is relatively high for a conventional p–n semiconductor diode as compared to that of a heterostructure type semiconductor device. This is why laser diodes are usually made of double heterojunction structures so as to have lower value of threshold current density.

Example 3.6 Threshold Gain

Find the threshold gain if the length of the cavity is 0.4 mm and the values of reflectivities on either ends of the cavity is 0.5. Assume loss coefficient = 3 mm^{-1}.

Solution:

We know that threshold gain is given by

$$g_{th} = \alpha + \frac{1}{2L}\ln\left(\frac{1}{R_1R_2}\right)$$

where, α represents the loss coefficient, L represents the length of the cavity, and R_1, R_2 denote the end reflectivities.

Using the given values of various parameters as $\alpha = 3$ mm^{-1}, $L = 0.4$ mm, $R_1 = R_2 = 0.5$, we get

$$g_{th} = 3 + \frac{1}{2 \times 0.4}\ln\left(\frac{1}{0.5 \times 0.5}\right) = 4.73\text{ mm}^{-1}$$ **Ans.**

Example 3.7 Threshold Current

If the threshold current density for a particular laser device having an active area of 0.2 × 0.5 mm^2 is specified as 3×10^6 Amp/m^2, then determine the threshold current.

Solution:

We know that threshold current, $I_{th} = J_{th} \times$ laser area

Using given values of $J^{th} = 3 \times 10^6$ Amp/m^2 and laser area = 0.2 × 0.5 mm^2, we get

$$I_{th} = \left(3 \times 10^6\text{ Amp/m}^2\right) \times \left(0.2 \times 10^{-3}\text{m} \times 0.5 \times 10^{-3}\text{m}\right)$$

$$I_{th} = 0.3\text{ Amp, or }300\text{ mA}$$ **Ans.**

In the laser cavity, oscillations usually occur over a very small range of frequencies (i.e., narrow spectral band). Moreover, the cavity gain is just adequate to compensate for transmission losses. It may be recalled that the laser structure forms a resonant cavity which acts as an amplifying medium for population inversion to exist and emissions to begin. Ultimately the light emission levels increases because the standing electromagnetic waves between the end mirrors occurs only at those frequencies for which L is an integral number of $\lambda/2$. Therefore, the resonance condition along the axis of the cavity is given by

$$L = \frac{\lambda}{2} . \frac{q}{n} \tag{3.12}$$

where, L is the optical spacing between mirrors, λ is the emission wavelength, q is an integer, and n represents the refractive index of amplifying medium.

Using $\lambda = c/f$, we have

$$L = \frac{c}{2f} \frac{q}{n} \tag{3.13}$$

Hence, the discrete emission frequency is given as

$$f = \frac{qc}{2nL} \tag{3.14}$$

Thus, we can say that desired frequency of oscillations that can occur within the laser cavity can be computed using various integer values of q. In fact, each frequency of oscillation corresponding to an integer value constitutes a propagation mode. The separation between different modes are described by a frequency interval δf, given by

$$\delta f = \frac{c}{2nL} \tag{3.15}$$

Assuming $\delta f << f$, we can write the spacing between modes as

$$\delta\lambda = \frac{\lambda \delta f}{f} \tag{3.16}$$

Using $\lambda = c/f$

$\Rightarrow$ $f = c/\lambda$

We can write

$$\delta\lambda = \frac{\lambda \delta f}{c/\lambda} = \frac{\lambda^2}{c} \delta f \tag{3.17}$$

Using $\delta f = \frac{c}{2nL}$, we get

$$\delta\lambda = \frac{\lambda^2}{c} . \frac{c}{2nL} = \frac{\lambda^2}{2nL} \tag{3.18}$$

It may be noted here that only the longitudinal modes are included in the laser emission within the specified spectral width of the gain curve, as shown in Fig. 3.15.

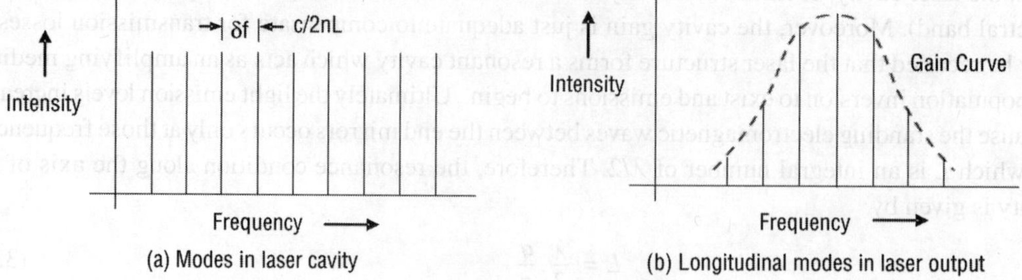

Fig. 3.15 Modes in laser cavity

It is possible that laser oscillations can take place in a direction that is exactly transverse to the cavity axis which may produce resonant modes. The laser modes consist of a sequence of wavelength peaks that correspond to different longitudinal modes. The spectral spacing between these modes (typically of the order of a few tenths of a nm) depends on the length of the cavity. This is why laser is known as a multimode optical source device.

But in single-mode operation, it is essential that the optical power output from a laser device must have one and only one longitudinal as well as transverse mode of propagation. Such a narrow spectral width can be achieved by employing smaller aperture of the resonant cavity which can be obtained by reducing the length of the cavity (approximately 50 μm). If the frequency interval is more than the transition linewidth of the laser, then only the single mode of propagation of light can oscillate within the transition linewidth of the laser. Therefore, rigid control of cavity parameter is essential, such as in double heterojunction structure with vertical width of wavelength region (< 0.4 μm). By definition, the lasing threshold is the minimum value of required current that is needed for stimulated emissions to occur. If the current happens to be below the specified lasing threshold value, then spontaneous emission will occur. This implies that laser diodes will function as ELEDs at currents below lasing threshold value. Fig. 3.16 shows the lasing characteristics, that is relative optical output power versus input drive current.

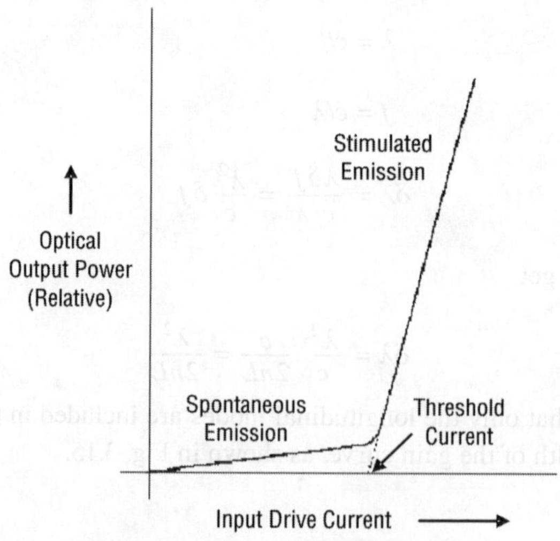

Fig. 3.16 Lasing characteristics

Laser diodes are much more sensitive to variations in operating temperature as compared to that of LEDs. They require higher current to operate which means quite complex drive circuits with possibility of more heat dissipation due to larger current. The selectivity of required longitudinal mode can be improvised by using suitable frequency-selective feedback arrangement. This will also ensure that various longitudinal modes undergo different cavity losses. These types of optical sources are commonly termed as dynamic single-mode lasers, or more precisely the single frequency lasers. These are also known as short cavity lasers, coupled cavity resonators, or distributed feedback lasers.

- Short cavity lasers can be obtained with shortened cavity length (i.e., by reducing cavity length from 250 μm to 25 μm). This, in turn, will tend to increase the mode spacing from 1 nm to about 10 nm. In order to obtain the desired single-mode operation, the peak of the gain curve may be adjusted accordingly. Alternatively, configurations employing resonators, either micro-cleaved or etched having short cavity of length 10–20 μm only are utilized.
- Coupled cavity resonators, also known as multiple element resonators or resonators within distributed reflectors, provide single frequency oscillators by coinciding longitudinal modes of each Fabry–Perot cavity, resulting in mode selectivity.
- Distributed feedback lasers (DFB lasers) can be obtained with distributed Bragg diffraction grating type structure. Basically, it contains periodic variations in the index of refraction along the direction of light wave propagation in heterostructure waveguide of the laser.

Fig. 3.17 shows the comparison between Fabry–Perot laser and distributed Bragg diffraction grating laser.

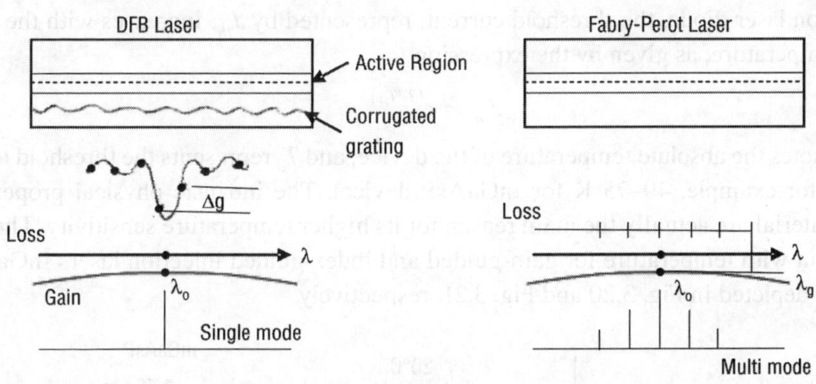

Fig. 3.17 Distributed Bragg diffraction grating vs Fabry–Perot laser

There are two distinct types of semiconductor lasers employing distributed feedback mechanisms. One of them is called distributed feedback (DFB) laser, whereas the other one is known as distributed Bragg reflector (DBR), as shown in Fig. 3.18 and Fig. 3.19, respectively.

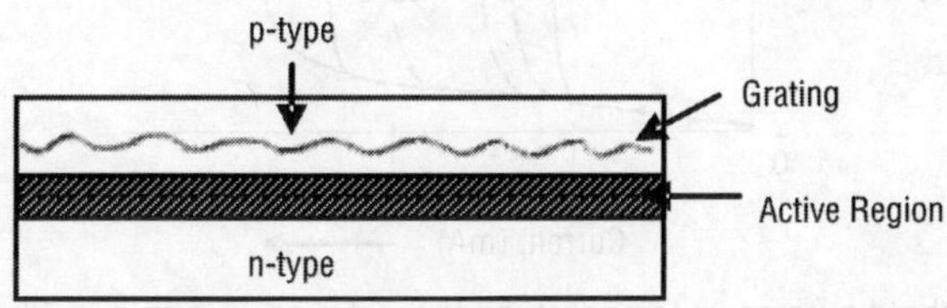

Fig. 3.18 Distributed feedback (DFB) semiconductor laser

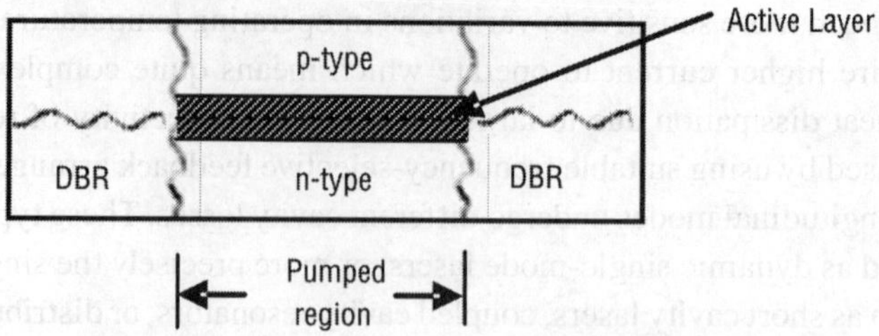

Fig. 3.19 Distributed Bragg reflector (DBR) laser

Table 3.5 provides a comparative study between DFB and DBR lasers in brief.

Table 3.5 DFB versus DBR lasers

S. No.	Distributed Feedback (DFB) Laser	Distributed Bragg Reflector (DBR) Laser
1.	In DFB laser, the optical grating is normally functional over the whole active region that is pumped.	In DBR laser, the optical grating is engraved near the cavity ends only.
2.	Distributed feedback occurs in the whole active region.	Distributed feedback occurs in the whole active region except in the central region.
3.	DFB lasers are well developed.	DBR lasers are less developed.

In injection laser diode, the threshold current, represented by J_{th}, increases with the increase in operating temperature, as given by the expression

$$J_{th} \propto e^{(T/T_o)} \tag{3.19}$$

where, T denotes the absolute temperature of the device, and T_0 represents the threshold temperature coefficient (for example, 40–75 K for InGaAsP device). The intrinsic physical properties of the InGaAsP material are actually the main reason for its higher temperature sensitivity. The variations in light output with temperature for gain-guided and index-guided injection lasers InGaAsP at λ = 1550 nm are depicted in Fig, 3.20 and Fig. 3.21, respectively.

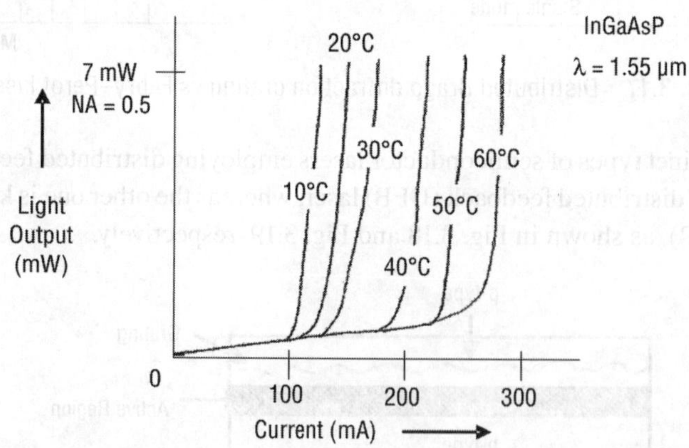

Fig. 3.20 Threshold current vs temperature characteristic for gain-guided injection laser

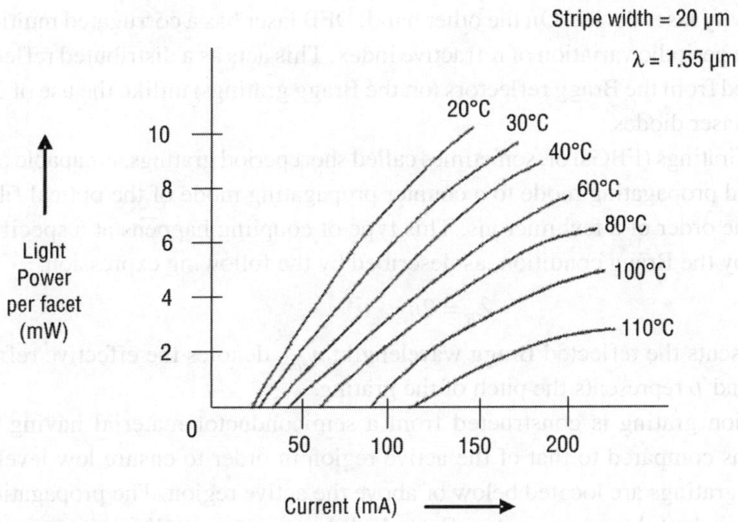

Fig. 3.21 Threshold current vs temperature characteristic for index-guided injection laser

It is observed that the optical power output of the laser diode strongly depends on the junction temperature. Obviously, the use of thermoelectric coolers may be necessary. It is recommended to use appropriate heat sink in order to ensure reliable operation of the laser devices over the desired range of the current.

The dynamic behavior of the injection laser diode is quite crucial in high transmission data rate applications. On applying a current step to injection laser diode, switch-on delay (approximately 0.5 ns) may be observed. It is often followed by high frequency of the order of 10 GHz damped oscillations, known as relaxation oscillations. Relaxation oscillator damping occurs due to gain non-linearities, lateral carrier diffusion, feeding of spontaneous emission into lasing mode, etc. It depends on the device structure and the operating conditions also. When an input current pulse is applied at the input of the laser having considerable amount of parasitic capacitance due to initial delay time, the pulse may be broadened as compared to the actual one. It may so happen that the optical output of the laser diode may contain several pulses because the electron density is repeatedly built up and received quickly. As a result of this phenomenon, there may be occurrence of relaxation oscillations. At transmission data rates exceeding 100 Mbps, this action may lead to a severe degradation in the shape of the transmitted optical pulse. It is imperative to say that switch-on delay and damping relaxation oscillations must be reduced.

- The switch-on delay can be reduced by biasing the laser diode near threshold (that is, pre-biasing).
- The damping relaxation oscillations can be reduced by having stripe widths less than 3 μm in Burrus type surface emitting LED (SLED) structure. This also helps in achieving fast response.

It is obvious that the gain difference in a laser diode is defined as the actual difference of the cavity gain and the cavity loss. It does not differ significantly among different longitudinal modes in Fabry–Perot laser diodes. It implies that several longitudinal modes can be supported. This will certainly degrade the spectral purity of the output optical signal. However, in a distributed feedback (DFB) laser diode; it is possible to increase the gain difference for a particular mode of particular interest. This can be achieved by presenting a uniform frequency-dependent loss over all modes in a conventional cavity laser device. On the other hand, DFB laser has a corrugated multi-layer structure that introduces a periodic variation of refractive index. This acts as a distributed reflector. The lasing action is achieved from the Bragg reflectors (or, the Bragg gratings) unlike the use of cleaved mirrors in Fabry–Perot laser diodes.

Fiber Bragg Gratings (FBGs) or, sometimes called short period gratings, is capable of coupling light from the forward propagating mode to a counter propagating mode of the optical fiber. The period of FBGs is of the order of a few microns. This type of coupling happens at a specified wavelength which is given by the Bragg condition, as described by the following expression:

$$\lambda_B = 2n_{core}\upsilon \tag{3.20}$$

where, λ_B represents the reflected Bragg wavelength; n_{core} denotes the effective refractive index of the fiber core, and υ represents the pitch of the grating.

The diffraction grating is constructed from a semiconductor material having greater energy bandgap value as compared to that of the active region in order to ensure low level of absorption. The corrugated gratings are located below or above the active region. The propagation mode which is closest to the pitch of the grating gets reflected while suppressing all other modes. The end facets of the cavity resonator are coated with anti-reflection material. In a quarter wave shifted DFB laser diode, a phase shift of $\pi/2$ radians that correspond to a path difference of exactly $\lambda/4$ is introduced at the center of the laser cavity. This will result into minimum threshold value and the maximum gain difference for the main mode corresponding to the Bragg wavelength. These structures exhibit higher mode stability, lower mode partition noise, and narrower linewidths (less than a few MHz). Longer cavity lengths or the use of extended cavities help reduce the line width. Longer the cavity, smaller is the linewidth. It is possible to make these devices tunable over a range of wavelengths by introducing frequency tunable elements such as diffraction gratings and electro optic/acousto-optic filters.

The refractive index, grating period, and the cavity length determine the extent of reflectivity (either high or low) of the grating over a narrow or broad spectrum. The other novel and important type of fiber grating is the long period grating (LPG). As the name suggests, it has a period, υ, which may be of the order of hundreds of microns and index modulation depth of the order of 10^{-4} or more. The LPG couples fundamental mode propagating in the fiber core into the cladding at specified wavelength as given by

$$\lambda_i = \left[n_{01} - n_{i(clad)}\right]\upsilon \tag{3.21}$$

where, n_{01} and $n_{i(clad)}$ represent the effective index of the core mode and the ith axially symmetrical cladding mode, respectively. It may be noted that the light propagating in the cladding rapidly decays because of transmission losses at the intersection of the cladding and external air. As a result, there may be a sequence of resonances, known as loss bands in the guided mode. Fig. 3.22 depicts coupling of guided mode in a LPG laser.

It is observed that there may be several resonances over a broad spectrum in an individual grating. The central wavelength of the LPG resonances depends analytically on the relative refractive index difference between the fiber core and the cladding of the fiber. It is obvious that any deviation in the refractive index due to strain or temperature variations can result into large shifts in the resonance wavelengths. The sensitivity to a specific measurand depends on many factors that include the order of the cladding mode, the composition of the fiber to which the guided optical mode is coupled, etc. Therefore, the sensitivity is different for various attenuation bands.

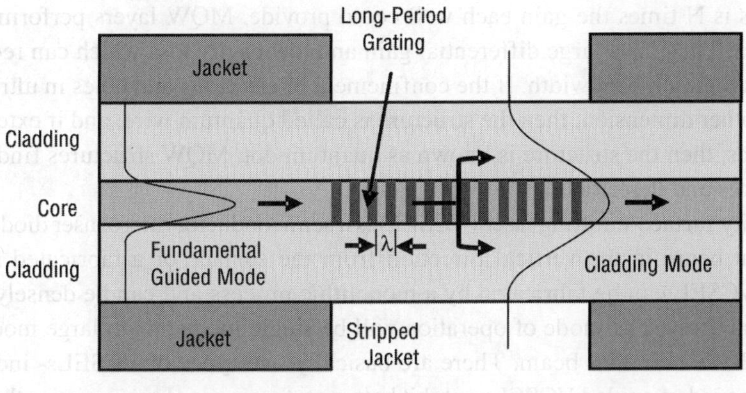

Fig. 3.22 Fundamental mode propagation in a long period grating laser

Multiple electrode DFB lasers are used to improve the tenability characteristics. Its operation depends on applying a larger amount of current to one electrode while a smaller value of current to the other electrode. The region near the output port is pumped at current densities near or just above threshold density. As stated earlier, this region serves as a Bragg reflector. Larger change in refractive index is possible at a lower current which helps in wavelength tuning, whereas the other region pumped at higher current provides the required gain. Tuning ranges of the order of 2–3 nm can be obtained with a linewidth of 15 MHz.

The active region in double heterojunction structure LDs has a typical thickness of the order of 150 nm which is sufficient for carrier confinement. When thickness of the active region is decreased by employing sophisticated fabrication techniques, the electronic and optical properties exhibit drastic changes. When active layer thickness is reduced further (much below 50 nm), may be up to 10 nm, then such type of laser structures are commonly known as quantum well structures. These can be of three types: quantum wells, quantum wires, and quantum dots, based on the actual number of dimensions of the quantum well structure in which thickness of the active region is reduced. The lasing action takes place by stimulating electron–hole recombination among discrete quantum well states. There are distinct advantages of using quantum structures in laser diodes as compared to bulk active layer devices such as

1. Higher differential quantum efficiency
2. Higher device efficiency
3. Narrower linewidth
4. Smaller value of threshold current
5. Improved noise performance
6. Lower frequency chirping, higher differential gain
7. Tunability

In a typical quantum well structure, a thin layer (<50 nm) of GaAs is surrounded by AlGaAs. The bandgap of GaAs is smaller than that of the surrounding layers. Rectangular potential wells are performed in conduction and valence bands in which electrons and holes are confined. When multiple layered structures (of the order of 100) of different semiconductor materials (arranged alternatively) are fabricated, a Multiple Quantum Well (MQW) structures is formed. The gain of the MQW laser having N layers is N times the gain each well could provide. MQW layers perform better at high current densities. They have large differential gain and low cavity loss which can reduce linewidth and increase modulation bandwidth. If the confinement of electrons and holes in ultra-thin layers is extended to another dimension, then the structure is called quantum wire, and if extended in all the three dimensions, then the structure is known as quantum dot. MQW structures find application in both laser sources and detectors.

Vertical-cavity surface-emitting laser (VCSEL) is a semiconductor micro laser diode that produces cylindrical light beam in the vertical direction from the surface of a fabricated semiconductor crystal wafer. VCSEL can be fabricated by a monolithic process and can be densely packed into a two-dimensional array. The mode of operation will be single mode due to large mode spacing and the output can have a circular beam. There are basically two types of VCSELs- index-guided and gain-guided. The index-guided VCSELs exhibit better performance characteristics than gain-guided VCSELs owing to the strong overlapping of the electromagnetic field with the gain region in the device and current independent modal control. A novel three-contact VCSEL with an interactivity absorber has been used in high-speed data communications.

There are certain advantages associated with VCSEL structure, such as:

1. It is possible to integrate the VCSEL structure in a specific two-dimensional array arrangement.
2. High-density arrays can be configured due to requirement of low threshold currents.
3. Easy alignment and packaging is possible without additional optics.
4. Low-cost due to testing at the wafer level.
5. High transmission data rate can be achieved.
6. Low power consumption.

Let us now turn our attention to another important phenomenon, known as frequency chirping. It refers to a dynamic shift of the weak wavelength emitted from the semiconductor laser device. It occurs due to direct injection current modulation of a single longitudinal mode. It causes dynamic linewidth broadening. One of the main reasons for frequency chirping to occur is the gain-induced variations in the refractive index of the laser device. It has been observed that minute variations in the carrier density may result in relaxation oscillations at the output of the laser device. This causes variations in the resonance frequency within the structure of Fabry–Perot resonator as well as DFB laser. In a single-mode fiber, the frequency chirping of the laser alongwith chromatic dispersion of the fiber can cause a significant shift from the specified zero-dispersion wavelength (for example, 0.05 nm in InGaAsP laser, corresponding to 6.4 GHz frequency shift), and subsequently may degrade the system performance.

There are certain techniques that can help to reduce frequency chirp. Some of them are mentioned below:

- Biasing the laser sufficiently above threshold
- Damping of relaxation oscillation by shaping the electrical current drive pulses
- Employing certain device structures like quantum-well lasers, Bragg wavelength detuned DFB lasers, multi-electrode DFB laser diodes
- Allowing laser diode to emit constantly
- Using the external modulator
- Impressing the user data on to the optical carrier

When we consider analog transmission, it becomes imperative to study the noise behavior of the injection laser diode operation, known as laser noise. Various types of noise sources in the laser device are as follows:

(a) *Frequency or phase noise*: It occurs due to discrete as well as random spontaneous or stimulated transitions. As a result, there may be fluctuations in the optical emission intensity. Upto 1 MHz, phase noise varies inversely as frequency. Beyond 1 MHz, the phase noise is independent of variation in frequency and it is termed as flat quantum noise, as shown in Fig. 3.23.

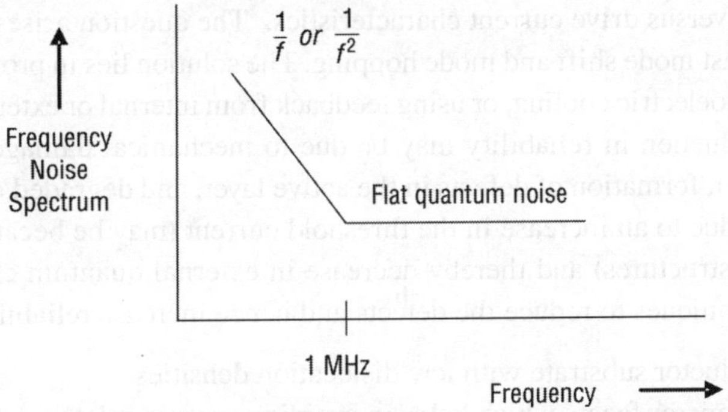

Fig. 3.23 Spectral characteristics showing laser phase noise

Quantum noise is a principal cause of linewidth broadening within semiconductor laser diodes. Single-mode laser diodes exhibit better noise immunity by more than 30 dB.

(b) *Relative intensity noise* (RIN): This is also known as instabilities in light output. These may be caused by temperature variations, or from spontaneous emission contained in laser output.
(c) *Self pulsation*: These are random intensity fluctuations which create a noise source.
(d) *Optical feedback*: It is reflection of light back into the laser device which occurs from undesired external reflections. It affects the frequency stability as well as the intensity of the emitted light of semiconductor lasers. However, this effect is significantly reduced with multimode lasers due to distribution of reflections among various propagation modes. As a result, the light is weakly coupled back into the laser mode. In single-mode fiber-optic systems operating at wavelength of 1550 nm, optical feedback may cause pulse broadening and reflection-induced frequency hops.
(e) *Mode partition noise*: It occurs when various propagation modes are not well stabilized in multimode laser diodes. Due to changes in temperature, the relative intensity of different longitudinal modes may vary significantly from one optical pulse to the next one, irrespective of maintaining constant output optical power from the laser device. These are also termed as *spectral fluctuations*. In conjunction with fiber dispersion, mode partition noise results in random distortion of received optical which may cause an increase in bit error rate (BER).

Can mode partition noise occur in single-mode lasers? The answer is yes. It happens due to occurrence of residual side modes in the output spectrum of the laser. The extent of effect of mode partition noise depends on the emitting wavelength. It varies for laser diodes emitting at $\lambda = 1300$ nm from those emitting at 1550 nm. For achieving better performance, the side modes should be suppressed.

> ***Note:*** All types of laser noise, including phase noise, relative intensity noise, self-pulsation, optical feedback, and mode partition noise can be minimized by employing optical isolators and mode stabilized devices.

There is another phenomenon, known as mode hopping, that occurs in a single-mode injection laser diode. It occurs mainly due to increase in temperature of the device junction. It is seen that the mode hopping is a discontinuous function of the drive current for the laser and normally occurs all of a sudden over 1–2 mA range only. Mode hopping results in variations (in the form of kinks) in the output optical power versus drive current characteristics. The question arises as to how one obtain stable operation against mode shift and mode hopping. The solution lies in providing either adequate heat sinking, or thermoelectric cooling, or using feedback from internal or external grating structures.

In general, the reduction in reliability may be due to mechanical damage of the cavity mirror facets used in the laser, formation of defects in the active layer, and degraded current limiting areas. All these effects are due to an increase in the threshold current (may be because of leakage current in Burrus type laser structures) and thereby decrease in external quantum efficiency with ageing. There are certain techniques to reduce the defects and hence increase reliability. These are:

1. Use of semiconductor substrate with low dislocation densities
2. Passivating the mirror facets which helps in avoiding surface-related defects
3. Mounting the soft solders in order to avoid possibility of external strains
4. Improving the crystal growth technique
5. Improvements in material selection and device fabrication techniques

Facts to Know

Mean lifetime for continuous wave injection laser diodes is reported to be 10^6 hours or more than 100 years for a wide range of GaAs/AlGaAs laser structures; 10^5 hours for InGaAsP/InP Burrus type lasers emitting at 1300 nm; 2000 hours for DFB lasers at 1550 nm subject to 60°C temperature.

Section Practice Problems

1. Calculate the number of longitudinal modes in a Ruby laser of a crystal length = 4 cm having a refractive index = 1.8. Assume that the peak emission wavelength is 550 nm. Also find the frequency spacing of longitudinal modes. *[Ans.: 2.62×10^5; 2.1 GHz]*
2. A GaAs laser is operating at 850 nm wavelength and has a length of 500 μm. If its refractive index is 3.7, find (a) the frequency spacing between longitudinal modes; (b) the wavelength spacing. *[Ans.: (a) 81 GHz; (b) 0.195 nm]*
3. The length and width of an optical cavity used for fabrication of a GaAs injection laser device are 250 μm and 100 μm, respectively. If the gain factor = 0.021 A/cm^3 and the loss coefficient = 10/cm, then determine the threshold density at normal operating temperature. Use the reflectivity for normal incidence of plane wave as 0.32. *[Ans.: 2.62 × 103 A/cm^2]*
4. Calculate the optical gain at threshold of a ruby laser diode having the following values: $R_1 = R_2 = 0.32$, α = 10 cm, L = 500 μm. *[Ans.: 33.7/cm]*

5. If the average operating wavelength of a laser is 0.5 μm with 1000 K temperature. The refractive index of laser is 1.38 with 1.3 × 104 longitudinal modes. Determine the optical spacing between mirrors.

[Ans.: 24 nm]

3.4 Optical Transmitter Block Diagram

In optical fiber communication system, it is the optical transmitter unit which is responsible for converting an electrical information signal into an optical one. So an optical transmitter essentially comprises of an electrical-to-optical converter and an encoder when needed. Typical features of an optical transmitter unit include

- narrow spectral linewidth
- single-mode cladding mode stripped fiber pigtail
- GaInAsP distributed feedback single-mode (1550 nm) laser chip
- InGaAs monitor photodiode
- internal thermo–electric cooling arrangement using precision NTC thermistor
- hermetically-sealed
- operates up to 2.5 Gbps over long distance single-mode fiber
- typical rated output power = 2 mW
- 14-pin butterfly package
- In-built optical isolator

The optical transmitter unit basically comprises the following:

1. A light source, i.e., LED or laser diode
2. Coupling optics
3. A signaling circuit, i.e., data conversion unit
4. A power control unit

All these transmitter components including a thermoelectric cooler (TEC), a heat sink and coupling optics are hermetically sealed inside the package to protect it from the environment. The major components of an optical transmitter are depicted in a functional block diagram shown in Fig. 3.24.

The main component in optical transmitter unit is an optical source—either an LED or an LD—both of which operate with almost similar requirements. For example, typical current levels needed are in the range of 20–300 mA, with a junction voltage drop of 1.2–2.5 V. Continuous emission optical sources provide a few mW of power while some pulsed lasers can deliver peak powers of a few watts. Therefore, the drive circuit required should be capable of delivering the needed current. There may also be a need for an additional linearising circuit with feedback control because the drive current versus output optical power behavior is non-linear. Transmitters using LED are generally limited to about 50 MHz, whereas laser transmitters can be in excess of 1 GHz due to the response time limitation of the respective devices. Thus, the circuit design and the selection of components will also be a function of the required response time or bandwidth. Fast direct modulation of LEDs with digital signal can be carried out by using a low impedance emitter follower circuit, a compensating matching network and a bipolar transistor switch while in saturation.

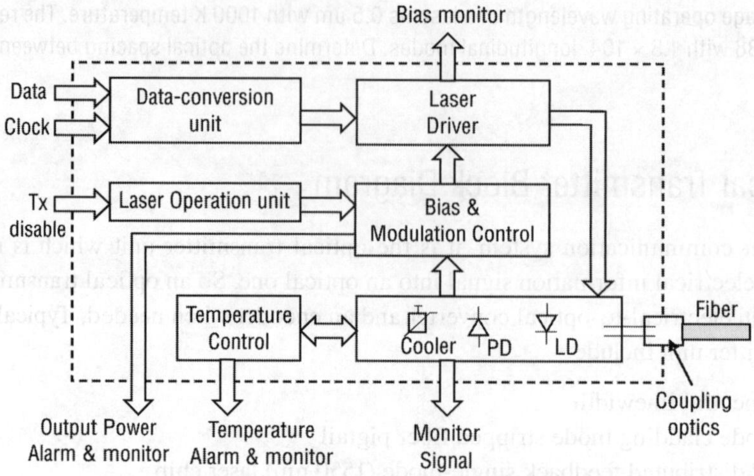

Fig. 3.24 Block diagram of optical transmitter unit

As far as coupling from optical source to single-mode or multimode fiber (i.e., source-to-fiber power launching) is concerned, it is desirable that the coupling efficiency must remain unchanged with time and there should be adequate mechanical stability. There are two methods of coupling light from LED source:

- Direct coupling, also known as butt coupling. When surface-emitting LED is coupled with a multimode fiber, then the coupling efficiency is typically 1%. However, when edge-emitting LED is coupled with a multimode fiber, then the coupling efficiency is about 10%. But when an edge-emitting LED is coupled with a single-mode fiber, the coupling efficiency may be less than 1%.
- Lens coupling. When using the lens coupling technique, coupling efficiency will be much higher (upto 30% for edge-emitting LED with multimode fiber coupling).

When a laser source is used with a multimode fiber, mode interference usually results in the formation of speckle patterns—bright and dark spots—within the fiber core because of the coherence of laser radiation, resulting in modal noise. It is not a problem with single-mode fiber. There are three distinct approaches for coupling light from a laser source into a single-mode fiber:

- Direct coupling or butt coupling. The efficiency of butt coupling with a laser source depends on the ratio of the laser-beam diameter to the fiber-core diameter. In single-mode optical fiber cable, the coupling efficiency is approximately 10–20%, whereas it is 50–70% in case of multimode fiber.
- Lens coupling. A focal lens can focus a laser beam at the fiber end-face, in order to match a laser beam to a fiber core for transmission. It increases the coupling efficiency.
- Collimating lens coupling. Collimating lens has a refractive index with a graded-index profile. Coupling efficiency up to 80% can be achieved.

The concept of lens coupling and collimated lens coupling (i.e., source-to-fiber power launching) is shown in Fig. 3.25 and Fig. 3.26, respectively.

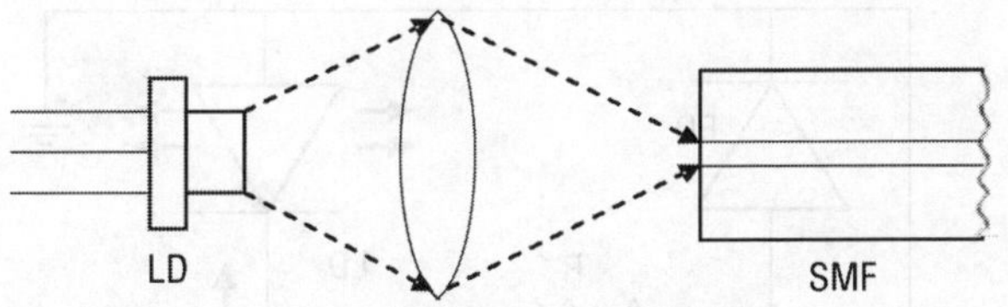

Fig. 3.25 Lens coupling

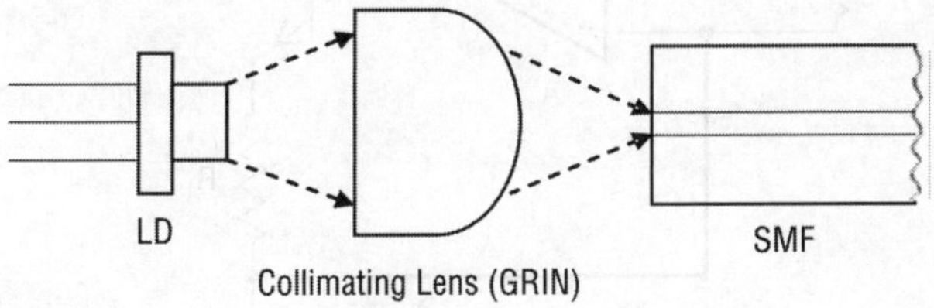

Fig. 3.26 Collimated lens coupling

An optical isolator is usually placed between a laser source and a fiber in order to transmit light from laser to fiber only.

The data conversion unit performs three major functions: encoding, parallel-to-serial conversion, and reshaping the electric format of the data, as shown in Fig. 3.27.

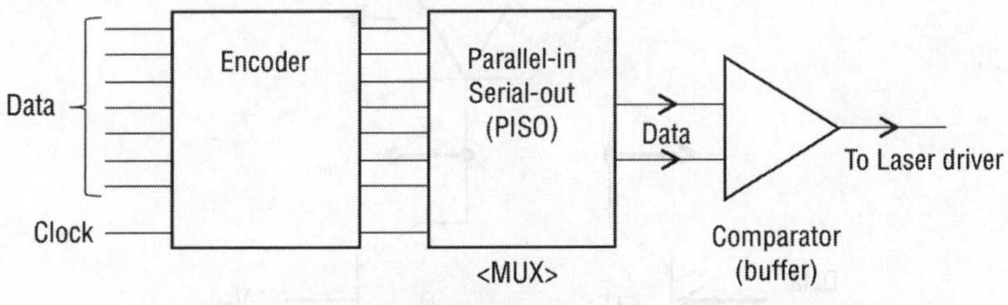

Fig. 3.27 Block diagram of data conversion unit

Laser driver circuit is required because a laser diode is a current-driven device and a laser driver converts outside voltage into the current needed to drive the laser. It also provides a stable bias current to a laser diode to speed the modulation process. The feedback signal from the temperature sensor that reached the laser driver through the bias-control circuit closes the current loop. A typical laser driver circuit is shown in Fig. 3.28.

As shown, control voltage, V_{bias}, is the input voltage for operational amplifier. V_{cc} is grounded and V_{EE} is tied to -5.2V. Current flowing through R depends only on the input voltage and does not depend on the load resistance which is laser diode here. Thus, by changing V_{bias}, one can control the bias current I_{bias}. When temperature rises, the feedback signal obtained from a photodetector (PD) helps to stabilize the average output power by changing the bias current. Modulation is controlled by simply changing the driving current from the bias level to maximum. A simple circuit which produces intensity modulation is shown in Fig. 3.29.

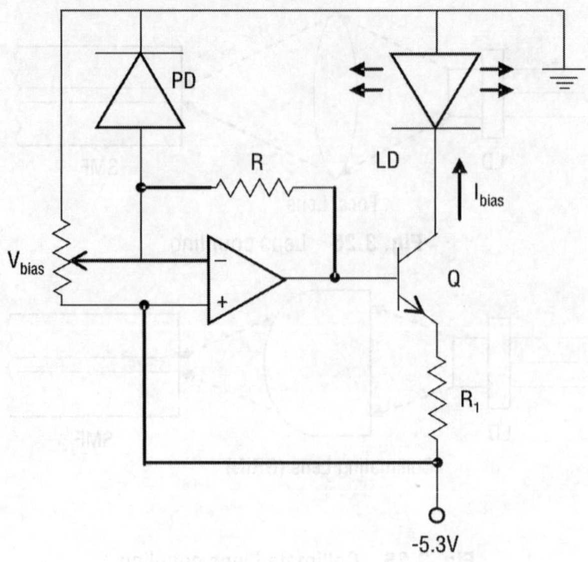

Fig. 3.28 Laser driver circuit

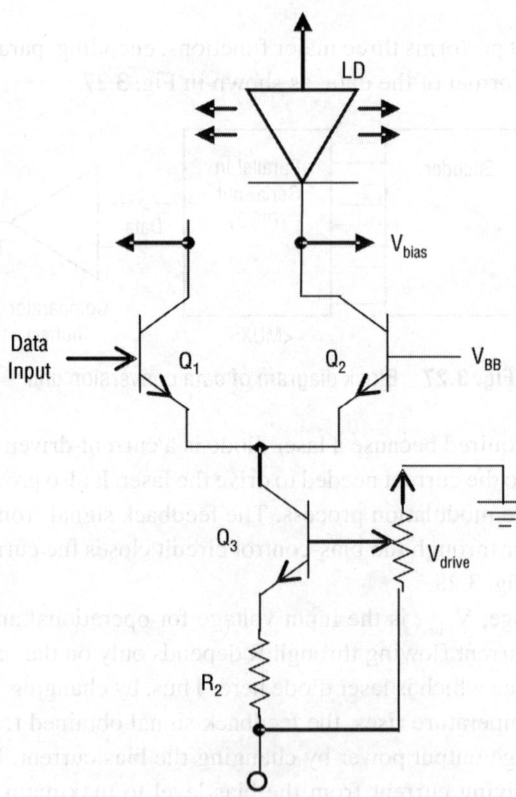

Fig. 3.29 Intensity modulation circuit

It is seen that when data input is greater than V_{BB}, Q1 conducts and hence laser diode (LD) is OFF. When data input is lesser than V_{BB}, Q2 conducts and as a result LD becomes ON. So this circuit produces intensity modulation. Thus, the driving circuit dynamically adjusts the bias level without changing the modulation current. As mentioned earlier, a thermoelectric cooler is quite often employed which helps to provide stability to the operating temperature of the laser device. A better method would be to design a driving circuit with dual-loop feedback configuration. In this way, both the modulation as well as the bias currents are automatically adjusted.

For high speed optical transmitters operating at transmission data rates of 10 Gbps or more, the laser device is normally biased at a constant driving current. This provides the continuous wave output. This is followed by an optical modulator to convert it into an equivalent data-coded optical pulse signal. The performance of an external optical modulator is measured by the parameters such as modulation bandwidth and extinction ratio (also called on-off ratio). In order to improve the performance of high-speed optical transmitters, we can use monolithic integration of the laser device with the current driver circuit. It implies that both electrical as well as optical devices are required to be fabricated on the same IC chip. This is why they are sometimes referred to as opto-electronic integrated circuit (OEIC). It is desirable that an optical transmitter provide reliable operation over a longer period, i.e., may be of the order of 10 years or more.

In an optical transmitter, the optical signal is applied at the input of an optical fiber in a digital form using binary line coding formats such as non-return-to-zero (NRZ), return-to-zero (RZ) and optical Manchester code. The minimum bandwidth and the average optical transmitted power mainly depend on the type of line coding format employed. In NRZ line codes, a transmitted data bit occupies the full bit period. RZ codes use pulse width with half bit period and each data bit can be encoded as two optical line code bits. Both full width and half width data pulses are present in phase encoded format. In Manchester line code, the data transition occurs at the center of each bit period. Moreover, generation and decoding of Manchester code is relatively simple and easy.

Facts to Know

The reliability of an optical transmitter is measured in terms of either failure-in-time (FIT) which is number of failures per billion hours (typically 300 at 25°C and 1500 at 50°C) as well as mean-time-to-failure (MTTF), which means an average time until a component misbehaves (typically 274 years at 25°C and 78 years at 50°C).

Points to Remember

- An LED has good linear response and is best suited for analog modulation.
- A laser diode is a threshold device and the response curve described between the output optical power and the forward current is small.
- Population inversion is the mechanism that is responsible for laser operation.
- Below the specified threshold level, there is spontaneous emission of light. It increases progressively until the specified threshold level. Thereafter the lasing action starts and the output optical power is produced. Hence, a laser device is normally used with digital modulation technique.
- The laser diode is quite sensitive to variations in the bias current and operating temperature. However, it is recommended to maintain the bias state above the specified threshold condition and perform an analog modulation for a very limited range.

- A reduction in broadening of emitted optical pulse and an increased power coupling into the fiber are readily achieved with the laser diode due to its narrow output light spectrum in the radiated cone form.
- Semiconductor laser diodes employ a double-hetero (DH) structure and a Fabry–Perot resonant cavity having active layer thickness of the order of a few hundred nanometers for optical oscillation and carrier confinement.
- The prominent laser devices include Distributed Feed Back (DFB), Distributed Bragg Reflection (DBR), and Multiple Quantum Well (MQW) lasers.
- In terms of circuit requirements, the laser diode differs from the LED mainly due to the fact that it is a threshold device and it has a limited linear range.
- Quite often the LD is pre-biased near but below its threshold to achieve minimum switch-on delay and reduction in relaxation oscillation.

Important Equations

The emission wavelength, $\lambda(\mu m) = \dfrac{1.24}{E_g(eV)}$, *where E_g (eV) represents the energy bandgap of the semiconductor material used in fabrication of optical source.*

The optical power output as a function of modulating frequency, $P(f) = \dfrac{P_o}{\sqrt{1+(2\pi f\tau)^2}}$, *where P_o is the output power at dc current , f is the modulation frequency, and τ is the carrier lifetime.*

In case of heterojunction LED structure, the carrier lifetime $\dfrac{1}{\tau} = \dfrac{1}{\tau_r} + \dfrac{1}{\tau_{nr}} + \dfrac{2V}{d}$, *where τ_r and τ_{nr} represents radiative and non-radiative lifetimes (i.e., recombination times), respectively,* V *represents the recombination velocity and* d *represents the active region thickness.*

The internal quantum efficiency, $\eta_{int} = \dfrac{1}{1+(\tau_r/\tau_{nr})}$, *where τ_r and τ_{nr} represent radiative and non-radiative lifetimes (i.e., recombination times), respectively.*

The optical power output, $P(f) = \eta iE$, *where η is the quantum efficiency, i represents the forward current in Amps, and E is expressed in eV.*

The threshold gain, $g_{th} = \alpha + \dfrac{1}{2L}\ln\left(\dfrac{1}{R_1R_2}\right)$, *where α represents the loss coefficient, L denotes the length of the cavity, and R_1, R_2 are the reflectivities on the ends.*

Threshold current density, $J_{th} = \dfrac{1}{\beta}\left[\alpha + \dfrac{1}{2L}\ln\dfrac{1}{R_1R_2}\right]$, *where β denotes a proportionality factor that can be measured experimentally and is known as slope of the gain curve.*

In the laser cavity, the optical spacing between mirrors, $L = \dfrac{\lambda}{2}.\dfrac{q}{n}$, *where λ is the emission wavelength, q is a positive integer, and n represents the index of refraction of the amplifying medium.*

The reflected Bragg wavelength, $\lambda_B = 2n_{core}\upsilon$ *where, n_{core} represents the effective core refractive index and υ represents the grating pitch.*

Key Terms with Definitions

Bragg Grating	Based on interferometric methods, a technique for constructing various functions of optical filters embedded within an optical fiber.
Diffraction grating	Transmitting or reflecting equally spaced lines in the form of an array that mutually enrich the effects of diffraction.
Distributed feedback (DFB) laser	An injection laser diode (ILD) having a Bragg reflection grating in its active layer with an objective of enhancing a single longitudinal mode while suppressing multiple longitudinal modes.
Edge-emitting Diode	A type of optical source such as an LED that emits focused light from its edge.
ELED	An edge-emitting LED double heterostructure with stripe contact geometry in which the light is taken out from a low refractive-index guiding layer edge present on either side of the active layer.
Extrinsic semiconductor	An intrinsic semiconductor is doped with dopant impurities which can donate or accept electrons (donors or acceptors) to the conduction band and thereby making electrons holes as the majority charge carriers.
Intrinsic semiconductor	It has no charge carriers at absolute zero temperature, but it develops equal number of two types of charge carriers – negative charge carriers known as electrons and positive charge carriers known as holes.
Heterojunction	A p–n junction in which each of the two sides are made of different type of semiconductor materials.
Homojunction	A p–n junction which is formed by doping same type of semiconductor material with p- and n-type impurities.
Internal quantum efficiency	The fraction of charges that recombine radiatively, and it is given as the ratio of the rate of radiative transitions and the rate of total transitions.
Laser	An acronym for Light Amplification by Stimulated Emission of Radiation. It is a coherent optical source that produces narrow spectrum light.
Laser diode (LD)	Same as laser. It can be considered a low-power optical source having almost identical construction to that of an LED.
Lasing	The stimulated emission is also called lasing.
LED	Semiconductor p–n heterojunction (single or double) device operated with forward bias condition.
p–n junction	It is a transition region between p- and n-type doped materials of the same single-crystal of a semiconductor.
SLED	Surface emitting LED structure in which a well is carved in the surface of the semiconductor substrate and a fiber is fixed such that the emitting surface is near to the fiber for an efficient power transfer.
Threshold gain	The minimum gain for which laser emissions can start.

Short Answer Type Questions

1. **What are the most commonly used optical sources?**
   Light-emitting diodes (LEDs) and laser diodes (LDs) are the most popular optical sources used in optical fiber communications. The difference between LEDs and laser diodes is that LEDs produce incoherent light, while laser diodes produce coherent light.

2. Why are LED and LD types of semiconductor diode sources the natural choice for fiber optic systems?
Optical sources for optical fiber communications should satisfy some desirable properties such as response time, emission wavelength, light intensity, radiation pattern, and output spectral characteristics. LED and LD types of semiconductor diode sources are the natural choice for fiber optic systems because they possess these desirable properties. In addition, these optical sources exhibit low power consumption, high efficiency, highly reliable operation, and compact size.

3. Define LED in its simplest form.
A light-emitting diode (LED) is basically a forward-biased p–n junction semiconductor device. It emits incoherent light with a relatively wider spectral width (of the order of 30–60 nm) using the principle of electroluminescence (i.e., spontaneous emission).

4. Compare the basic characteristics of LED and Laser Diode as optical sources.
A light emitting diode (LED) is primarily an incoherent optical source that is capable of emitting light radiations over a relatively wider angle and contains a broad spectrum of wavelengths. On the other hand, a laser diode (LD) is highly coherent, emits radiation in a narrow range of angles, has a narrow spectrum and fast response time. However, in terms of operating current requirement, cost and reliability, LED is preferred over LD.

5. Distinguish between homojunction and heterojunction.
A homojunction is similar to a normal p–n junction which is formed by adding relatively small concentrations of impurities into the same semiconductor. It is also possible to produce quite different semiconductor materials which, however, crystallize with nearly same lattice constant but differ in the bandgap, electron-affinity and such other properties. These can also be doped with impurities and the resulting junction is known as heterojunction. To achieve high efficiency, both LEDs and LDs are made with heterojunctions.

6. What are the advantages of employing double heterojunction devices?
A heterojunction may be either a single heterojunction (SH), or a double heterojunction (DH) type, depending on whether the junction property differs on one side or on both sides of the junction. Some of the advantages of deploying double heterojunction type devices include high injection devices, minority carrier confinement, photon confinement (also called optical guidance) and improved ohmic contact.

7. Compare and contrast LED and Laser Diode as optical sources.

1. LED emits light due to spontaneous emission of radiation, whereas laser diode emits light due to stimulated emission of radiation.
2. LED requires lower current densities for its operation as compared to higher current densities requirement for a laser diode.
3. Emitted photons in LEDs possess phases that are random in nature, whereas the emitted photons are in-phase with incident photons in laser diodes.
4. LED is incoherent optical source and laser diode is a coherent one.
5. Energy of emitted photons in LED gives a much wider spectral linewidth (about 100 times) than that of in laser diode.
6. LED can support several propagation modes within its structures, and hence is mainly used as a multimode optical source. On the other hand, laser diode supports single mode only.

8. What are the most common materials used for communication LEDs?
Gallium arsenide (GaAs) and Indium gallium arsenide phosphide (InGaAsP) are the two most commonly used semiconductor compound materials in communication LEDs. Because InGaAsP LEDs operate at longer wavelength than GaAs LEDs (1.3 μm vs 0.85 μm), their output spectrum is wider in wavelength terms by a factor of about 1.7.

9. Comment on the utility of LED as an optical source.

When an LED is utilized as an optical source, its light transmission is quite inefficient. It has been observed that about 1% of the input power is eventually converted into launched optical power which is coupled into the optical fiber. However, due to their relatively simple design, LEDs are very useful for low-cost optical fiber communication applications.

10. How is optical bandwidth related to the electrical bandwidth?

Optical bandwidth is considerably larger than the electrical bandwidth. Assuming Gaussian system response, we have

$$\text{Optical bandwidth} = \sqrt{2} \times \text{electrical bandwidth}$$

Optical bandwidth is also known as modulation bandwidth.

11. What are the factors on which modulation bandwidth of LEDs depend? How is modulation bandwidth related to output optical power?

The modulation bandwidth of LEDs depends on the following factors:

(a) The doping level in the active region of its structure
(b) Reduction in radiative lifetime
(c) Parasitic capacitance of the device

Modulation bandwidth is inversely proportional to output optical power.

12. Mention typical application of LEDs as optical source.

LEDs emit light having relatively large spectrum width. Therefore, the transmitted optical pulse will experience more fiber dispersion. This adversely affects the bit rate–distance product parameter which is a useful measure of its utility in various applications. Hence, LEDs find application mostly in local-area-networks with data rate capability limited to about 10–100 Mbps over transmission distances up to a few kilometers only. Recently, LEDs have also been developed that use several quantum wells to emit light at different wavelengths over a broad spectrum, and are currently in use for local-area WDM networks.

13. Why are LEDs not popular in modern optical fiber communications?

Nowadays, LEDs are being superseded by Vertical Cavity Surface Emitting Laser (VCSEL) devices as optical sources. VCSELs offer improved speed, power and spectral properties, at a similar cost. Common VCSEL devices couple well with multi-mode fiber.

14. What do you understand by a semiconductor laser?

A semiconductor laser is an optical source widely used in optical fiber communications that emits light based on stimulated emission instead of spontaneous emission, which results in high output power (~100 mW) as well as other benefits related to the nature of coherent light. The output of a laser is relatively directional, allowing high coupling efficiency (~50 %) into single-mode fiber.

15. List the most commonly used lasers.

Fabry–Perot, Distributed Feed Back (DFB) and Vertical Cavity Surface Emitting Laser (VCSEL) are the most commonly used semiconductor lasers as optical sources in optical fiber communications applications.

16. Which aspects of semiconductor lasers make these useful in optical fiber communications?

The narrow spectral width exhibited by semiconductor lasers helps to reduce the effects of chromatic dispersion that permits for higher transmission data rate. Moreover, it is possible to modulate them directly at high frequencies due to relatively small recombination time.

17. What are the desirable properties for semiconductor optical transmitters?

There are certain design aspects which must be considered for semiconductor optical transmitters. Their operation should be efficient as well as reliable over desired wavelength range. They should be capable to be modulated directly at high frequencies. Of course, their size should be quite compact.

18. How are laser diodes modulated?

Laser diodes are quite often modulated directly. This simply means that the optical output power is controlled by a current applied directly to the laser device. For very high data rates or very long distance links, a laser source may be operated continuous wave, and the light modulated by an external device such as an electro-absorption modulator or Mach–Zehnder interferometer. External modulation increases the achievable link distance by eliminating laser chirp, which broadens the linewidth of directly modulated lasers, increasing the chromatic dispersion in the fiber.

19. What are the general considerations for choosing an optical source?

The general considerations for an optical source for a particular application include physical dimensions to be suitable for the fiber used, narrow radiation pattern (beam width), linearity (output light power proportional to driving current), and ability to be directly modulated by varying driving current.

20. What are the specific aspects of an optical source to be considered for optical fiber communications?

There are several aspects of an optical source to be considered for optical fiber communications. Some of them pertain to coupling of sufficient optical power into the fiber, fast response time, narrow linewidth, stable and efficient operation, reliability, driving circuit issues, and cost.

21. Distinguish between the operation of LED and Laser.

LED is a p–n junction comprising of direct band gap semiconductor materials which acts as the active or recombination region. When the PN junction is forward biased, electrons and holes recombine either radiatively (emitting photons) or non-radiatively (emitting heat). This is simple LED operation. For fiber-optics, the LED should have a high radiance (light intensity), fast response time and high quantum efficiency. In an LASER, the photon is further processed in a resonance cavity to achieve a coherent, highly directional optical beam width.

22. Mention some applications of Fiber Bragg Grating laser structures.

The Bragg wavelength is independent of the type of laser structure used. This leads to versatility in the fabrication of Fiber Bragg Gratings (FBGs). This characteristic makes FBGs excellent all-optical sensors for applications such as detecting acoustic waves in the ocean, civil structural monitoring, down hole monitoring in the crude oil industry, monitoring of electrical power transformers and transmission lines, security and process control sensor networks. These sensors are easily multiplexed, very robust (they can be used even in extreme and hostile environments), reliable, immune to electromagnetic interference, and relatively inexpensive.

23. List the advantages of Fiber Bragg Grating (FBG) sensors.

Fiber Bragg Grating (FBG) sensors offer numerous distinct advantages:

1. They can give an absolute measurement that is insensitive to fluctuations in the radiance of the illuminating source, as the information is usually obtained by detecting the wavelength shift induced by the measurand,
2. They can be directly written into the fiber without changing the fiber diameter, making them compatible with a wide range of situations where small diameter probes are important, such as in advanced composite materials for strain mapping or the human body for temperature profiling
3. FBG sensors are small in diameter, lightweight and can be unintrusively embedded/ bonded in a wide variety of materials, including concrete and polymer based components
4. They can be mass produced at low cost, making them potentially competitive with conventional electrical sensors
5. They can be multiplexed.

24. Why are laser diodes more difficult to operate than LEDs in an optical transmitter unit?

In an optical transmitter unit, laser diodes are more difficult to operate than LEDs because their current must be carefully controlled, and the required current varies with temperature. With too little current, lasing stops, but too much will destroy the device. Laser diodes often use thermoelectric coolers for temperature control and have their light output monitored by a photodetector to allow the current to be adjusted automatically. Therefore, lasers are usually modulated by changing the current from a value just below the lasing threshold to some considerably higher value. This can be done at rates exceeding 10 GHz.

25. Are there any precautions to be taken while using lasers?

A safety note about lasers: since the light emitted is coherent and intense, and even a low-power laser such as a laser diode can damage the eye, which may focus the light onto a small spot on the retina, adequate care should be taken not to look into the laser beam. Infrared lasers are especially dangerous because the light is invisible and it is not known that the laser is operating. Anyone operating with lasers, even with very low-powered ones used in communications, should be very careful not to look into an operating laser.

26. When does a semiconductor diode work like a laser?

A semiconductor diode works like a laser when the conditions of stimulated emission, population inversion, and positive feedback are met. The spectral characteristics of a semiconductor laser are strongly influenced by the properties of the function medium such as doping.

Multiple Choice Questions

1. In a semiconductor if the charge carriers (electrons or holes) can make a transition from the conduction band to the valence band without any change in the momentum value, the semiconductor material is known as

A. Narrow band gap material. B. Wide band gap material.
C. Indirect band gap material. D. Direct band gap material.

2. The absolute power level produced by an optical source is 0 dBm. On a linear scale, it will be

A. 0 mW B. 1 mW
C. 0.1 W D. 1 W

3. If the band gap of a semiconductor material is 1 eV, its emission wavelength will be

A. 1300 nm B. 1240 nm
C. 1550 nm D. 685 nm

4. Which one of the following statements is true in case of an LED?

A. It is an incoherent optical source that emits light over a relatively wider angle with a broad spectrum.
B. It is a highly coherent optical source that emits light over a relatively wider angle with a broad spectrum.
C. It is an incoherent optical source and emits radiation in a narrow range of angles and contains a narrow spectrum of wavelengths.
D. It is a highly coherent optical source and emits radiation in a narrow range of angles and contains a narrow spectrum of wavelengths.

5. Which one of the following statements is true?

A. A laser diode is an incoherent optical source that emits light over a relatively wider angle with a broad spectrum.
B. A laser diode is highly coherent optical source that emits light over a relatively wider angle with a broad spectrum.
C. A laser diode is an incoherent optical source and emits radiation in a narrow range of angles and contains a narrow spectrum of wavelengths.
D. A laser diode is a highly coherent optical source and emits radiation in a narrow range of angles and contains a narrow spectrum of wavelengths.

6. ________ is an example of a direct bandgap semiconductor material which is suitable for emission of light.
A. Si B. Ge
C. GaAs D. GaAlSi

7. For the emission wavelength of 954 nm, the bandgap of a material of which a laser is made, is
A. 1 eV B. 1.24 eV
C. 1.3 eV D. 1.5 eV

8. When a particular semiconductor device is formed by adding
A. higher impurities concentrations into the same crystal, the resultant junction is known as homojunction.
B. lower impurities concentrations into the same crystal, the resultant junction is known as homojunction.
C. lower impurities concentrations into the same crystal, the resultant junction is known as heterojunction.
D. higher impurities concentrations into the same crystal, the resultant junction is known as heterojunction.

9. ________ are the advantages of employing double heterojunction type devices.
A. Low injection efficiency, improved ohmic contact, optical guidance, and minority carrier confinement
B. High injection efficiency, improved ohmic contact, optical guidance, and majority carrier confinement
C. Low injection efficiency, majority carrier confinement, optical guidance and improved ohmic contact
D. High injection efficiency, improved ohmic contact, minority carrier confinement, and optical guidance

10. When an LED is modulated by an electrical signal,
A. the output optical power is constant at low modulation frequency but falls off at high modulation frequency.
B. the output optical power is constant at high modulation frequency but falls off at low modulation frequency.
C. the output optical power falls off at low modulation frequency but is constant at high modulation frequency.
D. the output optical power rises at low modulation frequency but falls off at high modulation frequency.

11. The modulation response of an LED is described by
A. $P(f)=\dfrac{P_o}{\sqrt{1+(2\pi f\tau)^2}}$ B. $P(f)=\dfrac{P_o}{\sqrt{1+(2\pi f/\tau)^2}}$
C. $P(f)=\dfrac{P_o}{1+(2\pi f\tau)^2}$ D. $P(f)=\dfrac{P_o}{\sqrt{1+2\pi f\tau}}$

12. By definition, the internal quantum efficiency is related with radiative and non-radiative lifetimes as per following relationship
A. $\eta_{\text{int}}=\dfrac{1}{1-(\tau_r/\tau_{nr})}$ B. $\eta_{\text{int}}=\dfrac{1}{1+(\tau_r/\tau_{nr})}$
C. $\eta_{\text{int}}=\dfrac{1}{1+(\tau_{nr}/\tau_r)}$ D. $\eta_{\text{int}}=\dfrac{1}{1+(\tau_r\times\tau_{nr})}$

13. When an LED is modulated by an electrical signal,
A. the output optical power is constant at low modulation frequency but falls off at high modulation frequency.
B. the output optical power is constant at high modulation frequency but falls off at low modulation frequency.
C. the output optical power falls off at low modulation frequency but is constant at high modulation frequency.
D. the output optical power rises at low modulation frequency but falls off at high modulation frequency.

14. In a quarter wave shifted DFB laser diode, ________________ is introduced at the center of the laser cavity.

A. a phase shift of $\pi/4$ radians for a path difference of $\lambda/2$
B. a phase shift of $\pi/2$ radians for a path difference of $\lambda/2$
C. a phase shift of $\pi/2$ radians for a path difference of $\lambda/4$
D. a phase shift of $\pi/4$ radians for a path difference of $\lambda/4$

15. An LED is applied with an input electric power of 170 mW. If the optical power launched into the fiber by this LED is typically 25 μW, then the percent efficiency is

A. 1.5 B. 0.15
C. 0.015 D. 0.0015

16. Statement I - Laser operation occurs only above a certain threshold current.
Statement II – Below threshold current, the laser diode acts as a conventional LED.

A. Statement I and Statement II are correct.
B. Only statement I is correct.
C. Statement I is incorrect but Statement II is correct.
D. Both statements I and II are incorrect.

17. In a heterostructure geometry for semiconductor optical source,

A. The active region has a relatively *smaller* refractive index as compared to that of adjoining p- and n-type semiconductor cladding layers due to a *smaller* bandgap.
B. The active region has a relatively *larger* refractive index as compared to that of adjoining p- and n-type semiconductor cladding layers due to a *larger* bandgap.
C. The active region has a relatively *smaller* refractive index as compared to that of adjoining p- and n-type semiconductor cladding layers due to a *larger* bandgap.
D. The active region has a relatively *larger* refractive index as compared to that of adjoining p- and n-type semiconductor cladding layers due to a *smaller* bandgap.

18. A laser device as optical source has a very narrow angular light beam spread as compared with LEDs, which permits coupling efficiency of the order of ________________ into single mode fibers.

A. 100% B. 70%
C. 50% D. 30%

19. A relatively narrow emitted light spectral width by semiconductor laser diodes enables operation at approximately ________________ data rates.

A. 10 Gbps B. 1 Gbps
C. 500 Mbps D. 100 Mbps

20. DFB laser structures exhibit

A. lower mode partition noise, lower mode stability and narrower linewidths.
B. lower mode partition noise, higher mode stability and narrower linewidths.
C. lower mode partition noise, higher mode stability and wider linewidths.
D. lower mode partition noise, lower mode stability and wider linewidths.

21. Assuming the effective refractive of the waveguide as 3.3, the grating period in a DFB laser to obtain single-mode operation at 1550 nm will be

A. 85 nm B. 177 nm
C. 235 nm D. 360 nm

22. The main function of an optical fiber transmitter is

A. to amplify optical signals, followed by launching them to an optical fiber.
B. to convert an electrical signal into an optical signal, followed by launching them to an optical fiber.

C. to convert an optical signal into an electrical signal.
D. to convert an electrical signal into an optical signal, followed by launching them through the air.

23. In comparison with a surface-emitting LED (SLED), an edge-emitting LED (ELED) possesses all the properties except that it
A. Emits light in a relatively narrower angle of emission.
B. Couples relatively more power into optical fibers having small numerical aperture.
C. Emits optical signal over a relatively wider spectral range.
D. Is relatively more sensitive to operating temperature.

24. The term lasing threshold means
A. the minimum possible value of the drive current at which the laser output will be primarily from stimulated emissions.
B. the front step of the laser chip.
C. the shortest wavelength that the laser emits.
D. the cut or polished surfaces at either edge of the active region in the laser.

25. In order to operate the lasers under varying thermal conditions, one of the following devices is generally included in the standard laser package
A. An optical isolator
B. Automatic overload protection circuitry
C. Thermoelectric cooler
D. Variable attenuator

Keys to Multiple Choice Questions

1. D	**2.** B	**3.** B	**4.** A	**5.** D	**6.** C	**7.** C	**8.** B	**9.** D	**10.** A
11. A	**12.** B	**13.** C	**14.** C	**15.** C	**16.** A	**17.** D	**18.** C	**19.** A	**20.** B
21. C	**22.** B	**23.** B	**24.** A	**25.** C					

Review Questions

1. Outline desirable properties of an optical source to be used for optical fiber communications.

2. Compare and contrast important properties of LED and LD as optical sources. Also comment on their suitability for various applications of optical fiber communications.

3. What are the two primary types of LEDs? Briefly describe the operation of an LED as optical source.

4. Draw a cross-section of an LED structure, and describe its operation.

5. Sketch at least two different structures each for LED and laser diode.

6. Draw a cross-section of a laser diode and describe its operation briefly.

7. For what is the term *laser* an acronym? Compare the emission linewidth of a typical laser diode with the emission linewidth of an LED.

8. What do you mean by lasing?

9. What do you understand by population inversion? Illustrate population inversion mechanisms in a three or four level system.

10. Describe the theory of semiconductor lasers and derive an expression for the threshold current.

11. List the conditions necessary for a semiconductor to function as a gain medium in a laser.

12. Discuss the construction details and operation of a laser diode.
13. Explain the structure and properties of a DFB laser.
14. Compare the performance of laser diode versus LED in optical fiber communications.
15. Give an expression for the internal quantum efficiency of an injection laser diode.
16. Linewidth is an important parameter in the case of optical sources for communication. Compare the typical source linewidths normally found in RF, microwave and optical communication.
17. Discuss how the linewidth of a semiconductor laser can be controlled.
18. Why are laser diodes preferred for use with single-mode fiber?
19. Where do you use distributed feedback (DFB) laser structure?
20. What do you understand by single longitudinal mode (SLM) laser?
21. Write a short note on laser structures and their characteristics.
22. Where is the use of DFB lasers preferred? List the main parameters which are exhibited by DFB laser structures.
23. Contrast the advantages and disadvantages of LEDs and ILDs.

Numerical Problems

1. Find the energy, in electron–volts, in one photon at a wavelength of 1 μm. *[Ans.: 1.24 eV]*
2. Calculate the energy in one photon of a light wave at a wavelength of 400 nm. Express the result in electron–volts. *[Ans.: 3.11 eV]*
3. An LED has a forward voltage of 1.5 volts at a current of 40 mA. It produces a usable light output of 200 μW. How efficient is this optical source? *[Ans.: 0.333%]*
4. An LED is applied with an input electric power of 170 mW. If the optical power launched into the fiber by this LED is typically 25 μW, then find its efficiency. *[Ans.: 0.015%]*
5. A GaAs injection laser (with refractive index of GaAs = 3.6) has length of an optical cavity = 250 μm, width = 100 μm, the gain factor = 21×10^{-2} A/cm^3, and the loss coefficient per μm = 10 at normal operating temperature. Determine the threshold current density and the value of current required to operate this laser device. Assume uncoated cleaved mirrors and current restricted within the cavity only. The refractive index of GaAs is 3.6. *[Ans.: 2.65×10^3 A/cm^2; 663 mA]*
6. For a double heterojunction LED, the minority carriers have radiative recombination life times of 60 ns in the active region. If the non-radiative recombination life times is given as 100 ns, then determine the total carrier recombination life time. *[Ans.: 37.5 ns]*
7. The peak emission wavelength is 870 nm at a drive current = 40 mA for a particular optical source. Determine the power internally generated within the device. *[Ans.: 35.6 mW]*
8. For a double heterojunction InGaAsP LED, the minority carriers have radiative recombination life times of 30 ns and non-radiative recombination life times of 100 ns in the active region, when operated at λ = 1310 nm. For the injected current of 40 mA, find the following parameters:
   (a) The bulk recombination life time
   (b) The internal quantum efficiency
   (c) The internal power level *[Ans.: (a) 23.1 ns; (b) 0.77; (c) 2.9 mW]*

9. A planar LED structure is manufactured using GaAs semiconductor material having refractive index = 3.6. For a transmission factor at the interface of semiconductor crystal and external air of 0.68, determine the emitted optical power as a percent of internal optical power. *[Ans.: 1.31%]*

10. When optical power generated internally is 50% of electrical power supplied, determine the external efficiency. *[Ans.: 0.66%]*

11. Determine the power radiated by an LED if its internal quantum efficiency is 3% and it has an operating wavelength of 670 nm with a device current of 50 nm. *[Ans.: 2.8 mW]*

12. A 10 km fiber with a bandwidth distance product of 1000 MHz–Km is used in an optical fiber communication system. Calculate the optical bandwidth and the electrical bandwidth of the system. *[Ans.: 100 MHz; 70.7 MHz]*

13. Compute the threshold gain for a GaAs laser diode having the following parameters: cavity length L = 500 μm, loss coefficient $\alpha = 10\ cm^{-1}$, the facet reflectivities $R_1 = R_2 = 0.32$. *[Ans.: 33 cm^{-1}]*

14. A GaAs injection laser device has a cavity length = 300 μm and a loss coefficient $\alpha = 10\ cm^{-1}$. The mirror reflectivity at the end is 0.8 and 0.5 respectively. Find the gain coefficient. *[Ans.: 25.3 cm^{-1}]*

15. Show that for an LED, f_{3dB} (optical) = $\sqrt{3}\ f_{3dB}$ (electrical).

16. The 3-dB modulation bandwidth f_{3dB} is related with carrier lifetime τ_c by the expression $f_{3dB} = \dfrac{\sqrt{3}}{2\pi\tau_c}$.

Determine the corresponding modulation bandwidth for the typical range of τ_c = 2 – 5 ns for InGaAsP LED. *[Ans.: 50–140 MHz]*

17. The light output from the GaAs LED structure is coupled into a step-index profile fiber having its core refractive index = 1.4, the core diameter slightly greater than that of LED device, and numerical aperture = 0.2. Determine the coupling efficiency and the optical loss in dB. *[Ans.: 0.04 and 14 dB]*

18. A surface-emitting LED (SLED) couples 150 μW of optical power into a multimode step-index profile fiber. Calculate the overall power conversion efficiency if forward voltage of 2.5 V is applied across the device and 25 mA current flows through it. *[Ans.: 0.24 %]*

19. When a constant dc drive current is applied at the input of an LED, it generates an output optical power of 300 μW. If the minority carrier recombination life time for the device is 5 ns, then calculate the output optical power when it is externally modulated at 20 MHz. *[Ans.: 254 μW]*

20. For a AlGaAs injection laser diode, find the ratio of threshold current density at 20°C and threshold current density at 80°C. Assume T_0 = 160° K. *[Ans.: 1.46]*

21. A typical GaAs laser is emitting a spectral line centered at 632.8 nm whose gain curve has a half-width of 3.003×10^{-3} nm. If the cavity length of the laser is 20 cm, calculate the number of longitudinal modes excited. *[Ans.: 1.001×10^{-3} nm]*

CHAPTER

Optical Receivers

4

Chapter Objectives

After studying this chapter, you should be able to

- ✓ know essential requirements for a photodetector
- ✓ describe various types of semiconductor photodetectors
- ✓ understand the basic operation and characteristics of photodetectors
- ✓ explain the design parameters of an optical receiver
- ✓ analyze sensitivity and noise performance aspects of optical receivers

The purpose of a receiver in an electronic communication system is to extract the information sent by the corresponding transmitter with as minimum a carrier power level as possible. The primary function of an optical receiver in an optical fiber communication link is to convert the received optical signal into an equivalent electrical signal and recover the data. One of the main components of an optical receiver is a photodetector that converts incident optical signals into electric signals using photoelectric effects. High Sensitivity, dynamic range, fast response (i.e., acquisition time), high reliability, low noise, compatible size with that of fiber, and low cost are some of the important requirements of a photodetector. These requirements are best met by semiconductor photodetectors that convert an optical signal transmitted via optical fiber cables to equivalent electrical signals for further processing to achieve the desired output. The type of photodetectors suitable for three optical spectrum ranges of 800–900 nm, 900–1100 nm, and 1100–1600 nm vary in the material used for their fabrication as well as assembly techniques. A p–i–n photodiode is an ideal semiconductor photodetector device, because it can provide high quantum efficiency, fast response and capability to operate at higher modulation frequencies. The minimum received optical power that can be detected by a photodetector is limited by noise. A fully integrated single beam optical receiver comprises of a semiconductor photodiode, preamplifier in the electric domain, digital logic circuits, and an off-chip electronic driver circuit. This chapter discusses all the important aspects of photodetectors and optical receivers. The discussion begins with basic concepts behind the photo detection process, followed by description of different types of photodetectors usually used by optical receivers. Next, the components used in an optical receiver unit are explained. Finally, different types of noise sources in optical receivers that limit the signal-to-noise ratio, the receiver sensitivity parameter and its degradation are covered in sufficient detail.

4.1 Requirements for a Photodetector

In an optical fiber communication system, a semiconductor photodetector is the basic component in an optical receiver, as shown in Fig. 4.1.

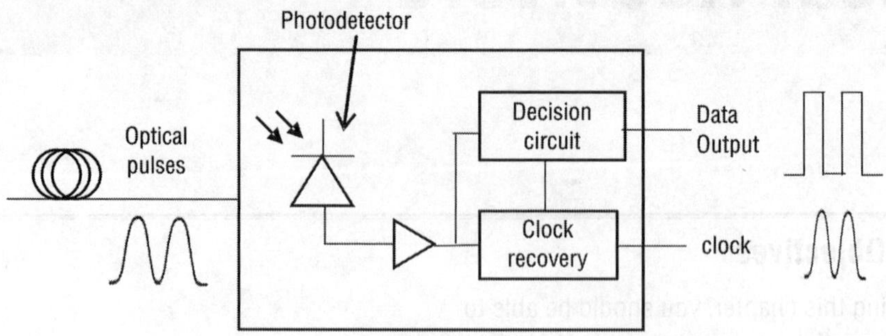

Fig. 4.1 A photodetector in an optical receiver

A photodetector converts the received optical signal into its equivalent electrical signal through the photoelectric effect. Photodetectors made of semiconductor materials absorb incident photons in the form of optical pulses and produces electrons, i.e., photocurrent. In general, the major requirements for a photodetector are listed below:

- Higher degree of sensitivity at the operating wavelengths
- Fast response to obtain a suitable bandwidth of the order of few GHz
- Linear response with regard to wide range of input optical signals so as to reproduce the received signal with high fidelity
- Low noise –minimum introduction of noise due to leakage and dark currents
- High quantum efficiency– A large electrical response to the given optical power
- Stability of performance characteristics against changes in ambient conditions such as operating temperature
- Small in size for efficient coupling with the optical fiber
- Compact for easy packaging
- Requirement of low bias voltage or currents
- Highly reliable operation at room temperature for several years
- Economical to use

Facts to Know

To convert an optical signal into an electrical signal, a number of physical effects such as photo-voltaic, photo-conductive, and the photo-emissive effects can be utilized in photodetectors.

There are different types of semiconductor photodetectors which meet these requirements. These are described in the next section.

4.2 Semiconductor Photodetectors

Broadly speaking, the semiconductor photodetectors are classified as:

1. Semiconductor photodiodes without internal gain
   (a) p–n junction semiconductor photodiode
   (b) p–i–n semiconductor photodiode
2. Semiconductor photodiodes with internal gain
   (a) Avalanche photodiode (APD)
   (b) Silicon-reach through Avalanche photodiode
3. Metal–Semiconductor–Metal (MSM) photodetector

4.2.1 p–n Junction Photodiode

We are aware that an electron–hole pair is created in a p–n semiconductor photodiode provided the energy of the incident photon is more than semiconductor band gap energy level, E_g. Moreover, an indirect band gap semiconductor material is more suitable for making a p–n diode. For an ordinary p–n junction to function as a p–n photodiode, it must be adequately reverse-biased such that adequate absorption of incident photons can take place at operating wavelength. This phenomenon is illustrated in Fig. 4.2.

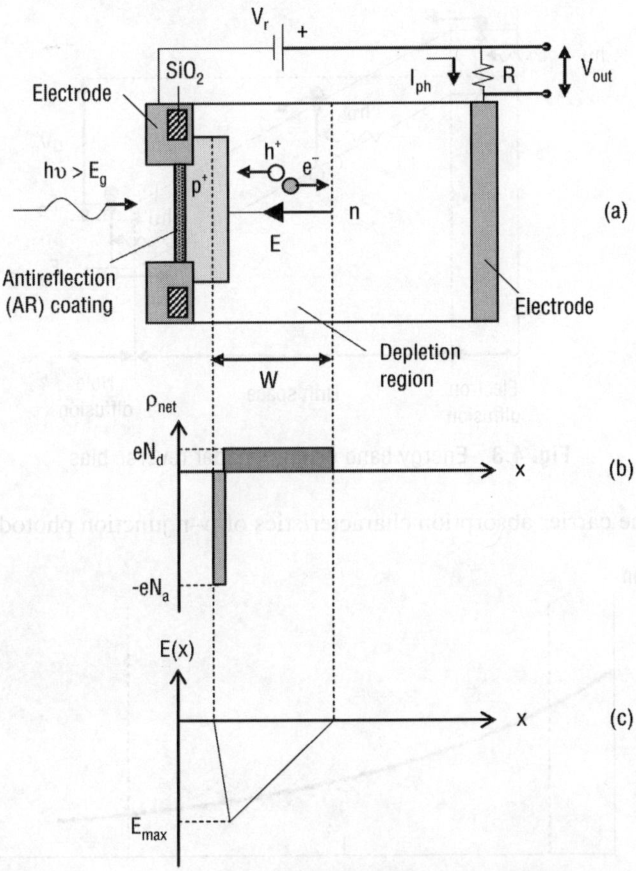

Fig. 4.2 (a) A reverse-biased p–n junction semiconductor photodiode, (b) Net space–charge distribution, (c) The E-field distribution across depletion region

As seen, the depletion region is formed by immobile holes in the n-type and immobile electrons in the p-type semiconductor materials. The width of the depletion region, denoted by *W*, depends on the doping concentration (lower the doping concentration, wider will be the depletion region) for a specified applied reverse bias voltage. When an optical pulse (in the form of light signal) is incident on its p-region (say), then new electron–hole pairs are created due to absorption of photons by the semiconductor material. There is a large built-in electric field because of external reverse bias voltage across the p–n junction. This results in an acceleration of movement of additionally generated electrons and holes in opposite directions within the depletion region. So these charge carriers tend to drift to the n-side and p-side of the p–n junction semiconductor, respectively. This leads to net flow of current due to movement of charge carriers across the junction. It is quite obvious that the amount of current is dependent on the amount of the incident optical power. Therefore, we can say that a reverse biased p–n junction semiconductor device can function as a photodetector.

It may be noted that as the incident optical signal is absorbed within the depletion region, the optical power that contributes to the flow of current decreases exponentially. Fig. 4.3 depicts a typical energy-band diagram that illustrates the movement of charge carriers (electrons and holes) through the process of drift and diffusion.

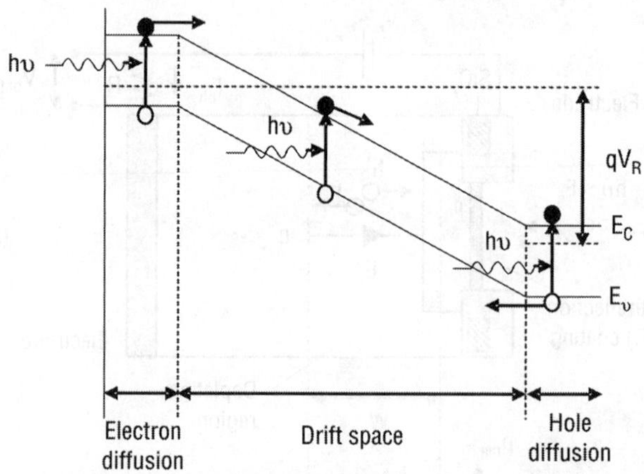

Fig. 4.3 Energy band diagram under reverse bias

Fig. 4.4 depicts the carrier absorption characteristics of p–n junction photodiode.

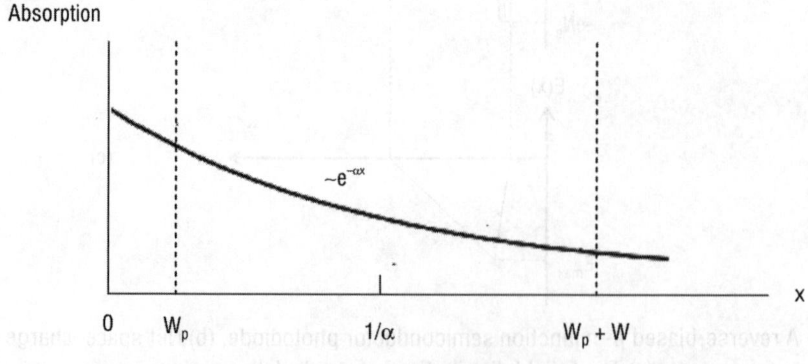

Fig. 4.4 Carrier absorption characteristics of p–n photodiode

The photocurrent I_p is directly proportional to the incident optical power levelP_{in}, that is,

$$I_p \propto P_{in}$$

$$\Rightarrow \quad I_p = R \times P_{in} \tag{4.1}$$

where the proportionality constant R is known as the *responsivity* of the photodetector, and is expressed in Amp/Watts.

Typical values of the responsivity of a p–n photodiode, R lies in the range 0.5 to 1.0 Amp/Watts Due to high quantum efficiency, the responsivity is generally high ($R \sim 1$ Amp/Watts).

Facts to Know

Due to mature technology, silicon is the most common semiconductor material for manufacturing of p–n photodetector for the wavelength region 800–900 nm. For wavelength region more than 900 nm, however, semiconductor materials such as Ge, GaInAs and InGaAsP are the preferred ones for p–n photodetectors.

(a) Input–output characteristics:

Fig. 4.5 depicts the output electric current, I_p with respect to variations in input reverse bias voltage at various light levels.

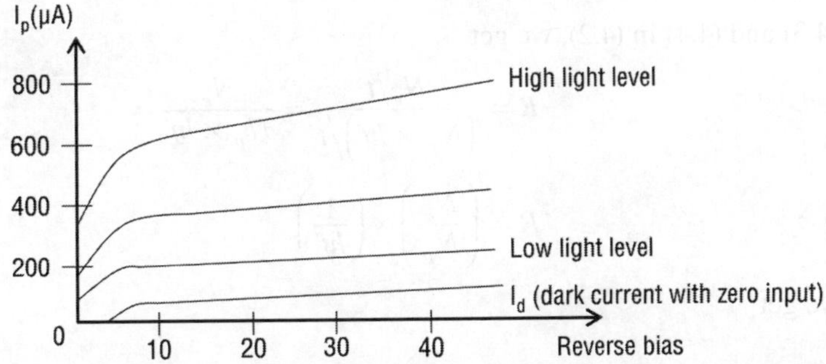

Fig. 4.5 Typical p–n photodiode output characteristics

For a good photodetector, dark current (the output electric current, I_p with no input light), denoted by I_d should be negligible (less than 10 nA). Fig. 4.6 shows the input optical power versus electric current output characteristics of a p–n photodiode.

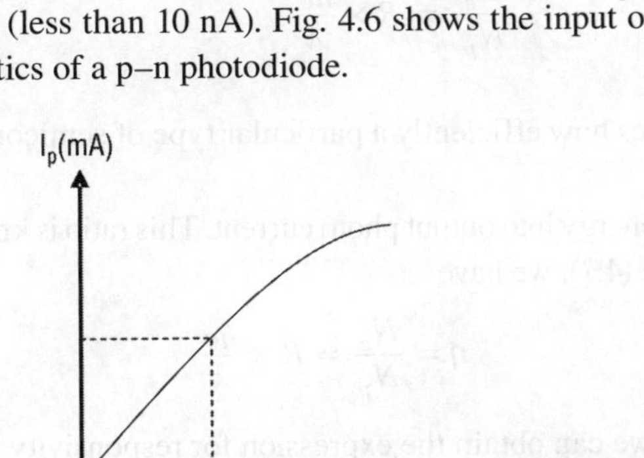

Fig. 4.6 Input–output characteristics of a photodiode

It is seen that at higher optical power, i.e., when there are very large number of photons that are incident on the photodetector per unit of given time, all generated electron–hole pairs (EHPs) contribute to resulting photocurrent. Thus, beyond a certain level of input optical power, a saturation effect is observed, as indicated by dotted line on the characteristic curve.

(b) Responsivity vs wavelength characteristics:

By definition, responsivity is given by

$$R = \frac{I_p}{P_{in}} \tag{4.2}$$

We know that photocurrent is generated due to number of electrons, N_e that flows per unit time, i.e.,

$$I_p = \frac{N_e}{t} \tag{4.3}$$

Similarly, the optical power, P_{in} is equivalent to the incident light energy per unit time. It is quite obvious that the light energy is same as the energy of a photon ($hc/\lambda = hf$) multiplied by the number of photons (N_p). That is,

$$P_{in} = \frac{N_p \times (hf)}{t} \tag{4.4}$$

Substituting (4.3) and (4.4) in (4.2), we get

$$R = \frac{N_e/t}{\left(N_p \times hf\right)/t} = \frac{N_e}{N_p \times hf}$$

$$\Rightarrow \qquad R = \left(\frac{N_e}{N_p}\right) \times \left(\frac{1}{hf}\right) \tag{4.5}$$

Using $f = (c/\lambda)$, we get

$$R = \left(\frac{N_e}{N_p}\right) \times \left(\frac{\lambda}{hc}\right) \tag{4.6}$$

$$\Rightarrow \qquad \frac{N_e}{N_p} = R \times \frac{hc}{\lambda} \tag{4.7}$$

The ratio $\left(\frac{N_e}{N_p}\right)$ indicates how efficiently a particular type of semiconductor material is capable of converting incident light energy into output photo current. This ratio is known as *quantum efficiency* of a photodiode. Then from (4.7), we have

$$\eta = \frac{N_e}{N_p} = R \times \frac{hc}{\lambda} \tag{4.8}$$

Rearranging the terms, we can obtain the expression for responsivity as

$$R = \eta \times \frac{\lambda}{hc} \tag{4.9}$$

Putting the standard values of Planck constant, $h = 6.626 \times 10^{-34}$ and velocity of light in air, $c = 3 \times 10^8$ m/s, we get

$$R\,(A/W) = \eta \times \frac{\lambda(\mu\text{m})}{1.24} \tag{4.10}$$

This shows that responsivity is directly proportional to the operating wavelength. It implies that the graph between responsivity and wavelength should be a straight line for an ideal photodiode, as indicated by dotted line in Fig. 4.7.

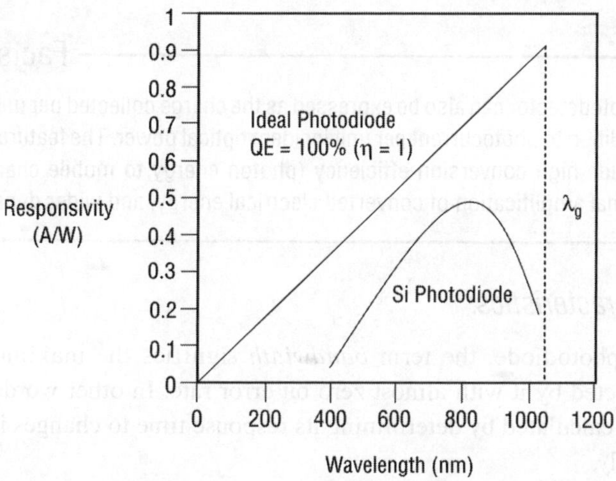

Fig. 4.7 Responsivity vs wavelength curve

The slope of responsivity versus wavelength curve is equal to $\eta/1.24$ where λ is specified in μm. It can be seen that for longer wavelength of the incident light, large number of photons will be required to obtain a specific amount of optical power. Increased number of photons also generates more number of electrons that produces larger photocurrent. Thus, the larger the wavelength, greater value of photo current will be produced for the given amount of optical power. For a particular semiconductor material having specified band gap energy level, the p–n semiconductor photodiode can detect only those wavelength for which $\lambda < \lambda_c = hc/E_g$ where λ_c is the cutoff wavelength which is the longest wavelength that can be detected by a photodetector. That is the reason that at the longer wavelength the responsivity approaches zero. Of course. InGaAs semiconductor compound material is mostly used for p–n semiconductor photodiodes, but Si semiconductor material can also be used for $\lambda < \lambda_c$.

Moreover, light incident directly on the active region of a p–n semiconductor photodiode is partially transmitted and partially absorbed. In that case, the quantum efficiency is also given by

$$\eta = \frac{P_{abs}}{P_{in}} = 1 - e^{-\alpha W} \tag{4.11}$$

where, P_{abs} represents the absorbed optical power, P_{in} represents the incident optical power, α denotes the absorption coefficient, and W is the width of depletion region (active region). It may be noted that the quantum efficiency is constant, not variable, and moreover, the absorption coefficient α is a function of wavelength.

$$\Rightarrow \qquad P_{abs} = P_{in}\left(1 - e^{-\alpha W}\right) \tag{4.12}$$

Case I: When α approaches zero; $P_{abs} = P_{in}(1 - e^0) = P_{in}(1 - 1) = 0$. The corresponding wavelength is the cutoff wavelength, λ_c.

Case II: When α approaches infinity; $P_{abs} = P_{in}(1 - e^{\infty}) = P_{in}(1 - 0) = P_{in}$

This means that each absorbed incident photon can produce an additional electron in p–n semiconductor photodiode.

> **Facts to Know**
>
> Responsivity of a photodetector can also be expressed as the charge collected per unit amount of incident photon energy, in addition to photocurrent per unit incident optical power. The features of opto–electronic photodetectors include high conversion efficiency (photon energy to mobile charge carriers), higher response time, internal amplification of converted electrical energy, and wider dynamic range.

(c) Bandwidth characteristics:

In the context of a photodiode, the term *bandwidth* signifies the maximum frequency, or bit rate, that can be detected by it with almost zero bit error rate. In other words, the bandwidth of a photodetector can be calculated by determining its response time to changes in the incident optical power. Mathematically,

$$BW_{PD} = \frac{1}{2\pi\left(\tau_{t_r} + \tau_{RC}\right)} \tag{4.13}$$

where, τ_{t_r} is the transit time given by the relationship $\tau_{t_r} = \frac{W}{v_d}$, W being the width of the depletion region of the reverse-biased photodiode, and v_d represents the drift velocity of the charge carriers. Both the parameters: W and v_d need to be optimized in order to obtain minimum value of τ_{t_r}, thereby maximizing bandwidth. We know that the width of the depletion region W is dependent on the concentrations of donor and acceptor impurity atoms in the p–n photodiode, whereas the drift velocity v_d depends on the externally applied reverse bias voltage V_R. Depending on the type of semiconductor material used for fabrication of photodiode, the drift velocity reaches a maximum value (known as the saturation velocity) of 10^5 meters per second approximately. Fig. 4.8 shows a typical plot between drift velocity of charge carriers and electric field across reverse-biased p–n junction for electrons and holes in silicon material.

τ_{RC} signifies the time constant of equivalent RC circuit which is given by the relationship $\tau_{RC} = (R_S + R_L)C_p$, where R_s is the internal series resistance, R_L is the external load resistance, and C_p represents the value of electrical parasitic capacitance.

For higher bandwidth, τ_{t_r} should be reduced by decreasing width of depletion region (W). But the quantum efficiency, represented by η, starts decreasing by a significant amount in case the depletion region width W is reduced, and is given as

$$\eta = 1 - e^{-\alpha W} = R \times \frac{hc}{\lambda}$$

It implies that the bandwidth can be traded off with the responsivity of a photodetector. The parameter τ_{RC} limits the bandwidth due to presence of electrical parasitic capacitance.

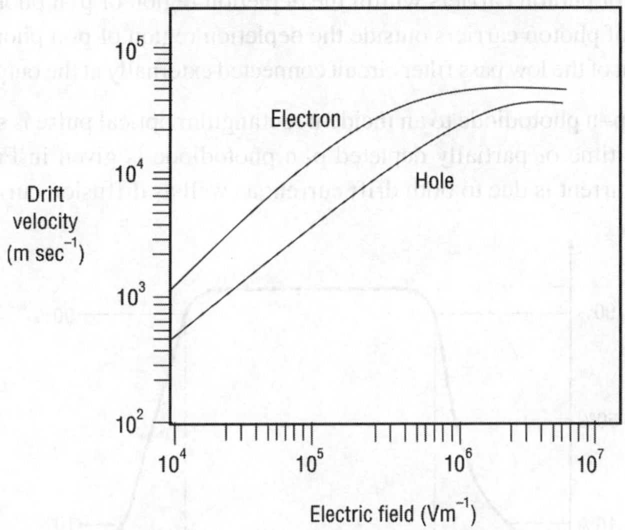

Fig. 4.8 Drift velocity vs electric field

(d) Photodiode response:

Fig. 4.9 depicts different types of optical response curves of p–n semiconductor photodetectors that illustrate clearly trade-off between quantum efficiency and response time.

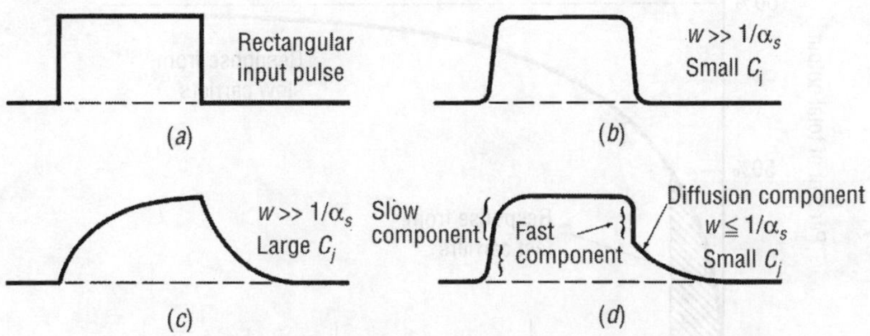

Fig. 4.9 Photodiode response to an optical pulse

In an ideal photodiode, in order to obtain high quantum efficiency, the width of depletion region across p–n junction must be much larger than $1/\alpha$ for type of semiconductor material used in the p–n photodiode so as to ensure the absorption of most of the incident light. The bandwidth that can be obtained with a p–n photodiode is quite often restricted by the value of transit time τ_{t_r}. It may be noted that p–n photodiodes have the capability to operate up to 40 Gbps bit rate.

The bandwidth of a p–n photodiode is also limited due to the presence of diffusion component in the resulting output photocurrent because the incident light which falls outside the depletion region is absorbed. Diffusion is an inherently slow process, distorting its response characteristics. The response of a photodetector depends on three major factors as given below:

1. Transition time of photon carriers within the depletion region of p–n photodiode.
2. Diffusion time of photon carriers outside the depletion region of p–n photodiode.
3. RC time constant of the low pass filter circuit connected externally at the output of the photodiode.

The response of a p–n photodiode to an incident rectangular optical pulse is shown in Fig. 4.10(a), and typical response time of partially depleted p–n photodiode is given in Fig. 4.10(b) under the situation that photo current is due to both drift current as well as diffusion current components.

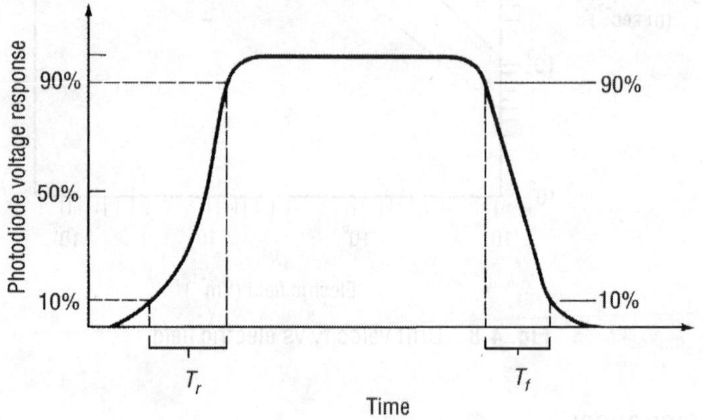

Fig. 4.10 (a) Photodiode response to optical pulse

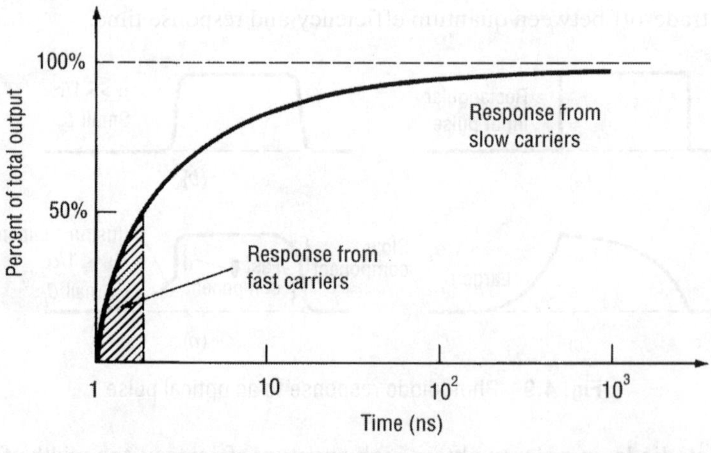

Fig. 4.10 (b) Typical response time of undepleted photodiode

In order to achieve a higher value of quantum efficiency of a p–n photodiode, the width of the depletion region should be greater than that given by the inverse of the absorption coefficient. This condition will ensure that most of the incident light will be absorbed. However, a larger depletion width means a smaller value of junction capacitance. This results in a smaller value of RC time constant. This, in turn, will lead to faster response. So far the situation is quite a favorable one and as desired. But, on the other hand, larger depletion width would also mean greater transit time in the depletion region. Thus we conclude that there may be a trade-off between depletion region width and quantum efficiency. However, the contribution in photodetector current output due to diffusion component may be minimized by smaller widths of the p- and n- regions themselves. This results in an increase in the width of the depletion region. This will ensure that most of the incident optical power gets absorbed within it. In fact, this principle is followed in the development of another type of photodetector, known as *p–i–n* semiconductor photodiode.

Example 4.1 Band Gap and Photodetection

(a) Find the maximum energy band gap a semiconductor material being used as a photoconductor so that it is quite sensitive to an incident light having $\lambda = 600$ nm.

(b) A photodetector having an active area of 0.05 cm^2 is illuminated with the incident light with an intensity of 20 mW/cm^2. Assume that each incident photon generates one electron–hole pair. Determine the number of photons emitted per second.

Solution:

(a) Given $\lambda = 600$ nm

We know that the maximum value of energy gap required, $E_g = \dfrac{hc}{\lambda}$

Therefore, $$E_g = \frac{\left(6.626 \times 10^{-34} Js/1.6 \times 10^{-19} C\right) \times \left(3 \times 10^8 m/s\right)}{600 \times 10^{-9} m} = 2.07 \text{ eV}$$ **Ans.**

(b) We know that the incident power for conversion is obtained by multiplying the given incident light intensity of 20 mW/cm^2 with cross-sectional area, $A = 5 \times 10^{-2}$ cm^2. That is,

$$P_{in} = \left(20 \text{ mW/cm}^2\right) \times \left(5 \times 10^{-2} \text{ cm}^2\right) = 1 \text{ mW}$$

We know that the number of photons generated per second, $N_p = \dfrac{P_{in}}{E_g}$

$$\therefore \quad N_p = \frac{1 \times 10^{-3} W}{(2.07 \, eV) \times \left(1.6 \times 10^{-19} \text{ J/eV}\right)} = 3 \times 10^{15} \text{ photons/s}$$ **Ans.**

Example 4.2 Responsivity of Photodiode

The radius of an active light receiving region of a p–n photodiode is specified as 0.02 cm. It has been observed that when light intensity of 0.1 mW/cm^2 is incident on it, it produces a photocurrent of 56.6 nA. Calculate the responsivity.

Solution:

The incident light intensity of 0.1 mW/cm^2 means that the incident power for conversion is obtained by multiplying it with cross-sectional area, $A = \pi r^2 = \pi\,(0.02)^2$ cm^2. That is,

$$P_{in} = \left(0.1 \text{ mW/cm}^2\right) \times \left(\pi \times 0.02^2 \text{ cm}^2\right) = 1.26 \times 10^{-7} \text{ W}$$

We know that responsivity, $R = \dfrac{I_p}{P_{in}}$

For given value of photocurrent I_p = 56.6 nA, we have

$$R = \frac{56.6 \times 10^{-9}\ A}{1.26 \times 10^{-7}\ W} = 0.45\ \text{A/W}$$ **Ans.**

Example 4.3 Quantum Efficiency of Photodiode

A photodiode has specified responsivity of 0.45 A/W. When radiation intensity of wavelength 700 nm is incident on it, find the quantum efficiency.

Solution:

We know that quantum efficiency, $\eta = R \times \frac{hc}{\lambda}$

For given values of responsivity R = 0.45 A/W and wavelength λ = 700 nm, we have

$$\eta = (0.45\ \text{A/W}) \times \frac{(6.626 \times 10^{-34}\ \text{Js}/1.6 \times 10^{-19}\ \text{C})(3 \times 10^{8}\ \text{m/s})}{(700 \times 10^{-9}\ \text{m})}$$

Hence, η = 0.80, or 80% **Ans.**

Example 4.4 Transit Time

We know that the transit time is one of the main factors that limit the bandwidth of a p–n photodiode. If the depletion region width = 10 μm and drift velocity of charged carriers = 10^5 m/s, then determine the transit time of the photodiode.

Solution:

We know that transit time, $\tau_{t_r} = \frac{W}{v_d}$; where W represents the depletion region width, and v_d denotes the drift velocity.

For given W = 10 μm and v_d = 10^5 m/s, we get

$$\tau_{t_r} = \frac{10 \times 10^{-6}\ m}{10^5\ m/s} = 100\ ps$$ **Ans.**

4.2.2 p–i–n Photodiode

Before we discuss another type of photodetector, let us recall that the p–n junction semiconductor photodiode has two drawbacks—one is depletion layer capacitance which is not sufficiently small to allow photo detection at high modulation frequencies (RC time constant limitation), and narrow space–charge layer (at most a few microns). This implies that the long wavelengths incident photons are absorbed outside space–charge layer which leads to low quantum efficiency. The p–i–n photodiode can significantly reduce these problems. To increase the depletion region width such that almost the whole incident optical power is absorbed in it, a region of very lightly-doped, or may be undoped (i.e., pure) semiconductor material is inserted between the p-type and n-type semiconductor materials. It is the property of an intrinsic layer that it has less doping and wider region (typically 5–50 μm). Since the middle layer comprises of an intrinsic semiconductor material, such a structure is often referred to as the p–i–n photodiode, as shown in Fig. 4.11.

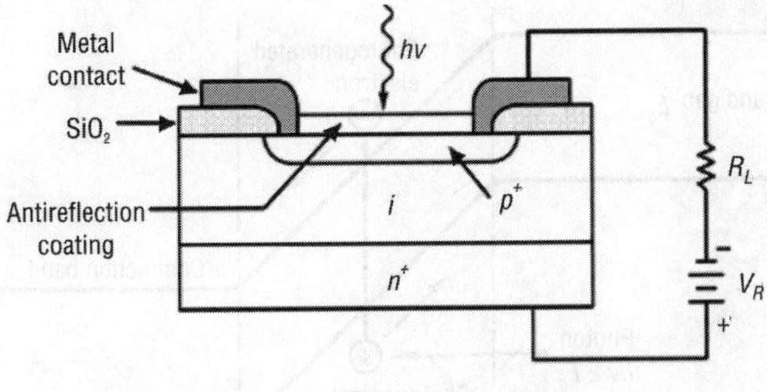

Fig. 4.11 Operation of a p–i–n photodiode

An important feature of the operation of the p–i–n photodiode is that it allows deep penetration of incident light at operating wavelength. A relatively shorter wavelength photon is incident near the surface of an externally reverse-biased p–i–n photodiode. This photon is completely absorbed by it. As a result, the photo-generated electron tends to diffuse within the depletion region to be swept into intrinsic-layer and then finally to be drifted across the photodiode. One of the main benefits of p–i–n photodiode is that small depletion region capacitance enables operation at relatively higher modulation frequencies. Moreover, the p–i–n photodiode offers reasonably higher quantum efficiency.

Now the question arises that what is the major difference between p–n photodiode and p–i–n photodiode as far as the photocurrent generation mechanism is concerned. In p–i–n photodiode, the drift component of the output photocurrent is more prominent than the diffusion component. Why is that so? The answer is that most of the incident optical power gets absorbed within the intrinsic region of a p–i–n photodiode. This, in turn, also results in higher quantum efficiency. Both power efficiency and bandwidth efficiencies are also high as there is no need to increase the reverse bias voltage V_R for increasing the width of the depletion region. Fig. 4.12 illustrates the basic operation of the p–i–n photodiode.

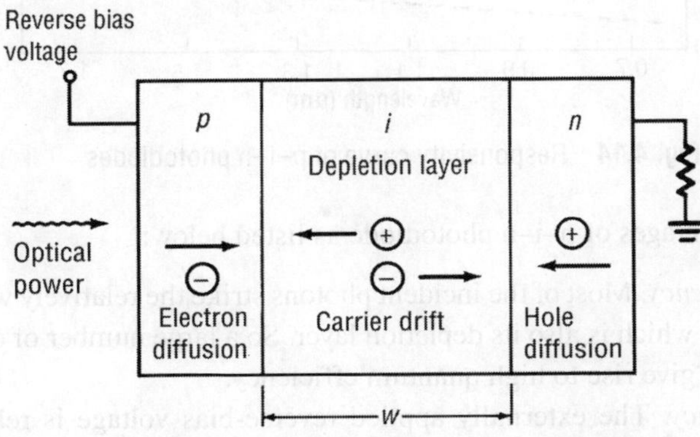

Fig. 4.12 Reverse-biased p–i–n photodiode

Fig. 4.13 depicts the energy–band structure of a typical p–i–n photodiode.

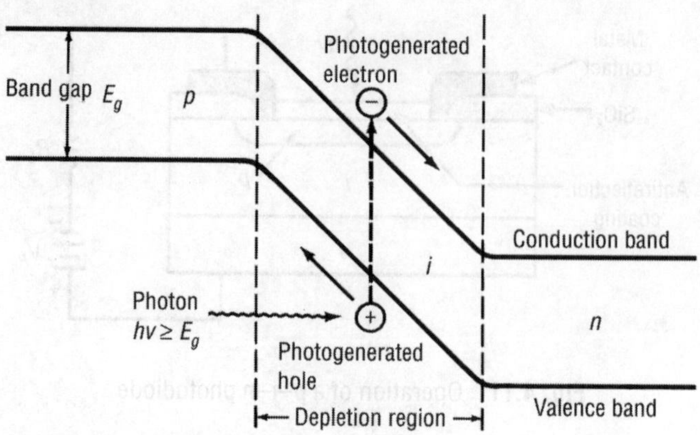

Fig. 4.13 Energy–band structure of p–i–n photodiode

Fig. 4.14 presents the responsivity curve for Si, Ge, and InGaAs p–i–n photodiodes.

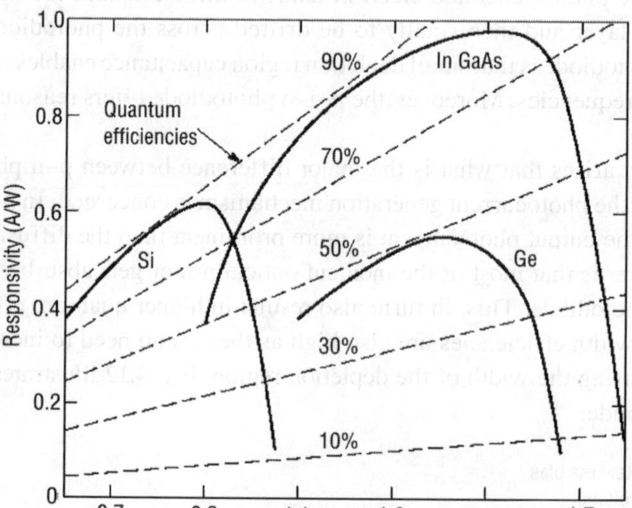

Fig. 4.14 Responsivity curve of p–i–n photodiodes

There are certain advantages of p–i–n photodiode as listed below:

1. *High quantum efficiency.* Most of the incident photons strike the relatively wider intrinsic layer of p–i–n photodiode which is also its depletion layer. So a large number of electron–hole pairs are generated which give rise to high quantum efficiency.
2. *High power efficiency.* The externally applied reverse-bias voltage is relatively very small (typically of the order of 5V). This is so because the depletion region width is precisely controlled by width of the intrinsic region, not by the amount of reverse-bias voltage.
3. *High bandwidth efficiency.* The incident photons strike the intrinsic region which is also the active region than the p- or n- regions of p–i–n photodiode, thereby resulting into high bandwidth efficiency.

4. Requirement of small reverse-biasing applied voltage (usually, 5V)
5. Decrease in dark current due to low thermally generated charge carriers
6. Small diffusion current due to very small width of the p- and n- regions as compared to relatively large width of the i-layer.

However, there are some disadvantagesof p–i–n photodiode as follows:

1. Rise in transit time, τ_{t_r} (= W/v_d) due to high W (width of i-layer). This results in decrease in bandwidth efficiency.
2. Increase in response time because charge carriers takes longer time to be drifted across relatively wider depletion region.

Facts to Know

A p–i–n type photodiode structure is an ideal solid-state photodetector device which is widely used in optical fiber communications. It has several features such as higher quantum efficiency, relatively faster response, capability to operate at high modulation frequencies and withstand high electric field.

Double–heterostructure p–i–n Photodiode

In order to improve the performance of p–i–n photodiode considerably, the middle intrinsic region is sandwiched between the p-and n-type regions of entirely different semiconductor materials having their energy band gap levels chosen in such a way that the whole incident light gets absorbed within the i-region only, as shown in Fig. 4.15.

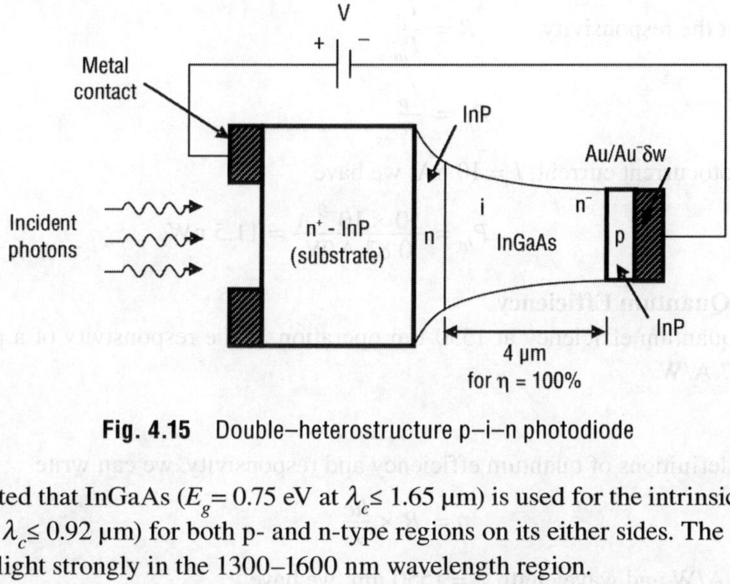

Fig. 4.15 Double–heterostructure p–i–n photodiode

It may be noted that InGaAs (E_g= 0.75 eV at $\lambda_c \leq$ 1.65 μm) is used for the intrinsic region and InP (E_g= 1.35 eV at $\lambda_c \leq$ 0.92 μm) for both p- and n-type regions on its either sides. The middle intrinsic region absorbs light strongly in the 1300–1600 nm wavelength region.

Double–heterostructure p–i–n photodiode has unique advantages such as requirement of relatively low reverse-bias voltage, capable of operating at higher bandwidth (i.e., up to 100 GHz), low noise, highly reliable operation, easy and economical in manufacturing, etc.

Example 4.5 p–i–n Photodiode

The responsivity curve of a commercial InGaAs p–i–n photodiode is given in Fig. 4.16.

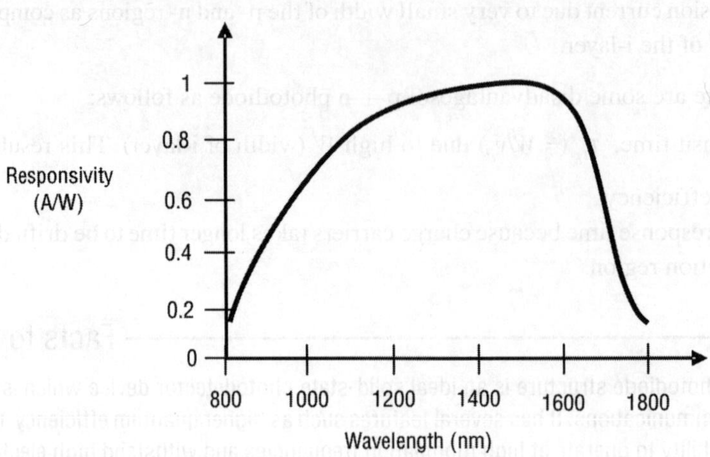

Fig. 4.16 Responsivity curve of InGaAs p–i–n photodiode

If the dark current is 5 nA, then what would be the incident optical power at $\lambda = 1550$ nm so as to deliver 10 nA photocurrent (which is twice that of dark current)?

Solution:

From the given responsivity curve of the p–i–n photodiode, we have R = 0.87 A/W at specified wavelength $\lambda = 1550$ nm.

We know that the responsivity, $R = \frac{I_p}{P_{in}}$

$$\Rightarrow \quad P_{in} = \frac{I_p}{R}$$

For given photocurrent current, $I_p = 10$ nA, we have

$$P_{in} = \frac{10 \times 10^{-9} A}{0.87 \text{ A/W}} = 11.5 \text{ nW}$$ **Ans.**

Example 4.6 Quantum Efficiency

Determine the quantum efficiency at 1550-nm operation, if the responsivity of a photodetector is specified as 0.87 A/W.

Solution:

From the basic definitions of quantum efficiency and responsivity, we can write

$$\eta = R \times \frac{hc}{\lambda}$$

For $R = 0.87$ A/W and wavelength $\lambda = 1550$ nm, we have

$$\eta = (0.87 \text{ A/W}) \times \frac{(6.626 \times 10^{-34} \text{Js}/1.6 \times 10^{-19} \text{C})(3 \times 10^{8} \text{m/s})}{(1550 \times 10^{-9} \text{ m})}$$

Hence, $\eta = 0.697$, or 69.7% **Ans.**

Example 4.7 Responsivity of p–i–n Photodetector

Determine the responsivity of a p–i–n photodetector at 0.85 μm operation for specified quantum efficiency of 60%.

Solution:

We know that quantum efficiency, $$\eta = R \times \frac{1.24}{\lambda(\mu m)}$$

$$\Rightarrow \quad R = \eta \times \frac{\lambda(\mu m)}{1.24}$$

Substituting the given values of $\eta = 60\%$ or 0.6, and $\lambda = 0.85$ μm, we have

$$R = 0.6 \times \frac{0.85}{1.24} = 0.41$$ **Ans.**

Example 4.8 Photocurrent of a p–i–n Photodiode

Determine the resultant photocurrent of a p–i–n photodiode having responsivity of 0.41 if an incident optical power = 10×10^{-6} W. Also find the voltage across load resistor of 50 ohms.

Solution:

We know that current, $$I_P = R \times P_{in}$$

For given values of $R = 0.41$ and $P_{in} = 10$ μW, we get

$$I_P = 0.41 \times 10 \times 10^{-6} = 4.1 \, \mu A$$ **Ans.**

Voltage across load resistor = $I_P \times R_L$

For calculated $I_p = 4.1$ μA and given $R_L = 50\Omega$, we get

Voltage across load resistor = $4.1 \times 10^{-6} \times 50 = 2.05 \times 10^{-4} V$, or 0.205 mV **Ans.**

Example 4.9 Time Constant of p–i–n Photodiode

It is specified in the data sheet of a p–i–n photodiode that it has a parasitic capacitance value of 3 pF. If the source and load resistances are 50 Ω each, then find the time constant of this p–i–n photodiode considering its equivalent RC circuit.

Solution:

We know that the time constant, $\tau_{RC} = (R_L + R_S)\, C_P$

For given values of $R_L = R_s = 50$ Ω, $C_p = 3$ pF, we have

$$\tau_{RC} = (50 + 50) \times 3 \times 10^{-12} = 300 \times 10^{-12} \text{s}$$

Hence, $\tau_{RC} = 0.3 \, ns$ **Ans.**

Example 4.10 Calculation of Bandwidth

It is specified in the data sheet of a p–i–n photodiode that it has a transit time of 2 ns. Considering its equivalent RC circuit, the time constant has been computed to be 0.3 ns. Comment on the available bandwidth of this photodiode whether it is limited by parasitic capacitance value of 3 pF. Also calculate the 3-dB bandwidth of the photodiode as given by the relationship $f_{3dB} = 0.35/\tau_{t_r}$.

Solution:
Given $\tau_{t_r} = 2\ ns$ and $\tau_{RC} = 0.3\ ns$

Since $\tau_{t_r} \gg \tau_{RC}$, the bandwidth of the photodiode is limited in this case.

For a given value of $\tau_{t_r} = 2\ ns$, we have

$$f_{3dB} = 0.35/\left(2 \times 10^{-9}\right) = 175 \text{ MHz}$$ **Ans.**

Example 4.11 Characteristics of a p–i–n Photodiode
Tabulate the operating characteristics of three commonly used p–i–n photodiodes made of Si, Ge, and InGaAs semiconductor materials. Also show that InGaAs photodiodes are more suitable for most of the relevant practical applications.

Solution:
Table 4.1 depicts a comparative study of operating characteristics of p–i–n photodiodes made of Si, Ge, and InGaAs semiconductor materials.

Table 4.1 Operating characteristics of various p–i–n photodiodes

S. No.	*Parameter*	*Si*	*Ge*	*InGaAs*
1.	Wavelength, λ	0.4–1.1μm	0.8–1.8μm	1.0–1.7μm
2.	Bias voltage, V_R	50–100 V	6–10 V	5–6 V
3.	Responsivity, R	0.4–0.6 A/W	0.5–0.7 A/W	0.6–0.9 A/W
4.	Dark current, I_d	1–10 nA	50–500 nA	1–20 nA
5.	Quantum efficiency, η (%)	75–90	50–55	60–70
6.	Rise time, τ_r (ns)	0.5–1	0.1–0.5	0.02–0.5
7.	Bandwidth, *BW*	0.3–0.6 GHz	0.5–3 GHz	1–10 GHz

Facts to Know

The dark current of a photodetector is defined as the current produced in the absence of any input optical signal. It may be originated either from thermally generated pairs of electrons and holes, or from stray light. Dark current represents the noise level in a photodetector, so a smaller dark current corresponds to a higher signal-to-noise ratio. The dark current should be as small as < 10 nA for a good photodetector.

4.2.3 Avalanche Photodiode (APD)

For reliable operation, it is imperative to say that all semiconductor photodetectors do need a certain minimum value of output photocurrent. This translates into requirement of a minimum input optical power, given by the relationship $P_{in} = I_P/_R$. This implies that semiconductor photodetectors having large value of responsivity, R, will always be a better choice because they would then need less amount of input optical power, P_{in}.

The responsivity of a p–i–n photodiode is limited by $R \approx \eta\lambda(\mu m)/1.24$. Obviously, the maximum value, $R_{max} \approx \lambda(\mu m)/1.24$ will be obtained at $\eta = 1$.

There is another type of photodetector, known as avalanche photodiodes (APDs) which can achieve a much larger value of R because they have been designed to offer an *internal current gain*. APDs are used when the amount of available optical power is limited at the input of an optical receiver.

The phenomenon, known as *impact ionization*, is responsible for internal current gain in APDs. It is the process of creating many secondary carriers, known as avalanche multiplication. For this, an additional layer is included in the structure of standard p–i–n photodiode. Fig. 4.17 illustrates the basic structure of APD alongwith its operation under reverse-biased condition.

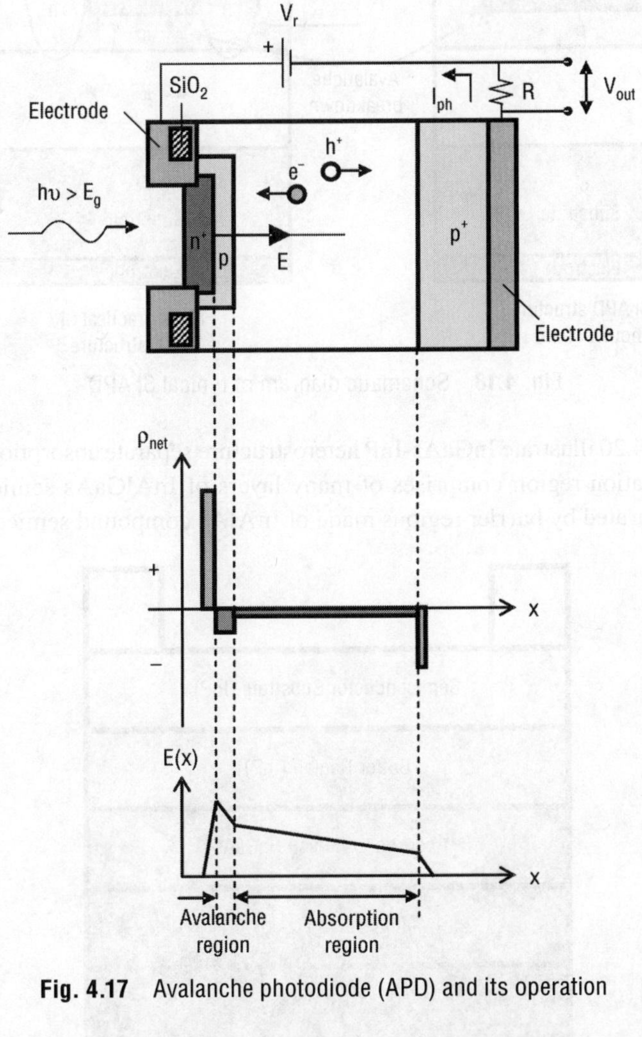

Fig. 4.17 Avalanche photodiode (APD) and its operation

An external reverse-bias supply voltage is applied such that a relatively higher electric field is developed in the p-region which is sandwiched between *i*-region and *n*-region of standard p–i–n photodiode structure. This region is often referred to as multiplication region because of generation of secondary electron–hole pairs (EHPs) through the phenomenon of impact ionization. The *i*-region still functions as the depletion region. In this region, not only most of the incident photons are absorbed but the primary electron–hole pairs are also produced. The electrons produced in the *i*-region cross the gain region which results in a generation of secondary electron–hole pairs. This, in turn, is responsible for the net current gain. Fig. 4.18 depicts a schematic diagram of typical Si APD without a quad ring and practical Si APD.

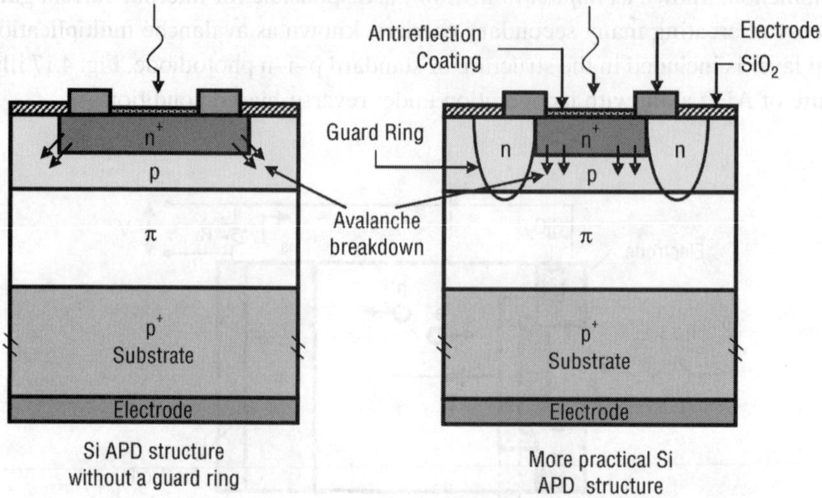

Fig. 4.18 Schematic diagram of typical Si APD

Fig. 4.19 and Fig. 4.20 illustrate InGaAs-InP heterostructure separate absorption and multiplication APD. The multiplication region comprises of many layers of InAlGaAs semiconductor quantum wells which are separated by barrier regions made of InAlAs compound semiconductor material.

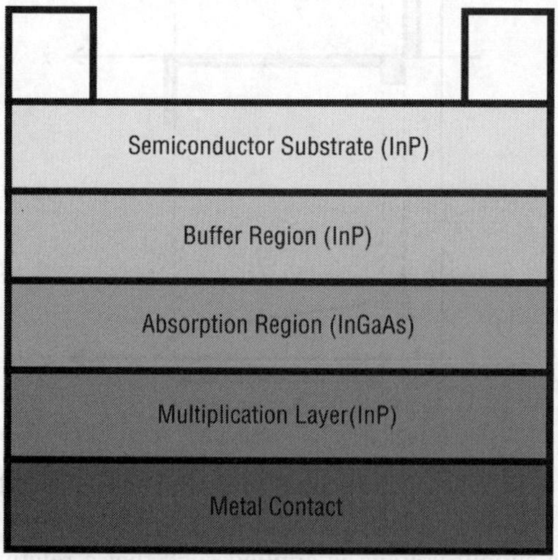

Fig. 4.19 InGaAs APD super lattice structure

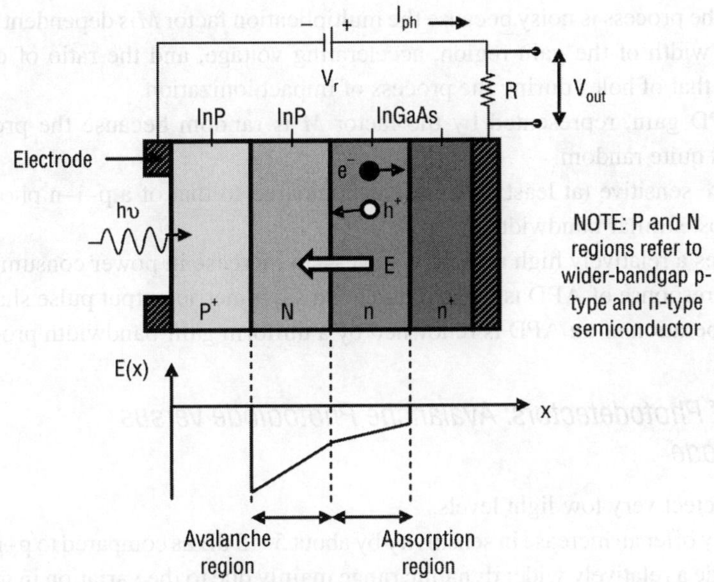

Fig. 4.20 Heterojunction APD

As indicated, an interpolating grading region (InGaAsP) with a transitional band gap energy level breaks and makes it convenient for the holes to cross through the InP region. As mentioned earlier, APDs multiply the primary photocurrent internally using impact ionization before it is processed further. For carrier multiplication to take place, it is very necessary that the photo-generated carriers (electrons and holes) must travel along the region having high electric field in which they gain sufficient energy so as to ionize bounded electrons available in the valence band to collide among themselves. The newly generated charged carriers due to high electric field lead to more ionization known as *avalanche effect*.

Some of the characteristics of the APD are as follows:

1. The responsivity of an APD is augmented by the multiplication factor M, also known as average APD gain, and is given by

$$R_{APD} = M \times R = M \times \left(\frac{\eta\lambda}{1.24}\right) \tag{4.14}$$

where, λ is the operating wavelength in μm.

2. The intrinsic bandwidth of an APD is dependent on the multiplication factor M because the transit time τ_{t_r} rises significantly due to additional time required for creation and gathering of secondary electron–hole pairs, as given by

$$M \times BW = \frac{1}{2\pi k_A \tau_{t_r}} \tag{4.15}$$

where, $k_A = \dfrac{\text{No. of electrons}}{\text{No. of holes}}$

3. The APD gain reduces at relatively higher frequency due to rise in transit time that limits the bit rate or bandwidth.

4. The avalanche process is noisy because the multiplication factor M is dependent on many factors such as the width of the gain region, accelerating voltage, and the ratio of concentration of electrons to that of holes during the process of impact ionization.
5. Average APD gain, represented by the factor M is random because the process of impact ionization is quite random.
6. APD is more sensitive (at least 10 times) as compared to that of a p–i–n photodetector, both having almost similar bandwidth.
7. APD requires a relatively high reverse voltage—an increase in power consumption.
8. The overall response of APD is limited due to an asymmetric output pulse shape.
9. The basic operation of an APD is renowned by a uniform gain-bandwidth product.

Comparison of Photodetectors: Avalanche Photodiode versus p–i–n Photodiode

1. APDs can detect very low light levels.
2. APDs usually offer an increase in sensitivity by about 5–15 dB as compared to p–i–n photodiodes.
3. APDs provide a relatively wider dynamic range mainly due to the variation in gain with respect to response time as well as external applied reverse bias voltage, as given by Fig. 4.21.

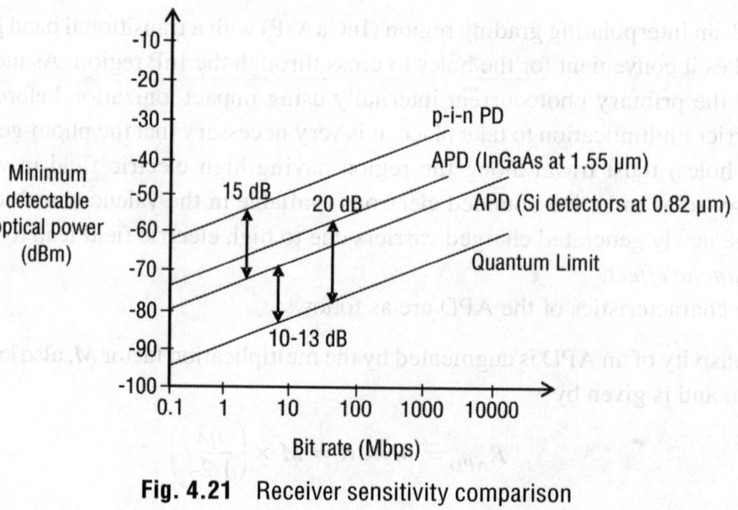

Fig. 4.21 Receiver sensitivity comparison

Drawbacks of APD

1. Difficulty in fabrication process because of quite complex structure
2. Additional noise contribution due to random nature of gain mechanism
3. Requirement of relatively high erreverse-bias voltage of the order of 50-400 V, and too depends on the operating wavelength
4. Effect of temperature on avalanche gain– In APD, the electron and hole ionization rates depend on temperature at high bias voltage. This causes avalanche gain to vary with temperature. For a Si APD, the ionization rates (and hence APD gain) increases as operating temperature decreases at constant applied bias voltage.

The gain-bandwidth product of InGaAs APDs is limited because of the use of InP semiconductor material for creation of secondary electron–hole pairs. We can adopt a hybrid approach—an additional Si multiplication region being introduced immediately next to an InGaAs absorption region. Generally the thickness of an absorbing region in most of APDs is quite thick (about 1 μm) which ensures more than 50% quantum efficiency. But it affects the bias voltage as well as the transit time which can be considerably minimized by using a relatively thinner absorbing region (as thin as approximately 0.1 μm only). This results in improved APDs as long as higher quantum efficiency can be maintained.

Facts to Know

In the modified design of APD, usually called reach-through APD, the depletion regional most stretches up to the contact region through absorption region as well as the multiplication region. It offers several unique features such as relatively higher gain ($M >> 100$) with larger bandwidth and low noise.

Example 4.12 Responsivity of an APD

The output current of an APD is measured as 100 nA corresponding to an incident optical power of 5 nW. The operating wavelength is 1.5 μm. Find its responsivity.

Solution:

We know that responsivity of an APD, $R_{APD} = \dfrac{I_p}{P_{in}}$

For given values of $I_p = 100$ nA and $P_{in} = 5$ nW, we get

$$\Rightarrow \quad R_{APD} = \frac{100 \times 10^{-9}\ \text{A}}{5 \times 10^{-9}\ \text{W}} = 20\ \text{A/W}$$ **Ans.**

Example 4.13 Multiplication Factor of an APD

The data sheet of an APD specifies that responsivity = 20 A/W and quantum efficiency = 70%. Determine the avalanche multiplication factor for operating wavelength of 1.5 μm.

Solution:

We know that the responsivity, $R_{APD} = M \times \left(\dfrac{\eta\lambda(\mu m)}{1.24}\right)$

$$\Rightarrow \quad M = \frac{R_{APD}}{\left(\dfrac{\eta\lambda(\mu m)}{1.24}\right)}$$

Using given values of $R_{APD} = 20$ A/W, $\eta = 70\%$ or 0.7, $\lambda = 1.5$ μm, we get

$$M = \frac{20}{\left(\dfrac{0.7 \times 1.5}{1.24}\right)} = 23.62$$ **Ans.**

Example 4.14 Quantum Efficiency of an APD

Consider a silicon p–i–n photodetector and an APD to detect light at $\lambda = 850$ nm. For an incident light intensity of 0.1 mW/mm^2, the photocurrent generated by the p–i–n photodetector and APD are

10 μA and 500 μA, respectively. In both cases, the active area is 0.2 mm². Compute the quantum efficiency and the avalanche multiplication factor.

Solution:

By definition, the quantum efficiency $\eta = \frac{\textbf{carrier generation rate}}{\textbf{photon generation rate}} = \frac{N_e}{N_p}$

The carrier generation rate in p–i–n photodetector, $N_e\big|_{p-i-n} = \frac{I_p\big|_{p-i-n}}{q}$

For the given value of $I_p\big|_{p-i-n} = 10\ \mu A$, we get

$$N_e\big|_{p-i-n} = \frac{10 \times 10^{-6} A}{1.6 \times 10^{-19} C} = 6.25 \times 10^{13}$$

The photon generation rate, $N_P = \frac{P_{in}}{(hc/\lambda)}$

where, P_{in} = Incident light intensity × Active area

For the given values of incident light intensity = 0.1 mW/mm² and active area = 0.2 mm², we get P_{in} = 0.1 mW/mm² × 0.2 mm² = 0.02 mW. Therefore,

$$N_p = \frac{0.02 \times 10^{-3}}{\left[\left(6.62 \times 10^{-34} \times 3 \times 10^{8}\right)/\left(0.85 \times 10^{-6}\right)\right]} = 8.56 \times 10^{13}$$

Hence, the quantum efficiency, $\eta = \frac{6.25 \times 10^{13}}{8.56 \times 10^{13}} = 0.73$, or 73% **Ans.**

Example 4.15 Multiplication Factor of an APD

Consider a silicon p–i–n photodetector and an APD to detect light at λ = 850 nm. For an incident light intensity of 0.1 mW/mm², the photocurrent generated by the p–i–n photodetector is 10 μA and that of APD is 500 μA respectively. Find the avalanche multiplication factor.

Solution:

We know that avalanche multiplication factor, $M = \frac{I_p\big|_{APD}}{I_p\big|_{p-i-n}}$

For given values of $I_p\big|_{p-i-n} = 10\ \mu A$ and $I_p\big|_{APD} = 500\ \mu A$, we get

$$M = \frac{500 \times 10^{-6} A}{10 \times 10^{-6} A} = 50$$ **Ans.**

Example 4.16 Characteristics of APDs

Tabulate a comparative study of various characteristics of different types of APDs such as Si, Ge, and InGaAs APDs.

Solution:

Table 4.2 depicts a comparative study of operating characteristics of common APDs made of Si, Ge, and InGaAs semiconductor material.

Table 4.2 Operating characteristics of APDs

S. No.	*Key Parameter*	*Si*	*Ge*	*InGaAs*
1.	Wavelength, λ	0.4–1.1 μm	0.8–1.8 μm	1.0–1.7 μm
2.	Bias voltage, V_R	200–250 V	20–40V	20–30 V
3.	Responsivity, R_{APD}	80–130 A/W	3–30 A/W	5–20 A/W
4.	APD Multiplication Factor, M	100–500	50–200	10–40
5.	k-factor, k_A	0.02–0.05	0.7–1.0	0.5–0.7
6.	Dark current, I_d	0.1–1nA	50–500 nA	1–5nA
7.	Rise time, τ_r	0.1–2 ns	0.5–0.8 ns	0.1–0.5 ns
8.	Bandwidth, BW	0.2–1 GHz	0.4–0.7 GHz	1–10 GHz

As $k_A << 1$ for Si, these APDs find applications in 800 nm optical band at bit rates ~ 100 Mbps. Ge or InGaAs APDs are used in 1300–1600 nm optical bands.

4.2.4 MSM Photodetector

In metal–semiconductor–metal (MSM) photodetector, a semiconductor absorbing region is inserted between two regions of a metal. Each metal–semiconductor interface forms a Schottky barrier that does not allow flow of electrons from metal side to semiconductor side. Electron–hole pairs (EHPs) are created through the phenomenon of photo-absorption. They flow towards metal contacts, which results in a photocurrent. This is, in fact, a measure of the incident optical power ($I_P \propto P_{in}$). Fig. 4.22 depicts the primitive structure of MSM photodetector.

Fig. 4.22 Basic structure of MSM photodetector

A set of uniform metal contacts, known as fingers, is placed on the flat surface of a semiconductor material. These fingers are biased interchangeably so that a relatively higher electric field occurs between them. Photons are incident on the surface of the semiconductor material that generate selectron–hole pairs. It has distinct advantages of higher bandwidth (of the order of 300 GHz) due to low capacitance, ease of fabrication, and planar structure. However, it has drawbacks of low responsivity (0.4–0.7 A/W) and reduction in active area. The progressive response of MSM photodetectors is usually different when illuminated from the top and back sides. A graded super lattice structure can be used to improve the performance of a MSM photodetector further. They offer responsivity of 0.6 A/W at 1300 nm, a relatively low dark-current density, and a rise time of approximately 16 ps. The planar structure of MSM photodetectors is the most appropriate structure for the purpose of monolithic integration.

Table 4.3 depicts typical values of important parameters of some commercially available photodiodes.

Table 4.3 Properties of commercially available photodiodes

Parameter	*UDT PIN5D*	*Epitaxx ETX75FJ*	*Fujitsu FID13Z81PZ*	*GAV30*	*Fujitsu FPD13U51KS*
Type	Silicon PD	InGaAs PIN PD	InGaAs PIN PD	Ge APD	InGaAs APD
Operating wavelength	850 nm	1300 nm	1550 nm	1300 nm	1300 nm
Responsivity	0.5 A/W	0.8 A/W	0.6 A/W	0.63 A/W	0.78 A/W
Rise time/cut-off frequency	15 ns	1.5 GHz	2 GHz	2 GHz	1 GHz
Dark current	10 nA	0.5 nA	1 nA	50 nA	20 nA
Capacitance	15 pF	1.5 pF	1 pF	1 pF	1 pF

Section Practice Problems

1. The quantum efficiency of a *p–n* photodiode is 70% for photons having energy value as 1.52×10^{-19} J. Determine the following:
(a) The operating wavelength of incident photons
(b) The optical power in order to obtain photocurrent = 3 μA at the operating wavelength determined in part (a).

[Ans.: (a) 1300 nm; (b) 4.1 μW]

2. The number of incident photonsper second on a *p–n* photodiode is 5×10^{10} at $\lambda = 900$ nm. On an average, the rate of collection of electrons at the output of photodiode is 2×10^{10} per second. Determine the following:
(a) The quantum efficiency
(b) The responsivity *[Ans.: (a) 0.4; (b) 0.3 A/W]*

3. A *p–i–n* photodetector produces one electron–hole pair for every two photons incident on it on an average at $\lambda = 850$ nm. Assume that all the photo-generated electrons are collected. Determine the following:
(a) The quantum efficiency of the photodiode
(b) The maximum band gap energy of the semiconductor in eV (assuming the incident wavelength to be a long-wavelength cut-off)
(c) The average photocurrent when the incident optical power is 10 μW.

[Ans.: (a) 50%; (b) 1.46 eV; (c) 3.43 μA]

4. The quantum efficiency of an avalanche photodiode (APD) is 0.4 at operating wavelength of 1300 nm. When illuminated with an optical power of 0.3 μW, it generates photocurrent = 6 μA due to avalanche gain. Compute the multiplication factor. *[Ans.: 47.6]*

5. A silicon reach-through APD, operating at a wavelength of 800 nm, exhibits a quantum efficiency of 90%, a multiplication factor of 800. Determine the rate at which photons should be incident on the device so that the output photocurrent due to avalanche gain is more than the specified dark current of 2 nA.

[Ans.: 1.74×10^7 per second]

4.3 Optical Receiver Block Diagram

The main function of an optical receiver is to detect the incoming optical signal (either analog or digitally modulated) and extract analog or digital information signal that is transmitted. The output of photodetector should be capable of driving the electronic circuit that follows in the system. Thus, the format of the modulation used by an optical transmitter decides the design of an optical receiver—an analog receiver or a digital receiver. Optical receivers that employ binary intensity modulation technique are known as digital optical receivers. A basic functional block diagram of digital optical receiver is shown in Fig. 4.23.

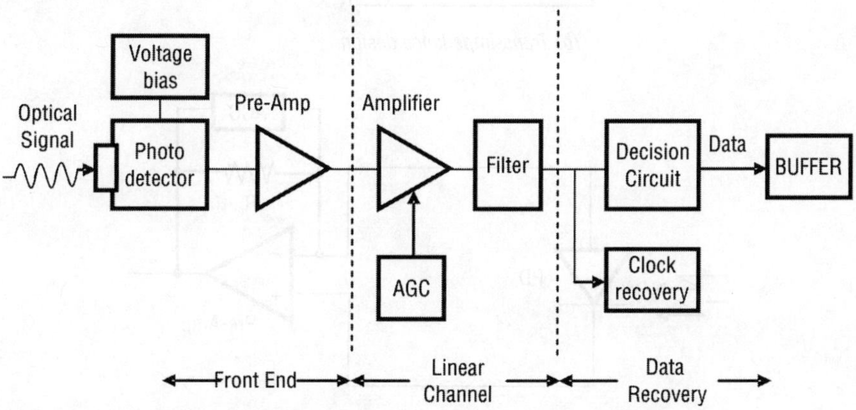

Fig. 4.23 Functional block diagram of digital optical receiver

The main entities of optical receiver can be arranged into three main groups:

1. *Front End*– It comprises of a photodetector and a pre-amplifier. Light from an optical fiber passes through coupling optics and falls on the sensitive area of a photodetector. A photodetector converts an incoming light into photocurrent. This photocurrent is converted into voltage and amplified by pre-amplifier.

   As far as digital receiver performance is concerned, there is a trade-off between sensitivity and data rate in the design of the front-end. The bandwidth of an optical receiver gets restricted due to component having the slowest response. If available bandwidth is significantly less than the desired bit rate, then it is not recommended to use a high-impedance front-end. In order to accomplish increased bandwidth, an equalizer (a type of high-pass filter) may be included in the front-end of the optical receiver. The noise sensitivity of the receiver limits its bandwidth. The equivalent circuit of an optical front end is shown in Fig. 4.24 (a) High impedance design; (b) Trans-impedance design; (c) Trans-impedance design with AGC.

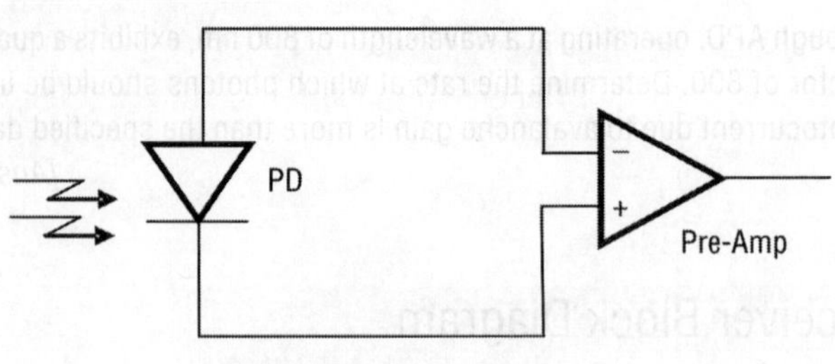

(a) High-impedance design

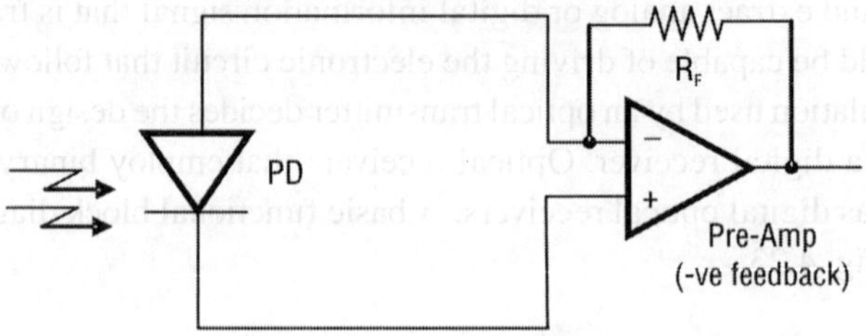

(b) Trans-impedance design

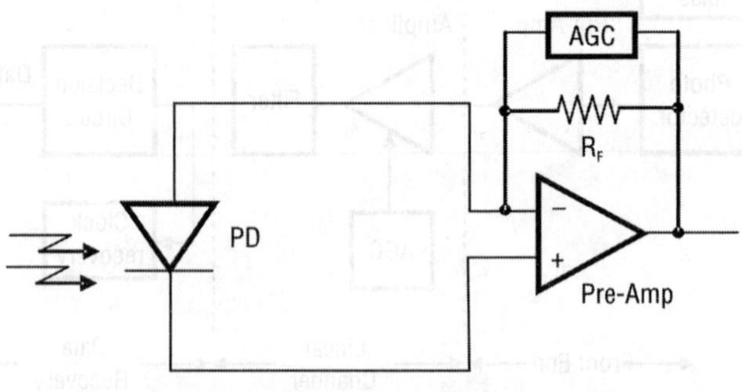

(c) Trans-impedance design with AGC

Fig. 4.24 An optical front-end–equivalent circuit

We know that in a p–i–n photodetector, the receiver sensitivity is directly proportional and thermal noise is inversely proportional to the external load resistance. Hence, in order to increase sensitivity and to decrease thermal noise of a photodetector, load resistance needs to be high. It is achieved by directly connecting a photodetector to an amplifier, as shown in Fig. 4.24(a). The bandwidth of a photodetector is inversely proportional to the load resistance. Therefore, in order to increase the bandwidth, one needs to reduce the load resistance. The function of a pre-amplifier is to convert current into voltage before amplifying the signal. The conversion is performed by an amplifier with negative feedback, as depicted in Fig. 4.24(b). To obtain a larger dynamic range (35–40 dB), trans-impedance pre-amplifier with AGC in feedback loop is used.

2. *Linear Channel Section–* It comprises of a pre-amplifier, a high-gain amplifier, followed by a low-pass filter. In certain situations, an equalizer is included prior to amplifier in order to enhance the limited bandwidth of front-end of an optical receiver. The gain of the amplifier is automatically adjusted so that the output voltage level on an average is limited and maintained

to a fixed level notwithstanding the average level of incident optical power at the photodetector. The use of low-pass filter after amplifier ensures that the output voltage pulse is shaped properly by reducing noise due to inter symbol interference. It also determines the bandwidth of the receiver, besides improving the sensitivity, or the signal-to-noise ratio (SNR).

3. *Data Recovery Section*– It mainly comprises of a decision making device and a clock recovery circuit. Its main function is to determine the logical meaning of the received signal. Typically, a decision circuit is a comparator driven by the input signal, as shown in Fig. 4.25.

Fig. 4.25 The function of decision circuit of a data recovery section

When the received signal is above threshold, the comparator's output is high. This means that the decision is made that this signal carrier logic HIGH, or 1 . When the signal is lower than threshold, the comparator's output is low, or almost zero. The clock recovery circuit provides the required timing synchronization of the received bit duration to the decision-making device.

In an analog optical receiver, the incident optical signal is detected by the photodetector, followed by a low-noise pre-amplifier and a post-amplifier having limiting action or automatic gain control (AGC). The signal is passed through an analog demodulator to recover the transmitted information. The performance of analog receiver is determined by the signal-to-noise ratio.

We know that when data is transmitted through any medium such as an optical fiber in optical fiber communication link, errors are likely to occur in the received data. Therefore, it is desirable to detect (or even correct) these errors in the receiver. There area number of error control (error detection and correction) techniques available which can be used. Examples of such error control codes are linear block codes (e.g., Hamming codes, cyclic redundancy check codes), convolutional codes, etc.

Facts to Know

Most of digital optical receivers have been designed to achieve the bit-error-rate (BER) of less than 10^{-9}.

4.4 Receiver Noise

We may recall that the objective of an optical receiver is to produce photocurrent, I_p corresponding to an incident optical power P_{in} on a photodetector. Assuming noise-free conversion of incident optical power into photocurrent, $I_p = (R \times P_{in})$, where R is the responsivity of the photodetector. However, in an optical receiver, there are various sources of noise which lead to continuous variations in the photocurrent, I_p, inspite of constant input optical power. These variations in the photocurrent do affect the overall performance of the optical receiver. Fig. 4.26 depicts various sources of noise present in the front-end of a typical optical receiver.

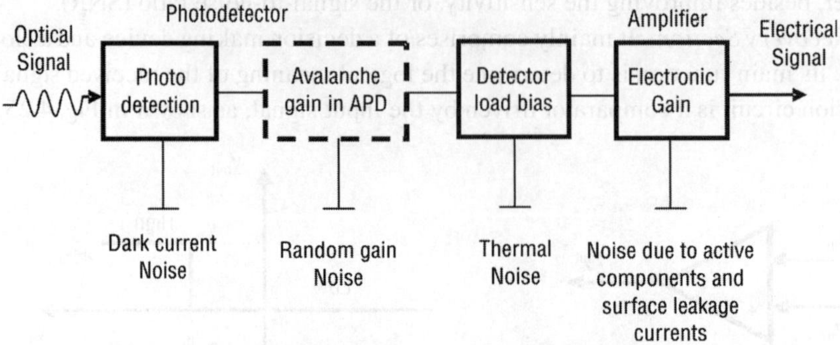

Fig. 4.26 Optical receiver front-end showing noise types

In almost all types of optical receivers, there are two prominent noise mechanisms, i.e., the shot noise and thermal noise.

(a) *Shot Noise*– Ideally an input optical power is constant. It implies that on an average, the number of photons per unit time does not vary. But, in fact, the actual number of photons arriving at a particular time is unknown and so is a completely random variable. Hence, the number of photo-generated electrons at any particular instance is a random variable. In addition, the number of electrons producing photocurrent will vary because of their random recombinations and absorptions. Generally, the actual number of electrons are not exactly same as the average number of electrons. This contributes to noise, known as shot noise. Thus, the photocurrent produced by a constant incident optical power signal can be expresses as

$$I(t) = I_p + i_s(t) \tag{4.16}$$

where, $I_P = R \times P_{in}$ signifies the average value of the photocurrent produced, and $i_s(t)$ represents the variations in the photocurrent due to shot noise. It may be noted that the spectral density of shot noise does not change at all. So, we can say that shot noise resembles the white noise whose spectral density does not depend on frequency.

The root-mean-square (RMS) value of shot-noise current, $i_s(t)$, is the representative characteristic of a shot-noise phenomenon. In practice, noise is also represented by its current RMS value per unit of bandwidth, i.e., the bandwidth normalized RMS value of noise current, i_N. For shot-noise current,

$$i_{sN}\left(A/\sqrt{Hz}\right) = \frac{i_s}{\sqrt{BW_{PD}}} = \sqrt{2qI_p} \tag{4.17}$$

The dark current I_d also generates shot noise. In fact, the dark current represents the value of the photocurrent which is produced even in the absence of any incident optical power. Surface leakage currents and currents due to generation–recombination of electron–hole pairs within the depletion region even in the absence of incident optical power contribute to dark current in photodetectors. Therefore,

$$I(t) = I_p + I_d + i_s(t) \tag{4.18}$$

The RMS value of the shot noise current can be expressed as

$$i_{sN}\left(A/\sqrt{Hz}\right) = \sqrt{2q\left(I_p + I_d\right)} \tag{4.19}$$

(b) *Thermal noise*– Due to change in operating temperature of photodetectors, the actual number of electrons may not be identical to the average number of electrons, T, any instant of time. This gives rise to noise, known as thermal noise. Electron motion due to temperature occurs in a random manner, which gives rise to variations in the photocurrent even when no external bias voltage is applied to the photodetector. There is another type of variation in the photocurrent due to the load resistor, which forms part of the front-end circuit of an optical receiver. This gives rise to an additional noise component which is often referred to as Johnson noise (sometimes known as Nyquist noise), which is similar to thermal noise.

Therefore, we can say that the total photocurrent, *I(t)*, is the linear sum of the actual photocurrent, shot noise current, and thermal noise current. That means,

$$I(t) = I_p + i_s(t) + i_{Th}(t) \tag{4.20}$$

where, $i_{Th}(t)$ represents the variations in the photocurrent due to thermal noise. Thermal noise current can be modelled as a stationary Gaussian random process having its spectral density expressed as

$$S_{Th}(f) = \frac{2kT}{R_L} \tag{4.21}$$

where, $S_{Th}(f)$ represents the spectral density of $i_{Th}(t)$, k = 1.38 × 10-23 J/K, T represents the absolute temperature expressed in K, and R_L represents the load resistor.

We know that the bandwidth-normalized RMS value of thermal noise current,

$$i_{ThN}\left(A/\sqrt{Hz}\right) = \sqrt{(4kT/R_L)} \tag{4.22}$$

(c) *Dark-current noise*– The RMS value of darkcurrent noise,

$$i_d\left(A/\sqrt{Hz}\right) = \sqrt{2qI_d BW_{PD}} \tag{4.23}$$

The RMS value per unit bandwidth is given by

$$i_{du}\left(A/\sqrt{Hz}\right) = \sqrt{2eI_d} \tag{4.24}$$

The dark current noise is essentially at the low bit rate (around 100 Mbps) while at high speed (> 1 Gbps) even 100 nA dark current does not produce an essential contribution to total noise.

(d) *1/f noise*– A photodiode also generates another type of noise that occurs in complete darkness (absence of incident optical power) other than the dark current noise in a photodetector. Known as 1/f noise, its RMS value per unit bandwidth is inversely proportional to frequency, which means this is not white noise.

$$i_{1/fN}\left(A/\sqrt{Hz}\right) = \left(K_{1/f}I^{\alpha}\right)/f^{\beta} \tag{4.25}$$

where, $K_{1/f}$, α, β are found empirically (α ~ 2, β ~ 1–1.5. This noise is important only in low frequency range and can be neglected when modulating frequency is greater than 100 Hz.

Fig. 4.27 depicts generalized equivalent circuit of noise in a photodetector.

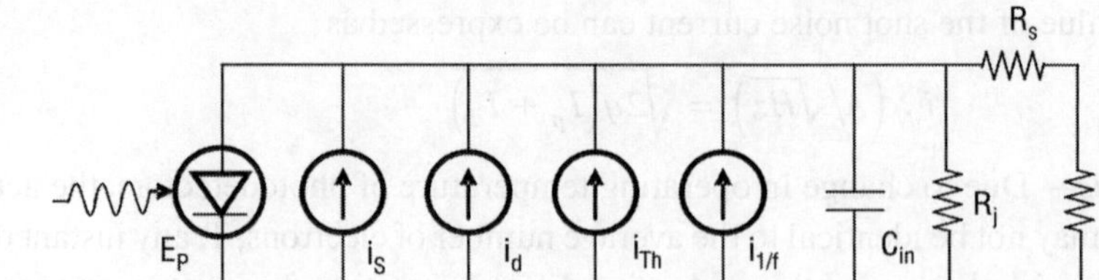

Fig. 4.27 Generalized equivalent circuit of noise in photodetector

It may be noted that each noise component is an independent random process approximately by Gaussian statistics. The overall noise is given by

$$i_{noise} = \sqrt{\left(i_s^2 + i_d^2 + i_{Th}^2 + i_{1/f}^2\right)} \tag{4.26}$$

The signal-to-noise ratio (SNR) is one of the performance determining parameter of an optical receiver. It is defined as the ratio of average signal power and the noise power, i.e.

$$SNR = \frac{\text{average signal power}}{\text{noise power}} = \frac{{I_p}^2}{i_{noise}^2} \tag{4.27}$$

Since $I_p = RP_{in}$, therefore

$$SNR = \frac{R^2 {P_{in}}^2}{i_{noise}^2} \tag{4.28}$$

where, R is the responsivity of a photodiode.

As stated earlier, it is thermal noise that contributes maximum to the overall noise present in a p–i–n photodiode as given by the expression

$$i_{noise}^2 = i_s^2 + i_d^2 + i_{Th}^2 + i_{1/f}^2 \tag{4.29}$$

Neglecting all other terms except thermal noise term, SNR limited by thermal noise, SNR_{Th} is given by

$$SNR_{Th} = \frac{R^2 {P_{in}}^2}{i_{Th}^2} \tag{4.30}$$

Substituting $i_{Th} = \sqrt{\dfrac{4kT\left(BW_{PD}\right)}{R_L}}$, (4.31)

$$SNR_{Th} = R^2 {P_{in}}^2 \times \frac{R_L}{4kT\left(BW_{PD}\right)} \tag{4.32}$$

where, BW_{PD} represents bandwidth of photodetector.

To reduce thermal noise and its influence on SNR, the load resistance R_L should be increased. If we wish to quantify the effect of thermal noise, then we define a term called **noise-equivalent power (NEP)**, which is the minimum incident optical power per unit bandwidth so as to give unity value of signal-to-noise ratio (SNR). Mathematically,

$$NEP = \frac{P_{in}}{\sqrt{BW_{PD}}} = \sqrt{\left(\frac{4kT}{R^2 R_L}\right)} \tag{4.33}$$

By definition, the optical receiver parameter detectivity is the inverse of NEP.

SNR in p–i–n Receiver

Consider the case in which the performance of an optical receiver is governed by its shot noise only, i.e., $i_s^2 \gg i_{in}^2$. Since i_s^2 increases linearly as the value of incident optical power P_{in} increases, the limit for shot-noise can be achieved with large value of P_{in}. Neglecting the dark current, I_d, we can write the expression for SNR due to shot-noise as

$$SNR_S = \frac{RP_{in}}{2qBW_{PD}} = \frac{\eta P_{in}}{2hcBW_{PD}} \tag{4.34}$$

It may be noted that SNR_s is dependent upon quantum efficiency η, bandwidth of the photodetector, and the photon energy *hc*.

SNR in APD Receiver

APD-based optical receivers generally offer a relatively higher value of signal-to-noise ratio (SNR) for identical level of incident optical power. We may recall that in APDs internal gain increases the photocurrent which is then given as

$$I_P = M \times R \times P_{in} = R_{APD} \times P_{in} \left(\because R_{APD} = M \times R\right) \tag{4.35}$$

where, M represents the multiplication factor of APD.

There is no change in thermal noise in APD-based optical receivers from other type of optical receivers. But amplification of photocurrent due to impact–ionization process raises the shot noise. Shot noise increase due to random generation of secondary electron–hole pairs which makes its multiplication factor (M) also random.

$$i_s^2(APD) = M^2\left[2eF_s R_{APD} BW_{APD}\right] \tag{4.36}$$

where, F_s is excess noise factor of APD.

Case I: Let $i_s^2 \gg i_{Th}^2$, i.e. shot noise is much greater than thermal noise. Then

$$SNR(APD)_s = \frac{R_{APD} P_{in}}{2eF_s BW_{APD}} \tag{4.37}$$

Case II: Let $i_s^2 \ll i_{Th}^2$, i.e. the mean-square value of thermal-noise current is many times greater than mean-square value of shot-noise current. Then

$$SNR(APD)_{Th} = \frac{M^2 R^2 P_{in}}{(4kT/R_L)BW_{APD}} \tag{4.38}$$

That is, $SNR(APD)_{Th} \propto M^2$.

Example 4.17 Shot-noise Current of Photodiode

A semiconductor photodetector has responsivity $R = 0.5$ A/W for an incident optical power of 10 μW. It has dark current of 2 nA. Determine the mean-square value of the shot-noise current for a specified bandwidth of 1 MHz.

Solution:

We know that mean-square shot-noise current, $i_{sN}^2 = 2q\left(I_p + I_d\right)BW$

where, $q = 1.6 \times 10^{-19}$C, $I_p = RP_{in}$, and $I_d = 2$ nA.

For the given $R = 0.5$ A/W and $P_{in} = 10$ μW, we get

$$I_p = (0.5A / W) \times \left(10 \times 10^{-6} W\right) = 5 \times 10^{-6} A$$

Hence, $i_{sN}^2 = 2 \times 1.6 \times 10^{-19} \left(5 \times 10^{-6} + 2 \times 10^{-9}\right) \times 1 \times 10^6 = 1.6 \times 10^{-18} A$

Example 4.18 Dark Current and Thermal Noise Current

An optical receiver has 20 MHz bandwidth operating at a wavelength of 1100 nm. It uses an InGaAs p–i–n photodiode producing a photodiode current of 4 nA with quantum efficiency of 90%. The load resistor of the circuit is 1 kΩ. Assuming negligible surface leakage current, find the value of dark current and thermal noise current if the incident optical power is 300 nW.

Solution:

We know that dark current, $i_d = \sqrt{2eBI_p}$

For the given $B = 20$ MHz and $I_p = 4$ nA, we have

$$i_d = \sqrt{2 \times 1.6 \times 10^{-19} \times 20 \times 10^6 \times 4 \times 10^{-9}} = 0.16\ nA \qquad \textbf{Ans.}$$

We know that thermal noise current, $i_{TH} = \sqrt{\dfrac{4kTB}{R_L}}$

For the given $B = 20$ MHz and $R_L = 1$ kΩ, we have

$$i_{TH} = \sqrt{\frac{4 \times 1.38 \times 10^{-23} \times 300 \times 20 \times 10^6}{1 \times 10^3}} = 18\ nA \qquad \textbf{Ans.}$$

Section Practice Problems

1. The maximum 3-dB bandwidth permitted by an InGaAs photoconducting detector is 450 MHz when the electron transit time in the device is 6 ps. Assuming quantum efficiency of 75%, determine
   (a) the gain, G
   (b) the output photocurrent when an optical power of 5 μW at a wavelength of 1300 nm is incident on it.
   [Ans.: (a)59; (b) 232 μA]
2. An InGaAs *p–i–n* photodiode operates at room temperature at a wavelength of 1300 nm. Its quantum efficiency is 70% and the incident optical power is 500 nW. Assume that the primary dark current of the device is 5 nA, load resistor is 1 kΩ, and the effective bandwidth is 25 MHz. Determine the rms values of shot noise current, dark current, and thermal noise current.
   [Ans.: 1.715 nA; 0.2 nA; 20.34 nA]

3. Calculate the signal-to-noise ratio at the input of an optical receiver amplifier if the photodiode current is 0.3662 μA, and the sum of mean square noise currents is 4.17×10^{-16} A^2. *[Ans.: 324]*

4.5 Receiver Sensitivity

By definition, the optical receiver sensitivity is the minimum average input received optical power that a photodetector can detect for it to operate at a given bit-error rate (BER) usually of the order of 10^{-9}.

We know that BER is defined as the ratio of the number of bits received in errors and the total number of bits transmitted. It signifies the probability of incorrect detection of a received bit by the decision device of the optical receiver. BER is one of the key performance indicators for digital optical receivers. Fig. 4.28 illustrates the inconsistent signal received at the input of the decision device which is sampled at the instant determined by the local clock recovery signal.

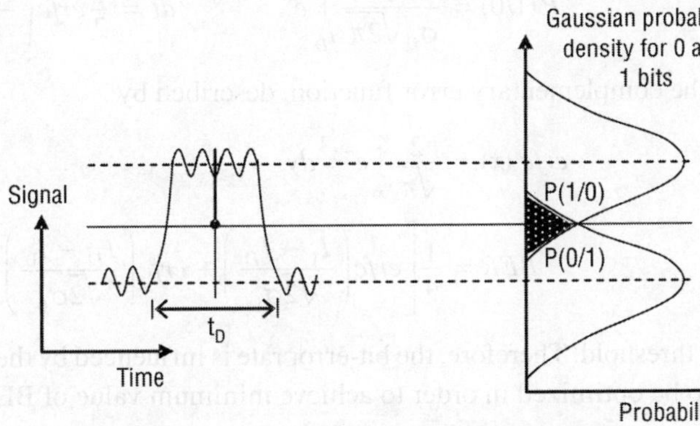

Fig. 4.28 Illustration of BER concept

In this figure, the condition probability *P*(1/0) denotes the probability of deciding 1 when 0 is received (incorrect decision), and likewise the conditional probability *P*(0/1) denotes the probability of deciding 0 when 1 is received by the decision device. This also depicts the dependence of *P*(1/0) and *P*(0/1) on the probability density function, denoted by *p*(I), where *p*(I) is dependent on the noise source statistics that contribute to variations in the output photocurrent. Let *p*(0) and *p*(1) represent the corresponding probabilities of receiving bits 0 and 1. Therefore,

$$BER = p(0)P(1/0) + p(1)P(0/1) \tag{4.39}$$

Since $p(0) = p(1) = 1/2$ as bits 0 and 1 are equally likely to occur.

$$\therefore \quad BER = \frac{1}{2}\left[P(0/1) + P(1/0)\right] \tag{4.40}$$

As we know that net photocurrent, $I(t) = I_p + i_s(t) + i_{Th}(t)$, where thermal noise $i_{Th}(t)$ is completely defined by Gaussian statistics having zero mean value as well as zero variance, given by $\sigma_{Th}^2 = \left(\frac{akT}{R_L}\right) \times BW$. The statistics of shot noise contribution $i_s(t)$ is also approximately Gaussian for p–i–n photodetectors as well as APDs having different expression for variance σ_s^2 as defined below:

$$\sigma_s^2\Big|_{p-i-n} = 2qI_p BW \tag{4.41}$$

$$\sigma_s^2\Big|_{APD} = 2qM^2F_A\left(RP_{in}\right)BW \tag{4.42}$$

where, F_A represents the excess noise factor in case of APD.

So, $\sigma^2 = \sigma_s^2 + \sigma_{Th}^2$ is also a Gaussian random variable. If σ_1^2 and σ_0^2 are the corresponding variances for bit 1 and 0 respectively, then the conditional probabilities can be written as

$$P(1/0) = \frac{1}{\sigma_1\sqrt{2\pi}}\int_{-\infty}^{I_D} e^{\left(-\frac{(I-I_1)^2}{2\sigma_1^2}\right)}dt = \frac{1}{2}erfc\left(\frac{I_1 - I_D}{\sqrt{2}\sigma_1}\right) \tag{4.43}$$

$$P(1/0) = \frac{1}{\sigma_0\sqrt{2\pi}}\int_{I_D}^{\infty} e^{\left(\frac{(I-I_D)^2}{2\sigma_0^2}\right)}dt = \frac{1}{2}erfc\left(\frac{I_D - I_o}{\sqrt{2}\sigma_0}\right) \tag{4.44}$$

where, *erfc* represents the complementary error function, described by

$$erfc\,(x) = \frac{2}{\sqrt{\pi}}\int_x^{\infty} e^{-y^2}\,dy \tag{4.45}$$

$$\therefore \qquad BER = \frac{1}{4}\left[erfc\left(\frac{I_1 - I_D}{\sqrt{2}\sigma_1}\right) + erfc\left(\frac{I_D - I_0}{\sqrt{2}\sigma_0}\right)\right] \tag{4.46}$$

where, I_D is the decision threshold. Therefore, the bit-error rate is influenced by the decision threshold value I_D, which needs to be optimized in order to achieve minimum value of BER. It is given by

$$I_D\,(opt) = \frac{\sigma_0 I_1 + \sigma_1 I_0}{\sigma_0 + \sigma_1} \tag{4.47}$$

When $\sigma_1 = \sigma_0$, then

$$I_D = \frac{\sigma_0 I_1 + \sigma_0 I_0}{\sigma_0 + \sigma_0} = \frac{I_1 + I_0}{2} \tag{4.48}$$

This implies that the decision threshold value must be set almost in the middle. It is applicable for p-i-n photodetector based optical receivers where thermal noise is the dominating noise source $\left(i.e.,\ \sigma_T \gg \sigma_s\right)$ which does not depend on the average value of the photocurrent.

Due to linear variation of σ_s^2 with average value of the photocurrent, the shot noise is relatively more for received bit 1 than for received bit 0. For APD-based optical receivers, it is possible to minimize the value of the BER by choosing the decision threshold value as per the expression given above.

Let $$Q \equiv \frac{I_D - I_0}{\sigma_0} = \frac{I_1 - I_D}{\sigma_1} \tag{4.49}$$

By solving these equations, we get

$$\sigma_0 Q = I_D - I_0$$

and $$\sigma_1 Q = I_1 - I_D$$

$$\Rightarrow \quad (\sigma_1 + \sigma_0)Q = I_1 - I_0$$

$$\Rightarrow \quad Q = \frac{I_1 - I_0}{\sigma_1 + \sigma_0} \tag{4.50}$$

The minimum value of BER thus obtained depends solely only on Q parameter as

$$BER = \frac{1}{2} erfc\left(\frac{Q}{\sqrt{2}}\right) \approx \frac{e^{-Q^2/2}}{\sqrt{2\pi}Q} \tag{4.50}$$

Fig. 4.29 shows BER versus Q parameter variations.

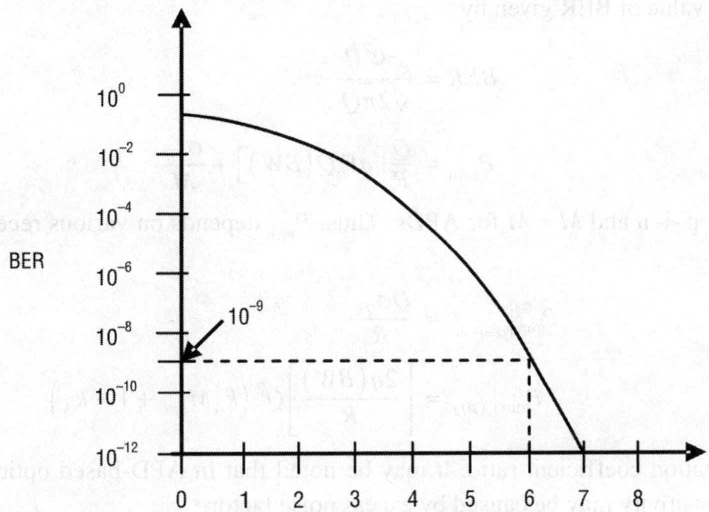

Fig. 4.29 BER vs Q

It can be seen that as the Q parameter increases, the BER improves significantly. The BER value can be achieved even less than 10^{-12} for Q greater than 7. It is also observed that at Q = 6, BER ~ 10^{-9}, and the sensitivity of the optical receiver relates to the average incident optical power for which Q ~ 6.

Facts to Know

The performance of the optical receiver is generally characterized by considering the value of BER as a function of average received optical power. For example, a measure of sensitivity of an optical receiver corresponds to the average received optical power for which BER = 10^{-9}.

Minimum Received Optical Power

The minimum received power (P_{min}) is the average optical power received in both bit 1 and bit 0, that is,

$$P_{\min} = \frac{(P_1 + P_0)}{2} \tag{4.51}$$

Assume $P_0 = 0$ and neglecting i_d, we have

$$I_1 = RP_1 = 2RP_{\min} \qquad (\because \quad P_1 = 2P_{\min}) \tag{4.52}$$

The RMS values of noise currents, represented by σ_1 and σ_0, include the contributions made by shot noise and thermal noise, respectively. These can be expressed as

$$\sigma_1 = \sqrt{\sigma_s^2 + \sigma_{Th}^2} \text{ and } \sigma_0 = \sigma_{Th} \tag{4.53}$$

where, $\sigma_s^2 = 2qM^2F_A(2RP_{\min})BW$ and $\sigma_{Th}^2 = (4kT/R_L)F_n(BW)$

$$\therefore \qquad Q = \frac{I_1}{\sigma_1 + \sigma_0} = \frac{2RP_{\min}}{\sqrt{\sigma_s^2 + \sigma_{Th}^2} + \sigma_{Th}} \tag{4.54}$$

For a specific value of BER given by

$$BER = \frac{e^{-Q^2/2}}{\sqrt{2\pi}Q} \tag{4.55}$$

$$P_{\min} = \frac{Q}{R}\left[qF_nQ(BW)\right] + \frac{\sigma_{Th}}{M} \tag{4.56}$$

where, $M = 1$ for p–i–n and $M = M$ for APDs. Thus, P_{min} depends on various receiver parameters. Accordingly,

$$P_{\min}\big|_{p-i-n} \approx \frac{Q\sigma_{Th}}{R} \tag{4.57}$$

$$P_{\min}\big|_{APD} = \left[\frac{2q(BW)}{R}\right]Q^2\left(k_A M_{opt} + 1 - k_A\right) \tag{4.58}$$

where, k_A is ionization coefficient ratio. It may be noted that in APD-based optical receivers, the degradation in sensitivity may be caused by excess noise factor.

$$P_{\min}\big|_{ideal} = \left[\frac{q(BW)}{R}\right]Q^2 \tag{4.59}$$

Quantum Limit

Let us consider the probabilities P(1/0) and P(0/1), assuming that no photons have been received with bit 0 and one photon is received with bit 1. The probability of identifying a bit as 1 when bit 0 arrives as P(1/0) is zero because no photons have been received in this case. The probability of identifying a bit as 0 when bit 1 arrives as P(0/1) is equal to $P(0) = e^{-N_p}$ where N_p is number of photons because $n = 0$.

$$\therefore \qquad BER = \frac{1}{2} \times \left[P(0/1) + P(1/0)\right]$$

$$\Rightarrow \qquad BER = 0.5\left[0 + e^{-N_p}\right] = 0.5e^{-N_p} \tag{4.60}$$

Thus, the BER depends on absolute minimum number of photons that are essential for detection of a bit. It is known as the quantum limit.

If one photon is received, i.e., $N_p = 1$, then

$$BER = 0.5e^{-1} = 0.18$$

This means that 18 out of 100 bits received are incorrectly interpreted.

If $N_p = 10$, then $\quad BER = 0.5e^{-10} = 2.3 \times 10^{-5}$

If $N_p = 20$, then $\quad BER = 0.5e^{-20} = 1.03 \times 10^{-9}$

If $N_p = 26$, then $\quad BER = 0.5e^{-10} = 0.5e^{-26} = 10^{-12}$

Using these values, a plot between BER and sensitivity P_{min} is shown in Fig. 4.30.

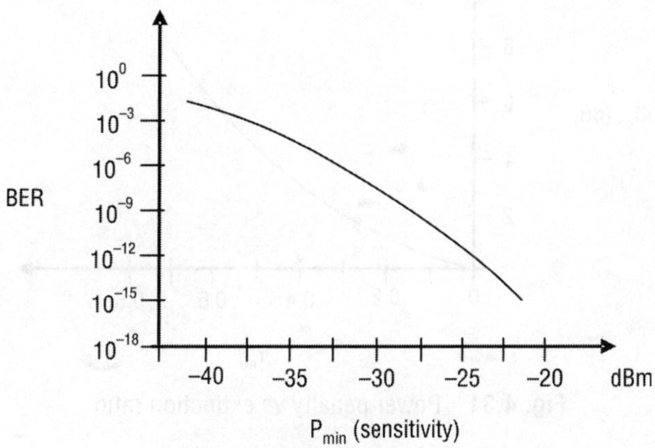

Fig. 4.30 BER vs P_{min} (dB)

Sensitivity Degradation

In an ideal situation, the incident optical signal at the input of an optical receiver may be considered as a bit stream in which no energy is contained in bit 0 and an optical rectangular pulse of constant energy is contained in bit 1. The degradation in sensitivity energy is then only due to receiver noise. However, in practice,

- The optical signal produced by an optical transmitter itself departs from an ideal bit stream as described above
- In addition, noise is introduced at various stages of optical amplifiers when a stream of optical pulses representing bit stream is transmitted via the optical fiber link.

As a result of these practical aspects, the minimum average optical power required by the optical receiver increases. This increase in average received optical power is often termed as power penalty, defined as

$$\delta_{ex} = 10 \log\left(\frac{1 + r_{ex}}{1 - r_{ex}}\right) \tag{4.61}$$

where, r_{ex} is known as extinction ratio which is defined as the ratio of the OFF-state power (P_0) and the ON-state power (P_1). That is,

$$r_{ex} = \frac{P_0}{P_1} \tag{4.61}$$

In semiconductor lasers, P_0 depends on the bias current I_b and the threshold current I_{Th}. Thus, one of the prominent sources of power penalty can be attributed to the energy carried by 0 bits. It simply implies that some optical power is always transmitted by most of the optical transmitters even in the OFF state. Generally, $P_0 << P_1$, therefore, $r_{ex} << 1$. Fig. 4.31 shows a plot between power penalty and extinction ratio.

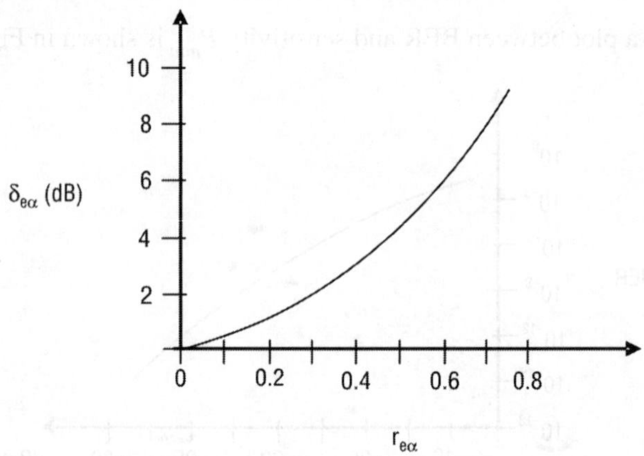

Fig. 4.31 Power penalty vs extinction ratio

It is seen that there is an increase in power penalty with corresponding increase in the extinction ratio. For an APD-based optical receiver, the power penalty is twice for same value of extinction ratio.

Mainly, there are two causes of degradation in optical receiver sensitivity even when there is no transmission of optical signal through the optical fiber. These are:

1. *Intensity noise*– In all practical situations, the optical signal generated by any optical transmitter suffers from variations in output optical power, called intensity noise. The optical receiver converts power fluctuations with current fluctuations which add to shot noise and thermal noise. As a result, the receiver SNR is degraded. We know that

$$\therefore \quad \sigma^2 = \sigma_s^2 + \sigma_{Th}^2 + \sigma_I^2 \tag{4.62}$$

where, $\sigma_I = R \times P_{in} \times r_I$; r_I being the amount of noise level present in the received optical signal. It is correlated with the relative intensity noise (RIN) of an optical transmitter by the following expression

$$r_I^2 = \frac{1}{2\pi}\int_{-\infty}^{\infty} RIN(\omega)\,d\omega \tag{4.63}$$

The relative intensity noise, r_I of the optical transmitter is given as the inverse of the SNR of optical signal generated by it. Normally, the value of SNR is greater than 20 dB, and the value of relative intensity noise r_I is less than 0.01. Accordingly,

$$Q = \frac{2RP_{\min}}{\sqrt{\left(\sigma_s^2 + \sigma_{Th}^2 + \sigma_I^2\right)} + \sigma_{Th}} \tag{4.65}$$

where, $\sigma_s^2 = 4qRP_{\min}(BW)$, $\sigma_{Th}^2 = (4kT/R_L)F_n(BW)$, $\sigma_I = 2r_I RP_{\min}$

Fig. 4.32 shows a plot between power penalty versus intensity noise parameter r_I.

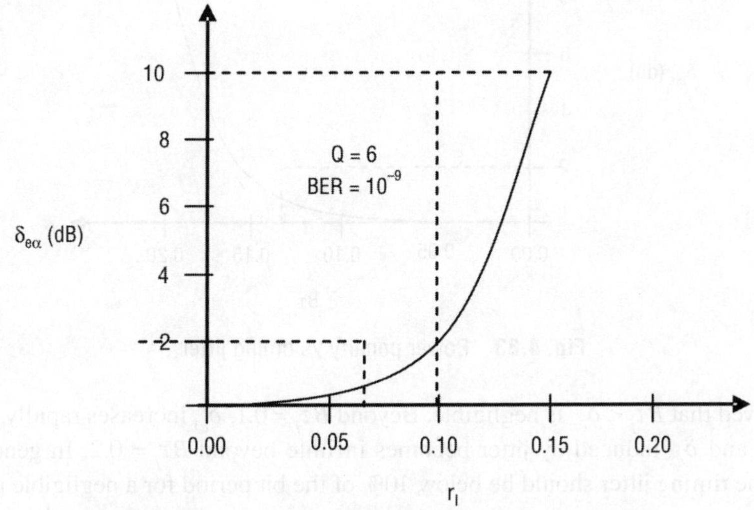

Fig. 4.32 Power penalty vs intensity noise parameter

From the plot, the following observations can be made:

- For $r_I < 0.01$, $\delta_{ex} < 0.02$ dB. Thus, in practice, the effect of noise due to optical transmitter is almost negligible.
- For $r_I = 0.1$, $\delta_{ex} = 2$ dB
- For $r_I = 0.167 = \dfrac{1}{Q}$, $\delta_{ex} = \infty$

What is the significance of power penalty to be infinity? Obviously, it simply means that the optical receiver is just unable to operate at specified BER inspite of increasing the received optical power indefinitely.

Other sources of intensity noise are as follows:

(i) Intensity noise introduced by in-line optical amplifiers
(ii) Mode partition noise due to fiber dispersion in a multimode semiconductor laser
(iii) Optical feedback due to parasitic reflections which might occur throughout the length of the optical fiber link.

2. *Timing jitter*– Timing jitter is defined as the fluctuations in the sampling time from one bit to another bit in the received bit stream because of the noisy nature of the input to the clock–recovery circuit in an optical receiver. The SNR is degraded because fluctuations in the sampling time lead to additional fluctuations in the signal. If the bit is not sampled at the bit center, the sampled value is reduced by an amount that depends on the timing jitter Δt. Since Δt is a random variable, the reduction in the sampled value is also random. The SNR is reduced as a result of such additional fluctuations, and the receiver performance is degraded. The SNR can be maintained by increasing the received optical power. This increase is the power penalty induced by time jitter. Fig. 4.33 shows power penalty vs the timing jitter parameter $B\tau_j$.

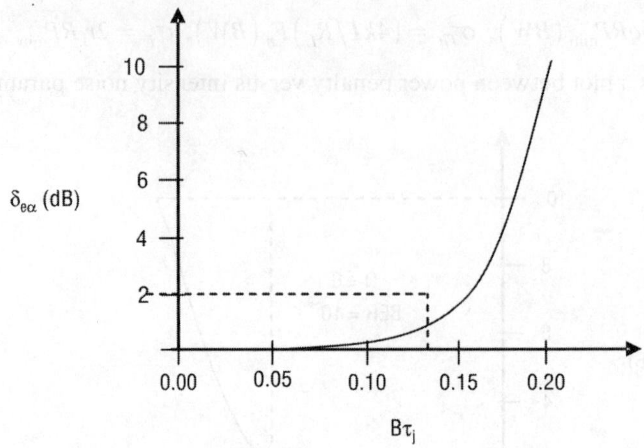

Fig. 4.33 Power penalty vs timing jitter

It is observed that $B\tau_j < \delta_{ex}$ is negligible. Beyond $B\tau_j = 0.1$, δ_{ex} increases rapidly. At $B\tau_j = 0.16$, $\delta_{ex} = 2dB$ and δ_{ex} induced by jitter becomes infinite beyond $B\tau_j = 0.2$. In general, the RMS value of the timing jitter should be below 10% of the bit period for a negligible power penalty.

Facts to Know

There is a substantial degradation (25–30 dB) in the optical receiver sensitivity from the specified quantum limit for optical fiber communication systems that operate at relatively higher bit rate. It is mainly because of dispersion which depends on the bit rate R_b and the length of the optical fiber link. It increases with ($R_b \times L$) product.

Points to Remember

- Photodetector is the core component of an optical receiver which functions on the principle of photoelectric effect to converts optical signals into electrical signals.
- A photodetector is a semiconductor photodiode which is mostly fabricated using InGaAs material.
- There are different types of semiconductor photodiodes based on their structure; for example, p–n photodiodes, p–i–n photodiodes, and avalanche photodiodes (APD).
- Metal–semiconductor–metal (MSM) photodetectors are mostly suitable for fabrication of monolithic circuit integration in several applications such as wavelength–division multiplexers (WDMs) and optical regenerators.
- The responsivity is the key performance parameter of the photodiode which depends upon the energy band gap of the semiconductor material used, the doping concentrations, the operating wavelength, and the thickness of the p, i and n regions.
- To obtain higher quantum efficiency and maximum optical receiver sensitivity, the width of the depletion region must be increased so as to maximize the absorption of photons within it.
- In the wavelength regions 1330 nm and 1550 nm, InGaAs p–i–n photodiodes exhibit relatively higher value of responsivity and quantum efficiency.
- Photodetectors in fiber-optics systems are usually p–i–n photodiodes or avalanche photodiodes (APDs).
- Germanium or Silicon APDs have operating wavelength region of 800 nm, whereas InGaAs APDs have operating wavelength of 1550 nm.

- The avalanche photodiodes offer relatively higher responsivity as well as sensitivity due to the avalanche multiplication.
- In photodetector, there are mainly three sources of noise—shot noise due to random arrival of photons (optical signals); thermal noise due to thermally induced random motion of charge carriers; and dark current due to reverse leakage current flow even when there is no light.

Important Equations

The photocurrent of a photodiode, $I_p = RP_{in}$; *where R represents the responsivity expressed in Amp/Watts, and P_{in} represents the optical power expressed in Watts that is incident on it.*

The quantum efficiency of a photodiode, $\eta = \frac{N_e}{N_p} = R \times \frac{hc}{\lambda}$; *where N_e is the number of electrons flowing per unit time, N_p represents the number of photons, R is the responsivity, and λ is the wavelength.*

The bandwidth of a photodetector, $BW_{PD} = \frac{1}{2\pi\left(\tau_{t_r} + \tau_{RC}\right)}$; *where τ_{t_r} is the transit time given by the relationship $\tau_{t_r} = \frac{W}{v_d}$, W being the depletion region width and v_d being the drift velocity; and $\tau_{RC}\tau_{RC}$ represents the time constant of equivalent RC circuit given by the relationship $\tau_{RC} = (R_L + R_s)C_p$, R_L being the external load resistance, R_s the internal series resistance, and C_p the parasitic capacitance.*

The responsivity of an APD, $R_{APD} = M \times R = M \times \left(\frac{\eta\lambda}{1.24}\right)$; *where M denotes the multiplication factor, also known as average APD gain.*

The overall noise in a photodetector, $i_{noise} = \sqrt{\left(i_s^2 + i_d^2 + i_{Th}^2 + i_{1/f}^2\right)}$; *where i_s represents the fluctuations in the photocurrent due to shot noise, i_d represents the RMS value of the dark-current noise, i_{Th} represents the fluctuation in the photocurrent due to thermal noise, and $i_{1/f}$ represents the noise per unit bandwidth.*

The signal-to-noise ratio (SNR) of a photodetector receiver, $SNR = \frac{R^2 P_{in}{}^2}{i_{noise}^2}$; *where the notations have the usual meanings as given above.*

Key Terms with Definitions

Avalanche photodiode (APD) A type of photodiode in which the amplification of output photocurrent takes place internally due to avalanche multiplication of charged carriers within the junction regions.

Bit-error rate (BER) The probability of incorrect identification of a bit (0 or 1) by the decision devise of the digital optical receiver.

Dark current In a reverse biased photodiode, the photocurrent that flows even when there is no incident optical power.

Extinction ratio The ratio of the OFF-state to ON-state optical power levels corresponding to received bit 0 and 1 at the photodetector.

Intensity noise Fluctuations in the optical power emitted by transmitter which gets converted into current fluctuations by the optical receiver which adds to photocurrents due to shot noise and thermal noise, causing degradation in receiver SNR.

Jitter Signifies the small and fast changes in the timings of an optical pulse waveform, mainly due to noise, external supply voltages, variations in the characteristics of the components used, and deficient synchronizing circuits.

Noise-equivalent power(NEP) The minimum received optical power per unit bandwidth which can result in unity signal-to-noise ratio.

Photodetector An opto-electronic transducer that converts optical signal into electrical signal.

Photodiode A semiconductor device that forms a photodetector.

Photon Represents a particle of light, or a quantum of electromagnetic energy.

Power penalty Signifies the requirement of increasing the average received optical power due to addition of noise at optical amplifiers which may result in degradation in receiver sensitivity when an optical signal is transmitted through the optical fiber.

Quantum efficiency The ratio of generation rate of primary charged carriers such as electron–hole pairs to the rate of incident photons in a photodetector.

Receiver Same as optical receiver. An optical device comprising of a photodetector for converting an optical-to-electrical signal as well as electronic circuitry for signal processing.

Receiver sensitivity The minimum value of received optical power needed to achieve an acceptable BER.

Responsivity The ratio of an output electrical current of the photodetector to its input optical power, expressed in Amp/Watts (A/W).

Shot noise A type of noise created by variations in the photocurrent due to basic nature of electrons.

Signal-to-noise ratio (S/N, or SNR) A measure of received signal quality that is defined as the ratio of available signal power to the total noise power at the same point in the device. It signifies the amount by which the signal level is higher than the noise level.

Timing jitter The sampling time fluctuations from one received bit to another received bit mainly due to noisy clock-recovery circuit.

Short Answer Type Questions

1. **What do you mean by optical detectors?**

   An optical detector is an opto–electronic transducer that has the capability to convert an incident optical signal (in the form of light) into an electrical signal (output photocurrent) In general, transducers are devices that convert input energy of one form into output energy of another. An optical detector does so by generating an electrical current output proportional to the intensity of the incident optical light.

2. **List at least four main requirements of an optical detector.**

   (1) Compact in size for efficient packaging as well as compactible in order to provide an efficient coupling of the output from optical fibers to its input.

   (2) High sensitivity level at the operating wavelength.

   (3) Low noise impact.

   (4) Stable operation even in varying environmental surroundings.

3. What are the factors on which current generated by semiconductor photodiode depends?
When semiconductor photodiodes absorb photons (optical light), they generate photocurrent. The extent of generated photocurrent depends on operating wavelength, size of the active area of the photodiode irrespective of the size of the fiber core, responsivity parameter of the photodiode, alignment of the photodiode with the optical fiber cable.

4. What are the major factors responsible for response time of a photodetector?
The factors responsible for response time of a photodetector includes the thickness of the active area which is directly related to the amount of time required for the electrons generated to flow out of the detector active area, and detector RC time constant which depends on the capacitance of the photodiode and the resistance of the load.

5. Illustrate the operation of generation of photocurrent in a p–i–n photodiode.
Fig. 4.34 depicts the simplified operation of generating photocurrent in a p–i–n photodiode.

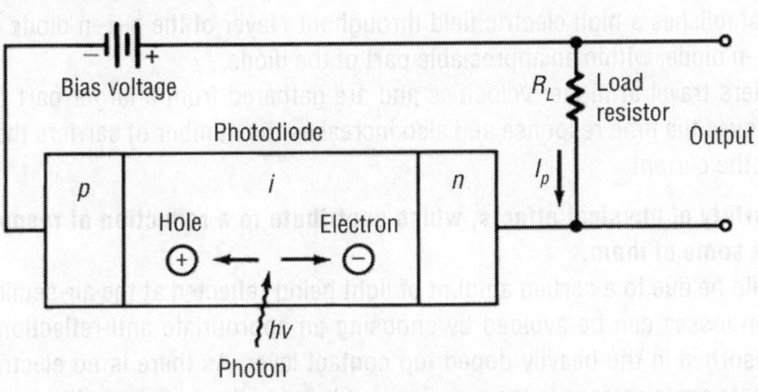

Fig. 4.34 Schematic of p–i–n photodiode operation

As shown, the high electric field present in the depletion region causes photo-generated carriers (electrons and holes) to get separated and be collected across the reverse–biased junction. This gives rise to a current flow in an external circuit, known as photocurrent.

6. Draw the energy-band diagram for a semiconductor p–i–n photodiode.
Fig. 4.35 illustrates the energy-band diagram for a semiconductor p–i–n photodiode.

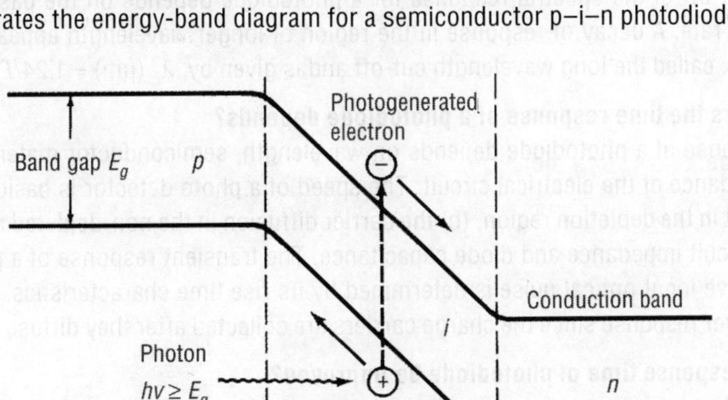

Fig. 4.35 Energy–band diagram of a p–i–n photodiode

7. Mention at least five advantages of p–i–n photodiode.

(1) The output photocurrent is directly proportional to the incident optical power having a linear relationship between them.
(2) It requires a relatively lower value of biasing voltage (typically less than 4V).
(3) The contribution to net photocurrent due to various noise sources is quite less.
(4) It has very small dark current.
(5) It provides operation at higher bit rate.

8. What are specific advantages of reverse biasing of photodiodes?

There are certain advantages of reverse biasing of photodiodes such as

(1) The photocurrent is a linear function of the generated charged carriers (electron–hole pairs) even at relatively higher optical powers.
(2) The capacitance and series resistance of the diode are minimized because the depletion region within the photodiode widens.
(3) The bias establishes a high electric field throughout *i* layer of the p–i–n diode or, in the case of an ordinary p–n diode, within an appreciable part of the diode.

Thus, carriers travel at higher velocities and are gathered from a larger part of the diode without bias. This improves the time response and also increases the number of carriers the number of carriers contributing to the current.

9. There are a variety of physical effects, which contribute to a reduction of responsivity or quantum efficiency. List some of them.

Firstly, this could be due to a certain amount of light being reflected at the air-semiconductor interface. These reflection losses can be avoided by choosing an appropriate anti-reflection coating. Secondly, light can be absorbed in the heavily doped top contact layer. As there is no electric field in this layer, the generated carriers (electrons in the p-region and holes in the n-region), diffuse toward the depletion region of the photodiode.

10. Define spectral response of the photodiode. What are the factors on which it depends?

The spectral response of the photodiode is defined as the variation in the quantum efficiency with respect to the operating wavelength of the incident optical signal. A decrease in the responsivity with shorter wavelengths is attributed to an enhanced energy absorption level in the thin surface layer. Hence the short wavelength cut-off of the spectral response for a photodiode depends on the base width and surface recombination rate. A decay of response in the region of longer wavelength appears at the absorption band edge (E_g), called the long wavelength cut-off and is given by, λ_c (μm) = 1.24/E_g (eV).

11. On what factors the time response of a photodiode depends?

The time response of a photodiode depends on wavelength, semiconductor material, diode geometry, and load impedance of the electrical circuit. The speed of a photo detector is basically governed by (a) the carrier drift in the depletion region, (b) the carrier diffusion in the non-depleted region and (c) the RC constant of circuit impedance and diode capacitance. The transient response of a practical photodiode to a square wave input optical pulse is determined by its rise time characteristics. A diode without any bias has a slower response since the charge carriers are collected after they diffuse to the diode surface.

12. How can the response time of photodiode be improved?

The response time can be improved by designing the device to withstand voltages which are greater than the voltages required for full depletion, creating a higher field gradient and accelerated photo carrier collection. At a sufficiently high reverse biasing voltage, the potential gradient exists throughout the device depth and a uniformly fast response is achieved.

13. What limits the bandwidth of indirect–bandgap and direct–bandgap semiconductor photodiodes?

In order to achieve higher quantum efficiency, the deletion region width should lie within 20–50 μm for indirect–bandgap semiconductor materials (Si and Ge). Due to this, their bandwidth is often limited because of relatively longer transit time (τ_{t_r} greater than 200 ps). By contrast, *W* can be as small as 3–5 μm for photodiodes that use direct-bandgap semiconductors, such as InGaAs. The transit time for such photodiodes is τ_{t_r} ~10 ps. Such values of τ_{t_r} correspond to a detector bandwidth of approximately 10 GHz if τ_{t_r} is much greater than τ_{RC} .

14. How does an avalanche photodiode (APD) produce large photocurrent?

An APD produce large photocurrent due to its internal amplification through an avalanche process which takes place when a large amount of reverse-bias voltage is applied across the active region. It is possible to control the APD gain by changing the external reverse-bias voltage.

15. Why are p–i–n photodiodes preferred in most applications?

Although higher dark currents and the excess noise due to multiplication are problems in APD's, yet they can provide 3 to 5 dB higher sensitivities than p–i–n photodiodes above 100 Mbps. At lower bit rates, low leakage currents of p–i–n results in better sensitivity. Employing APDs call for precise bias and temperature control, if their noise performance is to be controlled. Therefore, p–i–n photodiodes are preferred in most applications.

16. What are the two most important specifications of a photodetector?

Responsivity and dark current are the two most important specifications of a photodetector. Responsivity signifies an indication of the sensitivity of the photodetector to incident optical signal. Its basic units are amperes per watt of light. For example, in case of the Si PIN photodiode Honeywell HFD3876, a typical value of responsivity is 0.33 A/W. The leakage current, or the normal reverse current produced by the photodiode under conditions of no input optical signal is often termed as the dark current. It basically sets a sensitivity floor. Since the dark current is due to electron–hole pairs created by thermal activity, it increases with temperature, just as for any diode. For HFD3876, typical dark-current value is 0.05 nA at 25°C.

17. Illustrate a simplified block diagram of an optical receiver, depicting the essential components only.

An optical receiver is basically an electro-optical device that converts optical signals in the form of light pulses coupled from an optical fiber into useable electrical signals. It essentially comprises of a photodetector, low-noise electronic signal amplifier, and other associated electronic circuitry, as shown in Fig. 4.36.

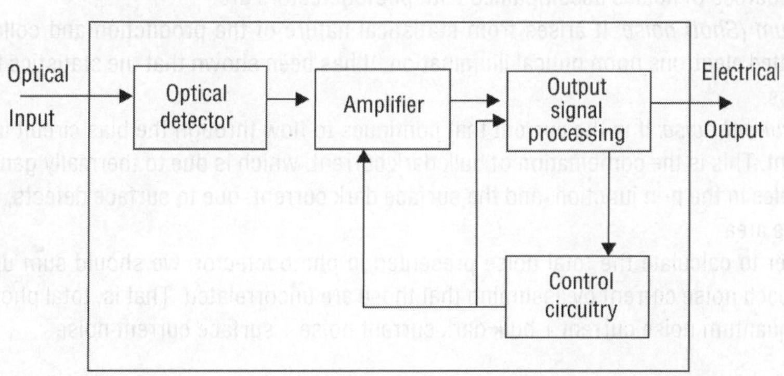

Fig. 4.36 A simplified block diagram of an optical receiver

18. Define the terms: Sensitivity and Dynamic range.

Sensitivity– It signifies the lowest value of the optical power required at the input of the optical receiver in order to achieve a reliable operation, for example, a specified bit error rate (BER) in digital optical receivers.

Dynamic range– It is basically the capability of the optical receiver to accommodate a wide range of input optical power levels so that it can be used for both long and short repeater spacings. It is very critical requirement in LAN applications wherein the receiver may be at varying distances from the probable transmitters, as in the case of Ethernet LAN.

19. What do you understand by bit rate transparency and bit pattern independence in case of optical receiver design?

Bit rate transparency is the ability of the receiver to operate reliably over a specified range of bit rates from a few Mbps to a few Gbps or more, whereas bit pattern independence is the ability of the optical receiver to perform with any type of signalling encoding format.

20. Distinguish between a high-impedance and a trans-impedance optical amplifier configurations widely used in the design of the optical receiver front-design.

A high-impedance optical amplifier used at the front-end of the optical receiver offers maximum detection sensitivity with minimum noise level. But, because of high input load impedance, the frequency response is limited by this high RC time constant. An equalizing circuit is needed at the output to extend the receiver bandwidth. In addition, a dc balanced coding scheme is necessary to remove the baseline wander effect. Such a receiver is not bit pattern independent and it has a limited dynamic range. On the other hand, a trans-impedance amplifier alleviates the problem of limited dynamic range and bandwidth by employing a negative feedback. But it results in higher noise level and degradation in receiver sensitivity.

21. What are different measures which can be considered for improving the performance of optical receivers?

The performance of optical receivers in 1300–1600 nm optical band is severely limited by thermal noise. The use of APD receivers improves the performance to a limited extent because of the excess noise factor associated with InGaAs APDs. The effect of thermal noise can be considerably reduced by using coherent-detection techniques in which the received signal is mixed coherently with the output of a narrow-linewidth laser diode. The receiver performance can also be improved by amplifying the optical signal before it is incident on the photodetector.

22. State the principal noises associated with photodetectors.

The main sources of noises accompanied with photodetectors are:

(1) *Quantum (Shot) noise:* It arises from statistical nature of the production and collection of photo-generated electrons upon optical illumination. It has been shown that the statistics follow a Poisson process.

(2) *Dark current noise:* It is the current that continues to flow through the bias circuit in the absence of the light. This is the combination of bulk dark current, which is due to thermally generated electrons and holes in the p–n junction, and the surface dark current, due to surface defects, bias voltage and surface area.

In order to calculate the total noise presented in photodetector, we should sum up the root mean square of each noise current by assuming that those are uncorrelated. That is, total photodetector noise current = quantum noise current + bulk dark current noise + surface current noise

Multiple Choice Questions

1. There is an increase in the responsivity of a photodetector with increase in operating wavelength due to which one of the following reasons?
   A. Presence of more number of electrons for the same optical power
   B. Presence of more number of photons for the same optical power
   C. Presence of less number of photons for the same optical power
   D. Presence of less number of electrons for the same optical power
2. Which statement is true?
   A. Indirect–bandgap semiconductor materials (e.g., Ge and Si) can be employed for fabrication of photodetectors inspite of the fact that the absorption edge is not as sharp as that for direct-bandgap materials such as GaAs.
   B. Indirect–bandgap semiconductor materials (e.g., Ge and Si) cannot be employed for fabrication of photodetectors inspite of the fact that the absorption edge is not as sharp as that for direct-bandgap materials such as GaAs.
   C. Indirect–bandgap semiconductor materials (e.g., Ge and Si) can be employed for fabrication of photodetectors inspite of the fact that the absorption edge is as sharp as that for direct-bandgap materials such as GaAs.
   D. Indirect–bandgap semiconductor materials (e.g., Ge and Si) cannot be employed for fabrication of photodetectors inspite of the fact that the absorption edge is as sharp as that for direct-bandgap materials such as GaAs.
3. Optical detector materials commonly used at operating wavelength of 850 nm are
   A. Ge, InP, InGaAs. B. Si, Ge, InP.
   C. GaAs, InGaAs, GaAlAs. D. Si, GaAs, GaAlAs.
4. Optical detector materials mostly used at operating wavelength of 1300 and 1550 nm are
   A. Ge, InP, InGaAs. B. Si, Ge, InP.
   C. GaAs, InGaAs, GaAlAs. D. Si, GaAs, GaAlAs.
5. The semiconductor material used in photodetectors determines the responsivity, defined as
   A. the ratio of the input optical power and the output photocurrent, and is a linear function of the operating wavelength and the quantum efficiency.
   B. the ratio of the output photocurrent, expressed in Amps and the input optical power (W), and is independent of the operating wavelength and quantum efficiency.
   C. the ratio of the output photocurrent, expressed in Amps and the input optical power, and varies linearly with the operating wavelength.
   D. the ratio of the input optical power and the output photocurrent,and is independent of the operating wavelength and quantum efficiency of the device.
6. In the p–n photodiode, the width of the depletion region
   A. does not vary with the junction voltage, doping densities and profiles.
   B. varies with the junction voltage, doping densities and profiles.
   C. varies with the junction voltage but does not vary with doping densities and profiles.
   D. varies with the junction voltage and doping densities, but remains independent with profiles.
7. Statement I*:* Photodiodes behave like constant current generators, the amount of current being proportional to the incident light energy.
   Statement II*:* In photodiodes, the deviation from linearity occurs at high light intensities, where the saturation comes into effect.

A. Only Statement I is correct. B. Only Statement II is correct.
C. Both statements are correct. D. Both statements are incorrect.

8. The equivalent circuit of a photodiode consists of
A. the photo current term and the depletion layer capacitance only.
B. terms related to the photocurrent and noise-current, and the depletion region capacitance only.
C. terms related to the photocurrent and noise-current, the depletion layer capacitance and a parallel resistance only.
D. terms related to the photocurrent and noise-current, the depletion layer capacitance, a parallel resistance, and the diode series resistance.

9. The shot noise in photodiode is due to one of the following:
A. the photo-generated currents only.
B. the background light current only.
C. the dark current only.
D. the photo-generated currents, the background light signal and the dark current.

10. In case of a p–i–n photodiode, typical value of the noise-equivalent power is
A. 10^{-12} to 10^{-15} watt. B. 10^{-9} to 10^{-11} watt.
C. 10^{-6} to 10^{-8} watt. D. 10^{-18} to 10^{-20} watt.

11. The signal-to-noise ratio (SNR) of an optical receiver can be expressed as
A. $SNR = \dfrac{R^2}{P_{in}{}^2 i_{noise}^2}$ B. $SNR = \dfrac{P_{in}}{R^2 i_{noise}^2}$
C. $SNR = \dfrac{R^2 P_{in}{}^2}{i_{noise}^2}$ D. $SNR = \dfrac{P_{in}}{i_{noise}}$

12. In general, APD's can provide ________ higher sensitivity than p–i–n photodiodes above 100 Mbps.
A. 3 to 5 dB B. 6 to 8 dB
C. 10 to 12 dB D. 1 to 2 dB

13. Which statement is true?
A. Photodetectors made of semiconductor materials absorb incident photons and produces electrons.
B. Photodetectors made of semiconductor materials absorb electrons and producesphotons.
C. Photodetectors help in conversion of an input electrical signal into an output optical signal.
D. Photodetectors does not transform an input optical signal into an output electrical signal.

14. A ________ p–i–n photodiode is usually excited with a photon having ________ wavelength for its absorption near the surface.
A. forward biased; short B. forward biased;long
C. reverse biased; short D. reverse biased; long

15. The performance of optical receivers that operate in 1300-1600 nm optical band is very much limited by
A. shot noise. B. 1/*f* noise.
C. dark current noise. D. thermal noise.

16. Which of the following is an inherent property of an optical signal and cannot be eliminated even in principle?
A. shot noise B. thermal noise
C. environmental noise D. background noise

17. The responsivity of a *p–i–n* semiconductor photodiode is specified as 0.5 A/W for λ = 1000 nm. What is the output photocurrent when optical power of 0.2 μW at this wavelength is incident on it?
A. 0.1 μA B. 1 μA
C. 10 μA D. 100 μA

18. A(n) ________ device gives an amplified photocurrent at its output.
A. photovoltaic detector B. *p–n* photodiode
C. avalanche photodiode (APD) D. *p–i–n* photodiode

19. The term receiver sensitivity means
A. The lowest optical power needed to achieve a specified performance of the optical receiver.
B. The ratio of output photocurrent to the input optical power.
C. The range of wavelengths for which the receiver functions properly.
D. The range of input optical power levels for which the receiver functions properly within specified values.

20. ________ does not decide how much photocurrent would be produced by a photodiode.
A. Responsivity at the operating wavelength
B. The bit rate of the incident optical signal
C. The size of active area in relation to the size of the fiber core
D. The alignment of the optical fiber cable with which the photodiode is coupled

21. The exact meaning of detector responsivity is given by one of the following statements.
A. The ratio of the output photocurrent to the input optical signal power
B. The ratio of the input optical signal powerto the output photocurrent
C. The minimum amount of input optical power essential for obtaining specified performance of the optical receiver
D. The complete range of input optical power levels for which the receiver functions properly meeting desired specifications

22. The gain of an APD can be varied by varying
A. the bit rate of the input optical signal
B. the input power of the optical signal
C. the external reverse-bias voltage applied across the photodetector
D. the modulation format of the input optical signal

23. Which statement is true?
A. Minimizing the receiver noise will maximize the receiver sensitivity.
B. Maximizing the receiver bandwidth will maximize the receiver sensitivity
C. Maximizing the receiver dynamic range will maximize the receiver sensitivity
D. Minimizing the reverse-bias voltage of photodetector will maximize the sensitivity of an optical receiver.

Keys to Multiple Choice Questions

1. B	**2.** A	**3.** D	**4.** A	**5.** C	**6.** B	**7.** C	**8.** D	**9.** D	**10.** A
11. C	**12.** A	**13.** A	**14.** C	**15.** D	**16.** A	**17.** A	**18.** C	**19.** D	**20.** B
21. A	**22.** B	**23.** C							

Review Questions

1. Define the terms responsivity and quantum efficiency of a photodiode, and derive expressions for the same. How are they related to each other?
2. Distinguish between intrinsic and extrinsic photoconduction. Which one is expected to be more sensitive and why?
3. Discuss the advantages of the structures used for p–i–n photodiode over that used for p–n photodiode.
4. Illustrate the functional block schematic of a digital optical receiver. Describe briefly the role of each functional block.
5. State the criteria for using the received signal by the decision devicein the optical receiver correlated with the incident optical power.
6. Explain the various noise mechanisms in a photodiode. Give the significance of each one of them.
7. Describe different mechanisms that limit the frequency response of a photodiode. Derive the necessary expressions.
8. Discuss the effect of transit times on the frequency response of photodiodes.
9. Highlight how the speed can be improved by proper choice of parameters.
10. Let thermal noise as well as shot noise affect the sensitivity of an APD-based optical receiver. Ignoring the effect of dark current, derive a general expression for the sensitivity by considering a finite value of extinction ratio.
11. Derive a general expression for power penalty (induced by intensity noise) of a p–i–n photodetector based optical receiver by considering a finite value of extinction ratio. Neglect contributions due to shot-noise and intensity-noise as compared to the thermal noise in the OFF state but not in the ON state.
12. Draw a characteristic curve depicting power penalty versus intensity-noise for different values of the extinction ratio. Highlight the point for power penalty to be infinite and comment.
13. Discuss noise and bandwidth considerations for the case of high-impedance and trans-impedance optical receivers. Comment on their merits and demerits.
14. Name and compare two types of photodetectors that are used in optical fiber communications.
15. From the basic concepts, show that responsivity of a photodetector is expressed as $R = \eta \dfrac{\lambda(\mu m)}{1.24}$; where η represents the quantum efficiency and λ denoted the operating wavelength of the incident light, expressed in μm. Use $h = 6.6 \times 10^{-34}$, $q = 1.6 \times 10^{-19}$C, $c = 3 \times 10^{14}$ μm/s.
16. Distinguish between the salient features of three types of photodetectors - p–n photodiode, p–i–n photodiode, and an APD. Is it possible to make these three types of photodiodes using the same semiconductor material?
17. What are the factors responsible for making the responsivity versus wavelength curve for a practical Si photodiode deviate from an ideal curve? How can the quantum efficiency of such a photodiode be improved?

Numerical Problems

1. Given that the semiconductor material Ge has an energy band gap value of 0.67 eV. What is the maximum wavelength that will be absorbed by it? *[Ans.: 1850 nm]*

2. The highest wavelength that Si can absorb is 1200 nm. What is the approximate band gap of Si. *[Ans.: 1.1 eV]*

3. Using the energy bandgap value of GaAs semiconductor material as E_g = 1.42 eV, determine the primary wavelength of photons generated due to electron–hole recombinations. Verify whether this wavelength falls in the visible spectrum region? *[Ans.: 873 nm; No]*

4. Will a silicon photodetector (E_g = 1.1 eV) be sensitive to emissions produced by a GaAs laser (E_g = 1.42 eV)? Justify your answer. *[Ans.: λ of GaAs = 873 nm is shorter than λ of Si =1120 nm; Yes]*

5. On an average, a p–n photodiode generates only one electron–hole pair (EHP) for every five number of incident photons at λ = 900 nm. Determine the quantum efficiency if all the photo-generated electrons are gathered. *[Ans.: 20%]*

6. The quantum efficiency of a p–i–n photodiode is specified as 80%. Determine its responsivity at λ = 1300 nm. *[Ans.: 0.84]*

7. Find the responsivity of a p–i–n photodiode at λ = 1550 nm if its specified quantum efficiency is 80%. Give sufficient reasons to justify that the photodiode at λ = 1550 nm is much more responsive as compared to that of at 1300 nm for the same quantum efficiency. *[Ans.:1; $R_{\lambda=1.55} > R_{\lambda=1.3}$]*

8. A photodiode circuit has responsivity R = 0.5 A/W and incident optical power of 10 μW. Find the resultant photodiode current. *[Ans.:5 × 10^{-6} A]*

9. Fig. 4.37 depicts the responsivity curve as function of its operating wavelengths for a commercial InGaAs p–i–n photodiode. It has dark current value of 5 nA.

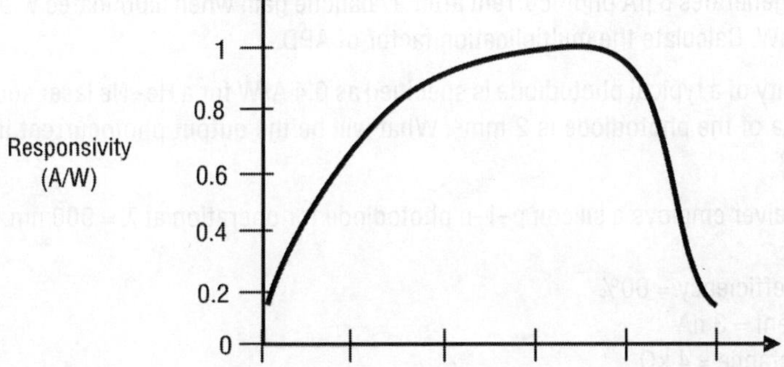

Fig. 4.37 For P9

(a) Calculate the incident optical power at λ = 1300 nm that would generate photocurrent which will be two times of specified dark current.

(b) Determine its quantum efficiency at λ = 1300 nm. *[Ans.: 9.43 nA; 78%]*

10. The Honeywell's HFD 3876/1 PIN photodiode has typical value of responsivity $R = 0.33$ A/W. Calculate the photodiode current for an incident optical power of 500 nW. *[Ans.: 165 nA]*

11. In the data sheet of a photodiode, the responsivity is specified as 0.4 A/W. What optical power is required to produce a photocurrent of 500 nA? *[Ans.: 1.25 μA]*

12. In the data sheet of a photodiode, the responsivity is specified as 0.4 A/W and the value of dark current as 50 pA. To what light input power does this dark current level correspond? *[Ans.: 0.125 nW]*

13. A laser diode with a power output of 4 mW is connected to a fiber length of 25 km having fiber loss specification of 0.2 dB/km. The receiver at the other end of the fiber has a responsivity of 0.2 A/W. How much current flows through the photodetector? *[Ans.: 252 μA]*

14. Determine the approximate bandwidth of the photodetector if the transit time and time constant of the equivalent RC circuit are given as $\tau_{t_r} = 100\ ps$ and $\tau_{RC} = 100\ ps$, respectively. What should be done for lightwave system to operate at bit rates of 10 Gbps or more? *[Ans.: 795.8 MHz; $\tau_{t_r}, \tau_{RC} < 10\ ps$]*

15. What would be the approximate bandwidth of InGaAs direct-bandgap semiconductor photodiode having transit time $\tau_{t_r} \sim 10\ ps$ and time constant of the equivalent RC circuit is $\tau_{RC} \ll \tau_{t_r}$. *[Ans.: 16 GHz]*

16. Photons at λ = 850 nm strike a *p–i–n* photodiode at the rate specified as 4×10^{10} photons per second and, on an average, the number of electrons gathered is 2×10^{10} electrons per second. Find the responsivity of the photodiode at this wavelength. *[Ans.: 0.34 A/W]*

17. In the data sheet of a p–n photodiode, it is specified that the quantum efficiency = 50% at λ = 900 nm. Determine the following:
(a) Responsivity
(b) Incident optical power for average photocurrent of 1 μA
(c) The corresponding number of received photons
[Ans.: (a) 0.36 A/W; (b) 2.76 μW; 1.25×10^{13} per second]

18. In the data sheet of an avalanche photodiode (APD), it is specified that the quantum efficiency = 50% at λ = 1300 nm. It generates 8 μA photocurrent after avalanche gain when illuminated with an incident optical power of 0.4 μW. Calculate the multiplication factor of APD. *[Ans.: 38]*

19. The responsivity of a typical photodiode is specified as 0.4 A/W for a He–Ne laser source (λ = 632.8 nm). The active area of the photodiode is 2 mm^2. What will be the output photocurrent if the incident flux is 100 μW/mm^2? *[Ans.: 80 μA]*

20. An optical receiver employs a silicon p–i–n photodiode for operation at λ = 900 nm. It has the following specifications:
- Quantum efficiency = 60%
- Dark current = 3 nA
- Total resistance = 4 kΩ
- Incident optical power = 200 nW
- Post-detection bandwidth = 5 MHz at operating temperature of 20°C.

Determine the mean square values of shot noise and thermal noise. *[Ans.: 1.44×10^{-19} A; 2×10^{-17} A]*

21. The post detection bandwidth of an optical receiver used in an optical fiber communications system is 5 MHz at λ = 1000 nm. Assuming the use of an ideal photodetector and the effect of quantum noise only, determine the optical power (in dBm) required at the input of the receiver so as obtain SNR = 50 dB. *[Ans.: -37 dBm]*

CHAPTER

5

Optical Amplifiers

Chapter Objectives

After studying this chapter, you should be able to

- ✓ know various types of optical amplifiers based on their functions
- ✓ describe principle of operation, basic structure types, and performance parameters of semiconductor optical amplifiers (SOAs)
- ✓ understand the concept of amplification, performance parameters and uses of Raman fiber amplifiers (RFAs)
- ✓ explain amplification mechanism, characteristics, and advantages of Erbium-doped fiber amplifiers (EDFAs)
- ✓ compare the major features and applications of SOAs, RFAs and EDFAs

An optical amplifier is a widely used device in optical fiber communications for the purpose of amplification of the optical signal generated by an optical transmitter. The optical signal is directly amplified to yield optical signal without any conversion to electrical signal first and then restoring it to amplified optical signal. This implies that optical amplifiers operate on photons in all-optical domain. In essence, optical amplifiers do require source and photodetector as optoelectronic devices and additional electronic circuits for other operations such as shaping of optical pulses including retiming. But why do we need optical amplifiers? We know that typical fiber loss around 1500-nm wavelength is approximately 0.2 dB/km. Let us examine what happens after the optical signals travel about 100 km down the fiber. The signals are attenuated by 20 dB. So they need to be amplified, otherwise the signal-to-noise ratio of detected signals is too low and bit-error-rate becomes too high (typically desired value of BER is $<10^{-9}$). One way to resolve this issue is to detect the weak signals, followed by modulating a new laser (optical-to-electrical-to-optical conversions) which requires high-speed (>10 GHz) electronic circuitry. Therefore, the best way to amplify the signal is only optically and the preferred method is fiber amplifier which is the most efficient, the most stable and the one with the lowest loss.

This chapter focuses on optical amplifiers. The discussion begins with functional types of optical amplifiers such as power amplifiers or boosters at the transmitter end, or in-line amplifiers used along with the fiber link, or as a pre-amplifier at the input of the optical receiver. This is followed by a detailed discussion on different aspects of semiconductor optical amplifiers (SOAs), fiber-based Raman optical amplifiers, and Erbium-doped fiber amplifiers (EDFAs). The principles of operation, amplification mechanisms, performance characteristics, advantages and drawbacks of each one of them are elaborated. Finally, a comparative study of major features offered by these three categories of optical amplifiers is presented.

5.1 Functional Types of Optical Amplifiers

An optical amplifier is nothing but a laser diode with anti-reflection coatings in place of end mirrors. The signal light is passed through a semiconductor which is essentially single-mode waveguide having ~ 0.5–2 mm length and 1–2 μm transverse dimensions. Fig. 5.1 shows a simplified functional block schematic of a generic optical amplifier.

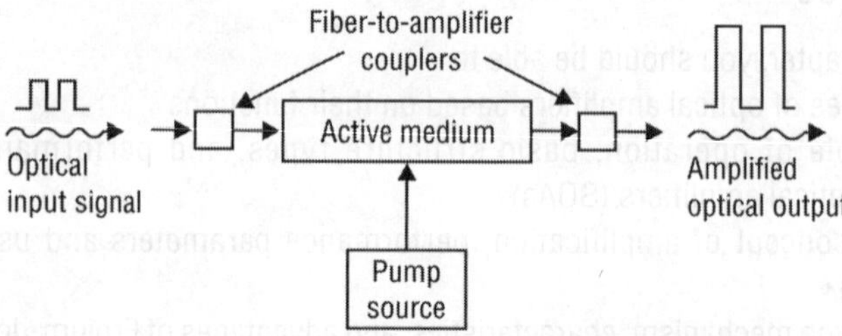

Fig. 5.1 A simplified functional schematic of generic optical amplifier

The optical input signal flowing through the optical fiber is applied to an active medium (an amplifying region) through fiber-to-amplifier coupler. The active medium is pumped with an electric current from a pump source. This results into optical transitions from the conduction band to the valence band because the injection current from the pump source creates a certain carrier density in the conduction band. Since the photon energies are slightly above the bandgap energy, the maximum optical gain occurs.

The distance between a transmitter and receiver link is extended by the use of optical amplifiers. They can compensate for signal attenuation mainly due to fiber loss. But this may reduce the optical SNR (and hence noise figure) by a small amount. Moreover, dispersion (and crosstalk in DWDM systems) cannot be compensated by optical amplifiers. In terms of the functions performed in a typical optical fiber communications link, optical amplifiers can be broadly classified in three basic types as given below:

(i) Power Amplifier or Booster (at Optical Transmitter)
(ii) In-line Optical Amplifier (along the fiber)
(iii) Pre-amplifier (at Optical Receiver)

5.1.1 Power Amplifier or Booster

It is a power amplifier that raises the power of an optical signal available at the output of an optical transmitter to the highest level before sending it down the optical fiber. Fig. 5.2 shows an arrangement of deploying optical amplifier as power amplifier.

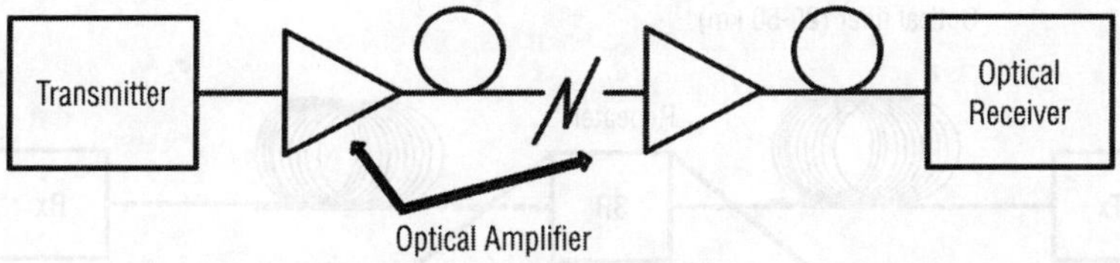

Fig. 5.2 An optical amplifier as Tx power amplifier or booster

Depending on the optical gain of the power amplifier and fiber losses, it is possible to increase the transmission distance by 100 km or more. For example, most of the DFB lasers used as optical sources deliver very small output power of the order of 2 mW only. A power amplifier used immediately after the optical transmitter can boost the optical signal.

> Note: Typical Erbium-doped Fiber Amplifier (EDFA) used as power amplifiers have an output of around 100 mW. In fact, an optical fiber can handle aggregate optical power levels of up to about 100–200 mW before nonlinear effects appear.

5.1.2 In-line Optical Amplifier

An in-line optical amplifier operates with a signal in the middle of a fiber–optic link. Fig. 5.3 shows the use of optical amplifiers as in-line amplifiers along the fiber link between optical transmitter and optical receiver.

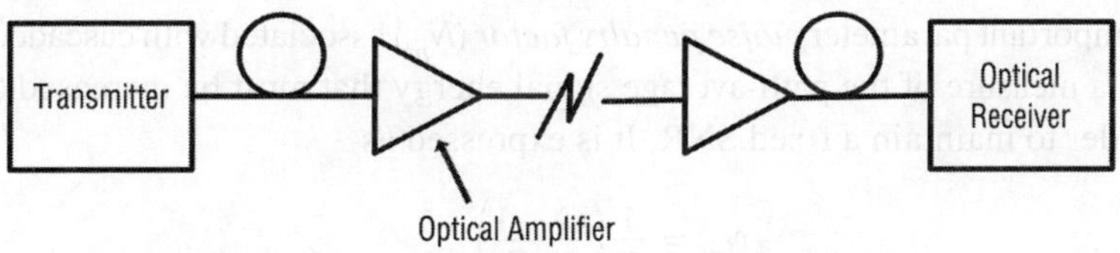

Fig. 5.3 An optical amplifier as in-line amplifier

The main function of an in-line optical amplifier is to compensate for signal losses caused by fiber attenuation, losses due to interconnections and signal distribution in WDM networks. In a typical application, an opto–electronic repeater (comprising of photodiode, timing and shaping circuits, and laser) can be replaced with an appropriate in-line optical amplifier within a long-haul optical fiber communication link, as shown in Fig. 5.4.

Many in-line optical amplifiers can be cascaded along the fiber link in a long-haul fiber–optic communication link, as shown in Fig. 5.5.

In this arrangement, we have to ensure that the system performance is not degraded significantly by cumulative effects of amplified spontaneous emission (ASE) noise, dispersion, non-linearity, and stability over entire WDM bandwidth. To compensate for the accumulated ASE noise level, there should be almost linear increase of optical signal power with the length of the optical fiber link. This helps to maintain a constant SNR. Moreover, for a very low noise figure, high power output as well as high optical gain is needed.

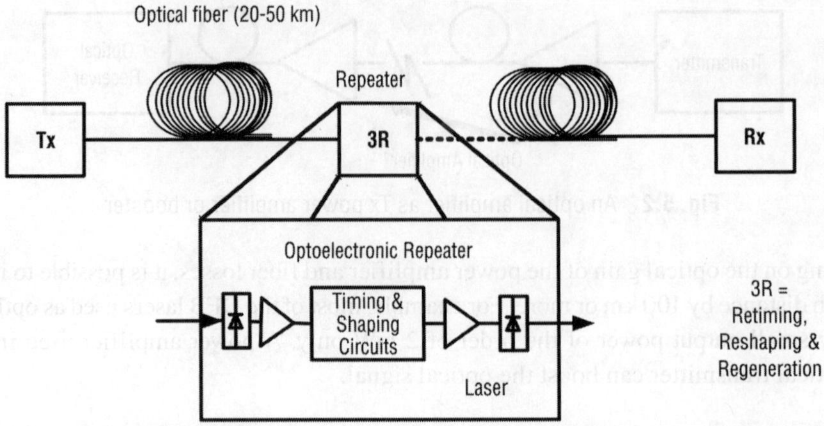

Fig. 5.4 An in-line optical amplifier as repeater

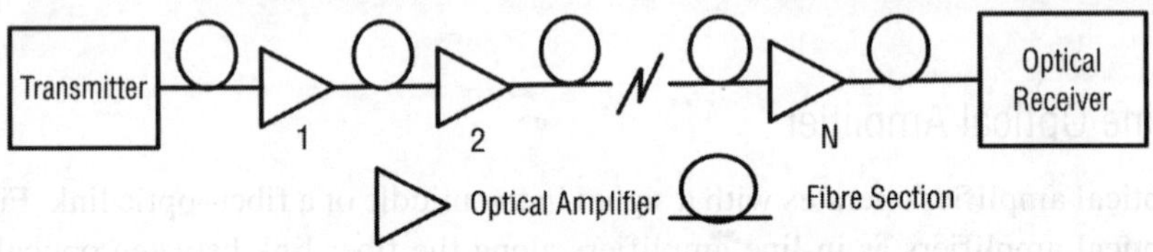

Fig. 5.5 Cascaded in-line optical amplifiers

There is an important parameter, *noise penalty factor* (N_{pf}) associated with cascaded in-line optical amplifiers. It is a measure of the path-average signal energy that must be increased (as optical gain increases) in order to maintain a fixed SNR. It is expressed as

$$N_{pf} = \frac{1}{G}\left(\frac{G-1}{\ln G}\right)^2 \tag{5.1}$$

where, G represents the optical gain (in ratio) of an in-line amplifier.

It is imperative to say here that in order to yield the best combination of overall optical gain and output SNR, the location of cascaded in-line optical amplifiers should be uniform along the fiber–optic communication link. Typical values of input optical power level for these in-line optical amplifiers ranges from -26 dBm to -9 dBm (i.e., 2.5 μW to 125 μW), with optical gains of more than 15 dB.

> ***Note:*** Nowadays, sophisticated in-line optical amplifiers are made as a multi-section optical amplifiers separated by an isolator. In fact, an in-line optical amplifier serves as a pre-amplifier employed within the fiber link, cascaded with a power amplifier at the transmitter end.

5.1.3 Pre-amplifier

When an optical amplifier is used as pre-amplifier, it amplifies an optical signal just before it reaches the optical receiver, as depicted in Fig. 5.6.

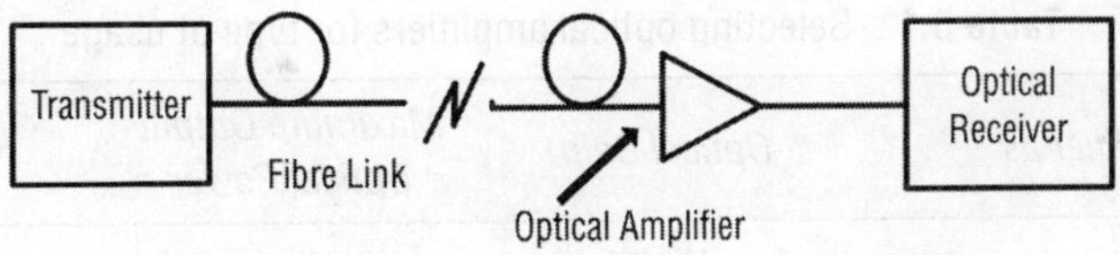

Fig. 5.6 An optical amplifier as pre-amplifier

By using an optical amplifier as a pre-amplifier just before an optical receiver, what exactly do we achieve? Obviously, the sensitivity of direct-detection optical receivers is significantly improved which has been limited by thermal noise. Another advantage is to compensate for distribution losses in LANs. The next question arises: what is the most essential requirement of an optical amplifier to be used as a pre-amplifier? It is, of course, a low noise characteristic because the input signal level is usually very low. The optical gain requirement may not be very high because the received optical signal is applied directly into an optical receiver. Moreover, a pre-amplifier can operate well below saturation; it will not have any feedback control. So, a pre-amplifier should have good sensitivity, high gain, and low noise.

The detector sensitivity or the improvement of minimum detectable signal of an optical amplifier employed as a pre-amplifier at the receiver end can be defined as the ratio of minimum value of the electrical signal power, S_{min}, that is required for it to perform with an acceptable bit-error rate (BER) to the new minimum detectable electrical signal level S'_{min} that is needed to maintain the same signal-to-noise ratio. It is given as

$$\frac{S_{\min}}{S'_{\min}} = G^2 \frac{N}{N + N'} > 1 \tag{5.2}$$

where, G represents the gain of an optical pre-amplifier, N represents the receiver's noise power level, N' is the spontaneous emission from the optical pre-amplifier that gets converted by the photodiode in the receiver to an additional background noise.

Facts to Know

Optical amplifiers used as power amplifiers amplify transmitter output and can deliver up to 50 mW (equivalent to +17 dBm) optical power. They find applications in cable TV systems just before a star coupler for signal distribution. On the other hand, optical amplifiers used as in-line amplifiers have medium output power and good noise figure. Along a fiber–optic communication link, they are normally installed after every 30–70 km distance. An optical pre-amplifier is a low-noise optical amplifier in front of an optical receiver.

Now the question arises how to select an optical amplifier to be used either as power amplifier at the optical transmitter end, or in-line amplifier along a fiber link, or a pre-amplifier at the optical receiver end. There are three major performance parameters: optical gain, maximum output power and noise figure which decides its usage. Table 5.1 shows the requirements for these three types of usage of optical amplifiers in optical fiber communication link applications.

Table 5.1 Selecting optical amplifiers for type of usage

Usage of Optical Amplifier as	*Optical Gain*	*Maximum Output Optical Power*	*Noise Figure*
Power Amplifier	High	High	Not significant
In-Line Amplifier	Medium	Medium	Good NF
Pre-Amplifier	Low	Low	Low < 5 dB

Table 5.2 shows the possible improvement of system gain with the use of optical amplifiers in optical fiber communication link applications.

Table 5.2 Improvement of system gain with optical amplifiers

Usage of Optical Amplifier as	*Improvement in Optical Gain (dB)*	*Improvement in Distance (km)*	*Main Requirement*
Power Amplifier	10–15	40–60	High efficiency
In-Line Amplifier	15–30	60–120	Low noise supervisory
Pre-Amplifier	5–10 (APD); 10–15 (PIN)	20–40 40–60	Low noise

Example 5.1 **Output Optical Power Level**

A typical optical amplifier can amplify an input optical signal of 1 μW level to the 1 mW level at its output. If a 1 mW signal level is incident at its input, then what would be the output optical power level? Assume no saturation happens at the output.

Solution:

We know that gain of the optical amplifier, $G(dB) = 10\log\left(\frac{P_{out}}{P_{in}}\right)$

For the given input power level, P_{in} = 1 μW, and output power level P_{out} = 1 mW, we have

$$G(dB) = 10\log\left(\frac{1\times10^{-3}\,W}{1\times10^{-6}\,W}\right) = 30$$

Now specified input power level P_{in} = 1 mW

Then, output power level can be computed using $G(dB) = 10\log\left(\frac{P_{out}}{P_{in}}\right)$

For G = 30 dB, we have $$30 = 10\log\left(\frac{P_{out}}{1\times10^{-3}\,W}\right)$$

$\Rightarrow$ $$P_{out} = \left(1\times10^{-3}\,W\right)\times \text{anti}\log\left(\frac{30}{10}\right) = 1\text{ W}$$ **Ans.**

Example 5.2 Spacing between In-line Amplifiers

Consider a fiber–optic communication link containing four cascaded in-line optical amplifiers, each having a 20 dB optical gain. If the optical fiber used has a loss specification of 0.2 dB/km, then what would be the spacing between adjacent optical amplifiers (assuming no other system impairments)? Determine total length of the communication link.

Solution:

Given optical gain of in-line optical amplifier = 20 dB

Specified fiber loss = 0.2 dB/km

Since there are no other system impairments, the spacing between adjacent optical amplifiers

= 20 dB / (0.2 dB/km) = 100 km **Ans.**

Given number of cascaded in-line optical amplifiers in the link = 4

Total length of the communication link = 100 km + (4 × 100 km) = 500 km **Ans.**

Example 5.3 Noise Penalty Factor

A fiber–optic transmission link uses eight cascaded optical amplifiers, each having an optical gain of 10 dB. A 50-km fiber length is used with 0.2 dB/km specification of fiber loss. Calculate the noise penalty factor (dB) over the total path.

Solution:

We know that noise penalty factor,

$$N_{pf} = \frac{1}{G}\left(\frac{G-1}{\ln G}\right)^2$$

where, G is the optical gain (in ratio) of an in-line amplifier.

Given optical gain of in-line amplifier = 10 dB

Converting it into ratio, G = antilog (10/10) = 10

Therefore,
$$N_{pf} = \frac{1}{10}\left(\frac{10-1}{\ln 10}\right)^2 = 1.53$$

$$N_{pf}(dB) = 10\log(1.53) = 1.85 \text{ dB}$$ **Ans.**

Section Practice Problems

1. A typical optical amplifier can amplify an input optical signal of 1 μW level to the 10 mW level at its output. If a 1 mW signal level is incident at its input, then what would be the output optical power level? Assume no saturation happens at the output. *[Ans.: 100 mW]*
2. A typical optical amplifier can amplify an input optical signal of 1 μW level to the 10 mW level at its output. If the saturation power level at the output of this optical amplifier is specified as 10 mW, then what would be the output optical power level if an input optical signal of 1 mW is applied to it? *[Ans.: 1 mW]*
3. A fiber–optic transmission link uses five cascaded optical amplifiers, each having an optical gain of 30 dB. A 150-km fiber length is used with fiber loss specification of 0.2 dB/km. Determine the noise penalty factor (ratio) over the total path. *[Ans.: 4.62]*

There are different versions of optical amplifiers, such as semiconductor optical amplifiers (SOAs), Raman fiber amplifiers (RFAs) and Erbium-doped fiber amplifiers (EDFAs). These are described in the following sections.

5.2 Semiconductor Optical Amplifiers

A semiconductor optical amplifier (SOA), also known as laser amplifier, is an active medium of a semiconductor laser. In other words, an SOA is a laser diode without or with very low optical feedback. An electric current is externally applied to the laser device that excites electrons in the active region. When photons travel through the active region it can cause these electrons to lose some of their extra energy in the form of more photons that match the wavelength of the initial ones. Therefore, an optical signal passing through the active region is amplified and is said to have experienced optical gain. So we can say that semiconductor laser can act as a semiconductor optical amplifier when operating quite close to its threshold value.

Principle of Operation: The principle of stimulated emission is primarily used by a semiconductor optical amplifier for amplification of an optical information signal, as the case with laser operation. Figure 5.7 depicts the principle of operation of SOA.

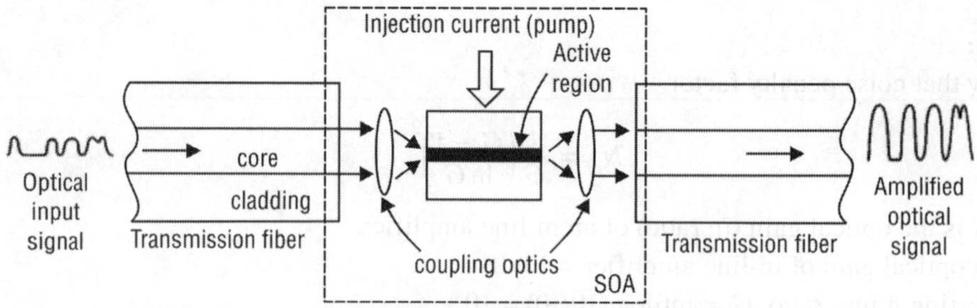

Fig. 5.7 Principle of operation of SOA

As shown, the injection current (also termed as pump signal) in the active region to achieve population inversion is actually responsible for the desired optical gain. The coupling optics is used at the input and output of the active region to couple it efficiently with transmission fiber on the either end of the active region. The optical gain depends on the following factors:

- the wavelength of the optical input signal
- the type and characteristics of the amplifier medium (active region)
- the local beam intensity at any point within the active region

5.2.1 Types of SOAs

Based on the structure, there are two types of semiconductor optical amplifier such as

1. Fabry–Perot Laser Amplifier (FPLA)
2. Traveling–Wave Semiconductor Laser Amplifier (TWSLA)

Fabry–Perot Laser Amplifier (FPLA): It has the same configuration as that of Fabry–Perot laser, as shown in Fig. 5.8.

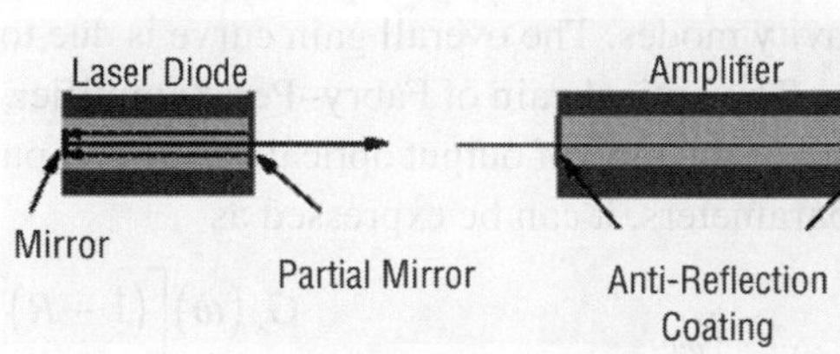

Fig. 5.8 Laser diode vs Fabry–Perot laser amplifier

Fabry–Perot laser amplifiers are almost identical to regular index-guided Fabry–Perot lasers. Either edges (or "facets") of the SOA are designed to have very low reflectivity so that there are no unwanted reflections of the signal within the semiconductor itself. The main difference from regular lasers is that they have reflective facets in order to build up the intensity of light within the semiconductor material. Whereas in the FPLA, the back facet is pigtailed. Fig. 5.9 shows the basic configuration depicting principle of operation of FPLA.

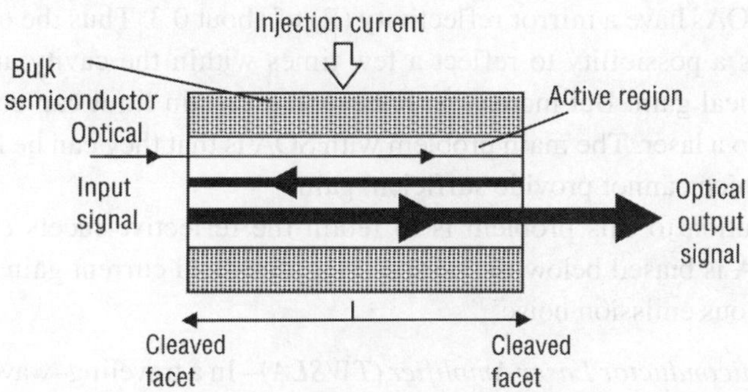

Fig. 5.9 Principle of operation of Fabry–Perot laser amplifier

As seen, the input optical signal that enters the active region is reflected several times from cleaved facets on both sides of the active region. When the signal leaves the cavity, it is amplified. The FPLA (or, FPA) is biased below the normalizing threshold current gain. Fig. 5.10 shows typical characteristic curves of optical gain versus frequency for various values of facet reflectivity, *R*.

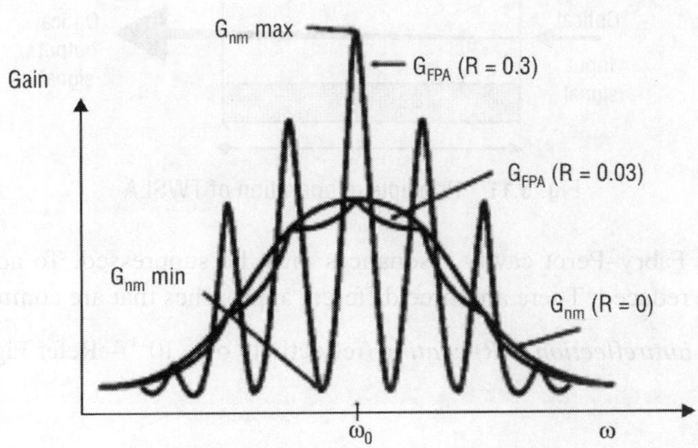

Fig. 5.10 Optical gain of FPA vs frequency

Ripples are caused by the cavity modes. The overall gain curve is due to the width of the atomic transition in the semi-conductor. The optical gain of Fabry–Perot amplifier, as a function of angular frequency, $G_{FPA}(\omega)$, is expressed as the ratio of output optical power to input optical power. In terms of facet reflectivity and other parameters, it can be expressed as

$$G_{FPA}(\omega) = \frac{P_{out}}{P_{in}} = \frac{G_s(\omega)\left[(1-R)^2\right]}{\left[1 - RG_s(\omega)\right]^2 + 4RG_s(\omega)\sin^2\left[\frac{(\omega-\omega_0)L}{\upsilon}\right]} \tag{5.3}$$

where, $G_s(\omega)$ is a single-passage amplification factor assumed to have a Gaussian-shape dependence on angular frequency, R is power-reflection coefficients of cleaved facets, ω is current angular frequency, ω_0 is the center angular frequency, L represents the length of the active region, and $\upsilon = c/n$ is the speed of light within the active medium with refractive index n.

The use of FP resonator which provides optical feedback can significantly increase the gain of a semiconductor optical amplifier. Higher the value of R, the higher will be the optical gain at resonant frequency. Typical SOAs have a mirror reflectivity (R) of about 0.3. Thus the optical signal entering the active region has a possibility to reflect a few times within the cavity and thereby providing reasonably good optical gain. But increasing R beyond a certain limit can create oscillations, i.e., amplifier turning into a laser. The main problem with SOA is that they can be fabricated up to about 450 μm long only which cannot provide sufficient gain.

One possible solution to this problem is to retain the reflective facets characteristic of laser operation. The FPLA is biased below the normalizing threshold current gain which also raises the problem of spontaneous emission noise.

Traveling–Wave Semiconductor Laser Amplifier (TWSLA)– In a traveling–wave semiconductor laser amplifier (TWSLA), or simply TWA, an input optical signal is amplified by a single passage through the active region. There is no optical feedback possible as it does not have any reflective facets. Fig. 5.11 shows the basic configuration depicting principle of operation of TWSLA.

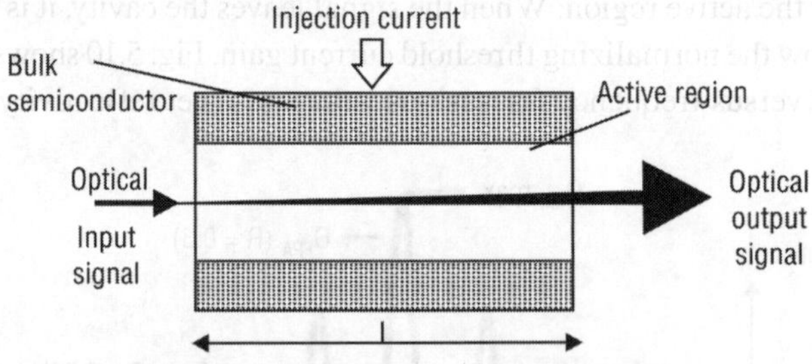

Fig. 5.11 Principle of operation of TWSLA

In TWSLA, the Fabry–Perot cavity resonances must be suppressed. To accomplish this, the reflectivity must be reduced. There are three different approaches that are commonly used:

(a) *TWSLA using antireflection (AR) coating* (reflectivity of ~ 10^{-4})– Refer Fig. 5.12.

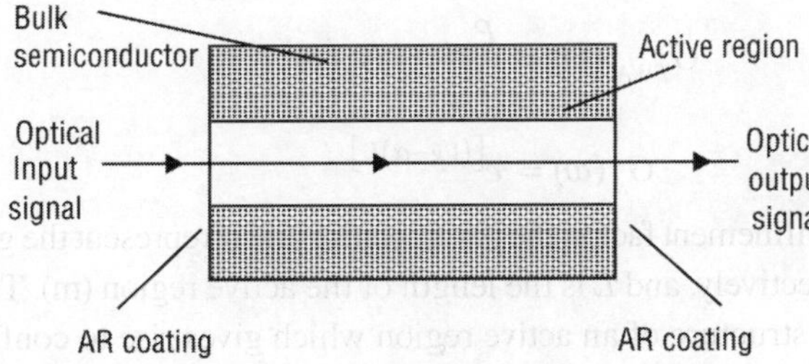

Fig. 5.12 TWSLA using AR coating

(b) *TWSLA using tilting active region*– Refer Fig. 5.13.

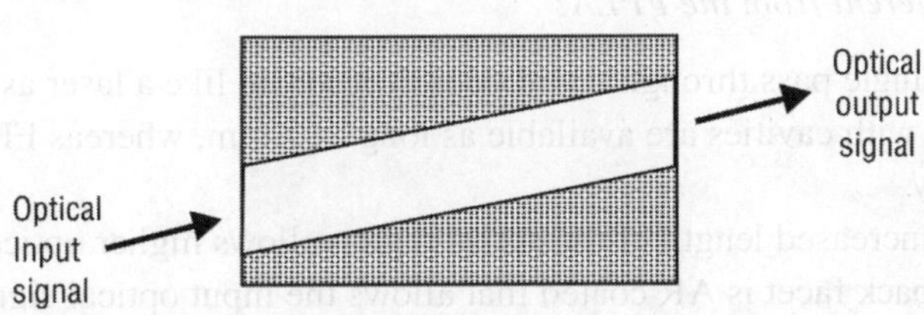

Fig. 5.13 TWSLA using tilting active region

The optical input signal enters the angled facet at one end of the active region, follows the path and leaves it at the other end as optical output signal. In this way, the reflected beam is physically separated from forward beam, thereby achieving reflectivity figure of ~ 10^{-3} to 10^{-4}.

(c) *TWSLA using transparent window region*– Refer Fig. 5.14.

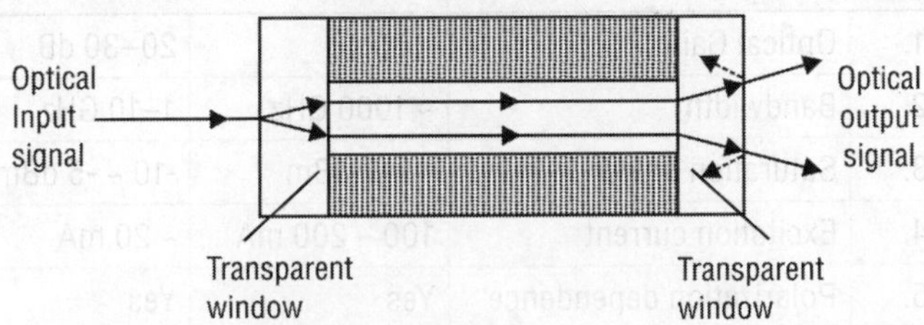

Fig. 5.14 TWSLA using transparent window region

Here the optical input signal (light beam) at one end of the active region spreads in the first transparent window region and then enters at the semiconductor–air interface. It travels straight through the medium (thin active region surrounded by bulk semiconductor material) up to the second transparent window region. Then the spread optical output signal is available at the other end of the medium. Due to further spreading of the reflected beam on the return trip, there may not be much coupling of light into the thin active region.

The gain of traveling-wave amplifier G_{TWA} (ω) is, in fact, the gain of a Fabry–Perot amplifier with $R = 0$. Therefore,

$$G_{TWA}(\omega) = \frac{P_{out}}{P_{in}} = G_s(\omega) \tag{5.4}$$

$$G_s(\omega) = e^{[(\Gamma g - \alpha)L]} \tag{5.5}$$

where, Γ represents the confinement factor, the parameters g and α represent the gain and loss coefficients per unit length (1/m) respectively, and L is the length of the active region (m). The radiated photons are guided by the waveguide structure of an active region which gives rise to confinement factor.

From this expression, it is obvious that the optical gain of traveling-wave amplifier without reflective facets can be increased either by increasing the value of the confinement factor (Γ), the gain coefficient (g), the length of the active region (L), or by decreasing the value of the loss coefficient (α).

How is TWSLA different from the FPLA?

- TWSLA uses a single pass through it and doesn't resonate like a laser as in FPLA.
- TWSLA devices with cavities are available as long as 2 mm, whereas FPLA are limited up to 450 μm long only.
- In TWSLA, the increased length of the active region allows higher optical gain.
- In TWSLA, the back facet is AR coated that allows the input optical signal.
- In TWSLA, the exit facet is also AR coated otherwise it is just the same as for a laser.
- The TWSLA can be operated above the lasing threshold level because no optical feedback is present.

Table 5.3 presents a comparative review of typical characteristic parameters of TWSLA and FPLA.

Table 5.3 Comparison between TWSLA and FPLA parameters

S. No.	*Parameter*	*TWSLA*	*FPLA*
1.	Optical Gain	~ 30 dB	20–30 dB
2.	Bandwidth	~ 1000 GHz	1–10 GHz
3.	Saturation power	~ +10 dBm	-10 ~ -5 dBm
4.	Excitation current	100 – 200 mA	~ 20 mA
5.	Polarization dependence	Yes	Yes

Note: Because of the absence of optical feedback, the TWSLA can yield higher optical gain per unit of length as compared to that of the FPLA. Typical optical gains of the order of 30 dB over 40-nm bandwidth range can be achieved in TWSLA.

5.2.2 Performance Parameters of SOAs

Semiconductor optical amplifiers saturate and typical saturation output power for SOAs is around 5–10 mW, as shown in Fig. 5.15.

The performance of semiconductor optical amplifiers can be measured in terms of gain, noise, bandwidth, and dependence on polarization. The detailed analysis of these performance parameters as well as comparison between Fabry–Perot and Traveling–Wave semiconductor optical amplifiers is presented here.

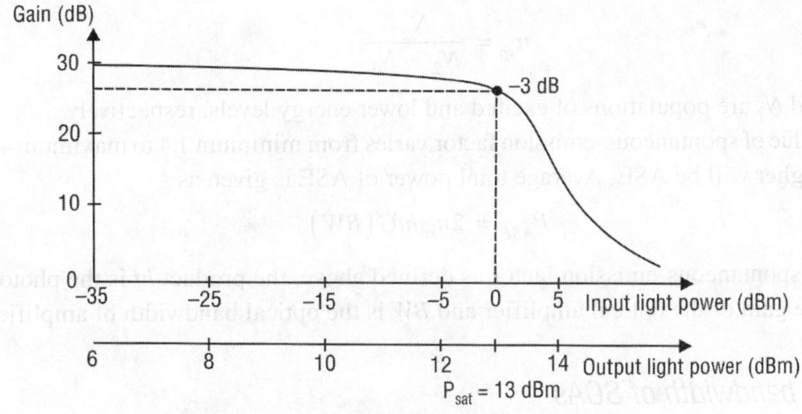

Fig. 5.15 Gain vs input power of SOA

(a) Gain as a function of angular frequency

We know that

$$G_{FPA}(\omega) = \frac{G_s(\omega)\left[(1-R)^2\right]}{\left[1-RG_s(\omega)\right]^2 + 4RG_s(\omega)\sin^2\left[\frac{(\omega-\omega_0)L}{\upsilon}\right]} \tag{5.6}$$

$$G_{TWA}(\omega) = G_s(\omega) = e^{\left[(\Gamma g-\alpha)L\right]} \tag{5.7}$$

It is worthwhile to mention here that the value of $G_s(\omega)$ depends on the state of polarization of the input optical signal. The confinement factor Γ and the gain coefficient g also depend on the state of polarization because of rectangular shape and crystal structure of active region.

It is evident that at $R = 0.3$ (corresponding to natural semiconductor–air interface), peaks of FPA gain are obtained at resonant frequencies. Smaller the values of R, the less pronounced gain peaks will be. The difference between $G_{FPA}(\omega)$ and $G_{TWA}(\omega)$ is the value of their reflectances. The ratio of maximum and minimum values of optical gain of FPA is given as

$$\Delta G = \frac{G_{FPA}(\max)}{G_{FPA}(\min)} = \left[\frac{(1+RG_s)}{(1-RG_s)}\right]^2 \tag{5.8}$$

For $\Delta G < 2$, $G_sR < 0.17$; and for given value of $G_s = 30$ dB or 1000, $R < 0.00017$.

(b) Noise generated by SOA

Spontaneous emitted and amplified photons constitute amplified spontaneous emission (ASE) in an optical amplifier. This phenomenon generates the noise in its active medium. Since they are random in phase and direction, they generate noise within the signal's bandwidth. *Spontaneous-emission factor* (n_{sp}), also termed as *population-inversion factor*, is defined as the ratio of population of excited energy levels and the difference between the populations of excited and lower energy levels. Mathematically,

$$n_{sp} = \frac{N_2}{N_2 - N_1} \tag{5.9}$$

where, N_2 and N_1 are populations of excited and lower energy levels, respectively.

Typical value of spontaneous-emission factor varies from minimum 1.4 to maximum 4. The higher this value, higher will be ASE. Average total power of ASE is given as

$$P_{ASE} = 2n_{sp}hfG(BW) \tag{5.10}$$

where, n_{sp} is spontaneous-emission factor as defined above, the product hf is the photon energy, G represents the gain of the optical amplifier and BW is the optical bandwidth of amplifier.

(c) Optical bandwidth of SOAs

By definition, the bandwidth of an amplifier is the difference between the maximum and minimum frequency at which the gain falls by 3 dB from its maximum value. The bandwidth of FPLA and TWSLA are given as

$$BW_{FPLA} = (\omega - \omega_0) = \frac{v}{L}\sin^{-1}\left(\frac{1 - RG_s}{2\sqrt{RG_s}}\right) \approx \frac{c}{L}\sqrt{\frac{(1-R)^2}{R}} \times \sqrt{G_{FPA}(\max)} \tag{5.11}$$

$$BW_{TWSLA} \approx \frac{c}{L}\sqrt{\frac{(1-R)^2}{R}} \times \sqrt{G_s} \tag{5.12}$$

So, there is trade-off between gain and optical bandwidth in case of FPLA as well as TWSLA. Since R for TWSLA is quite small, therefore

$$BW_{TWSLA} \times \sqrt{G_s} \gg BW_{FPLA} \times \sqrt{G_{FPLA}(\max)} \tag{5.13}$$

Because $G_{FPLA} > G_{TWSLA}$; it implies that $BW_{TWSLA} \gg BW_{FPLA}$

Figure 5.16 shows bandwidth comparison of FPLA (or, simply FPA) and TWSLA (or, simply TWA).

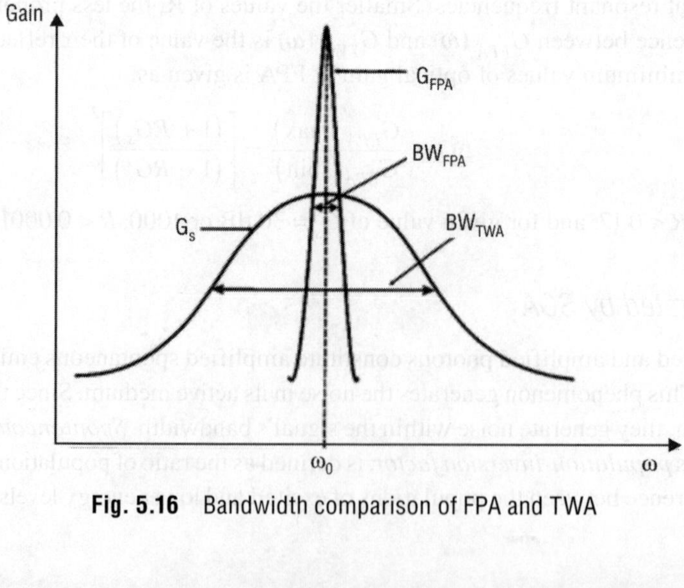

Fig. 5.16 Bandwidth comparison of FPA and TWA

(d) Polarization dependence of SOAs

The optical gain of SOAs depends on the state of polarization of the input optical signals, i.e., the amplification of TE and TM modes is different. This is due to the rectangular shape and the crystal structure of the active region, which make gain-coefficient (g) and confinement factor (Γ) dependent on polarization. The difference in gain between two orthogonal polarizations can be 5–7 dB. The polarization dependence in SOAs can be reduced in following ways:

(i) Make active region as square as possible in cross-section.
(ii) Connect two SOAs in series or in parallel to compensate for unequal gain in ortho–polarization.
(iii) A double pass through same active region.

5.2.3 Advantages of SOAs

- The optical gain provided by SOAs is relatively independent of wavelength of the incident optical signal.
- The injection current serves as the pump signal for amplification, not another laser.
- Due to compact size, SOAs can be integrated with several waveguide photonic devices on a single planar substrate.
- They use the same technology as diode lasers.
- SOAs have the ability to operate at operating wavelengths of 1300 nm and 1550 nm with wider bandwidth (up to 100 nm).
- They can be configured and integrated to function as pre-amplifiers at the optical receiver end.
- SOAs can function as simple logic gates in WDM optical networks.

Note: About ten laser diodes and a coupler can be fabricated on the same substrate that can serve as a WDM transmitter device. Integrating an SOA into the output could reduce some part of the coupling losses.

5.2.4 Limitations of SOAs

- SOAs can deliver output optical power up to a few mW only which is usually sufficient for single channel operation in a fiber–optic communication link. However, a WDM system may require up to a few mW power per channel.
- Since coupling of the input optical fiber into the SOA integrated chip tends to induce signal loss, SOA must provide additional optical gain in order to minimize the impact of this loss on the input facet of the active region.
- SOAs are highly sensitive to polarization of the input optical signal.
- They generate higher noise level in the active medium.
- In case multiple optical channels are amplified as required in WDM applications, SOAs can produce severe crosstalk.

Facts to Know

There is another undesired effect in an SOA. Cross saturation can cause undesired coupling between channels. However, this can be used for wavelength conversion and "controlling light with light" application. If used for multiple channels in a switched optical network, gain must be adjusted as channels are added and dropped. Four-wave mixing is also quite pronounced in SOAs that causes undesired coupling of light between channels. This can, however, also be used to advantage in wavelength converters.

Example 5.7 Pumping Rate of SOA

Consider an InGaAsP SOA having length of the amplifier = 500 μm, thickness of active layer = 0.3 μm, and width of the active area = 3 μm. If a 10 mA bias current is applied to it, then what would be the pumping rate? [Use $q = 1.6 \times 10^{-19}$C].

Solution:

We know that the pumping rate in SOA is given by

$$R_p = \frac{I}{qdwL}$$

where I is the bias–current in Amp , d is thickness of active region, w is width of active region, and L is the length of the amplifier.

For given I = 10 mA or 0.01 Amp , d = 0.3 μm, w = 3 μm and L = 500 μm, we have

$$R_p = \frac{0.01}{\left(1.6\times10^{-19}C\right)\left(0.3\times10^{-6}\right)\left(3\times10^{-6}\right)\left(500\times10^{-6}\right)}$$

$$\Rightarrow \quad R_p = 1.4\times10^{32}\,\text{electrons/m}^3 \text{ per second} \qquad \textbf{Ans.}$$

Section Practice Problem

1. Consider an InGaAsP SOA having length of the amplifier = 100 μm, thickness of active layer = 0.3 μm, and width of the active area = 3 μm. If a 10 mA bias current is applied to it, then what would be the pumping rate? *[Ans.: 9 × 10^{32} electrons/m^3 per second]*

5.3 Raman Fiber Amplifiers

A Raman fiber amplifier (RFA) is based on an intrinsic non-linearity present in the form of stimulated Raman scattering (SRS) mechanism. When a high-power optical signal (pump signal) propagates through the silica fibers, then SRS occurs.

How is SRS different from stimulated emission?

- In case of stimulated emission, an incident photon of the input optical signal stimulates emission of another identical photon without losing its energy.
- In case of SRS, the incident photon of the pump optical signal gives up some part of its energy to create another photon of reduced energy at a lower frequency. This phenomenon is termed as inelastic scattering. The remaining energy of the incident pump signal is absorbed by the medium in the form of optical phonons.

In a nutshell, Raman fiber amplifiers must be pumped optically in the optical silica-made fiber itself in order to provide optical gain. Fig. 5.17 illustrates the use of an optical fiber as a distributed Raman fiber amplifier.

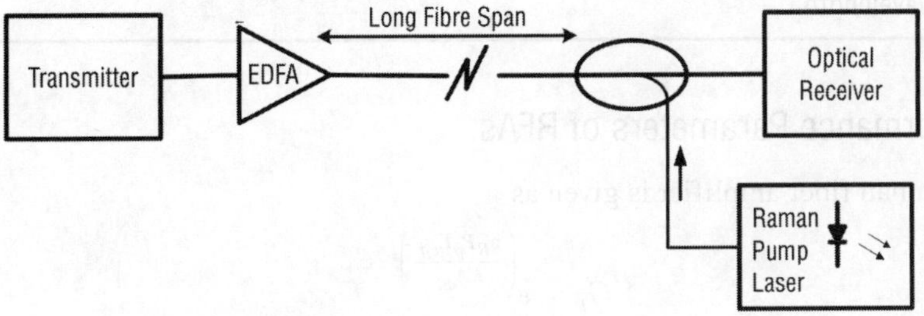

Fig. 5.17 Distributed Raman fiber amplifier

As shown, the pump signal from Raman pump laser diode at angular frequency ω_p is injected into the transmitted optical signal propagating into the long fiber span at angular frequency ω_s at a specific point through a fiber coupler. As these two optical signals co-propagate inside the fiber, the optical energy from Raman pump laser is transferred to the optical signal. It is important to note here that SRS Raman pumping takes place only in the backward direction (i.e., towards optical receiver side) over the fiber. Thus, the optical gain decreases in the direction of the transmitter, whereas it is maximum closer to the receiver end.

> ***Note:*** A new signal, known as a Stokes wave, is generated due to Stimulated Raman Scattering. The optical signal to be amplified must be longer in wavelength (or, lower in frequency) than that of the pump signal. This is the essential condition for amplification to occur due to SRS in Raman fiber amplifier. When the difference in these two frequencies is approximately 13.2 THz, then the optimal amplification occurs.

Fig. 5.18 shows a 1313 nm band Raman amplifier operation.

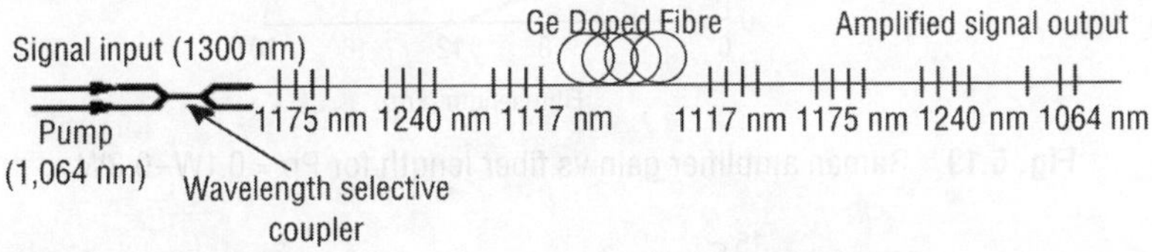

Fig. 5.18 Raman fiber amplifier operation at 1310 nm

As seen, the optical signal input (at 1300 nm wavelength) and the optical pump signal (at 1064 nm wavelength) enter the Ge doped fiber together through a wavelength selective coupler. A high level (around 20%) of Ge dopant is used in silica fiber in order to increase the SRS effect. The pump signal at 1064 nm is shifted to higher wavelengths in stages and then pumps the 1300 nm input signal by the SRS mechanism. In this way, sufficient optical gain is obtained.

> ***Note:*** A narrow fiber core size is used to increase the intensity of the light so as to achieve higher gain efficiency. This results in low noise process of amplification at small signal levels, yielding desired optical gain.

Facts to Know

It is easy to construct a Raman fiber optical amplifier. However, due to availability of specific wavelengths and limited power laser devices, it is difficult to manufacture very high power (about 0.5W) pump lasers at required wavelength.

5.3.1 Performance Parameters of RFAs

The gain of Raman fiber amplifier is given as

$$G_R = e^{\left(\frac{g_R P_p L_{eff}}{K A_{eff}}\right)} \tag{5.14}$$

where, g_R is Raman power gain coefficient, P_p is optical pump power, $L_{eff} = \dfrac{1 - e^{-\alpha_p L}}{\alpha_{p_2}}$ is the effective fiber core length (α_p being the fiber transmission loss at pump wavelength and L being the actual fiber length), K is constant (=2 in single-mode fiber), and $A_{eff} = \pi r_{eff}$ is the effective fiber core area (r_{eff} being the effective core radius).

Fig. 5.19 and Fig. 5.20 show Raman gain (G_R) versus fiber length (L) plot for different values of pump power in Raman optical amplifiers.

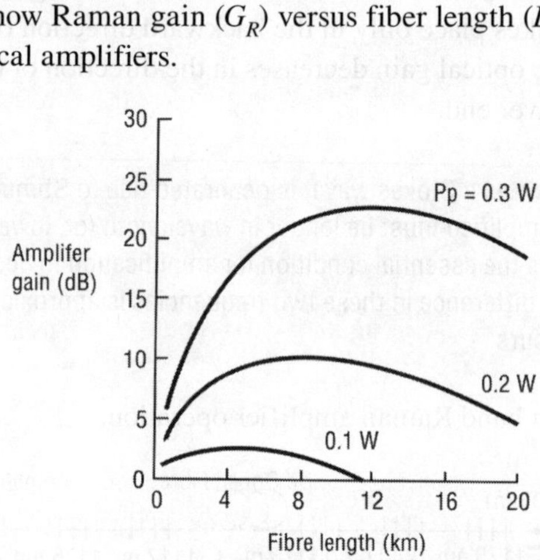

Fig. 5.19 Raman amplifier gain vs fiber length for Pp = 0.1W–0.3W

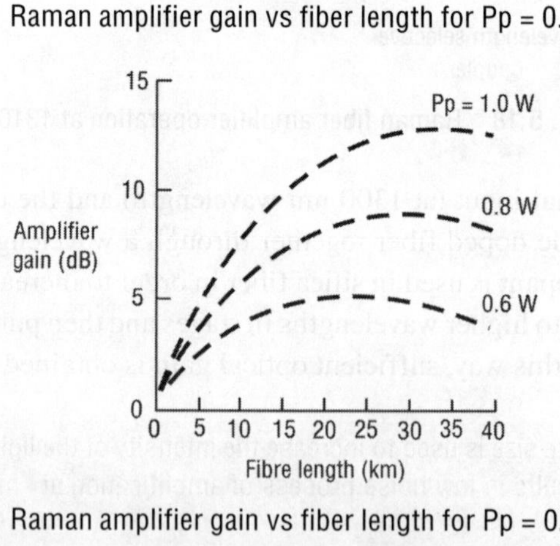

Fig. 5.20 Raman amplifier gain vs fiber length for Pp = 0.6W–1.0W

It is observed that Raman gain (G_R) becomes larger as fiber length L increases up to around 50 km where it reaches an almost constant value. Higher value of G_R can be achieved with low-loss fibers. Moreover, G_R is increased as fiber core diameter is decreased. However, there is requirement of high optical pump signal power.

Wide bandwidth Raman optical amplifiers can be realized using multiple pumps. Fig. 5.21 shows Raman optical gain versus wavelength characteristics with one pump and two pumps.

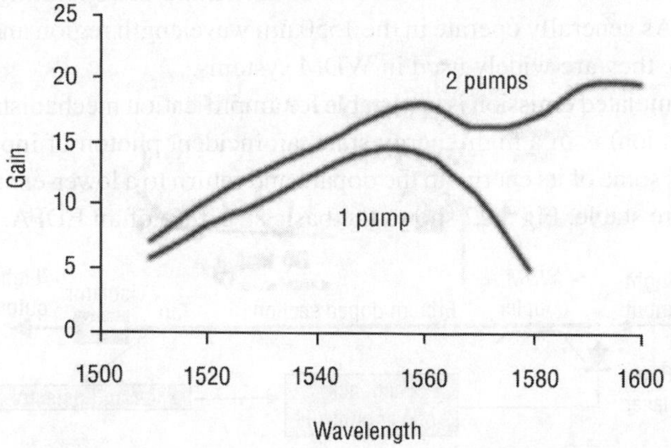

Fig. 5.21 Raman fiber amplifier gain vs wavelength

5.3.2 Advantages and Applications of RFAs

(i) Raman fiber amplifiers are mainly used as a preamplifier for improving receiver sensitivity as they can provide upto 20-dB gain for 1W of pump signal power.
(ii) Due to their low noise figure (approximately 4 dB), they can be employed as optical preamplifier for high speed optical receivers.
(iii) RFAs have broad bandwidth, so these are used for amplifying several channels simultaneously (WDM applications).

5.3.3 Drawbacks of RFAs

(i) High optical pump power is required.
(ii) Rayleigh crosstalk may be present due to backscattering.
(iii) Raman gain is sensitive to polarization of the input optical signal to some extent.

Facts to Know

In the 1980s, Raman fiber amplifier was demonstrated in 1270–1670 nm wavelength range. Due to non-availability of high-power diode laser pump source, any type of optical fiber can be used as the amplification medium. Raman amplification process itself provides high-power laser. However, its biggest disadvantage is cross-talk.

5.4 Erbium–Doped Fiber Amplifiers

Erbium–doped fiber amplifiers (EDFAs) make use of rare–earth element such as erbium (Er^{3+}) as amplification medium. It is doped into the fiber core during the manufacturing process. It consists of a short piece of fiber (typically 10 m or so) made of glass in which a small controlled amount of erbium is added as dopant in the form of an ion (Er^{3+}). Thus, the silica fiber acts as a host medium. It is the dopants (erbium) rather than silica fiber that determine the operating wavelength and the gain bandwidth. EDFAs generally operate in the 1550 nm wavelength region and can offer capacities exceeding 1 Tbps. So, they are widely used in WDM systems.

The principle of stimulated emission is applicable for amplification mechanism of the EDFA. When the dopant (an erbium ion) is in a high-energy state, an incident photon of input optical signal will stimulate it. It releases some of its energy to the dopant and return to a lower-energy state ("stimulated emission") that is more stable. Fig 5.22 shows the basic structure of an EDFA.

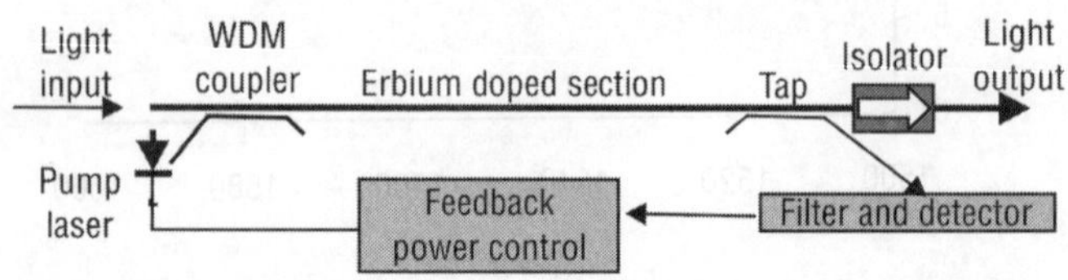

Fig. 5.22 Basic structure of an EDFA

The pump laser diode normally produces an optical signal of wavelength (at either 980 nm or 1480 nm) at high power (~ 10–200 mW). This signal is coupled with the light input signal in the erbium-doped section of the silica fiber through WDM coupler. The erbium ions will absorb this pump signal energy and jump to their excited state. A part of the output light signal is tapped and fed back at the input of pump laser through optical filter and detector. This serves as feedback power control mechanism so as to make EDFAs as self-regulating amplifiers. When all the metastable electrons are consumed then no further amplification occurs. Therefore, the system stabilizes automatically because the output optical power of the EDFA remains almost constant irrespective of the input power fluctuation, if any.

> ***Note:*** One of the distinguishing features of EDFA is that its active medium is a small section of regular silica fiber which is heavily doped with ions of erbium element. The external pump signal is supplied by a high-power laser diode radiating either at 1480-nm wavelength directly, or at 980-nm wavelength indirectly. Another distinguished feature is that its external energy is also delivered directly in the optical domain.

Fig. 5.23 shows the simplified functional schematic of an EDFA in which a pump signal from the laser is added to an input optical signal (at 1480 nm or 980 nm) through a WDM coupler.

This diagram shows a very basic EDF amplifier. The wavelength of the pump signal (with pump power of about 50 mW) is 1480 nm or 980 nm. Some part of this pump signal is transferred to the input optical signal by stimulated emission within a short length of Erbium-doped fiber. It has typical optical gain of about 5–15 dB and less than 10 dB noise figure. For 1550 nm operation, it is possible to obtain 30–40 dB optical gain.

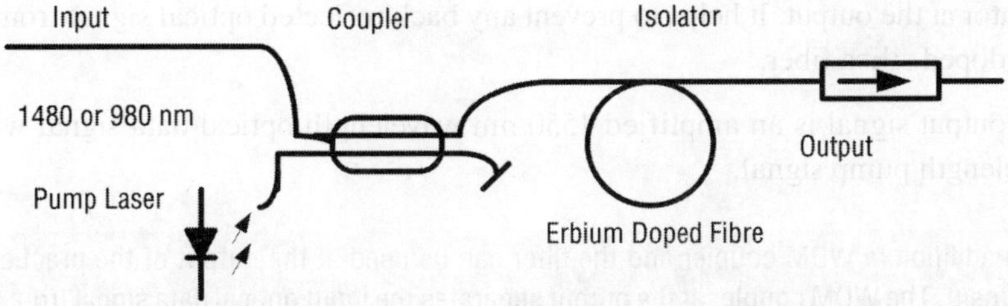

Fig. 5.23 Simplified functional schematic of an EDFA

> ***Note:*** High performance commercial designs provide output powers from 10 to 23 dBm (10 mW to 200 mW) and noise figures between 3.5 and 5 dB (the physical limit is 3.01 dB). EDFAs have been deployed in terrestrial and submarine links and now are considered as standard components using a well understood technology.

Fig. 5.24 depicts a simplified operation of an EDFA with its practical structure when used in WDM application.

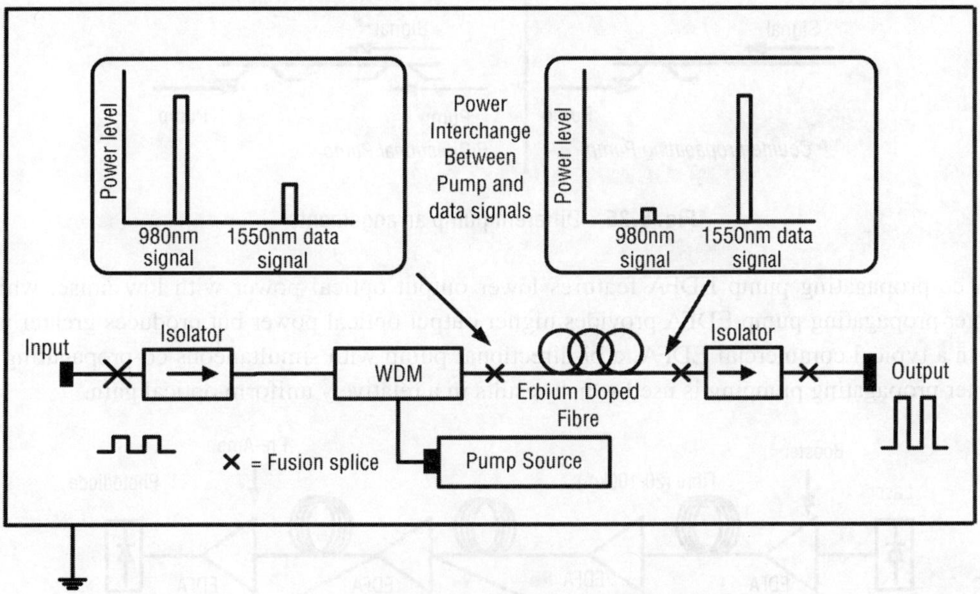

Fig. 5.24 Practical realization of an EDFA

As shown, it includes the following major parts:

- An isolator at the input. This keeps the noise generated by an EDFA from propagating towards the transmitter end.
- A WDM coupler. It combines the low-power 1550 nm optical input data signal with high-power pumping optical signal (from pump source such as laser) at 980 nm wavelength.
- A small section of erbium-doped silica fiber. In fact, this serves as the active medium of the EDFA.

- An isolator at the output. It helps to prevent any back-reflected optical signal from entering the erbium-doped silica fiber.

The final output signal is an amplified 1550 nm wavelength optical data signal with a residual 980 nm wavelength pump signal.

> ***Note:*** In addition, a WDM coupler and the filter can be used at the output of the practical EDFA arrangement. The WDM coupler at the output separates the input optical data signal from residual pump optical signal. The filter at the output further separates the residual pump optical signal from the optical data signal.

There are two types of structures of EDFAs:

- EDFA with co-propagating pump
- EDFA with counter-propagating pump

Fig. 5.25 shows counter-propagating pump and bidirectional pump arrangements that can be used in EDFA structures.

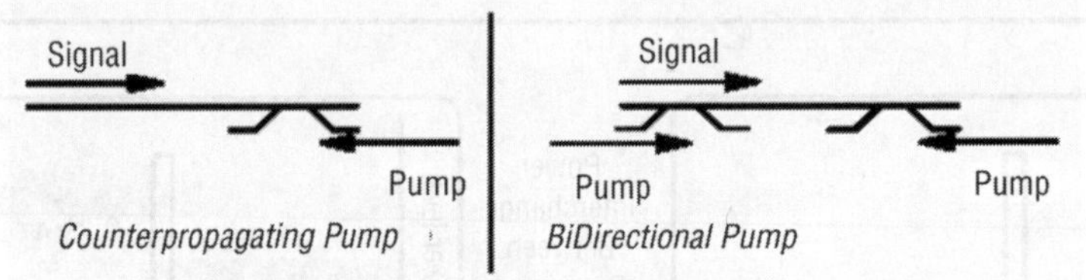

Fig. 5.25 Different pump arrangements

A co-propagating pump EDFA features lower output optical power with low noise; while a counter-propagating pump EDFA provides higher output optical power but produces greater noise too. In a typical commercial EDFA, a bi-directional pump with simultaneous co-propagating and counter-propagating pumping is used which results in a relatively uniform optical gain.

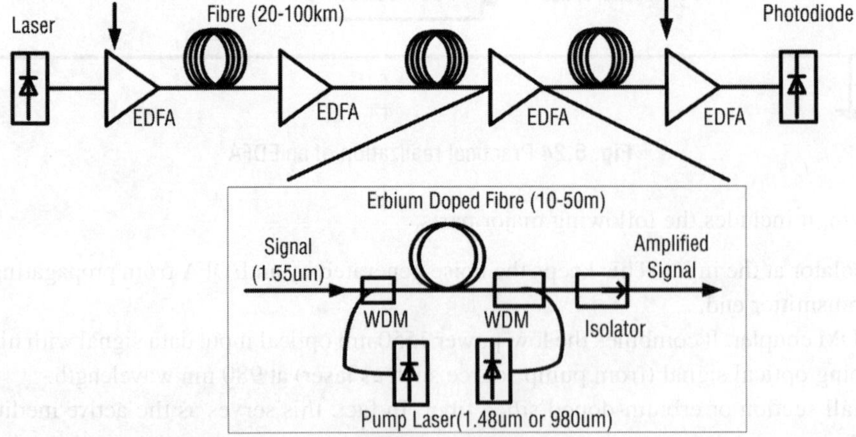

Fig. 5.26 Application of EDFA as booster, in-line, and pre-amplifier

In a long-haul application of an optical fiber communication link, EDFAs can be used as a booster amplifier at the output of the optical transmitter, an in-line optical amplifier along the optical fiber as well as a pre-amplifier just before the receiver, as shown in Fig. 5.26.

It may be noted that in-line EDFAs are placed at 20–100 km distance apart depending on the fiber loss. The optical input signal is at 1.55 μm wavelength, whereas the pump lasers operate at 1.48 μm or 980 nm wavelength. Typical length of Erbium-doped fiber is 10–50 m.

Facts to Know

EDFAs are commercially available since the early 1990s. They work best in the wavelength range of 1530–1565 nm. EDFAs are optically transparent and can provide optical gain up to 30 dB and virtually unlimited RF bandwidth. EDFAs have been deployed in terrestrial and submarine links and now are considered as standard components using a well understood technology.

5.4.1 Amplification Mechanism in EDFAs

As stated earlier, the amplification mechanism in an EDFA is based on stimulated emission similar to that of in laser. High energy from the optical pump signal (produced by another laser) excites the dopant erbium ions (Er^{3+}) in a silica fiber at the upper energy state. The input optical data signal stimulates the transition of the excited Erbium ions to the lower energy state and results in the radiation of photons with the same energy, i.e., the same wavelength as that of the input optical signal.

Energy-level Diagram: Free Erbium ions exhibit discrete levels of energy band. When Erbium ions are doped into a silica fiber, each of their energy levels splits into a number of closely related levels so as to form an energy band. See Fig. 5.27.

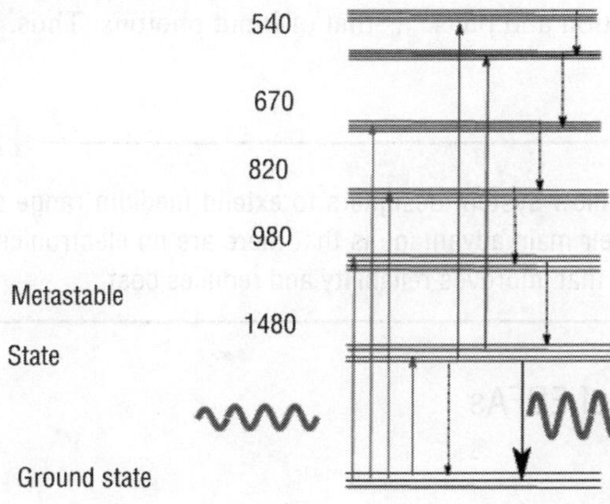

Fig. 5.27 Amplification mechanism in EDFA

To attain population inversion, Er^{3+} ions are pumped at the intermediate level 2. In indirect method (980-nm pumping), Er^{3+} ions are continuously moved from level 1 to level 3. It is followed by non-radiatively decay to level 2, from where they fall to level 1, radiating the optical signals in the desired wavelength of 1500–1600 nm. This is known as *3-level* amplification mechanism. The key to using this 3-level amplification mechanism is the lifetime of upper two levels, for example, τ_{sp} ~1 μs at level 3 and τ_{sp} ~10 ms at level 2 (known as *metastable level).* Therefore, Er^{3+} ions pumped at level 3 will descend to level 2 quickly and will stay there for a comparatively longer time. Thus, Er^{3+} ions will accumulate at level 2, creating population inversion. Thus, EDFAs operate on the basis of a three-level pumping scheme – excited state, metastable state and ground state.

When pumping is done directly at 1480-nm, due to longer τ_{sp} at this level, population inversion is created because of their accumulation. When an optical signal operating at one of the WDM wavelengths passes through such an inversely populated erbium-doped fiber, the transition of Er^{3+} ions from level 2 to level 1 will be stimulated. Fig. 5.28 depicts the flow of signals in EDFA causing amplification.

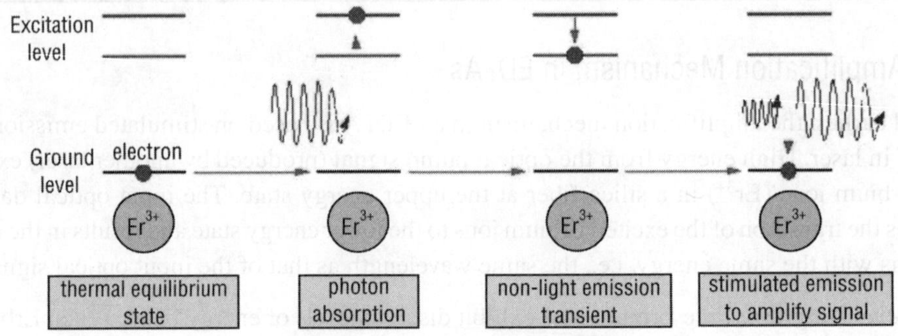

Fig. 5.28 Flow of signals in EDFA

This stimulated transition will be accompanied by the stimulated emission of photons having the same wavelength, direction and phase as that of input photons. Thus, amplification of the input signal occurs.

Facts to Know

Remotely pumped EDFAs allow system designers to extend medium range submarine links, such as those between islands. Their main advantage is that there are no electronics and therefore no power needs along the link, a fact that improves reliability and reduces cost.

5.4.2 Characteristics of EDFAs

(a) Optical Gain

By definition, the optical gain of an optical amplifier is the ratio of the output optical power to input optical power. That is,

$$\text{Optical gain} = P_{out}/P_{in} \tag{5.15}$$

Or,

$$\text{Optical gain (dB)} = 10 \log (P_{out}/P_{in}) \tag{5.16}$$

The output optical power includes the power of both the signal and the ASE (Amplifier Stimulated Emission) noise. It means

$$\text{Optical gain (dB)} = 10 \log [(P_{out} + P_{ASE})/P_{in}] \quad (5.17)$$

Typically, the optical gain can be obtained in the range of about 20–40 dB, depending on their functions such as boosters, in-line, or pre-amplifiers.

(b) Gain Spectrum

By definition, the gain spectrum of an optical amplifier means the characteristic curves that exhibit the variation of optical gain as a function of wavelength and gain flatness. Typically, the optical gain of an EDFA depends on number of factors like the concentration of dopant Er^{3+} ions, the length (L) of EDFA, the core radius (a) of the silica fiber, the power level (P_p) of the pump signal, and the wavelength of the input signal (λ_s). There is no optical gain outside the specific range of wavelengths. But even within this wavelength range, optical gain varies substantially as depicted in Fig. 5.29.

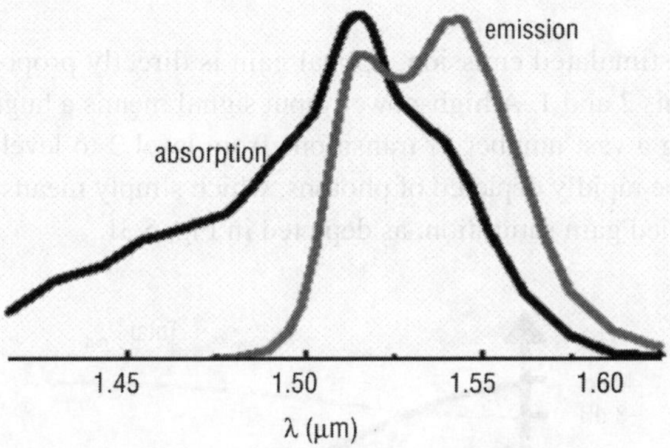

Fig. 5.29 Gain spectrum characteristics (optical gain vs wavelength)

There is significant fluctuation in the optical gain between 1.52 μm and 1.57 μm. The amorphous nature of silica and other co-dopants (Ge and Al) within the fiber core as well as the variations in pump signal power do affect the shape of the gain spectrum considerably.

The optical gain versus wavelength curve of the EDFA (as well as the ASE versus wavelength plot) can vary with input signal wavelength and power, as shown in Fig. 5.30.

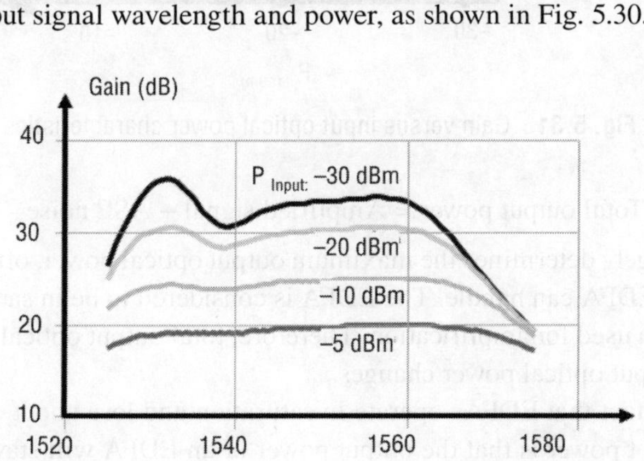

Fig. 5.30 Gain vs wavelength (nm) for various values of P_{in}

It is interesting to observe here that there is reduction in optical gain with increase in input power. If the input signal power is -20 dBm, then the gain is about 30 dB at 1550 nm, resulting in +10 dBm output. If the input is -10 dBm, then the gain is about 25 dB and the output about +15 dBm. In other words, when the input optical signal power varies by a factor of ten, then the output power varies only by a factor of three.

Above -10 dBm input power level, the amplifier is in full compression: at -5 dBm input power level, it has 20 dB gain, therefore, the 5 dB increase in input power has no effect on the output power (but it may have improved the noise figure). The saturation level can also be recognized by the fact that the traces become more flat when the input power increases. Saturation is a preferred point of operation because it stabilizes the system and reduces noise without causing nonlinear effects (like clipping) inside the amplifier for high speed modulation.

(c) Gain Saturation

As per the principle of stimulated emission, optical gain is directly proportional to the difference in the population of levels 2 and 1. A high-power input signal means a huge number of photons per unit of time, stimulating a vast number of transitions from level 2 to level 1. This implies that the intermediate level will be rapidly depleted of photons, which simply means decreasing optical gain. This phenomenon is called gain saturation, as depicted in Fig. 5.31.

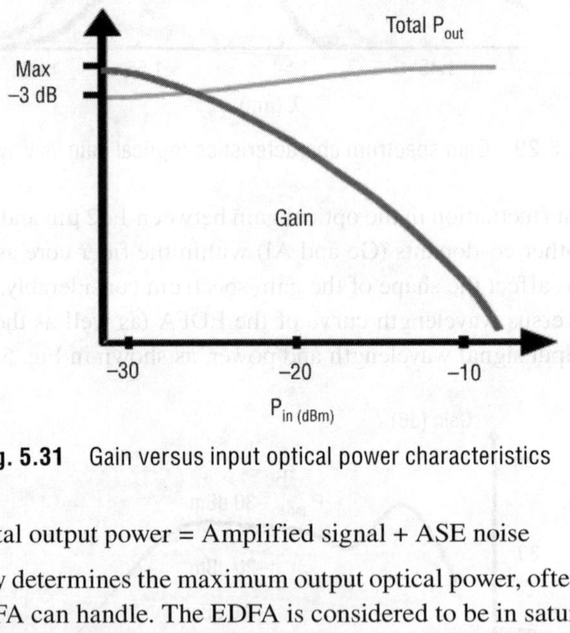

Fig. 5.31 Gain versus input optical power characteristics

Total output power = Amplified signal + ASE noise

Gain saturation largely determines the maximum output optical power, often called the saturated output power that an EDFA can handle. The EDFA is considered to be in saturation as if almost all Erbium ions have been used for amplification. Therefore, total output optical power remains almost same, regardless of input optical power changes.

One benefit of the fact that EDFAs operate in saturation and lose about one dB of gain for one dBm increase in output power is that the output power of an EDFA will stay fairly constant over a variety of operating conditions. This amplifier has an almost constant output power for a very wide input power range. At -30 dBm input optical power level, more than 50% of all metastable electrons are consumed for amplification. When the input optical power increases then this number approaches 100%. Because the pump power remains constant, the pool of excited electrons is limited.

When used with a single carrier, if the input power of the EDFA were to drop by one dB, the gain would increase by one dB to re-establish the previous output power operating level. If the input signal power is increased further, the optical gain would drop again, reestablishing the previous operating point. Again, we assume that the power fluctuations occur much slower than the time constant of the metastable state (~1 ms), and the modulation is much faster (20 kHz to many GHz).

Differences between Electronic and EDFAs at gain saturation (Refer Fig. 5.32)

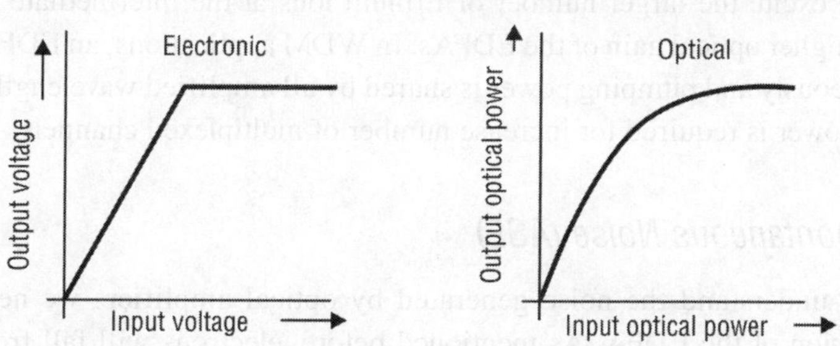

Fig. 5.32 Electronic amplifier vs EDFA at gain saturation

- An electronic amplifier operates relatively linearly until its gain saturates. Whereas in an EDFA, as the input power is increased, the gain increases slowly.
- An electronic amplifier operated near saturation just clips the peaks off and introduces significant distortions into the output signal. Whereas an EDFA at saturation yields less gain without any distortion of the output signal or crosstalk between WDM channels.

(d) Optimal Fiber Length

High-power pumping signal in an EDFA is provided along the length of an active silica fiber. At its input, a pumping signal has more power. As it propagates along an active optical fiber, the pumping power level decreases due to absorption. An amplified signal, on the other hand, becomes stronger while propagating along the active optical fiber. This implies that both the signal power level and the pump power level vary along the length of the amplifier. Due to available of only a finite number of Erbium ions, a finite gain (and a finite maximum power) per unit length of the amplifier can be obtained. This means there has to be optimal fiber length which would yield optimum optical gain.

- In an optical amplifier (EDFA) designed for single wavelength operation, the optimal amplifier length is a function of the data signal power level, the pump signal power level, the concentration of dopant Erbium ions, and the required optical gain.
- In an optical amplifier (EDFA) designed for multi-wavelength operation (for WDM application), the additional requirement of gain flatness over the desired range of amplified wavelengths demands a careful design and optimization of the amplifier's length.

Since the optical power of the pumping signal damps along an active fiber, an input optical signal will experience less and less gain and eventually begin to undergo loss. Thus, an active fiber has an optimal length which depends on doping concentration, gain flatness, gain bandwidth product, and shape of the gain characteristics.

> *Note:* Typical length of an active fiber in EDFA ranges from a few meters to 20–50 meters.

(e) Pumping Optical Power

Optimal fiber length of an EDFA depends on pumping optical power. Obviously, the higher pumping optical power will excite the larger number of Erbium ions at the intermediate level. This would yield, in turn, the higher optical gain of the EDFAs. In WDM applications, an EDFA amplifies many channels simultaneously and pumping power is shared by all amplified wavelengths, therefore more pumping optical power is required for increase number of multiplexed channels.

(f) Amplified Spontaneous Noise (ASE)

In order to better understand the noise generated by optical amplifiers we need to look at the spontaneous emission of the EDFA. As mentioned before, electrons will fall from the metastable state down to the ground stable either by stimulated emission due to an incoming photon (that is the amplification effect) or randomly with about a one millisecond time constant. Randomly emitted photons have a random phase, travel direction and wavelength within the amplifier's wavelength range. This is called the spontaneous emission. Those photons travelling along the fiber will trigger stimulated emission that of course will have their wavelength, phase, etc. At the end almost all energy pumped into an amplifier without any input signal reappears as amplified spontaneous emission (ASE). However, if an input signal consumes electrons in the metastable state, then fewer are left for spontaneous emission, therefore reducing the ASE. An EDFA can amplify many channels simultaneously because it has a relatively wide gain bandwidth product. Amplified optical signals (i.e., wavelength-separated channels) along with the noise associated with an EDFA are shown in Fig. 5.33.

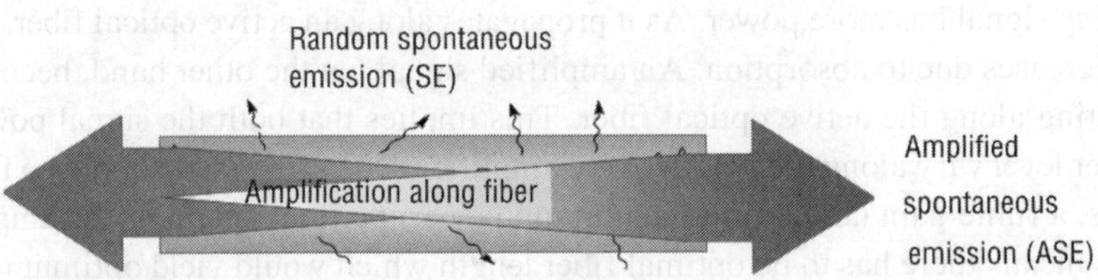

Fig. 5.33 Amplified Spontaneous Emission (ASE) mechanism

In EDFA, amplification of optical signal occurs along the Erbium-doped fiber. Due to this, random spontaneous emission occurs which is neither polarized nor coherent. This is inherent property of the Erbium ions that they randomly emit photons within the wavelength range of 1520–1570 nm. On the other hand, spontaneous emission stimulates emission of other photons.

- When input optical signal is not present, then all optical energy is eventually converted into amplified spontaneous emission (ASE) noise.

- When input optical signal is present, then it uses metastable electrons. So lesser amount of ASE noise is produced.

It implies that the optical noise in an EDFA, termed as amplified spontaneous noise (ASE) occurs due to noise–signal interference. This type of noise cannot be filtered because it is within the bandwidth of the optical signal. Thus, it contributes to the overall *noise figure of an EDFA*, denoted by F_n. It is expressed as

$$F_n = 2n_{sp} \tag{5.18}$$

where, n_{sp} is *spontaneous emission factor* which is the ratio of relative population of excited state (N_2) and the difference between population of excited state (N_2) and ground state (N_1). Mathematically, it can be expressed as

$$n_{sp} = \frac{N_2}{N_2 - N_1} \tag{5.19}$$

Since $N_1 \neq 0$, therefore $n_{sp} > 1$. Thus, noise figure (F_n) of EDFAs will be greater than 3 dB (typically 3.5–9 dB).

Fig. 5.34 gives a typical ASE output spectra of an EDFA with a stimulating input signal and with no input signal.

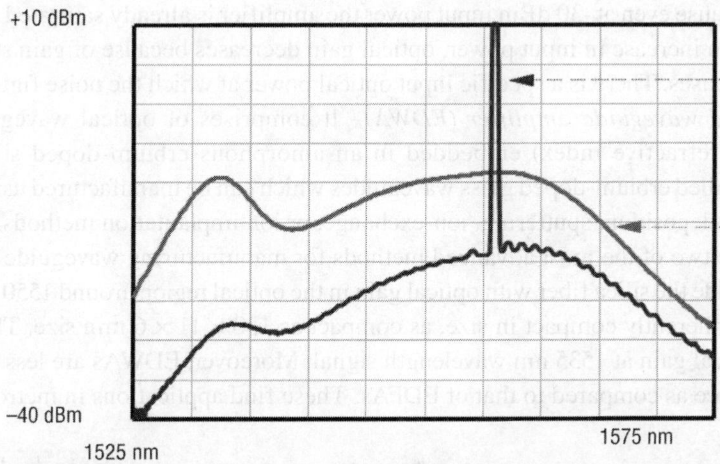

Fig. 5.34 ASE output spectra of an EDFA (input signal level vs wavelength)

From the ASE output spectra, the following observations may be made.

- Most of the pump signal power is present at the stimulating wavelength only.
- When the input signal is present, there is significant variation in the power distribution at the other wavelengths.

The basic question for characterizing EDFAs is how to measure its noise figure. If the input optical signal is turned off, then a large amount of ASE is present. And if the input optical signal is turned on, then a large amount of optical signal is present.

Noise figure describes how close an optical amplifier comes to an ideal optical amplifier that amplifies the input signal spectrum including noise but does not add any noise. According to quantum physics, it is impossible to build an optical amplifier with better than 3.0 dB noise figure. Fig. 5.35 shows the characteristic curves for noise figure versus wavelength of an EDFA.

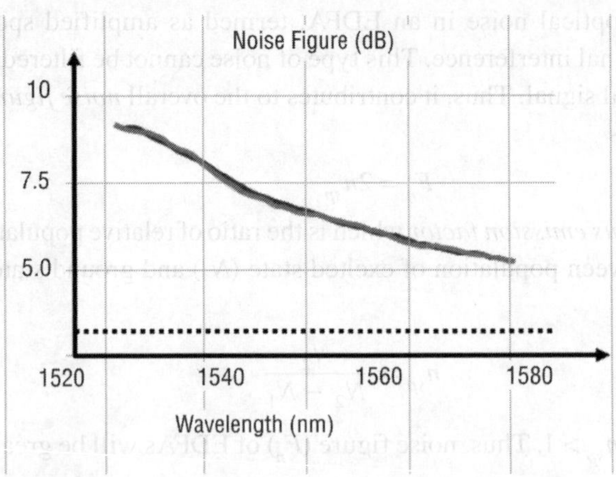

Fig. 5.35 Noise figure characteristics of an EDFA

As we can see, the noise figure becomes better with increasing wavelength. The traces overlap significantly because even at -30 dBm input power the amplifier is already saturated sufficiently. We know that with an increase in input power, optical gain decreases because of gain saturation, while noise figure increases. There is a specific input optical power at which the noise figure is minimum.

Erbium-doped waveguide amplifier (EDWA)– It comprises of optical waveguides (made of increased glass refractive index) embedded in an amorphous erbium-doped silica fiber. This arrangement is called erbium-doped glass waveguides which can be manufactured using PECVD and flame hydrolysis deposition, sputtering, ion-exchange, or ion implantation methods. Sputtering and ion-exchange are two of the most advanced methods for manufacturing waveguide amplifiers. The erbium ions provide the silica fiber with optical gain in the optical region around 1550 nm wavelength.

EDWAs are inherently compact in size, as compact as 130 × 11 × 6 mm size. They can provide up to 15-dB optical gain at 1535 nm wavelength signal. Moreover, EDWAs are less costly and offer better performance as compared to that of EDFAs. These find applications in metro area networks.

Facts to Know

The gain and noise figure of the first generations EDFAs were not flat over wavelength. The amplitude levels or signal-to-noise ratio (SNR) can degrade quicker than desired if several EDFAs are cascaded. Therefore, they may not be well suited for dense wavelength–division multiplexing (DWDM) applications. More recent designs compensate this effect with different fiber doping, compensation filters or simply by pre-emphasizing the channels at the transmitter side for improvements in SNR for all multiplexed channels.

5.4.3 Broadband EDFAs

Broadband EDFAs mean the utilization of the complete available bandwidth of 35–40 nm for amplification. This is possible if we employ tunable optical filters along with EDFAs to have wavelength-selective losses, thereby resulting in an almost flat gain spectrum. It implies that the transmission loss of an optical filter should be such higher loss occurs in the high-gain region and low loss occurs in the low-gain region. In other words, transmission loss should follow the gain profile. If such a type of optical filter is introduced after the Erbium-doped silica fiber, then the resultant output optical power will be more or less uniform for all input wavelengths. In order to make EDFAs broadband (i.e., flat gain over desired optical region), we can use various types of tunable optical filters such as thin-film optical interference-based Fabry–Perot and Mach–Zehnder interferometer filters, and optical diffraction-based long-period grating-based Michelson filters, acousto–optic filters, electro–optic filters.

It is desirable that in spite of considerable fluctuations of the input optical signal power, total output optical signal power of an EDFA should not vary. In a simple model, an amplifier has a pool of optical energy available for amplification. If a channel is dropped (or is added) in a DWDM system, then fewer (more) signals compete for this energy. As a result, the output power for other channels increases (decreases) accordingly. These unwanted power fluctuations must be taken care either by adjusting the power of the pump lasers within the EDFA, or by a high tolerance of the optical receiver.

Multistage EDFAs– One stage of EDFA simply means that it comprises of single section of erbium-doped silica fiber. Multistage EDFAs, or cascaded amplifiers may be needed for the following reasons:

- To enhance the total output optical power output without increasing amplified stimulated emission (ASE) noise.
- To achieve the gain flatness over the desired operating bandwidth
- To reduce the amount of ASE noise.

Fig. 5.36 shows a typical configuration employing two-stage EDFAs.

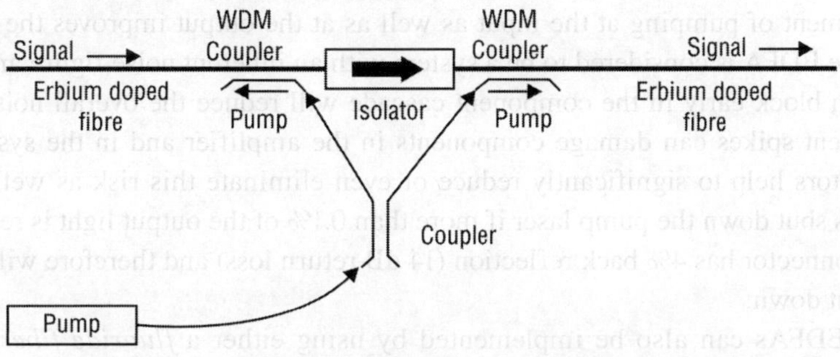

Fig. 5.36 Two-stage EDFA in-line amplifier configuration

Here, both EDFAs are used as in-line optical amplifiers. The pump signal is shared by these two EDFAs through optical couplers, having pump power ratio typically different from 50:50. The two sections of Erbium-doped fibers are isolated by an optical isolator after coupling pump power through WDM couplers. The two-stage EDFA in-line amplifier configuration helps to reduce the impact of ASE noise on the amplified optical signal. Usually, the first-stage EDFA is a low-noise device because of its low optical gain, whereas the second-stage EDFA is a high optical gain device and acts as a power amplifier. The overall noise figure is mostly determined by the first-stage low-noise EDFA.

Generally, in two-stage EDFA design, the first-stage EDFA is pumped using 980-nm laser with Erbium-doped fiber length of typically 20–30 meters. The second-stage EDFA is pumped bi-directionally using 1480-nm lasers with Erbium-doped fiber length of 200 meters. An optical isolator is used between these two stages. Its main function is to pass the amplified stimulated emissions from the first-stage EDFA to the second-stage EDFA for the purpose of providing necessary pumping to it while blocking the amplified stimulated emissions from backward-propagating towards the first-stage EDFA. The first-stage EDFA can provide optical gain in 1530–1570 nm wavelength range. Thus, cascaded design of multistage EDFAs are capable of providing flat gain over the desired wavelength region while maintaining relatively low level of ASE noise.

Commercial optical amplifiers are optimized for performance needed in a particular application (booster/in-line/pre-amplifier) as well as to optimize cost and functionality. Input and output monitors are added for safety and reflection monitoring reasons. Power sensors monitor overall system health and provide aging information. Fig. 5.37 shows a typical design of two-stage EDFA In-line amplifiers for telemetry and remote control application.

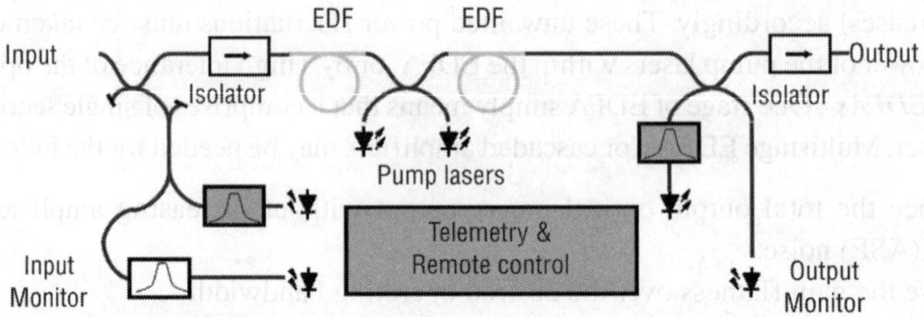

Fig. 5.37 EDFA in-line amplifiers for telemetry application

The arrangement of pumping at the input as well as at the output improves the noise figure of the EDFA. If the EDFA is considered to be a system with an inherent noise figure and a gain block, placing the gain block early in the component cascade will reduce the overall noise figure of the cascade. Transient spikes can damage components in the amplifier and in the system. Input and reflection monitors help to significantly reduce or even eliminate this risk as well. For example, some amplifiers shut down the pump laser if more than 0.1% of the output light is reflected back. A straight open connector has 4% back reflection (14 dB return loss) and therefore will cause such an amplifier to shut down.

Broadband EDFAs can also be implemented by using either a *fluoride fiber*, or a *telluride fiber* instead of silica fiber as the host medium. The Erbium atoms are heavily doped in this base material. Alternatively, Thulium-doped fiber amplifiers also can provide flat gain in 1480–1510-nm optical window. The total optical power (sum of all channels in WDM) coming out of an amplifier can cause non-linear effects in the fiber which may result into significant distortions. Therefore, system designers must carefully balance EDFA power levels, amplifier spacing and signal-to-noise requirements. Standards for long haul systems typically propose up to eight channels at 100 or 200 GHz spacing with up to twelve EDFAs to cover a span up to 600 km. Companies offering systems with more channels often target them for regional links with less or no amplifiers.

Conventional EDFAs are best used for single channel systems in the 1550-nm region. They can be designed to be deployed as power amplifiers, in-line amplifiers or pre-amplifiers. The bandwidth is not wide enough for DWDM, special EDFAs are needed. Table 5.4 gives a brief account of major specifications of three different types of conventional EDFAs.

Table 5.4 Parameters of conventional EDFAs

Parameter	*EDFA-19*	*EDFA-16*	*EDFA-14*
Wavelength Range	1525–1565 nm		
Maximum Gain	>40 dB	>38 dB	>35 dB
Optimum Gain Flatness	0.6 over 1545–1558 nm		
Saturation Power Level	19 dBm	16 dBm	14 dBm
Noise Figure	<6.0 dB	<6.5 dB	<6.0 dB
Minimum Isolation at Input and Output	30 dB		

Gain-flattened EDFA are now available for DWDM applications, with gain flatness within 1 dB in 1530–1560 nm wavelength range. Table 5.5 provides a comparative study of major specifications of three different types of gain–flattened EDFAs.

Table 5.5 Parameters of gain–flattened EDFAs

Parameter	*EDFA (FG18)*	*EDFA (FG15)*	*EDFA (FG13)*
Wavelength Range	1530–1560 nm		
Maximum Gain	>25 dB	>22 dB	>20 dB
Optimum Gain Flatness	<1 over 1530–1560 nm		
Saturation Power Levels	18 dBm	15 dBm	13 dBm
Noise Figure	<6.0 dB	<6.5 dB	<6.0 dB
Minimum Isolation at Input and Output	30 dB		

Facts to Know

The output gain, output power and noise figure of EDFAs can be tweaked by various design modifications. Using 980 nm pumps usually produce EDFAs with lower noise figures. EDFAs of this type make better preamplifiers. It is thought that the reason for this NF improvement is that the pump wavelength is farther out of the emission band and, for this reason, reduces ASE.

5.4.4 Advantages Offered by EDFAs

1. EDFAs can be used as power amplifier after optical transmitter, in-line optical amplifier along the fiber, and pre-amplifier before the optical receiver.
2. The optical gain is almost insensitive to polarization.
3. Inter-channel crosstalk does not occur in EDFAs due to large carrier lifetime (about 10-ms).
4. Inter-channel crosstalk due to cross-gain saturation can be minimized in the non-saturated regime operation.
5. EDFAs with a relatively high concentration of Erbium dopants (~ 500 parts per million) can be used as *lumped optical amplifiers.*

6. Most EDFAs are capable of providing optical gain as high as 20–25 dB gain using Erbium-doped fiber length of approximately 10 m.
7. The optical gain of EDFAs cannot be modulated at frequencies much larger than 10 kHz due to their relatively slow response.
8. The bandwidth of EDFAs is large which make it an ideal choice for point-to-point WDM networks.
9. EDFAs make WDM possible which increases the capacity of optical fiber communication links.

5.4.5 Limitations of EDFAs

1. The optical gain spectrum is not uniform.
2. EDFAs work only in the 1550-nm transparent window.
3. EDFAs cannot find application in CATV industry which operates exclusively at 1300-nm.
4. EDFAs are costlier as compared to other types of optical amplifiers.

Example 5.9 Maximum Input Power of EDFA

Consider that an EDFA pumped at 980 nm is being used as a power amplifier with a 10-mW pump power. When optical power of an input signal is 1 mW at $\lambda_s = 1550$ nm, the output of the amplifier is +20 dBm. Compute a) the optical gain of the amplifier (in dB), and b) the input pump power required to achieve this gain.

Solution:

(a) Given input optical signal power = 1 mW

Expressing it in dBm, we get input optical signal power = 10 log (1) = 0 dBm

Given output optical signal power = +20 dBm

Optical gain (dB) = Output optical power (dBm) – Input optical power (dBm)

Hence, Optical gain G (dB) = +20 – 0 = 20 dB **Ans.**

(b) We know that the maximum input power in an EDFA is given by

$$P_{s,in} \le \frac{(\lambda_p/\lambda_s) P_{p,in}}{G-1}$$

For the given $\lambda_p = 980$ nm, $\lambda_s = 1550$ nm, $P_{p,in} = 10$ mW and calculated G = 20 dB (i.e., 100 in ratio), we get

$$P_{s,in} \le \frac{(980/1550)(10 \times 10^{-3}\text{W})}{100-1} = 64\mu W$$ **Ans.**

Section Practice Problem

1. An EDFA, being used as a power amplifier, is pumped with a 60-mW power at 980 nm. If the optical gain of FDFA at $\lambda_s = 1550$ nm is 20 dB, then what would be the maximum input pump power required to achieve this gain? *[Ans.: 380 μW]*

5.5 Comparision of Optical Amplifiers

We have studied different aspects of semiconductor optical amplifiers, Raman fiber amplifiers and Erbium-doped fiber amplifiers in the preceding sections. Now we present a comparative study of various features offered by each one of them.

1. *Principle of operation:* Semiconductor optical amplifier (SOA) uses the principle of stimulated emission through population inversion in an active medium of a laser diode. Raman fiber amplifier (RFA) uses the principle of stimulated Raman scattering (SRS) that occurs in silica fibers when a high-power optical pump signal propagates through the fiber. Erbium-doped fiber amplifier (EDFA) uses the principle of stimulated emission in silica fiber which is heavily doped with Erbium (Er^{3+}) ions when pumping is done by a powerful laser diode.
2. *Types of basic structures:* SOAs are of two types– Fabry–Perot and Travelling–wave SOAs. Raman optical amplifier is fiber-based amplifier along with fiber coupler and optical coupler. EDFA is Erbium-doped fiber based amplifier along with WDM coupler, optical isolators and pump laser diodes.
3. *Optical Gain:* A typical optical gain of about 30 dB can be achieved with TWA type of SOA. Raman optical amplifier can provide 20–30 dB optical gain for pumped power of 1W with low-loss fiber. EDFA can provide 20–25 dB optical gain over EDF length of ~ 10 m through ~ 500 parts per millions of dopant density.
4. *Gain spectrum (operating wavelength region):* SOAs have the ability to operate in 1300-nm as well as 1550-nm wavelength band with wider bandwidth (upto 100-nm). Raman optical amplifier offers broad bandwidth suitable for WDM applications. EDFA operates in 1550-nm range only having 35–40 nm bandwidth. It is also widely used in WDM applications.
5. *Noise Characteristics:* In SOAs, amplified spontaneous emission (ASE) noise exists in active medium ($n_{sp} \approx 1.4 - 4$). Raman optical amplifier has low noise figure (≈ 4 dB). In an EDFA, the noise figure is typically in the range of 3.5 to 9 dB.
6. *Sensitivity to Polarization:* The optical gain of SOA is sensitive to the state of polarization of input optical signal. The optical gain of Raman optical amplifier is somewhat sensitive to polarization. The optical gain of EDFA is nearly insensitive to polarization.
7. *Inter-channel crosstalk in WDM applications:* The inter-channel crosstalk level is quite high in SOA. Rayleigh crosstalk is dominant in Raman optical amplifier mainly due to backscattering mechanism. There is no inter-channel crosstalk in EDFAs due to large carrier lifetime (about 10 ms).
8. *Application areas:* SOAs as well as EDFAs are used as power amplifiers in optical transmitter, in-line amplifiers and pre-amplifiers at optical receiver. Raman optical amplifiers are mainly used as pre-amplifiers due to low noise figure. EDFAs are widely used in WDM applications due to their large bandwidth, flat optical gain, insensitive to polarization, no inter-channel crosstalk at 1550-nm long-haul systems.

Table 5.6 gives a comparative study of typical specifications of SOA, Raman Optical Amplifier and EDFA.

Table 5.6 Comparison of optical amplifier parameters

S. No.	*Parameter*	*SOA*	*Raman Amplifier*	*EDFA*
1.	Optical Gain (dB)	>30	>25	>40
2.	Wavelength Range (nm)	1280–1650	1280–1650	1530–1560
3.	3 dB Bandwidth (nm)	60	Depends on pump	30–60
4.	Saturation Power Level (dBm)	15	0.5 × Pump Power	20
5.	Maximum Saturation Power Level (dBm)	18	0.75 × Pump Power	22
6.	Pump Signal Level	<400 mA	>30 dBm	25 dBm
7.	Noise Figure (dB)	8	5	5
8.	Polarization (dB)	<0.5	0	0
9.	Time Constant	0.2 ns	f_s	10 ms
10.	Switchable	Yes	No	No
11.	Cost Factor	Competitive	High	Medium

5.6 Applications of Optical Amplifiers

- Optical amplifiers can be used as in-line amplifiers along an optical fiber link. This enables linear amplification.
- Optical amplifiers can be bi-directional.
- Optical amplifiers are transparent for optical modulation type, signal format, or bit rate.
- A linear optical amplifier may allow multiplex operation of several optical signals at different wavelengths (e.g., wavelength–division multiplexing).
- The dispersion effects can be small with single-mode fiber for sufficient amplification of the optical signal.
- Optical amplifiers can be used as optical gain functional entities including optical receiver pre-amplifiers.
- When used in a non-linear mode, optical amplifiers can act as optical routers and cross-connect switches, optical pulse shapers and optical gates.
- Semiconductor optical amplifiers are widely used as transmitter amplifiers in 1300-mm window in CATV networks.
- Optical amplifiers can be employed in optical switches, filters, wavelength converters, etc.

Facts to Know

The recent advancements in WDM have created a need to extend the band useful for amplifications. New developments show that you can extend the range of the optical amplifier possibly up to 1625 nm (or more), effectively allowing designers to send more optical channels through the system. A requirement of better gain flatness in WDM can be achieved by using suitable optical filters along with or introducing appropriate modifications in doping of silica fibers.

Points to Remember

- An optical amplifier operates solely in the all-optical domain.
- Optical amplifiers are used to overcome fiber loss. They are used in basic applications such as power amplifier to increase the power to greater levels than possible from the optical source, in-line amplifiers for periodic power boosting, pre-amplifier to increase the received power sensitivity.
- Distribution loss in local area or cable networks can be compensated by the use of optical amplifiers.
- Various types of optical amplifiers include semiconductor optical amplifiers (lasers that aren't lasing), Raman fiber amplifiers and doped–fiber amplifiers.
- SOAs utilize stimulated emission from injected carriers.
- Fabry–Perot amplifier can provide high optical gain but it is non-uniform over wavelength spectrum.
- Traveling wave amplifier is broadband but very low facet reflectivities are needed.
- Optical reflections at the facets of the active region can severely affect the performance of the SOAs, especially when the single-pass gain is high.
- Several non-linear phenomena in SOAs lead to inter-channel crosstalk in multichannel operation.
- In fiber-based optical amplifiers, the optical gain is provided either by stimulated Raman scattering mechanism or by rare–earth dopants in the fiber.
- A fiber Raman amplifier is based on stimulated Raman scattering in which the pump energy at λ_p is transferred to the optical signal energy λ_s in a non-resonant process to provide optical gain at λ_s.
- An erbium-doped fiber amplifier (EDFA) is based on the principle of optical pumping of an Er^{3+} ions in silica fiber by low-attenuation region of silica-based fiber either at 980 nm or at 1480 nm optical source and stimulated emission at 1550 nm.
- The optical gain of an EDFA depends on several factors such as doping concentration, fiber length, pump power, etc.
- EDFAs are widely used in multichannel (WDM) systems.

Important Equations

The noise penalty factor of cascaded optical in-line amplifiers $N_{pf} = \frac{1}{G}\left(\frac{G-1}{\ln G}\right)^2$; *where G is the optical gain (in ratio).*

Gain of Fabry–Perot amplifier $G_{FPA}(\omega) = \frac{P_{out}}{P_{in}} = \frac{G_s(\omega)\left[(1-R)^2\right]}{\left[1-RG_s(\omega)\right]^2 + 4RG_s(\omega)\sin^2\left[\frac{(\omega-\omega_0)L}{\upsilon}\right]}$; *where $G_s(\omega)$ is a single-passage amplification factor assumed to have a Gaussian-shape dependence on angular frequency, R is power-reflection coefficients of cleaved facets, ω is current angular frequency, ω_0 is the center angular frequency, L is active medium length, and $\upsilon = c/n$ represents the speed of light within the active medium with refractive index n.*

Gain of traveling–wave amplifier, $G_{TWA}(\omega) = \frac{P_{out}}{P_{in}} = G_s(\omega) = e^{[(\Gamma g - \alpha)L]}$; *where Γ represents the confinement factor, 'g' and 'α' are the gain and loss coefficients, respectively, of an active region per unit length (1/m), and L is the length of the active medium (m).*

Average total power of ASE $P_{ASE} = 2n_{sp}hfG(BW)$; *where n_{sp} represents the spontaneous-emission factor, the product hf denotes the photon energy, G is the amplifier gain and BW is optical bandwidth of amplifier.*

Optical bandwidth of FPA $BW_{FPA} = \frac{\nu}{L}\sin^{-1}\left(\frac{1-RG_s}{2\sqrt{RG_s}}\right) \approx \frac{c}{L}\sqrt{\frac{(1-R)^2}{R}} \times \sqrt{G_{FPA}(\max)}$; *where $\upsilon = c/n$ represents the velocity of light within the active medium with refractive index n, L is the active medium length, R is power-reflection coefficients of cleaved facets, where $G_s(\omega)$ is a single-passage amplification factor assumed to have a Gaussian-shape dependence on angular frequency.*

Optical bandwidth of TWA $BW_{TWA} \approx \frac{c}{L}\sqrt{\frac{(1-R)^2}{R}} \times \sqrt{G_s}$; *where c represents the velocity of light, L is active medium length, R represents the power-reflection coefficients of cleaved facets, $G_s(\omega)$ is a single-passage amplification factor assumed to have a Gaussian-shape dependence on angular frequency.*

Relationship between optical bandwidth of FPA and TWA is given by

$$BW_{TWA} \times \sqrt{G_s} \gg BW_{FPA} \times \sqrt{G_{FPA}(\max)}$$

The gain of Raman optical amplifier, $G_R = e^{\left(\frac{g_R P_p L_{eff}}{KA_{eff}}\right)}$; *where g_R is Raman power gain coefficient, P_p is optical pump power, $L_{eff} = \frac{1-e^{-\alpha_p L}}{\alpha_{p_2}}$ is the effective fiber core length (α_p being the fiber transmission loss at pump wavelength, and L being the actual fiber length), K is constant (=2 in single-mode fiber), and $A_{eff} = \pi r_{eff}$ is the effective fiber core area (r_{eff} being the effective core radius).*

The overall noise figure of an EDFA, $F_n = 2n_{sp}$; *where* $n_{sp} = \frac{N_2}{N_2 - N_1}$ *is spontaneous emission factor and which N_1 and N_2 represents the relative populations of the ground and excited states, respectively.*

Key Terms with Definitions

ASE — Amplified Spontaneous Emission, comprising of spontaneous emitted photons as well as amplified photons in an optical amplifier.

Booster — A power amplifier that raises the output power of an optical transmitter before sending it down the optical fiber.

Detector sensitivity — The ratio of minimum value of the electrical signal power at the input of an optical receiver to deliver an acceptable bit-error rate to the new minimum detectable electrical signal level that is needed to maintain the same SNR after use of pre-amplifier at the input of optical receiver.

EDFA — Erbium-doped fiber amplifier– an optical amplifier that uses Erbium element as a gain medium for optical amplification by doping the silica fiber core.

EDWA — Erbium-doped waveguide amplifier– an optical amplifier that comprises of optical waveguides which are embedded in an amorphous erbium-doped glass substrate.

FPA — Fabry–Perot Amplifier, the same as regular index-guided FP lasers.

Gain spectrum — Optical gain as a function of wavelength and gain flatness of an optical amplifier.

In-line Amplifier An optical amplifier used to compensate for signal losses caused by fiber attenuation and operates in the middle of a fiber–optic link.

Noise penalty factor A measure of the path-average signal energy that must be increased (as optical gain increases) in a number of cascaded optical amplifiers used as in-line amplifiers to maintain a fixed SNR.

Optical bandwidth The frequency range at which the gain of an optical amplifier drops by 3 dB from its maximum value.

Optical gain Optical gain of an optical amplifier is the ratio of output to input optical power.

Pre-amplifier An optical amplifier that is employed just before it reaches the direct-detection optical receiver to improve its sensitivity.

Power penalty The increase in the average received power due to addition of noise at optical amplifiers, leading to sensitivity degradation.

Raman fiber amplifier An optical amplifier that exploits stimulated Raman scattering mechanism that occurs in silica fibers when a high-power optical signal from a pump source propagates through the fiber.

SOA Semiconductor Optical Amplifier, an active medium of a semiconductor laser without or with very low optical feedback.

TWSLA or TWA A traveling-wave semiconductor laser amplifier or traveling-wave amplifier, an active medium with no optical feedback and provides amplification of an input optical signal by a single passage through the active medium.

Short Answer Type Questions

1. What are common characteristics in all types of optical amplifiers?

All types of optical amplifiers are essentially based on the principle of population inversion. The input optical signal pulses propagating through the fiber induce stimulated emission. The output power levels of all optical amplifiers generally saturate above a certain level of optical signal power. They add amplified stimulated emission noise to the output optical signal.

2. List the major characteristics of semiconductor optical amplifiers (SOAs).

(1) SOAs are compact in size. In addition to fiber and electrical connections, an extremely small semiconductor chip area is needed.

(2) SOA devices can be made to operate in different wavelength regions although the gain bandwidth product is smaller.

(3) The optical gain responds very quickly to any instantaneous change in input signal level within nanoseconds due to small lifetime and stored energy of the photons.

(4) The slight variations in optical gain causes phase changes in the output optical signal. This phenomenon leads to linewidth enhancement factor.

(5) SOAs exhibit significant self-phase modulation and four-wave mixing types of non-linear distortions in the output optical signal.

(6) SOAs tend to have relatively higher noise figure.

(7) The amplification in SOAs is normally sensitive to polarization of the input optical signal.

3. Enumerate optical gain versus signal power in semiconductor optical amplifier.

(1) In SOAs, the carrier recombination lifetime is of the order of few hundred picoseconds that determine the optical gain dynamics. (This is in contrast to EDFAs, which have recombination lifetime of the order of few milliseconds).

(2) The optical gain responds fast to any variations in the input optical signal level.
(3) As the bandwidth of optically-modulated signal increases, the optical gain can cause severe distortions in the output optical signal.
(4) Semiconductor optical amplifiers have nonlinear characteristics which result in frequency chirping problem.
(6) SOAs can be configured to function as wavelength converters due to their nonlinear behavior.

4. What are typical applications of SOAs?
(1) Power booster (immediately after laser diode being used as optical source).
(2) In-line optical amplifier along the fiber.
(3) Wavelength converter in WDM networks.
(4) Optical switches in WDM networks.
(5) Pre-amplifier just before the detector at the optical receiver end.

5. What is the amplification process in semiconductor optical amplifier?
Semiconductors have valence and conduction bands. At thermal equilibrium, valence band has higher population of carriers. Under population inversion condition conduction band will have higher population. The semiconductor p–n junction is forward biased in order to achieve population inversion.

6. Why are Raman fiber amplifiers also known as distributed amplifiers?
Raman fiber amplifiers are topologically simpler to design because no special doping is required. They use intrinsic optical nonlinearity of fiber. The optical signal amplification takes place throughout the length of transmission fiber. Hence, these are also known as distributed amplifiers.

7. Mention the fundamental principle of amplification mechanism in Raman fiber amplifier.
Raman fiber amplifier uses intrinsic property of silica fiber which implies that medium of optical signal transmission itself combats signal losses. The mechanism behind this is called Stimulated Raman Scattering (SRS). Raman pumping is usually done backwards. High optical energy is co-launched into the fiber. The Raman gain is higher at the receiver end of the fiber than at the transmitter end which is very useful in compensating for the fiber losses.

8. List specific advantages of backward pumping in Raman amplifier.
(1) Raman amplification is an almost instantaneous process.
(2) Backward pumping in Raman amplifier averages out fluctuations in the output optical power.
(3) In case forward pumping is used in Raman amplifier, the pump noise will strongly affect the input WDM signals.
(3) If the pump has a slight fluctuation in time, individual bits are amplified differently, resulting into jitter in the output optical signal.

9. How does Raman gain in fiber depend on the frequencies of the input and pump signal, in addition to polarization of the input optical signal?
The usable gain bandwidth is about 48 nm and the maximum Raman gain occurs for a frequency offset of 13.2 THz in Raman fiber amplifier. The silica fiber structure has certain vibration modes. This results in the occurrence of maxima in the Raman gain spectrum curves. For example, a 1064-nm pump signal leads to the maximum Raman optical gain at a 1116-nm input signal.

10. Raman fiber has specific advantages and disadvantages. What are these?
The possibility of very broadband operation, compatibility with installed single-mode fibers, and amplification of variable wavelength are some of its major advantages. It can serve as an extension of EDFAs functionality. This may yield a lower average power and crosstalk. There are, however, certain disadvantages such as high pump power requirements, need of sophisticated gain control and noise.

11. EDFAs have certain promising technical characteristics. Mention them.

Efficient pumping, high output optical power, low noise and distortion, minimal inter-channel crosstalk, negligible sensitivity to polarization of input optical signal, and very high sensitivity are some of promising technical characteristics of EDFAs.

12. How does EDFA operate? What can cause it to become a laser?

A relatively high-powered pump optical signal at particular wavelength is coupled using a wavelength selective coupler with the input optical signal at standard wavelength different from that of pump signal. The mixed optical signals are guided into a section of standard silica fiber doped with erbium atoms. This high-powered pump optical signal excites the erbium atoms to their higher-energy state. When the signal photons meet the excited erbium atoms, they acquire some of their energy. This additional energy is in the form of extra photons which are coherent in nature (i.e., exactly in the same phase) and in the direction of the optical signal being amplified. An optical isolator is usually placed at the output to prevent reflections in the backward direction. Otherwise, these back-propagating reflections may cause the amplifier to become a laser in the extreme case.

13. State the fundamental properties of the element Erbium. How is it useful in providing amplification mechanism in EDFAs?

Erbium is a rare–earth element which can absorb and release light energy in the communications band around 1550 nm. When light at 980 or 1480 nm is applied to fiber doped with Er^{3+} ions, the fiber absorbs this energy, i.e., electrons are excited to the next higher energy level where they remain in a metastable state for some time. If left undisturbed, the Er^{3+} ions doped fiber will eventually release this energy in the band of frequencies from about 1530 to 1565 nm. If stimulated by an input signal in this band, the Er^{3+} ions doped fiber will emit the stored energy at the stimulated wavelength. The 980 nm absorption band is narrower than the 1480 nm band. In addition it is more difficult to make reliable 980 nm lasers. However, pumping the amplifier with 980 nm can result into a better noise figure of the amplifier.

14. Draw a suitable diagram to depict the functional properties of the element Erbium.

The element Erbium possesses phosphorescent properties, i.e., photons at 980 or 1480 nm wavelength activate electrons into a metastable state which is a temporary energy state. When they return to the ground state, they emit optical signals in the 1550 nm wavelength region, as shown in the Fig. 5.38.

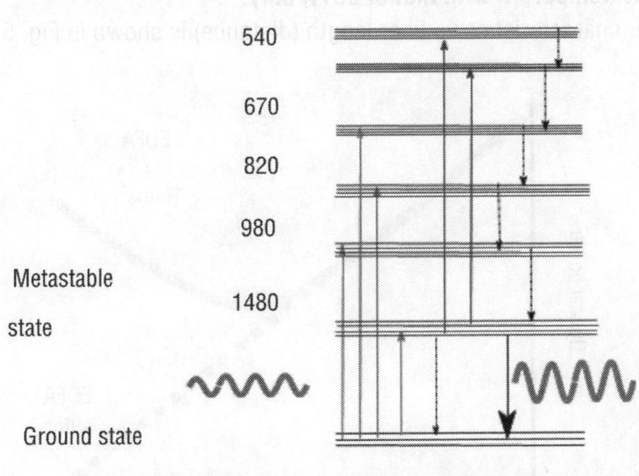

Fig. 5.38 Functional properties of Erbium

This phenomenon is known as stimulated emission and the emitted photons have the identical wavelength and phase as that of incident optical signals.

15. List some of attractive features of EDFAs for optical network use.

There are several attractive features for network use that include the following:

- Relatively simple construction
- Reliable, due to the number of passive components
- Allows easy connection to external fibers
- Broadband operation > 20 nm
- Bit rate transparency
- Ideally suited to long span systems
- Integral part of DWDM systems
- Undersea applications for OFAs are now common

16. What is the unique feature in distributed Raman amplification?

Distributed Raman amplification is usually used along with EDFA being employed as power amplifier in the transmitter section, as shown in the Fig. 5.39.

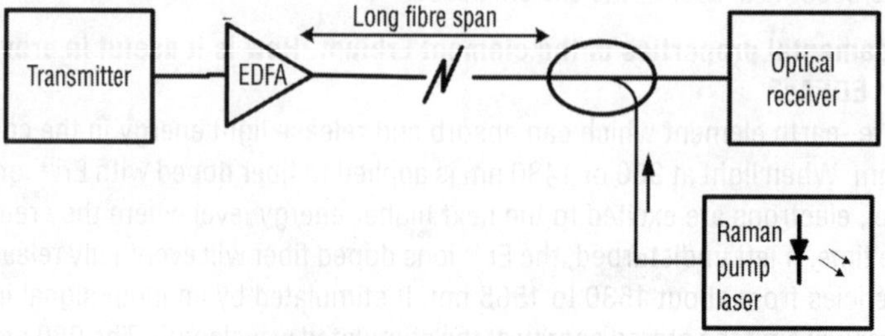

Fig. 5.39 Distributed Raman amplifier

Its unique feature is that Raman pumping happens in the direction towards the receiver end, not the transmitter end. This results in maximum optical gain towards the receiver end and minimum optical gain towards the transmitter end.

17. Plot the optical power versus fiber length (distance) characteristics for EDFA along with distributed Raman amplifier. Compare it with that of EDFA only.

The optical power characteristics vs fiber length (distance) is shown in Fig. 5.40.

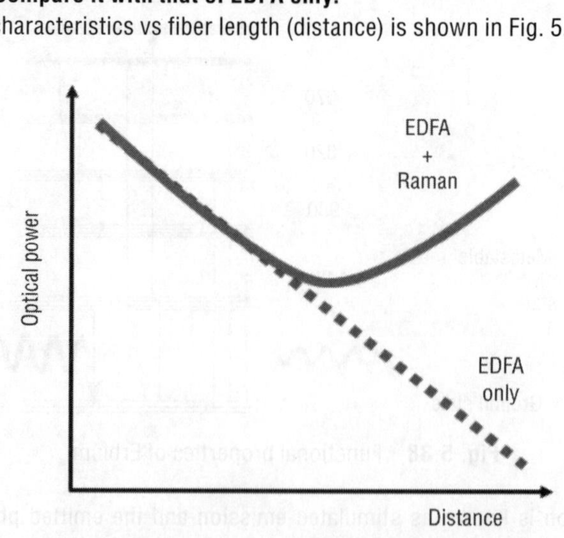

Fig. 5.40 Optical power vs fiber length (distance)

It is observed that as the length of the optical fiber (distance) increases, the optical power level of an EDFA located at the transmit end decreases. Whereas with an EDFA at the transmit end and Raman amplifier at the receive end (as in case of distributed Raman amplification), the minimum optical power level is obtained near the center of the fiber distance.

18. Can Raman amplification provide very broadband amplification? If yes, how?
Raman fiber amplifiers are capable of providing amplification over a wider range of wavelengths. In order to achieve very high gain over the desired wide range of wavelengths, usually multiple high-power pump signal sources such as wideband lasers are used. It has been successfully demonstrated that about 93 nm bandwidth amplification has been achieved with just two optical pump sources and 400 nm bandwidth amplification is possible by using more number of optical pump sources.

Multiple Choice Questions

1. Optical amplifiers can be used as
A. pre-amplifiers to precede the optical receiver.
B. in-line amplifiers to compensate for fiber loss.
C. power amplifiers to follow the optical transmitter.
D. all of these.

2. The difference between a regenerator and an optical amplifier can be best described by one of the following statements:
A. A regenerator amplifies as well as restores the optical signal transmitted.
B. A regenerator converts the optical signal into the electrical signal for amplification and then reconverts it into the optical domain, whereas an optical amplifier operates only in the optical domain.
C. An optical amplifier compensates for transmission loss.
D. Both are identical and as such there is no difference between the two.

3. The difference in the structure of a semiconductor laser and a semiconductor optical amplifier can be best described by one of the following statements:
A. The SOA is pumped electrically.
B. The end facets of the active region have 50% reflectivity.
C. The end facets of the active region have 100% reflectivity.
D. The end facets of the active region in the SOA have 0% reflectivity.

4. An optical amplifier when used as pre-amplifier has ________ optical gain and ________ maximum output power as compared to that of when used as in-line amplifier.
A. low; low B. high; high
C. low; high D. high; low

5. The use of FP resonator which provides optical feedback, can ________ the gain of a semiconductor optical amplifier.
A. significantly decrease B. significantly increase
C. marginally increase D. marginally decrease

6. The mirror reflectivity in typical SOAs is around
A. 100%. B. 50%.
C. 30%. D. 0%.

7. Statement I: In a Fabry–Perot semiconductor optical amplifier, the input optical signal that enters the active region is reflected several times from cleaved facets and leaves the cavity with optical gain.
Statement II: In traveling-wave semiconductor laser amplifier, an input optical signal is amplified by a single passage through the active region with no optical feedback.
A. Statement I is correct but Statement II is incorrect.
B. Statement I is incorrect but Statement II is correct.
C. Statement I as well as Statement II are correct.
D. Statement I as well as Statement II are incorrect.

8. Average total power of Amplified Spontaneous Emission (ASE) is given as
A. $P_{ASE} = n_{sp}hfG(BW)$ B. $P_{ASE} = 2n_{sp}hfG(BW)$
C. $P_{ASE} = (1/2)n_{sp}hfG(BW)$ D. $P_{ASE} = \left[2n_{sp}hfG(BW)\right]^2$

9. The optical gain and optical bandwidth in case of FPA as well as TWA are related to each other by the expression
A. $BW_{TWA} \times \sqrt{G_s} \gg BW_{FPA} \times \sqrt{G_{FPA}(\max)}$ B. $BW_{TWA} \times \sqrt{G_s} \leq BW_{FPA} \times \sqrt{G_{FPA}(\max)}$
C. $BW_{TWA} \times \sqrt{G_s} \ll BW_{FPA} \times \sqrt{G_{FPA}(\max)}$ D. $BW_{TWA} \times \sqrt{G_s} \geq BW_{FPA} \times \sqrt{G_{FPA}(\max)}$

10. The polarization dependence in SOAs can be reduced by
A. making the active region as square as possible in cross-section.
B. connecting two SOAs in series or in parallel to compensate for unequal gain in ortho–polarization.
C. providing a double pass through same active region.
D. Any of these methods.

11. Statement I: Raman gain (G_R) becomes larger as fiber length L increases upto around 50 km where it reaches an almost constant value.
Statement II: Raman gain (G_R) is increased as fiber core diameter is increased.
A. Statement I is correct but Statement II is incorrect.
B. Statement I is incorrect but Statement II is correct.
C. Both Statement I as well as Statement II are correct.
D. Both Statement I as well as Statement II are incorrect.

12. Raman fiber amplifiers are similar to Erbium-doped fiber amplifier in the sense that
A. both of them operate in all-optical domain.
B. both of them can be used around the 1550-nm optical window.
C. both of them can be employed for multichannel operation.
D. both of them operate in all-optical domain, can be used around the 1550-nm optical window as well as can be employed for multichannel operation.

13. EDFAs generally operate in the wavelength region near
A. 1550 nm and can offer capacities exceeding 1000 Gbps.
B. 850 nm and can offer capacities around 100 Gbps.
C. 1150 nm and can offer capacities around 1000 Gbps.
D. 1550 nm and can offer capacities around 100 Gbps.

14. Statement I: EDFAs are self-regulating amplifiers.
Statement II: The active medium in an EDFA is a small section of silica fiber which is heavily doped with rare-earth element erbium.
Statement III: EDFAs have been deployed in terrestrial and submarine links

A. Only statements I and III are correct. B. Only statement II is correct.
C. Only statements II and III are correct. D. Statements I , II as well as III all are correct.

15. EDFA differs from a Raman fiber amplifier (RFA) in the way that
A. An EDFA requires population inversion while the RFA does not.
B. An EDFA operates on the principle of stimulated emission.
C. A RFA operates on the principle of stimulated Raman scattering.
D. Both operate on the same principle.

16. Statement I: A co-propagating pump EDFA features higher output optical power with low noise.
Statement II: a counter-propagating pump EDFA provides lower output optical power but produces greater noise too.
A. Statement I is correct but Statement II is incorrect.
B. Statement I is incorrect but Statement II is correct.
C. Both Statement I as well as Statement II are correct.
D. Both Statement I as well as Statement II are incorrect.

17. Which wavelength is the most appropriate one for pumping an EDFA?
A. 850 nm B. 980 nm
C. 1300 nm D. 1550 nm

18. The optical gain in an EDFA depends on the following factors.
A. Doping concentration B. Pump power
C. Length of the doped fiber D. All of these

19. The main difference between an SOA and an EDFA is that
A. An SOA operates in the electrical domain, whereas the EDFA operates in the optical domain.
B. An SOA is pumped electrically, whereas the EDFA is pumped optically.
C. An SOA is pumped optically, whereas the EDFA is pumped electrically.
D. An SOA amplifies 1300 nm wavelength, whereas the EDFA amplifies 1550 nm.

20. Noise figure of EDFAs will be typically
A. < 3 dB. B. 2.5–4.5 dB.
C. 3.5–9 dB. D. > 9 dB.

Keys to Multiple Choice Questions

1. D	**2.** B	**3.** D	**4.** A	**5.** B	**6.** C	**7.** D	**8.** B	**9.** A	**10.** D
11. A	**12.** D	**13.** A	**14.** D	**15.** A	**16.** D	**17.** B	**18.** D	**19.** B	**20.** C

Review Questions

1. State the principle of operation and describe the structure of a semiconductor optical amplifier.
2. Compare the performance of FPA and TWA semiconductor optical amplifiers.
3. Describe the performance parameters of semiconductor optical amplifiers in terms of noise bandwidth, optical gain, and polarization dependence.
4. Show that the theoretical limit for noise figure is 3 dB in case of an optical amplifier system.
5. What is meant by gain ripple in a SOA?

6. The SNR of the amplified optical signal degrades by 3 dB even for an ideal optical amplifier. Why?
7. Describe the origin of gain saturation in Raman fiber amplifiers.
8. Distinguish between the amplification mechanisms in a Raman fiber amplifier and an erbium-doped fiber amplifier.
9. What are the flexibilities offered by Raman fiber amplifiers that are not offered by SOAs and EDFAs?
10. Illustrate the mechanism of amplification in an EDFA with a suitable energy level diagram.
11. Describe the optical gain and noise characteristics of EDFA.
12. Distinguish between the amplification mechanisms in a Raman fiber amplifier and an erbium-doped fiber amplifier.
13. How is EDWA different from EDFA?

Numerical Problems

1. An input optical signal of 1 μW level is applied to an optical amplifier. Its output optical power level is 1 mW. What is the output optical power level when a 1 mW signal is incident on it? If the saturation output power of this optical amplifier is specified as 10 mW, then what would be the output optical power level for an input optical power level of 1 mW? *[Ans.: 1 W; 10 mW]*
2. Consider a fiber–optic communication link containing N cascaded optical amplifiers, each having a 30 dB optical gain. If the optical fiber used has a loss specification of 0.2 dB/km, then the span between optical amplifiers is 150 km assuming there are no other system impairments. Determine
   (a) the number of in-line optical amplifiers needed for a 900 km link.
   (b) the noise penalty factor over the total path (in dB). *[Ans.: (a) Five; (b) 13.2 dB]*
3. An optical amplifier is operating at 1300 nm wavelength with input optical power of 0.5 mW and noise figure of 4 dB. What is the receiver bandwidth if the SNR at the output is 30 dB? *[Ans.: 9.2×10^{11} kHz]*
4. Determine the spontaneous emission factor for an optical amplifier having optical gain = 20 dB if ASE is 1 mW for a fractional bandwidth of 5×10^{-6}. *[Ans.: 3.8×10^{21}]*
5. Consider an InGaAsP SOA with thickness of active layer = 0.5 μm, width of the active area = 5 μm. If a 1 μW optical signal at 1550 nm wavelength is incident on it, then what would be the photon density? [Use group velocity of the incident light, $v_g = 2 \times 10^8$ m/s; Planck's constant, $h = 6.626 \times 10^{-34}$ J.s]. *[Ans.: 1.6×10^{16} photons/m^3]*
6. Consider a SOA with thickness of active layer = 0.3 μm, width of the active area = 3 μm, length of the amplifier = 500 μm. If a 100 mA bias current is applied to it, then what would be the pumping rate? [Use $q = 1.6 \times 10^{-19}$C]. *[Ans.: 1.4×10^{33} electrons/m^3 per second]*
7. For a 1300 nm InGaAsP SOA having parameter values as confinement factor = 0.3, gain coefficient = 2×10^{-20} m^2, time constant = 1 ns, threshold density = 1.0×10^{24} per m^3. If the pumping rate is 1.4×10^{33} electrons/m^3 per second, then determine the small-signal gain per unit length. *[Ans.: 23.4 per cm]*
8. Consider that an EDFA being used as a power amplifier with a 30-mW pump power at 980 nm. When an input optical signal power of 0 dBm at λ_s = 1550 nm is applied to it, the output power level is +20 dBm. Compute the following:
   (a) the optical gain of the amplifier (in dB).

(b) the input pump power required to achieve this gain. *[Ans.: a) 20 dB; b) 190 μW]*

9. An optical fiber has a loss specification of 0.2 dB/km. It is used in an optical fiber communication link. After light propagation for 50 km, an EDFA is used as in-line optical amplifier. Determine the optical gain which the EDFA should have so as to maintain no-loss-no-gain regeneration. *[Ans.: 10 dB]*

10. The optical gain of an EDFA is 20 dB at 1550 nm. It is pumped at 980 nm with a pump power of 30 mW. What would be the maximum input and output optical signal power? *[Ans.: 190 μW; 19.1 mW]*

11. Consider an EDFA (optical gain = 10 dB), which is used as a power amplifier after optical transmitter, is pumped at 980 nm wavelength. Assume that the amplifier input is a 0 dBm level from a laser diode transmitter. Determine the minimum required pump power for a 10 dBm output power level at 1540 nm. *[Ans.: 14 mW]*

12. An optical fiber has a loss specification of 0.2 dB/km. It is used in an optical fiber communication link. After light propagation for 50 km, a chain of three identical EDFAs are used in cascaded form. Let the optical gain and noise figure of each EDFA is 10 dB and 5 dB respectively, then what would be the effective noise figure of such an arrangement of EDFAs? *[Ans.: 5.45 dB]*

13. The input signal power to an EDFA is 200 μW at 1550 nm. It is pumped at 980 nm with pump power of 30 mW. Assuming that the fiber modes for λ_s and λ_p are fully confined, calculate

(a) the rate of absorption per unit volume from the Er^{3+} level E_1 to pump level E_3 due to the pump at λ_p (assume $N_2 \approx 0$).

(b) the rate of absorption per unit volume from the level E_1 to the metastable level E_2 and the rate of stimulated emission per unit volume from level E_2 to level E_1, both due to the signal at λ_s (assuming $N_2 \approx N_1$).

It is specified that the cross-sectional area of fully doped fiber core = 8.5 μm^2, doping concentration = 5×10^{24} per m^3, signal absorption cross-section = 2.57×10^{-25} m^2, pump absorption cross-section = 2.17×10^{-25} m^2, signal emission cross-section = 3.41×10^{-25} m^2.

[Ans.: a) 1.9×10^{28} m^{-3}s^{-1}; b) 1.2×10^{26} m^{-3}s^{-1}; 1.6×10^{26} m^{-3}s^{-1}]

14. Three identical EDFAs, each having optical gain = 10 dB and noise figure = 5 dB are cascaded to provide an improvement in the output SNR besides amplification to compensate for fiber loss in an optical fiber communication link. If 10 mW of input optical power is launched at its input with an SNR of 30 dB, then what would be the output SNR? *[Ans.: 24.5 dB]*

15. Consider a fiber–optic communication link containing N cascaded optical amplifiers, each having a 30 dB optical gain. If the optical fiber used has a loss specification of 0.2 dB/km, then the span between optical amplifiers is 150 km assuming there are no other system impairments. Determine

(a) the number of in-line optical amplifiers needed for a 900 km link.

(b) the penalty factor over the total path (in dB). *[Ans.: (a) Five; (b) 13.2 dB]*

16. Consider an optical transmission path containing eight cascaded optical amplifiers, each having a 20 dB optical gain. If the optical fiber used has a loss specification of 0.2 dB/km, then what would be the impairment-free transmission distance between two optical amplifiers used as in-line amplifiers to compensate for fiber loss? *[Ans.: 100 km]*

17. A fiber–optic transmission link uses 8 cascaded optical amplifiers, each having an optical gain of 20 dB. A 100-km fiber length is used with fiber loss 0.2 dB/km. Calculate the noise penalty factor (dB) over the total path. *[Ans.: 6.6 dB]*

18. The input power levels for in-line optical amplifiers nominally ranges from -26 dBm to -9 dBm, with optical gains generally greater than 15 dB. Express these input power levels in μW. *[Ans.: 2.5 μW to 125 μW]*

19. The input power levels for in-line optical amplifiers nominally ranges from -26 dBm to -9 dBm, with optical gains generally greater than 15 dB. What would be the range of output power levels for nominal flat gain of 15 dB? *[Ans.: -9 dBm to +6 dBm]*

20. Consider an EDFA which is used as a power amplifier with an optical gain of 10 dB). Assume that a laser diode transmitter delivers 1-mW output power level at the input of EDFA. If the pump wavelength is 980 nm, then find the minimum required pump power for a 10-mW output power level at 1540 nm. *[Ans.: 14 mW]*

21. Generally the input to the power amplifier is -8 dBm or greater, and the power amplifier gain must be greater than 5 dB in order to be more advantageous than using a preamplifier at the receiver. What would be the nominal output power level of power amplifier? *[Ans.: -3 dBm]*

Dispersion Management Techniques

CHAPTER **6**

Chapter Objectives

After studying this chapter, you should be able to

- ✓ understand the need of dispersion management in optical fiber communications
- ✓ describe pre- and post-compensation techniques for dispersion management
- ✓ know about dispersion compensating fibers
- ✓ explain fiber Bragg gratings
- ✓ describe different methods of fabrication of chirped fiber gratings

When all the spectral components are separated from an optical signal, it is termed dispersion. It usually occurs when optical signals travel along optical fiber from transmitter to receiver in an optic–fiber communication link. Dispersion causes distortion in the transmitted optical signal (analog or digital transmission) along the optical fiber. As the optical pulses travel along the optical fiber channel, when digital modulation is used in transmitting optical signals, the dispersion phenomenon causes the broadening of optical pulses. There are different types of dispersion effects such as modal dispersion (in multimode fiber, transmitted optical pulse tends to spread due to time delay between lower- and higher-order propagation modes, causing bandwidth limitation), chromatic dispersion (combination of material dispersion as well as waveguide dispersion that results in spreading of transmitted optical pulse as they travel through the optical fiber), and polarization mode dispersion. Material dispersion happens because of variations in the fiber core refractive index with respect to operating wavelength, and waveguide dispersion happens due to nature of the physical structure of the optical fiber. Due to variations in the fiber core refractive index, different wavelengths of the light beam would travel at somewhat different velocities of light. As a result, an optical pulse gets broadened, causing dispersion. With the introduction of optical amplifiers (as discussed in the previous chapter) as in-line amplifiers in an optic–fiber link, the signal attenuation due to fiber is no more the major concern for achieving desired performance for optical fiber communication systems. However, they aggravate the dispersion problems. In order to achieve the lowest attenuation, there is a need of implementing efficient dispersion management techniques. There are different varieties of optical fibers available including dispersion compensating fibers. With an objective of controlling the spread of transmitted optical pulse in optical fiber communications systems, the dispersion management (also known as dispersion compensation) techniques must be applied.

This chapter focuses on dispersion management in optical fiber communications. The discussion begins with the need of dispersion management because dispersion-induced pulse broadening imposes the severe limitations on the performance of the system. This is followed by detailed discussions on different techniques of dispersion management as pre-compensation as well as post-compensation. The discussion is carried forward by describing various types of dispersion-compensating fibers including fiber Bragg gratings. Finally, different methods of fabrication of chirped fiber gratings are covered.

6.1 Need for Dispersion Management

Fiber–optic communication systems are quite often limited in their performance by dispersive and non-linear effects. The use of optical amplifiers as in-line amplifiers along the fiber link do compensate for fiber losses but further worsen the dispersion problem. The degradation of the transmitted optical signals due to dispersion gets accumulated when cascaded arrangement of in-line optical amplifiers is employed along the fiber–optic transmission link. Dispersion-induced pulse-broadening imposes the serious limitations on the system performance. Let us first understand these limitations which include limiting bit rate and fiber bandwidth. For example, in 50 km length of the optical fiber cable, the transmitted bit rate is limited to about 2 Gbps due to dispersion. In fact, it is the available bandwidth of the fiber that delivers the dispersion-limited transmitted bit rate. The 3-dB fiber bandwidth is given by

$$f_{3dB} \approx \frac{0.188}{|D| L \sigma_\lambda} \tag{6.1}$$

where, $|D|$ denotes the dispersion parameter, expressed in ps/(km–nm); L represents the length of the in km, and σ_λ represents the root-mean-square value of the spectral width of the optical source in nanometers.

- When $\sigma_\lambda = 0$, the optical source spectral width very much smaller than the bit rate.
- At typical values of $\sigma_\lambda = 1$ nm or 5 nm, the bit rate reduces with fiber length for given value of $D = 16$ ps/(km–nm).

Let us denote $\sigma_D \equiv |D| L \sigma_\lambda$ which signifies the extent to which the optical pulse is broadened due to dispersion. Then, we can write

$$f_{3dB} \approx \frac{0.188}{\sigma_D} \tag{6.2}$$

The bit rate R_B is related with 3-dB fiber bandwidth by the relationship, $R_B \leq 1.33\, f_{3dB}$. This implies that the bandwidth of the optical fiber gives a rough estimate of the maximum transmitted bit rate that is possible in a typical optical fiber communication link which is dispersion-limited.

We know that if we use a laser having narrow linewidth as the optical source, then it is possible to minimize the effects of Group-Velocity Dispersion (GVD) considerably. Even if we operate it at the wavelength closer to the fiber's specified zero-dispersion wavelength, λ_{ZD}, then also the effects of GVD can be minimized. Practically, it is not always possible to operate the system at zero-dispersion wavelength. Why is that so? The 'standard' single-mode fiber (SMF) has $\lambda_{ZD} \approx 1310$ nm, whereas 3G terrestrial fiber–optic systems (which use distributed feedback laser as optical source) operate near $\lambda \approx 1550$ nm. In this optical band of standard single-mode fiber (SMF), the specified dispersion parameter is about 16 ps/(km–nm). In case the transmitted bit rate exceeds 2 Gbps, then the Group-

Velocity Dispersion severely limits the system performance. There are other reasons also that justify the need for dispersion management such as:

(a) For a directly-modulated distributed-feedback (DFB) laser, the relationship between the bit rate (R_B) and maximum transmission distance (L) can be expressed as

$$L < \frac{1}{4R_B|D|\sigma_\lambda} \tag{6.3}$$

where, σ_λ is the RMS spectral width of the dispersion-induced (mainly due to frequency chirping) optical pulse.

As an example, let $\sigma_\lambda = 0.25$ nm, $D = 10$ ps/(nm–km), then $L \approx 42$ km only is obtained at operating bit rate $R_B = 2.5$ Gbps. Since optical amplifiers can be spaced at a much larger spacing than $L \approx 42$ km, for dispersion-limited fiber–optic systems, dispersion management technique must be employed.

(b) The use of an external modulator along with DFB laser can significantly improve the system performance. This helps to minimize spectrum broadening of transmitted pulse, if any, due to frequency chirping. The $\sigma_\lambda = 0$ provides the upper limit of the dispersion provided standard fibers are used with this type of optical transmitters. The upper limit of the transmission distance L is then given by

$$L < \frac{1}{16|\beta_2|{R_B}^2} \tag{6.4}$$

where, β_2 represents the GVD coefficient and is related with D such that

$$\beta_2 = -D\frac{\lambda^2}{2\pi c} \tag{6.5}$$

Typical value of $\beta_2 = -20$ ps^2/km at $\lambda = 1550$ nm. Then, for required bit rate $R_B = 2.5$ Gbps, $L < 500$ km. This shows a considerable improvement over directly modulated DFB laser ($L \approx 42$ km only). However, the following observations can be made:

- When in-line optical amplifiers are used along with the optical fiber, then even this amount of dispersion is quite considerable.
- When the transmitted bit rate R_B is increased beyond 2.5 Gbps (10 Gbps, say), then the GVD-limited transmission distance decreases to 30 km only which is extremely less as envisaged for the use of in-line optical amplifiers.

We can conclude that the standard single-mode fibers have relatively considerable amount of GVD which may result into degradation in the performance of 1550 nm fiber–optic systems at higher bit rate (usually 10 Gbps or more). For improving the overall performance of optical fiber communications networks, dispersion-management techniques must be implemented.

Facts to Know

Dispersion management is the term used to refer to the management of dispersion compensation. Major tasks include choosing the right location for dispersion compensating fibers (DCFs), the sequence for placing DCF and regular fibers, and the right length of DCF. The ultimate objective is to maximize the bandwidth or the transmitted bit rate of the fiber–optic system.

Example 6.1 To calculate Dispersion
In a typical optic–fiber system, the specified dispersion parameter of the fiber used is 16 ps/(nm–km). Calculate the dispersion induced for 100 km fiber length.

Solution:

Total dispersion induced in the optical fiber = D × L

For the given D = 16 ps/(nm–km), and L = 100 km, we have
Total dispersion = 16 ps/(nm–km) × 100 km = 1600 ps/nm **Ans.**

Example 6.2 Maximum Transmission Distance
An optic–fiber system uses a directly-modulated DFB laser as an optical source at the transmitter. Determine the maximum transmission distance if the operating bit rate = 2.5 Gbps, the dispersion parameter = 10 ps/(nm–km), and RMS spectral width of the pulse = 0.15 nm.

Solution:
We know that in a directly-modulated distributed-feedback (DFB) laser, the maximum transmission distance,

$$L < \frac{1}{4R_B|D|\sigma_\lambda}$$

Given R_B = 2.5 Gbps, D = 10 ps/(nm–km), and RMS spectral width of the pulse σ_λ = 0.15 nm, we have

Therefore, $L < \frac{1}{4\times\left(2.5\times10^9\right)\times\left(10\times10^{-12}\right)\times0.15}\ km < 67\ \text{km}$ **Ans.**

Example 6.3 Dispersion-induced Limited Distance
An optic–fiber system uses an external modulator along with DFB laser as an optical source at the transmitter. Determine the maximum transmission distance if the required bit rate = 2.5 Gbps. Consider typical value of GVD coefficient β_2 = –20 ps²/km at λ = 1550 nm.

Solution:
We know that the maximum transmission distance for a DFB laser using an external modulator is given by

$$L < \frac{1}{16|\beta_2|R_B^{\,2}}$$

Given R_B = 2.5 Gbps, β_2 = -20 ps²/km, and λ = 1550 nm, we have
Therefore, $L < \frac{1}{16\times\left|-20\ \times10^{-12}\right|^2\left(2.5\times10^9\right)^2}\ km < 500\ \text{km}$ **Ans.**

Section Practice Problems

1. An optic–fiber system uses a directly-modulated DFB laser as an optical source at the transmitter. Determine the maximum dispersion-limited transmission distance if the operating bit rate = 10 Gbps, the dispersion parameter = 17 ps/(nm–km), and RMS spectral width of the pulse = 0.15 nm. *[Ans.: ~10 km]*

2. Determine the approximate transmission distance of a fiber–optic system operating at required bit rate = 2.5 Gbps. The system uses a directly-modulated DFB laser and affected by dispersion at specified dispersion parameter = 16 ps/(nm–km), and root-mean-square value of the spectral width of the optical pulse = 0.15 nm. *[Ans.: 42 km]*
3. Show that in a fiber–optic system, the maximum transmission distance using an external modulator along with DFB laser is more than ten times the maximum transmission distance using directly-modulated DFB laser, when both systems operate at required bit rate = 2.5 Gbps. Consider typical value of dispersion parameter D = 16 ps/(nm–km), RMS spectral width of the pulse σ_λ = 0.15 nm, GVD coefficient β_2 = –20 ps^2/km at λ = 1550 nm.

What is the main reason of degradation of the optical signal through the optical fiber due to dispersion phenomenon? When the optical signal propagates in the fiber, its spectral components acquire the phase factor. In order to restore the transmitted optical signal, this phase factor needs to be canceled. There are various dispersion management techniques for this purpose. Actual implementation of dispersion management techniques can be carried out in three ways:

- At the transmission, known as pre-compensation techniques.
- At the receiver, known as post-compensation techniques.
- Along the fiber link, known as dispersion-compensating fibers (DCFs).

All these dispersion management techniques are discussed next.

6.2 Pre-Compensation Dispersion Management

As the name suggests, pre-compensation dispersion management techniques are implemented prior to the occurrence of the dispersion. Since the dispersion occurs within the optical fiber as the optical signals propagate through them, it implies that these techniques are applied at the optical transmitter end. In pre-compensation, the characteristics of optical pulses are suitably modified (changing the spectral amplitude) before they are launched into the optical fiber. The nature and extent of modification is to counter the impact of GVD (within the fiber) exactly, and the shape of the output optical pulse will be retained as that of the input optical pulse. Fig. 6.1 shows a functional block schematic of pre-compensation dispersion management technique.

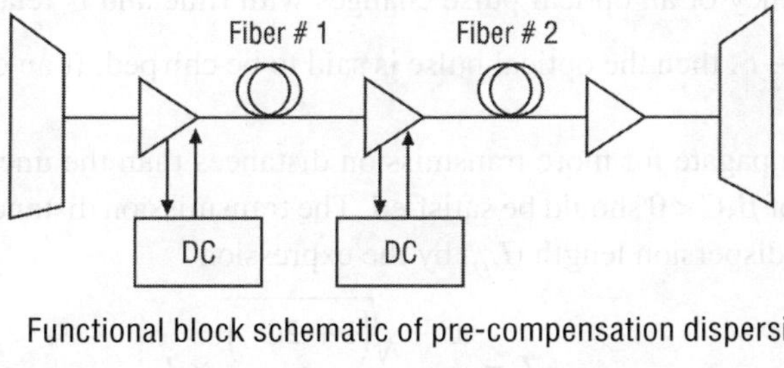

Fig. 6.1 Functional block schematic of pre-compensation dispersion (DC)

There are various pre-compensation dispersion management techniques.

6.2.1 Pre-chirp Pre-compensation

In pre-chirp pre-compensation dispersion management technique, the input optical pulse is frequency chirped at the transmitter end before its propagation down the fiber. The extent of frequency-chirping is such that the broadening of the pulse due to GVD within the fiber is minimized.

Let us choose the initial field for propagation of chirped Gaussian optical pulses within optical fibers as

$$\tilde{A}(0,\omega) = A_0\sqrt{\frac{2\pi T_0^2}{1+iC}}e^{\left[-\frac{\omega^2 T_0^2}{2(1+iC)}\right]} \quad \text{(In frequency domain) (6.6)}$$

$$\mathrm{A}(0, t) = A_0 e^{\left[-\frac{1+iC}{2}\left(\frac{t}{T_0}\right)^2\right]} \quad \text{(In time domain) (6.7)}$$

where, A_0 represents the maximum amplitude, C represents a parameter that manages the frequency chirp levied on the chirped Gaussian optical pulse, T_0 represents another parameter that signifies the half-width of the transmitted optical pulse (that is, width at *1/e* intensity point).

$$\Rightarrow \quad \tilde{A}(z,\omega) = \tilde{A}(0,\omega)e^{\left[\frac{i}{2}\beta_2 z\Delta\omega^2\right]} = A_0\sqrt{\frac{2\pi T_0^2}{1+iC}}e^{\left[-\frac{\omega^2 T_0^2}{2(1+C^2)}+\Delta\omega^2\frac{iCT_0^2+\beta_2}{(1+C^2)}\right]} \quad (6.8)$$

$$\Delta\omega_0 = \sqrt{1+C^2}\times\frac{1}{T_0} \quad (6.9)$$

$$\Rightarrow \quad A(z, t) = \frac{1}{2\pi}\int_{-\infty}^{\infty}\tilde{A}(0,\omega_0)e^{\left(\frac{i}{2}\beta_2 z\Delta\omega\right)}d\Delta\omega \quad (6.10)$$

$$\Rightarrow \quad A(z, t) = \frac{A_0}{\sqrt{Q(z)}}e^{\left(-\frac{1}{2T_0^2 Q(z)}\right)} \quad (6.11)$$

$$\text{where, } Q(z) = 1-\frac{(C-i)\beta_2 z}{T_0^2},\ T(z) = \sqrt{\left(1+\frac{C\beta_2 z}{T_0^2}\right)^2+\left(\frac{\beta_2 z}{T_0^2}\right)^2}\times T_0 \quad (6.12)$$

If the carrier frequency of an optical pulse changes with time and is related to phase derivative as $\delta\omega(t) = -\frac{\partial\phi}{\partial t} = \frac{C}{T_0^2}t$, then the optical pulse is said to be chirped. If an optical pulse is suitably chirped, then it can propagate for more transmission distances than the unchirped one. For this to happen, the condition of $\beta_2 C < 0$ should be satisfied. The transmission distance (*L*) is related to chirp parameter (*C*) and the dispersion length (L_D) by the expression

$$L = \frac{C+\sqrt{(1+2C^2)}}{(1+C^2)}\times L_D \quad (6.13)$$

The chirp parameter $C = 0$ corresponds to the unchirped Gaussian pulses, and therefore, $L = L_D$.

For $C = 1,\ L = \dfrac{1+\sqrt{(1+2\times 1^2)}}{1+1^2} L_D = \dfrac{1+\sqrt{3}}{2} L_D = 1.366 L_D$

That means that L is greater than L_D by 36.6 %.

For $C = 1/\sqrt{2},\ L = \dfrac{1+\sqrt{\left(1+2\times\left(\frac{1}{\sqrt{2}}\right)^2\right)}}{1+\left(\frac{1}{\sqrt{2}}\right)^2} L_D = \dfrac{1+\sqrt{2}}{\frac{3}{2}} L_D = 1.6 L_D$

That means, L increases by 60 %. In fact, this is the maximum improvement that occurs for $C = 1/\sqrt{2}$. However, for large values of C, $L < L_D$.

Thus, we can say that if the pre-chirp pre-compensation dispersion management is properly optimized, then the transmission distance can be increased by a factor of approximately 2.

- *Directly-modulated semiconductor lasers* (used as optical sources) chirp the output optical pulse automatically but the value of C is negative. For standard fibers, the value of β_2 is also negative at 1550-nm wavelength region, the required condition $\beta_2 C < 0$ is not satisfied.
- In *externally-modulated lasers* used as optical sources, the output optical pulses can be considered almost free of any frequency-induced chirp. The pre-chirp pre-compensation dispersion management technique imposes a frequency chirp having positive value of C such that the required condition given as $\beta_2 C < 0$ is fully met.

Fig. 6.2 depicts a basic functional block schematic diagram of the pre-chirp method used for pre-compensation dispersion management techniques.

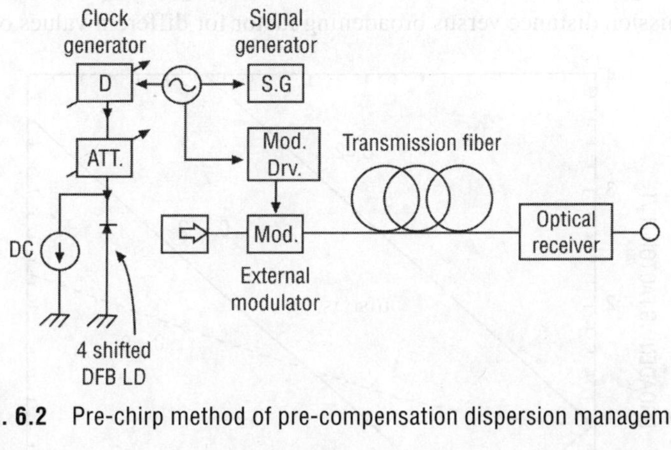

Fig. 6.2 Pre-chirp method of pre-compensation dispersion management

As seen, the frequency of the $\lambda/4$-shifted DFB laser used as optical source is first frequency modulated. Its output is then applied to an external amplitude modulator. Thus, the output modulated optical signal possesses frequency modulation as well as amplitude modulation simultaneously which is a pre-chirped optical pulse. It is propagated down the fiber link. Fig. 6.3 illustrates the input and output waveforms of pre-chirp method.

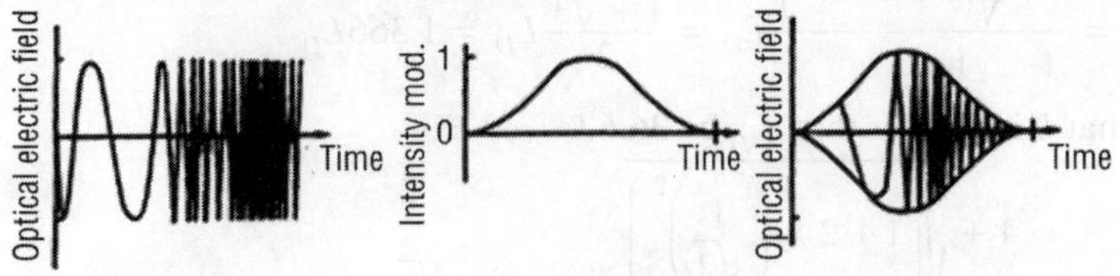

Fig. 6.3 Input and output waveforms of pre-chirp method

Consider the Gaussian-shaped optical pulse. Mathematically, the chirped optical pulse can be given as:

$$E(0, t) = A_0 e^{\left(-\frac{t^2}{T_0^2}\right)} e^{\left[-i\omega_0(1+\delta \sin \omega_m t)t\right]} \tag{6.14}$$

where, ω_0 represents the angular frequency of the carrier pulse which is frequency-modulated with a frequency-modulation depth δ at optical signal frequency ω_m.

Near to the center of the pulse, we have $\sin(\omega_m t) \approx \omega_m t$. Therefore,

$$E(0, t) \approx (0, t) e^{(-i\omega_0 t)} \approx A_0 e^{\left(\frac{1+iC}{2}\left(\frac{t}{T_0}\right)^2\right)} e^{-i\omega_0 t} \tag{6.15}$$

where the chirp parameter $C = 2\delta\omega_m\omega_0 T_0^2$ (6.16)

By changing the FM parameters such as modulation depth δ and modulating frequency ω_m, the magnitude and sign of the chirp parameter C can be changed. Fig. 6.4 shows the variation of normalized transmission distance versus broadening factor for different values of chirp parameter.

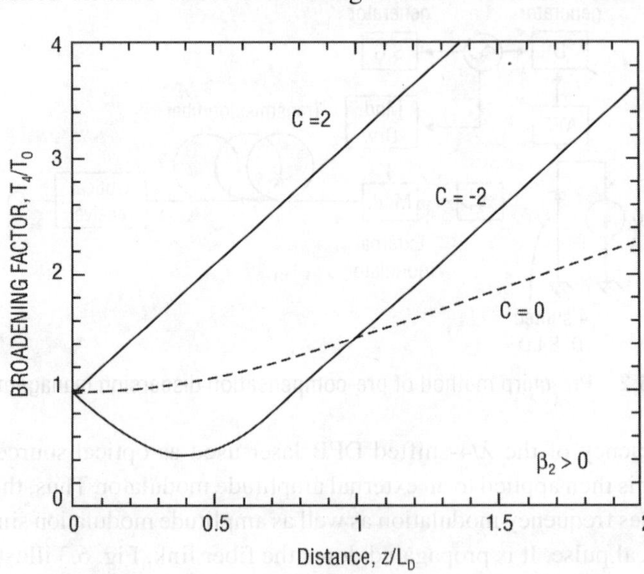

Fig. 6.4 Transmission distance vs broadening factor

With broadening factor $T_1/T_0 = \sqrt{2}$; the transmission distance is given as

$$L = \frac{C + \sqrt{\left(1 + 2C^2\right)}}{\left(1 + C^2\right)} \times L_D ;$$

and

$$L_D = \frac{{T_0}^2}{|\beta|} \tag{6.17}$$

Note: The refractive index of an external modulator should be varied electronically so that a frequency chirp with $C > 0$ can be imposed. For example,

- Using an electro–optic material lithium niobate $LiNbO_3$ modulator with $C \approx 0.6 - 0.8$, a transmission distance of 256 km was achieved for a 5-Gbps signal.
- Using an electro–absorption or a Mach–Zehnder modulator, an optical pulse with $C > 0$ can be chirped. Combined with DFB lasers, a transmission distance of over 100 km was achieved for a 10-Gbps NRZ signal was using standard fiber by implementing the pre-chirp dispersion management technique.

Facts to Know

When an optical carrier signal is phase modulated, it results in a positive chirp ($C > 0$). It has an advantage that an external modulator employed with DFB laser can modulate the phase of an optical carrier signal.

6.2.2 Novel Coding Pre-compensation

Novel coding pre-compensation dispersion management technique, also known as dispersion supported transmission technique, can either use frequency shift keying (FSK) format or duo–binary coding for transmission of optical signals.

(a) *Using FSK format for signal transmission:* When the laser wavelength (λ) is switched by a constant amount ($\Delta\lambda$) between binary signals 1 and 0, the FSK-modulated signal with constant power is produced. The wavelength offset $\Delta\lambda$ depends on the corresponding change in pulse duration $\Delta T = 1/R_B$, where R_B is the bit rate. We know that $\Delta T = |D| L\Delta\lambda$; where *D, L* and $\Delta\lambda$ represents the absolute value of the dispersion parameter, the fiber length, and the wavelength offset, respectively. When these two wavelengths ($\lambda + \Delta\lambda$ and $\lambda - \Delta\lambda$) propagate within the optical fiber, they tend to travel at slightly different velocities. This causes fiber dispersion that converts the FSK signal into an amplitude-modulated signal (equivalent to amplitude shift keying ASK). An integrator together with a decision-making electronic circuit is used at the receiver to decode this signal. The transmission distance can be improved significantly by using the FSK technique, e.g., 86 km @ 40 Gbps.

(b) *Using duo–binary coding for signal transmission:* In this technique, two consecutive bits are taken together, forming a 3-level duo–binary code in the digital data stream. The net bit rate as well as the signal bandwidth is reduced by 50%. This results in significant improvement in the transmission distance because signal degradation due to GVD depends on the signal bandwidth (transmission distance is inversely proportional to signal bandwidth). For example, an improvement over 30–40 km longer distance @ 10 Gbps data rate with duo–binary coding has been achieved as compared with binary coding. Combining it with an external modulator based pre-chirping pre-compensation dispersion management technique can produce a frequency chirp with $C > 0$, 160 km distance @ 10 Gbps data rate has been realized. Phase-shaped binary transmission takes advantage of phase reversal.

6.2.3 Non-linear Pre-chirp Pre-compensation

(a) In a simple non-linear pre-chirp pre-compensation dispersion management technique, the output signal of optical transmitter is amplified using SOA as power amplifier operating in the gain saturation region. This results in almost linear chirping over most of the pulse duration (the input pulse shape decides the extent of chirping) in the amplified pulse signal due to carrier–signal induced changes in the fiber core refractive index. An in-line optical amplifier amplifies the transmitted optical pulse as well as the chirp with $C > 0$. In a fiber meeting the condition $\beta_2 C < 0$, the chirped optical pulse can be compressed. The experiments show a five times increase in transmission distance. Moreover, the coupling and insertion losses can also be compensated prior to launching the optical signal into the fiber.

(b) A non-linear medium can also be employed that can pre-chirp the optical pulse before transmission. If the refractive index of the fiber core varies with the light intensity, an optical pulse is chirped due to self-phase modulation. Thus, in a simple pre-chirp pre-compensation dispersion management technique, the output optical signal of the optical transmitter is passed through an additional section of fiber having appropriate length and then it is launched into the fiber link. This results into positive value of the chirp parameter which makes it suitable for compensation of dispersion.

Example 6.7 Pre-chirp Dispersion

Calculate the value of L in terms of L_D for $C = 5.9$ and show that for such a large value of C, $L < L_D$.

Solution:

We know that the transmission distance, $$L = \frac{C + \sqrt{(1 + 2C^2)}}{1 + C^2} L_D$$

Given $C = 5.9$, we have $$L = \frac{5.9 + \sqrt{(1 + 2 \times 5.9^2)}}{1 + 5.9^2} L_D = 0.4 L_D$$ **Ans.**

This clearly shows that for $C = 5.9$, $L < L_D$.

Section Practice Problem

1. Using an electro–optic material lithium niobate $LiNbO_3$ modulator with $C \approx 0.6 - 0.8$, a 5-Gbps signal could be transmitted over transmission distance of 256 km. Show that transmission distance $L = 1.4 L_D$ for Gaussian chirped parameter $C = 0.6$, where L_D is the dispersion length.

6.3 Post-Compensation Dispersion Management

In post-compensation dispersion management techniques, actual implementation is carried out at the optical receiver end so as to cancel out the phase factor which is responsible for dispersion-induced degradation of the optical signal during its propagation in the fiber. Fig. 6.5 shows a functional block schematic of post-compensation dispersion management technique.

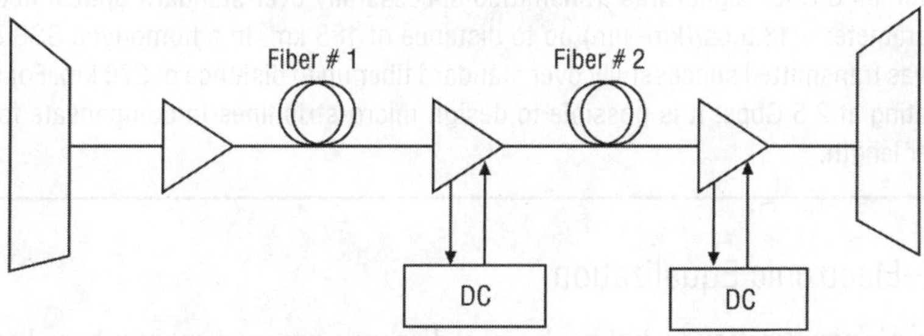

Fig. 6.5 Functional block schematic of post-compensation dispersion (DC)

Various post-compensation techniques are discussed below.

6.3.1 Electronic Equalization

Electronic equalization is the most practical dispersion compensation approach for coherent fiber–optic communication systems in which direct detection is used at the receiver end. Group-velocity dispersion (GVD) within the optical receiver can be compensated by employing electronic equalization techniques. It is assumed that the optical fiber behaves like a linear system. GVD-degraded optical signal can be equalized at the receiver. The compensation for dispersion is relatively easy when a heterodyne receiver detects the received optical signal. In a heterodyne receiver, the received optical signal is first converted into an intermediate frequency while preserving the amplitude and phase information of the signal. The original signal is recovered by passing it through a microwave bandpass filter.

Now the question is: Is it possible to compensate GVD using a linear electronic equalization technique? The answer is No! Why? We know that a photodetector reacts to an optical signal power only, and all phase information is lost. So, it is obvious that a linear electronic equalization circuit cannot retrieve a spread optical pulse.

Then what is the solution? The solution lies in the use of the non-linear equalization technique. The dispersion-degraded optical signal can be recovered by it. For example, the decision threshold at the receiver is changed as per the preceding bit. The analog waveform is examined over number of bit intervals around the bit under consideration before making a decision about a given bit.

The major drawbacks of electronic equalization techniques are:

- The requirement of electronic logic devices and circuits operating at the required bit rate.
- Exponential increase in their complexity as the number of bits representing spread optical pulse increases. This happens due to resultant GVD-induced spreading of the optical pulse.

Due to this, there is a restriction to operation for transmission distance (limited up to a few dispersion lengths) as well as transmission bit rate (relatively quite low) with electronic equalization post-compensation dispersion management techniques.

Facts to Know

When a length of 31.5 cm micro-strip line is used as an electronic dispersion equalizer at the heterodyne optical receiver, an 8 Gbps signal was transmitted successfully over standard optical fiber cable with dispersion parameter = 18.5 ps/(km–nm) up to distance of 188 km. In a homodyne SSB detector, a 6 Gbps signal was transmitted successfully over standard fiber up to distance of 270 km. For a fiber–optic system operating at 2.5 Gbps, it is possible to design micro-strip lines to compensate for GVD up to 4900-km fiber length.

6.3.2 Opto–Electronic Equalization

An opto–electronic equalization technique for post dispersion management is based on a *transversal filter.* In opto–electronic equalization, the received optical signal is split into several branches by a power splitter used at the optical receiver. Each branch uses fiber–optic delay lines that introduce variable delays. Photodetectors having variable sensitivity are used in each branch, which convert the optical signal into corresponding photocurrent. The photocurrent from each branch is then summed and applied to the decision-making circuit for final recovery of the signal. Using this technique, the transmission distance can be extended three-fold for a fiber–optic communication system operating at 5 Gbps.

6.3.3 Optical Equalization

We know that the optical signal is affected by GVD through the spectral phase. So, an optical equalization filter, having its transfer function such that it cancels the phase and suitable for compensating the GVD exactly, will be able to restore back the original optical signal. But there cannot be such an ideal optical filter. However, there can be an arrangement of using an optical filter along with an optical amplifier in such a way that both GVD as well as fiber attenuation can be compensated together. If the bandwidth of an optical filter is much smaller than bandwidth of optical amplifier, then amplifier noise can also be reduced. Interferometry optical filter and fiber gratings filters are two most popular methods of optical equalization for dispersion managements.

Interferometry Optical Filter– Optical filters designed using an interferometer have frequency dependent transmission characteristics. This means that it is sensitive to the input optical signal frequency. For example, it is possible to obtain the Fabry–Perot (FP) interferometer optical filter in view of number of round-trips of the optical signal travelling between two mirrors. Its transfer function is represented by

$$A(L,t) = \frac{1}{2\pi}\int_{-\infty}^{\infty} \widetilde{A}(0,\omega)\,H(\omega)\,e^{\left(\frac{i}{2}\beta_2 L\omega^2 - i\omega t\right)}d\omega$$

The design of a reflective FP interferometer is based on 100% reflective back mirror whose transfer function is expressed with the following expression

$$H_{FP}(\omega) = H_0 \times \frac{\left(1 + re^{(-i\omega T)}\right)}{\left(1 + re^{(i\omega T)}\right)} \tag{6.18}$$

where, H_0 is constant, considering all type of losses, $|r|^2$ represents reflectivity of the front-mirror, and T represents round-trip time in the Fabry–Perot cavity.

The Fabry–Perot filter modifies the spectral phase which is a periodic function. The spectral phase is maximum at the FP resonances. This type of Fabry–Perot interferometer optical filter can compensate the GVD which has been accumulated over a length of about 110–130 km of standard fiber. An optical amplifier can be used along with filter in order to compensate the relatively high insertion loss (about 6–8 dB). For separating the signal from the incident optical signal, an optical circulator can be used in place of a 3-dB fiber coupler, which will then reduce the insertion loss to about 1 dB.

For optical equalization, Mach–Zehnder interferometer (MZI) is normally used which is basically an optical filter. In MZI optical filter, two 3-dB directional couplers are connected in series. Out of these, the first 3-dB directional coupler divides the input optical signal equally. If arm lengths are different (one arm length will be longer than the other), then these two signals acquire different phase shifts before getting interference by the second 3-dB directional coupler. Design of the MZI optical filter ensures that higher frequency components of the input optical signal travel in its longer arm, thereby experiencing more delays as compared to lower frequency components that travel through the relatively shorter arm. Therefore, depending on its arm lengths and frequency components, the final signal output may be taken from either of the two output ports. The relative delay is just opposite to dispersion and thus is capable of compensating fiber dispersion. It is only a few cm long device that is required for 50 km fiber length. Practically, cascaded stages of many MZI optical filters are used in place of a single MZI optical filter to achieve better performance of optical equalization.

The main advantage of MZ interferometer optical filter is that by merely varying the length of the arm and selecting appropriate number of MZ interferometer filters, it is possible to control the dispersion-equalization characteristics quite effectively. In other words, it can serve as a field programmable optical equalization filter technique in which the GVD as well as the operating wavelength can be precisely controlled. The main limitations of MZ interferometer optical filters are sensitivity to input polarization and a relatively narrow bandwidth (of the order of 10 GHz).

Note: Fiber grating filters are discussed under Fiber Bragg Gratings in section 6.5.

Facts to Know

Pre-compensation at the optical transmitter or post-compensation at the optical receiver end for dispersion management can increase the transmission distance by almost two times in a dispersion-limited fiber-optic system. Hence, these techniques have been found more suitable mainly for short-haul networks.

6.4 Dispersion Compensating Fibers

For long-haul fiber–optic communications systems, there is a need to provide GVD compensation along the fiber length periodically. For this purpose, a special kind of fiber, known as dispersion compensating fiber (DCF) can provide an all-optical solution for compensation of the fiber GVD completely. However, this is possible at low value of average optical power so as to keep almost negligible non-linear effects inside optical fibers. The most commonly employed dispersion compensating fibers techniques for dispersion compensation are

- dispersion-shifted fibers (DSF)
- dispersion-flattened fibers (DFF)
- dispersion-compensating fibers (DCF)

Now all these kinds of dispersion management fibers are described briefly.

Dispersion-Shifted Fibers: In this type of fibers, the wavelength corresponding to zero dispersion is shifted toward the region of lowest fiber attenuation (usually in the 1550 nm optical band). Dispersion-shifted fibers are best suited for single-channel transmission because they can provide minimal dispersion over a very narrow range of operating wavelengths. As a result, repeater spacing can be increased in long-haul optical fiber systems which will enhance the overall efficiency of the system. Dispersion-shifted single-mode fibers can be designed by using multi-clad fibers with step-index or graded-index fiber cores.

Dispersion-Flattened Fibers: The index profile of a fiber can be manipulated so as to achieve total dispersion quite close to zero at two or more than two different adjacent operating wavelengths. In between these operating wavelengths, the total dispersion still remains close to zero. This is known as dispersion-flattening. How is dispersion-flattening achieved in fibers? If waveguide dispersion is partially cancelled by material dispersion in the operating optical band, then dispersion flattening is possible. Multi-cladding fiber cables are generally employed in the design of dispersion-flattened fibers.

Fig. 6.6 shows the variation of dispersion parameter, expressed in ps/(km–nm) with wavelength (nm) for standard (i.e., uncompensated) fiber, dispersion-shifted fiber, and dispersion-flattened fiber over 1200–1700 nm range for comparison purpose.

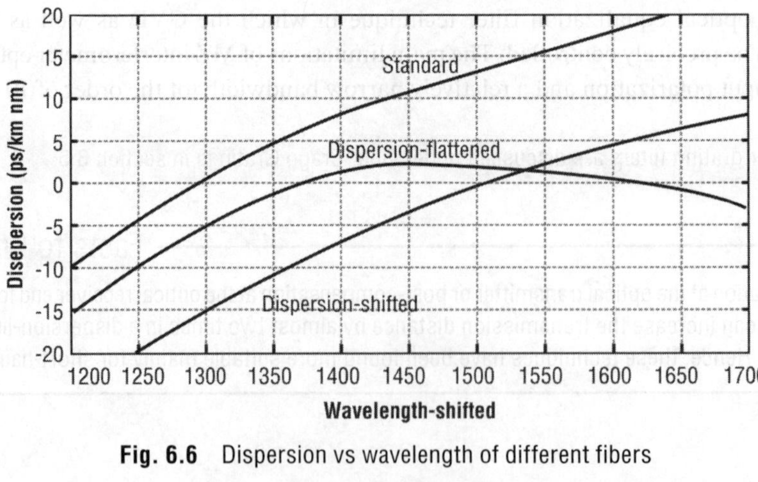

Fig. 6.6 Dispersion vs wavelength of different fibers

Facts to Know

The design of dispersion-flattened fibers has been reported where dispersion is less than 0.01 dB/km over the entire wavelength operating range of 1310–1670 nm. In optical fiber systems based on wavelength division multiplexing (WDM) applications, the use of dispersion-flattened fibers have resulted in manifold increase in information-carrying capacity.

Dispersion-Compensating Fibers (DCF): It is a relatively small closed loop of fiber cable which possesses a negative dispersion equivalent to the actual dispersion of the optical fiber used for transmission purpose. There are two ways to insert DCF in the main fiber cable - either at the optical transmitter end of the beginning of the fiber cable (pre-compensation dispersion management technique) or at the input of the optical receiver end (post-compensation dispersion management technique). Similar arrangement can be worked out between two in-line optical amplifiers used in the same fiber–optic communication link. A typical functional block diagram of fiber–optic communication link showing the use of dispersion compensating fiber is depicted in Fig. 6.7.

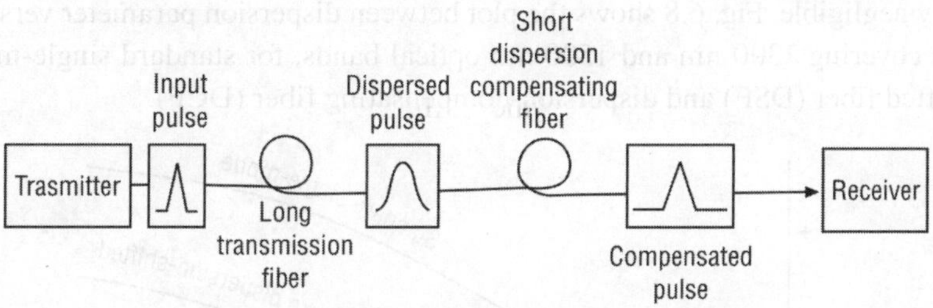

Fig. 6.7 Use of dispersion compensating fiber

It may be noted that the use of DCF gives large insertion loss. It has the advantages of simple construction, high reliability, and provides continuous compensation over a wide range of optical wavelengths. However, DCF has a small core size which may make it prone to certain types of nonlinearities.

In order to understand its functioning, let us consider the pulse-propagation equation as (Neglecting the 3rd-order dispersion term β_3 at $|\beta_2| > 0.1$ ps²/km)

$$\frac{\partial A}{\partial L} + \frac{i\beta_2}{2}\frac{\partial^2 A}{\partial t^2} = 0 \tag{6.19}$$

The solution can be expressed is

$$A(L, t) = \frac{1}{2\pi}\int_{-\infty}^{+\infty} \tilde{A}(0,\omega)\, e^{\left(\frac{i}{2}\beta_2 L\omega^2 - i\omega t\right)} d\omega \tag{6.20}$$

Let L_1 is the length of normal fiber and L_2 is the length of DCF such that $L = L_1 + L_2$, then

$$A(L, t) = \frac{1}{2\pi}\int_{-\infty}^{+\infty} \tilde{A}(0,\omega)\, e^{\left(\frac{i}{2}\omega^2(\beta_{21}L_1 + \beta_{22}L_2) - i\omega t\right)} d\omega \tag{6.21}$$

where, β_{2j} represents GVD parameter for given part of the fiber having length L_j $(j = 1, 2)$.

In case the dispersion-compensating fiber is chosen in such a way that ω^2 phase term is cancelled, then at the end of it the optical pulse will certainly retain its original rectangular shape. The criterion to achieve almost perfect dispersion compensation will be given by

$$\beta_{21}L_1 + \beta_{22}L_2 = 0 \tag{6.22}$$

$$\Rightarrow \quad D_1L_1 + D_2L_2 = 0 \tag{6.23}$$

$$\Rightarrow \quad L_2 = -\left(\frac{D_1}{D_2}\right)L_1 \tag{6.24}$$

Practically, the length L_2 should be as small as possible. As seen from this expression, this can be made possible with a sufficiently large negative value of D_2. Since $D_1 > 0$ for standard fibers, the DCF should possess normal GVD at 1550 nm ($D_2 < 0$).

Thus, we conclude that it is possible to compensate for the positive dispersion introduced by a conventional or standard fiber by introducing a specified section of a single-mode fiber having negative-dispersion characteristics in such a way so that the overall dispersion of the fiber–optic link becomes nearly negligible. Fig. 6.8 shows the plot between dispersion parameter versus wavelength over the range covering 1300 nm and 1500 nm optical bands, for standard single-mode, non-zero dispersion-shifted fiber (DSF) and dispersion-compensating fiber (DCF).

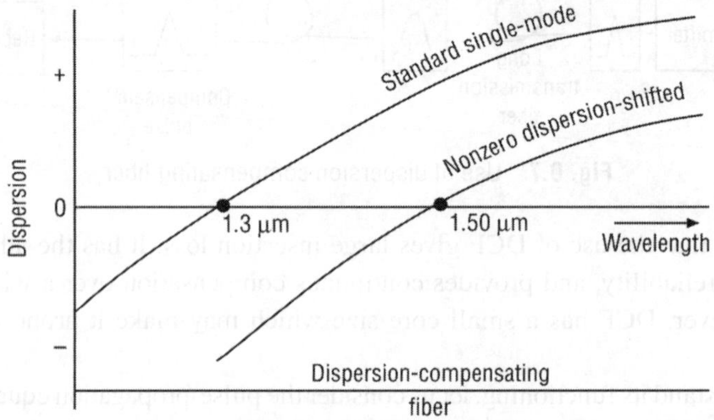

Fig. 6.8 Dispersion vs wavelength plots

It clearly shows that dispersion parameter for DCF is almost constant over the desired wavelength range. The first problem encountered in using DCF is its high attenuation. Thus, a new characteristics figure of merit (FOM) is used to describe the quality of DCF. It is defined as

$$FOM\left(ps/nm\right)\Big|_{dB} = \frac{dispersion\ co - efficient\left(ps/nm - km\right)}{attenuation\left(dB/km\right)} \tag{6.25}$$

FOM reveals the existence of a trade-off between the negative dispersion coefficient and attenuation of a DCF.

(a) *Depressed-cladding design of DCF:* To achieve a DCF with a high negative dispersion coefficient, the refractive-index profile as well as the relative difference of refractive index value, as may be necessary for a specific application, have to be manipulated. Typically, the refractive index of the inner part of the cladding is lowered by doping silica with fluorine. However, depressed-cladding design of DCF leads to an increase in Rayleigh scattering losses which is generally not desired.

(b) *Decreasing the core radius:* High negative dispersion may also be obtained by decreasing the core radius. Moreover, the power penalty in DCF is much higher than in regular fiber because of small core size. This leads to high level of non-linear effects, causing deterioration in system performance. In addition, FOM reaches at its maximum value at specific value of $\Delta = \Delta_{opt}$. However, extrinsic loss in DCF is bending loss.

(c) *Using a DCF module with optical amplifier:* A practical solution in using DCF lies to compensate for its high attenuation. We know that fiber attenuation can be compensated by using in-line optical amplifiers with physical separation of about 60–80 km between them, in addition to a DCF module having about 6–8 km of dispersion-compensating fiber in order to compensate for GVD. But this arrangement also has two problems:

- To compensate high attenuation of DCF, amplifier gain has to be increased. But this results into severe non-linear effects such as augmented ASE noise.
- The optical intensity happens to be more within a DCF for a specified input optical power due to its small mode diameter. This results in considerable increase in the non-linear effects.

(d) *Using two-mode DCF (TM–DCF):* Two-mode dispersion compensating fibers (TM–DCF) can be designed with values of $V \approx 2.5$ which results in higher-order mode to be near cut-off. We know that for higher-order mode, the dispersion parameter D has large negative value. The use of TM-DCF necessitates a mode conversion device which is capable of converting the fundamental mode energy to higher-order mode energy which is fully supported by DCF. A mode conversion device uses a two-mode fiber which has fiber gratings for providing necessary coupling between modes. It should operate over a wide bandwidth and must be polarization insensitive.

Major Drawbacks of DCFs

A relatively long enough (usually more than 5 km length) DCF may be needed in order to compensate for GVD which might have been accumulated over 50 km length of conventional fiber. Obviously, in long-haul applications, this would result in substantial addition of fiber loss to the existing optical fiber link loss. Table 6.1 shows link distances with various levels of dispersion slope matching with DCFs.

Table 6.1 Link distances versus bit rate and dispersion slope

Bit Rate	*Dispersion tolerance (power penalty = 1 dB)*	*Link distance without slope match*	*Link distance with 60% slope match*	*Link distance with nearly 100% slope match*
40 Gbps	30 ps/nm	~20 km	~40 km	300 km
10 Gbps	500 ps/nm	~300 km	~640 km	5000 km
2.5 Gbps	8000 ps/nm	~4700 km	~10250 km	80000 km

Facts to Know

The recent advancements in WDM have created a need to extend the band useful for amplifications. New developments show that you can extend the range of the optical amplifier possibly up to 1625 nm (or more), effectively allowing designers to send more optical channels through the system. Other WDM requirements include better gain flatness which can be achieved with doping modifications or filters.

Periodic Dispersion Maps

Periodic dispersion management technique provides the effective solution to the degradation in signal quality due to intra-channel effects (i.e., the non-linear interaction among various optical pulses available in the same optical fiber channel), and inter-channel effects (i.e., the non-linear interaction among optical pulses of adjacent channels in a wavelength-division multiplexing system). In this technique, fibers having positive GVDs are mixed with fibers having negative GVDs periodically. As a result, the net dispersion is almost negligible over each period. That is,

average dispersion, $$\overline{D} = \frac{\sum_{j=1}^{n} D_j L_j}{\sum_{j=1}^{n} L_j} \tag{6.26}$$

where, D_j represents dispersion of the fiber segment L_j (j = 1, 2, ….n) and $L_m = \sum_{j=1}^{n} L_j$ represents the dispersion map period that is selected in such a way that the required system performance is completely satisfied.

In the case of $\overline{D} \approx 0$, dispersion can be compensated for every map period. In practice, $L_m = L_A$ (the amplifier spacing) is typically 50 km for submarine systems and 80 km for terrestrial light wave system. However, in the presence of non-linear effects, it does not provide the best solution for perfect dispersion compensation of GVD in every dispersion map. The main problem occurs due to large broadening of the transmitted optical pulse travelling through the segment of the standard fiber of the specified dispersion map. It results in the non-linear interaction among the adjacent (may be overlapping also) optical pulses. Generally, it is desirable to keep relatively large local GVD so as to enable to suppress the resulting non-linear effects. Simultaneously, the average dispersion for all optical channels of a long-haul WDM light wave system should be minimized.

Example 6.9 Need of DCF

A 12 km DCF fiber having dispersion parameter = 100 ps/(nm–km) is used for compensating dispersion in a 80-km standard fiber having specified dispersion parameter = 17 ps/(nm–km). Can this DCF reduce the dispersion to zero? If not, what is the solution?

Solution:

We know that the total dispersion in the fiber = D × L

For DCF used, given D = 100 ps/(nm–km), and L = 12 km, we have

Dispersion provided by DCF = 100 ps/(nm–km) × 12 km = 1200 ps/nm

For the actual fiber used, given D = 17 ps/(nm–km) and fiber length L = 80 km, we have

Dispersion induced by fiber = 17 ps/(nm–km) × 80 km = 1360 ps/nm

Since dispersion induced by the fiber is more than the dispersion provided by DCF, so this DCF cannot reduce the induced dispersion to zero.

The solution is to use DCF which can provide dispersion equivalent to 1360 ps/nm. For specified D = 100 ps/(nm–km) of DCF, the length of DCF needed should be 1360/100 = 13.6 km. **Ans.**

Section Practice Problem

1. A 15.6 km DCF fiber having dispersion parameter = 100 ps/(nm–km) is used for compensating dispersion in a 80-km standard fiber having its specified dispersion parameter = 16 ps/(nm–km). Can this DCF reduce the dispersion to zero? Give reasons to support your answer. *[Ans.: NO]*

6.5 Fiber Bragg Gratings

'Grating' implies the periodic structure, i.e., a periodic change in the value of the fiber core refractive index. A very small portion of light gets reflected from the fiber core where there is a slight change in its refractive index. All these reflective portions of light combine into one reflected beam provided that the Bragg condition, as specified by the following expression, is satisfied.

$$\lambda_B = 2\Lambda n_{eff} \tag{6.27}$$

where, λ_B represents the Bragg wavelength, Λ represents the grating period (i.e., distance between two adjacent maximum points of the periodic refractive index), and n_{eff} represents the fiber core effective refractive index value.

Fiber Bragg grating works as a mirror, selectively reflecting the Bragg wavelength (λ_B) only, and thus transmitting all the other wavelengths of the optical signal. Chirped means the optical grating period, Λn_{eff}, changes linearly over the length of the grating. Thus, chirped FBG reflects not a single wavelength but a set of wavelengths. An optical circulator is used to direct pulses into and out of the FBG. Fig. 6.9 shows the principle of operation followed by a fiber Bragg grating.

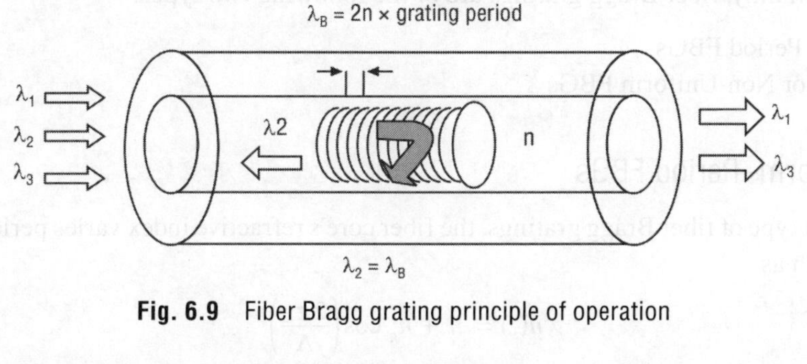

Fig. 6.9 Fiber Bragg grating principle of operation

An input optical pulse, dispersed after propagating along a fiber, is directed to the grating. The shorter wavelengths reflected almost immediately upon entering while the longer wavelengths penetrate deeper into the grating before they will be reflected. This effect is achieved by shortening the grating period at the grating entrance and lengthening it at the grating end. That's why it is called 'chirped'. Thus, the device ensures less delay for shorter wavelengths but creates more delay for longer delay. This is exactly the opposite of the delay introduced by a single mode fiber itself. Therefore, pulse spread caused by dispersion in fiber is compensated for by a chirped FBG. Chirped FBG works well for the λ_{Bragg} and its small variations. Generally, DCFs have been designed in such a way that dispersion parameter $|D|$ increases as operating wavelength increases, which plays an important role for WDM systems. This feature permits the broadband dispersion compensation. However, there is a trade-off between bandwidth of a fiber Bragg grating and its delay, i.e., its compensation ability.

Fabrication of an FBG

The fiber core refractive index can be changed under exposure to ultraviolet light. This phenomenon is known as fiber photosensitivity and is the physical basis for grating fabrication. There are two basic fabrication methods:

- *Directly exposing a fiber's core to a pair of interfering UV beams*– It provides radiation of both maximum and minimum intensity. The minimum intensity leaves the refractive index unchanged and the maximum intensity changes the refractive index.
- *The phase-mask technique*– It is based on the same interference principle but gives much better results because of the higher gratings precision it imposes.

FBGs as dispersion–compensation device

FBGs offer the most promising dispersion management solution. In fact, there exists a frequency region, known as the stop band, in which reflection of most of the incident light takes place. In this situation, a FBG can function almost equivalent to an optical filter with the stop band centered at the Bragg wavelength, $\lambda_B = 2\Lambda n_{eff}$ The index variations are periodic in nature and because of this the forward propagating waves as well as backward propagating waves are coupled together at wavelengths quite near to λ_B. This results into reflectivity to the incident optical signal which is frequency-dependent. The grating strength determines its bandwidth and functions as a reflection optical filter. Mainly, fiber Bragg gratings are of the following two types:

- Uniform Period FBGs
- Chirped or Non-Uniform FBGs

6.5.1 Uniform–Period FBGs

In the simplest type of fiber Bragg gratings, the fiber core's refractive index varies periodically along the fiber length as

$$n(z) = \bar{n} + n_g \cos\left(\frac{2\pi z}{\Lambda}\right) \tag{6.28}$$

where, $\bar{n}$ represents the average mode index, n_g denotes the modulation depth (typically ~ 10^{-4}), z is the fiber length and Λ represents the grating period.

The reflectivity of the grating becomes nearly 100% within the stop band. Outside the stop band, there is a possibility of the grating-induced dispersion as the phase is nearly linear there. For example, to compensate the GVD of 100-km fiber length, a single grating having 2-cm length may be adequate. Practically, the uniform fiber gratings are not possible for dispersion compensation. In case the wavelength of the optical signal happens to lie within the stop band, then the coupling coefficient can be tapered along the grating length for dispersion compensation and the gratings functions acts as a reflection filter.

Now let us move ahead with the analysis of the Bragg gratings. The coupled–mode equations are used to analyze the Bragg gratings. They describe the coupling between the forward propagating waves and backward propagating waves at a pre-defined angular frequency ω. The coupled-mode equations are generally expressed as:

$$\frac{dA_f}{dz} = i\delta A_f + iKA_b \tag{6.29}$$

$$\frac{dA_b}{dz} = -i\delta A_b - iKA_f \tag{6.30}$$

where, A_f represents the spectral amplitude of the forward propagating wave and A_b represents the spectral amplitude of the backward propagating wave.

$\delta = \frac{2\pi}{\lambda_0} - \frac{2\pi}{\lambda_B}$ represents the detuning from the Bragg wavelength λ_B; (6.31)

$K = \frac{\pi n_g \Gamma}{\lambda_B}$ denotes the coupling coefficient; (6.32)

The confinement factor, $\Gamma = \frac{P_{core}}{P_{total}} = 1 - e^{-\frac{2a^2}{\omega^2}}$; a being the core radius and ω being the field radius, also known as spot size.

The coupled-mode equations are linear in nature and can be solved analytically. The transfer function of the gratings which act as a reflective optical filter, can be written as

$$H(\omega) = r(\omega) = \frac{A_b(0)}{A_f(0)} = \frac{iK\sin(qL_g)}{q\cos(qL_g) - i\delta\sin(qL_g)} \tag{6.33}$$

where, $q^2 = \delta_2 - K_2$, and L_g represents the length of the grating.

The condition ($KL_g = 3$) means the stop band and the grating reflectivity becomes nearly 100%. There are two methods for dispersion compensation as specified below:

- *Apodization Method*– In uniform-period gratings, the change in refractive index, represented by Δn is maintained uniform throughout the grating. This results in z-dependent coupling coefficient K. Uniform-period gratings can be obtained by writing the grating holographically with the use of an ultraviolet Gaussian beam. Fig. 6.10 and Fig. 6.11 illustrate uniform Bragg grating period (typical) vis-à-vis Apodization of uniform-period FBG, respectively.

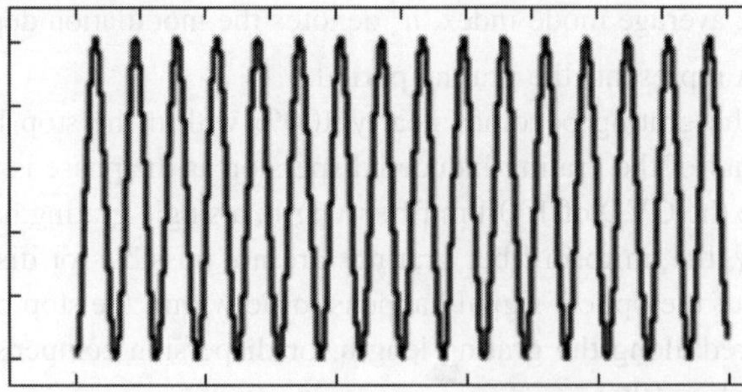

Fig. 6.10 Uniform fiber Bragg grating period

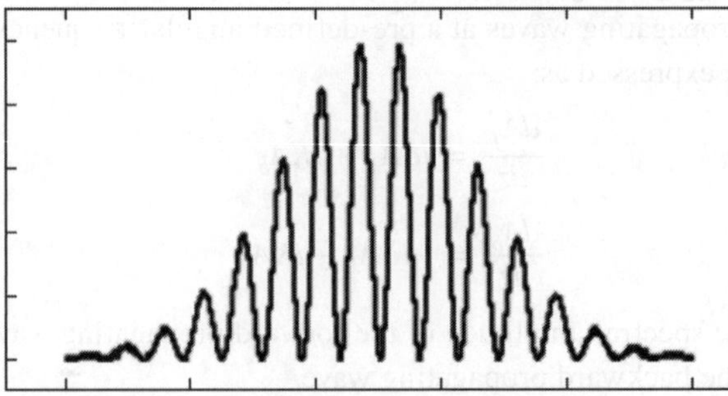

Fig. 6.11 Apodization for uniform–period FBG

For dispersion compensation, we can use V-shaped group-delay profile. This profile must be centered at Bragg wavelength λ_B so that the group delay varies linearly.

- *Tapering of Coupling Coefficient*– As mentioned previously, it is possible to obtain dispersion compensation if the coupling coefficient is tapered along the grating length. The necessary condition is that the wavelength of the optical signal must lie within the specified stop band, at which the grating is considered to function as a reflection optical filter.

6.5.2 Chirped FBGs

The most developed dispersion–compensating gratings (DCGs) are chirped fiber Bragg gratings (FBGs), also known as non-uniform FBGs. In a standard fiber, there may be long delays for those frequency components within an optical pulse which are on the lower side. This is mainly because of chromatic dispersion. So, we can design chirped Bragg grating fibers with longer delays for higher frequency components within an optical pulse. This would result into pulse compression.

Chirped or non-uniform fiber Bragg gratings have the following features:

- The variation of the Bragg wavelength $\lambda_B = 2\Lambda n_{eff}$ is all along the length of the grating.
- The optical period Λn_{eff} also varies over the fiber length.
- Whenever the Bragg condition is satisfied locally, there is every possibility of reflection of various frequency components present in an input optical pulse.
- As the specified Bragg wavelength alters along the length of the gratings, a chirped fiber grating may experience a stop band due to possible overlapping of several tiny stop bands.
- Chirped FBGs have a relatively wide stop band.

Fig. 6.12 shows how chromatic dispersion is compensated by chirped fiber Bragg grating.

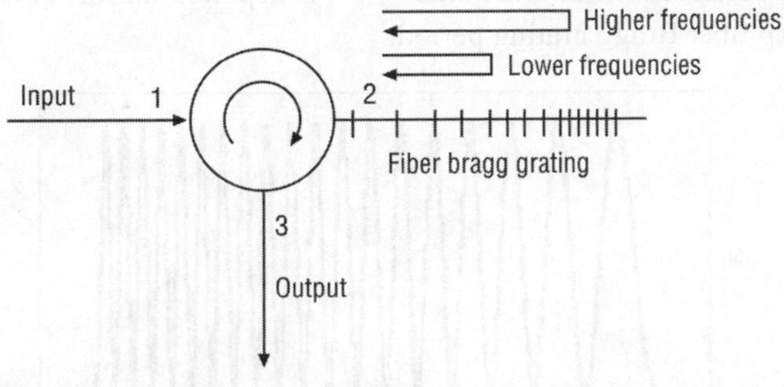

Fig. 6.12 Chromatic dispersion compensation with chirped fiber Bragg grating

Due to increasing Bragg wavelength and the optical period, there is delay in low frequency components in an optical pulse. Then the dispersion parameter D_g of a chirped grating of length L_g is given by

$$T_R = D_g L_g \Delta\lambda \tag{6.34}$$

$$\Rightarrow \quad D_g = \frac{T_R}{L_g \Delta\lambda} \tag{6.35}$$

where, T_R represents the round-trip time within the fiber Bragg gratings, $\Delta\lambda$ represents the difference in Bragg wavelengths at two extreme ends of the gratings.

Substituting $T_R = \frac{2n_{eff} L_g}{c}$ in the above equation, we get

$$\Rightarrow \quad D_g = \frac{2n_{eff} L_g}{c L_g \Delta\lambda} = \frac{2n_{eff}}{c\Delta\lambda} \tag{6.36}$$

For WDM systems, we are required to employ different chirped fiber Bragg gratings for each and every wavelength used in different channels for dispersion compensation. As an example, Fig. 6.13 shows chirped fiber Bragg gratings used for dispersion compensation of λ_1, λ_2, and λ_3.

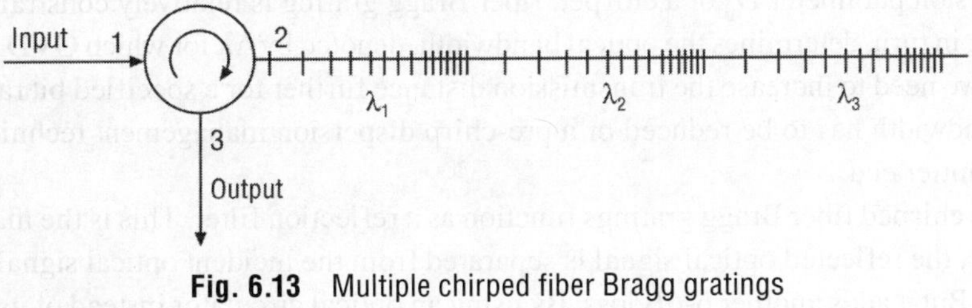

Fig. 6.13 Multiple chirped fiber Bragg gratings

Methods of fabrication of chirped fiber gratings:

- Dual-beam holographic technique
- Double exposure technique
- Tilting, or stretching of the fiber
- Phase mask technique

Apodization method can also be used for dispersion compensation for non-uniform Bragg grating period. The change in refractive index, denoted by Δn is kept non-uniform throughout the grating. Fig. 6.14 depicts chirp fiber Bragg grating period.

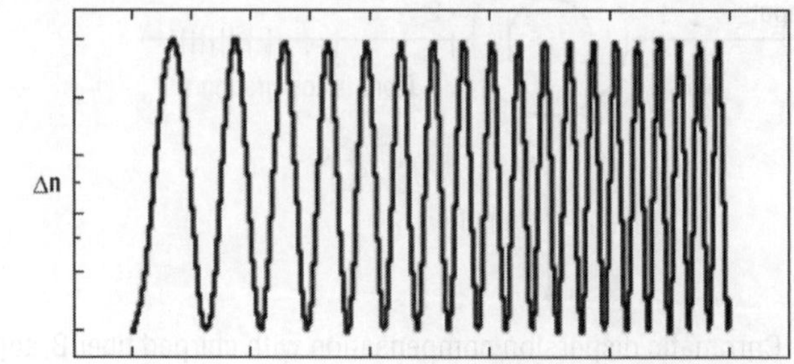

Fig. 6.14 Chirp fiber Bragg grating period

The slope of group delay actually signifies the dispersion–compensating ability of the chirped fiber Bragg grating. The chirped fiber Bragg grating should be apodized in order to achieve maximum coupling coefficient in the center and almost negligible coupling coefficient towards its extreme ends. Fig. 6.15 shows the concept of Apodization for chirp fiber Bragg grating.

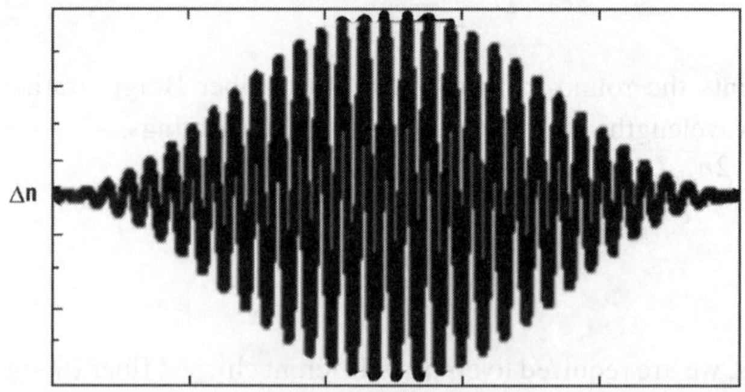

Fig. 6.15 Apodization for chirp fiber Bragg grating

The dispersion parameter D_g of a chirped fiber Bragg grating is relatively constrained by the bit rate R_B which, in turn, determines the optical bandwidth, denoted by $\Delta\lambda$ for which GVD compensation is needed. If we need to increase the transmission distance further for a specified bit rate, then either the signal bandwidth has to be reduced or a pre-chirp dispersion management technique is used at optical transmitter end.

In fact, the chirped fiber Bragg gratings function as a reflection filter. This is the main limitation. In some cases, the reflected optical signal is separated from the incident optical signal using a 3-dB fiber coupler. But it adds another 6-dB loss. By using an optical circulator instead of using an optical 3-dB fiber coupler, the insertion loss can be reduced to less than 2 dB. Fig. 6.16 shows an application of fiber Bragg grating for optical add/drop multiplexer.

A phase shift can be introduced in the middle of a single fiber Bragg grating in order to form a transmission filter having relatively small bandwidth. A transmission filter can also be formed by combining two or more fiber Bragg gratings. This can provide dispersion compensation having considerably low insertion loss.

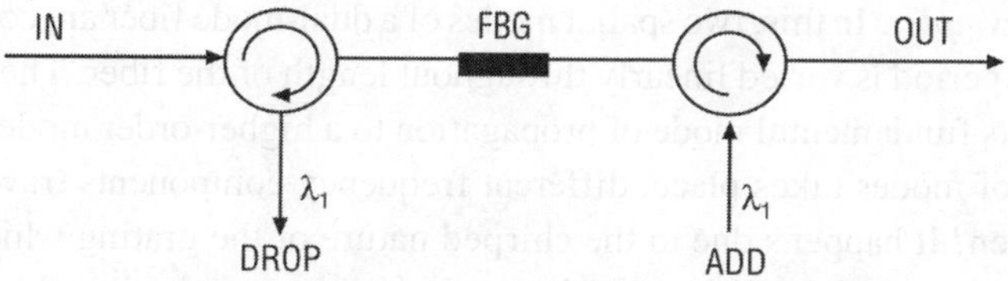

Fig. 6.16 Use of FBG in optical add/drop MUX

Application of FBG for dispersion compensation

Wavelength division multiplexing (WDM) networks employ a very large number of optical channels with an objective of providing the enhanced capacity of the order of more than 1 Tbps. Therefore, it becomes utmost necessary that dispersion should be compensated for each channel for reliable and efficient transmission. Chirp fiber gratings are often employed in WDM system for multiplexing of less than 10 number of different wavelength channels. If the resultant bandwidth of WDM signal is quite large, then a cascaded chirped fiber grating is used in series for dispersion management. Fig. 6.17 shows an arrangement of using cascaded fiber gratings in WDM system for dispersion management.

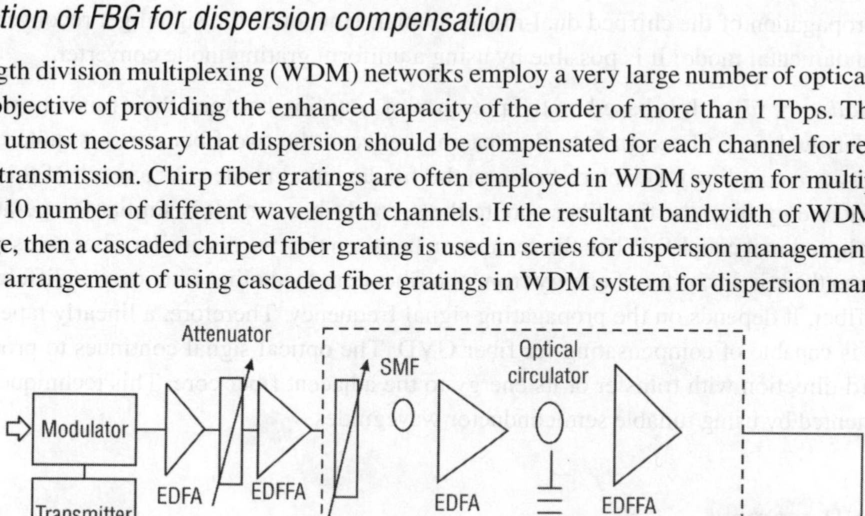

Fig. 6.17 Cascaded gratings in WDM systems for dispersion management

As shown, four number of fiber Bragg gratings are used for dispersion compensation of combined GVD of all the optical channels and two EDFAs along with an optical circulator are used to compensate for the fiber losses.

6.6 Chirped Mode Couplers

It is an all-fiber device which has been designed based on the principle of chirped distributed resonant coupling. Dispersion management is possible by using two fiber-based transmission fibers in the following two ways:

- Chirped dual-mode coupler
- Tapered dual-core fiber

Chirped dual-mode coupler: In this, two spatial modes of a dual-mode fiber are coupled by the chirped grating. The grating period is varied linearly throughout length of the fiber. The grating transfers an optical signal from its fundamental mode of propagation to a higher-order mode of propagation. But before this transfer of modes takes place, different frequency components traverse different paths. Why does this happen? It happens due to the chirped nature of the grating which is responsible for coupling the fundamental mode and higher-order modes. With an increase in the grating period along the length of chirped dual-mode coupler, there is a possibility to compensate for the fiber GVD. It may be noted here that the optical signal continues to propagate onward but results in a higher-order mode of propagation of the chirped dual-mode coupler. How does the signal get reconverted back into the fundamental mode? It is possible by using a uniform grating mode converter.

Tapered dual-core fiber: It is based on the presence of coupling between their fundamental modes of dissimilar dual-core fibers. In case the spacing between the two fiber cores is quite close, the evanescent-wave coupling takes place between the fundamental modes. As a result of this, there is a transfer of energy from the first fiber core to the second fiber core. It is similar to operation as in an optical directional coupler. When the separation between the two dissimilar fiber cores is tapered linearly, then there is a transfer of energy from one fiber core to another fiber core at different points along the fiber. It depends on the propagating signal frequency. Therefore, a linearly tapered dual-core fiber is capable of compensating for fiber GVD. The optical signal continues to propagate in the forward direction with transfer of its energy to the adjacent fiber core. This technique can also be implemented by using tunable semiconductor waveguides.

Points to Remember

- The performance of fiber–optic communication systems is quite often restricted more due to dispersion as well as non-linear effects rather than due to fiber transmission losses.
- The standard single-mode fibers exhibit relatively large GVD which limits the performance of 1550 nm fiber–optic systems at a bit rate exceeding 10 Gbps. This necessitates the extensive use of dispersion-management techniques.
- Dispersion management is the term used to refer to the management of dispersion compensation.
- Pre-compensation dispersion management deals with modifying the characteristics of optical pulses at the transmitter end prior to launching them into the fiber link.
- Electronic equalization scheme of post-compensation dispersion management techniques is the most practical approach in coherent fiber–optic communication links.
- An opto–electronic equalization technique for dispersion management is based on a transversal filter.
- As the optical pulse is affected by GVD through the spectral phase characteristics, an optical equalization filter (with transfer function that can cancel the phase) can restore the propagating optical signal.
- For long-haul fiber–optic communication system, it is essential to compensate for the GVD periodically along the fiber length.
- The use of dispersion compensating fiber (DCF) is an effective all-optical method for complete compensation of the fiber GVD provided that the average optical power is maintained reasonably low so as to have negligible non-linear effects within the optical fibers.
- By introducing a small segment of a single-mode fiber having appropriate negative dispersion characteristics so as to compensate for the positive dispersion introduced by a conventional fiber.

- Fiber Bragg gratings (FBGs) offer the most promising dispersion management solution. In fact, a stop band exists in FBGs over a frequency range when the incident light is mostly reflected back instead of propagation.
- The most developed dispersion-compensating gratings (DCGs) are chirped fiber Bragg gratings (FBGs).
- A transmission filter can be formed by combining two or more fiber gratings to provide dispersion compensation that results into relatively low insertion losses.
- An all-fiber device, known as a chirped mode coupler, is designed based on the fundamental principle of chirped distributed resonant coupling.

Important Equations

The 3-dB fiber bandwidth, $f_{3dB} \approx \dfrac{0.188}{|D|L\sigma_\lambda}$; *where D is dispersion parameter in ps/(km–nm), L represents the length of the optical fiber in km, and σ_λ represents the RMS spectral width in nm.*

For a directly-modulated DFB laser, the maximum transmission distance, $L < \dfrac{1}{4R_B|D|\sigma_\lambda}$; *where R_B is the bit rate, D represents the dispersion parameter given in ps/(km–nm), and σ_λ represents the RMS spectral width of the optical pulse broadened considerable due to frequency chirping.*

The transmission distance with chirped Gaussian pulse $L = \dfrac{C + \sqrt{(1+2C^2)}}{1+C^2} L_D$; *where C is the parameter that decides the frequency chirp imposed on the optical pulse, and L_D represents the dispersion length.*

The Bragg wavelength $\lambda_B = 2\Lambda n_{eff}$; *where Λ represents the grating period (that is, the distance between two adjacent maximum points of the periodic refractive index), and n_{eff} represents the fiber core's effective refractive index.*

In chirped fiber Bragg grating, dispersion parameter of a chirped grating $D_g = \dfrac{T_R}{L_g \Delta\lambda}$; where $T_R = \dfrac{2n_{eff}L_g}{c}$ *represents the round-trip time within the grating, L_g represents the length of a chirped grating, n_{eff} denotes the fiber core's effective refractive index, $\Delta\lambda$ represents the difference in the Bragg wavelengths at the two ends of the grating.*

Key Terms with Definitions

Chirped fiber Bragg gratings A type of grating that possess a relatively wide stop band. The optical period in a chirped fiber Bragg grating changes over the fiber length.

Chirped mode coupler An all-fiber device used for dispersion management, and is designed based on the basic principle of chirped distributed resonant coupling.

DCF Dispersion Compensating Fiber. It is a loop of a segment of special kind of fiber which has equal and opposite dispersion to that of transmission fiber so as to compensate the fiber GVD completely at low average optical power.

DFF	Dispersion-Flattened Fiber in which the index profile of a fiber is manipulated in such a way that the net dispersion is nearly zero at two or three different prominent wavelengths as well as in between.
Dispersion	Dispersion is a phenomenon by which all the spectral components are separated from a wave.
Dispersion management	Dispersion management is the term used to refer to the management of dispersion compensation.
DSF	Dispersion-Shifted Fibers in which there is a shift in zero-dispersion wavelength toward the region of lowest attenuation which mostly lies in the 1550-nm optical band.
Electronic equalization	A dispersion management technique used for GVD compensation based on the principle of heterodyning in the optical receiver.
FBG	Fiber Bragg grating works as a mirror, selectively reflecting Bragg wavelength only, and thus transmitting all the other wavelengths of the optical signal.
FOM	Figure of Merit– the ratio of fiber dispersion (ps/nm–km) to fiber attenuation (dB/km)
Grating	Grating implies the periodic structure, i.e., a periodic change in the value of the refractive index of the core.
Interferometry optical filter	Optical filters using an interferometer having frequency dependent transmission characteristics.
Optical equalization	An optical equalization filter having a transfer function which can cancel the phase in order to restore the GVD affected optical signal.
Opto-electronic equalization	A post-compensation dispersion management technique which uses a transversal filter at the optical receiver.
Pre-compensation	Pre-compensation dispersion management deals with modifying the characteristics of optical pulses at the optical transmitter. It changes the spectral amplitude of optical pulses in order to compensate GVD exactly, with restoration of original shape of the optical pulse at the output of the fiber.
Post-compensation	The actual implementation of dispersion management techniques is carried out at the optical receiver end so as to cancel out the phase factor which is responsible for dispersion-induced degradation of the transmitted optical signal through the optical fiber.
Uniform-period FBGs	A type of FBG in which the fiber core's refractive index changes periodically along the length of the fiber.

Short Answer Type Questions

1. **What are the different methods for reducing the effects of dispersion?**
   (1) Using an external modulation at optical transmitter to reduce frequency chirping
   (2) Use of small dispersion fiber
   (3) Use of dispersion compensation fiber

2. **Define chirp.**
   The chirp C is defined by the change in frequency $d\omega$ due to the rate of change of the phase. That is,

$$d\omega = -\frac{d\varphi}{dt} = \frac{Ct}{\tau^2}$$

where, τ is the initial *1/e* duration of the optical pulse.

3. List the desirable properties of an ideal dispersion compensation device.

(1) Large negative dispersion coefficient
(2) Low attenuation
(3) Minimal nonlinear contributions
(4) Wide bandwidth
(5) Corrects dispersion slope as well
(6) Minimal ripple
(7) Polarization independent

4. Illustrate the plot of dispersion vs wavelength for conventional (unshifted) fiber and dispersion-compensating fiber for comparison purpose.

Fig. 6.18 shows the plot of dispersion vs wavelength for conventional (unshifted) fiber and dispersion-compensating fiber.

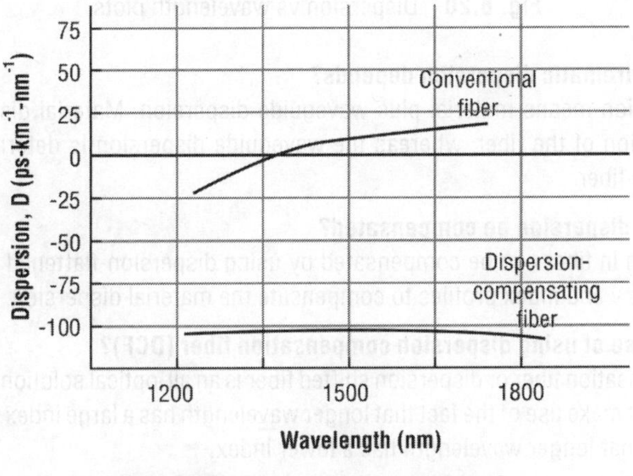

Fig. 6.18 Dispersion vs wavelength

5. Draw a functional block schematic diagram of a typical optical fiber communication link depicting the use of dispersion compensating fiber.

Fig. 6.19 shows the required functional block schematic.

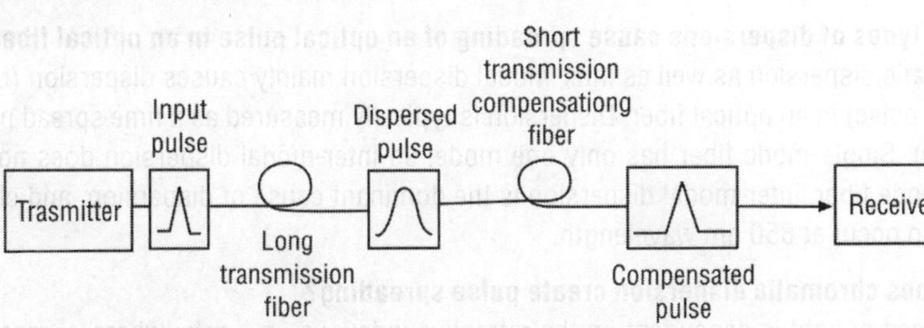

Fig. 6.19 Use of DCF in optical fiber communication link

6. With the help of plots between dispersion versus wavelength, show that dispersion compensating fiber (DCF) exhibits uniform dispersion over 1.3 μm – 1.5 μm wavelength region as compared to standard single-mode fiber and non-zero dispersion-shifted (at 1.5 μm wavelength).

Fig. 6.20 shows the plot between dispersion versus wavelength for standard single-mode, non-zero dispersion-shifted fiber as well as the dispersion-compensating fiber.

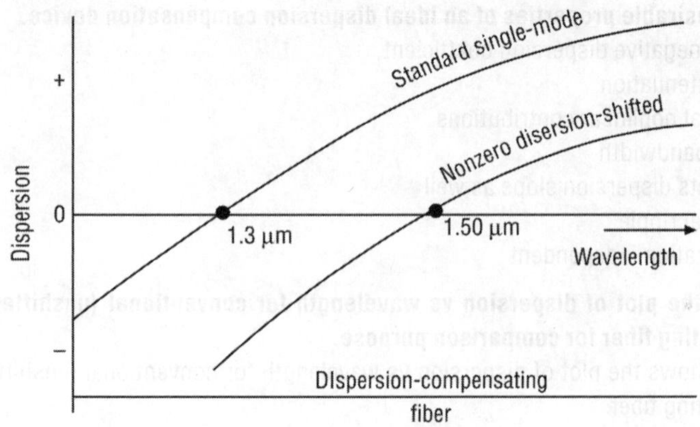

Fig. 6.20 Dispersion vs wavelength plots

7. **On what factors chromatic dispersion depends?**
Chromatic dispersion means material plus waveguide dispersion. Material dispersion depends on the material composition of the fiber, whereas the waveguide dispersion is determined by the waveguide index profile of the fiber.

8. **How can material dispersion be compensated?**
Material dispersion in fiber can be compensated by using dispersion-flattened fibers which manipulate waveguide geometry and index profiles to compensate the material dispersion.

9. **What is the purpose of using dispersion compensation fiber (DCF)?**
Dispersion compensation fiber or dispersion shifted fiber is an all-optical solution for long-haul fiber–optic communications. It make use of the fact that longer wavelength has a large index that make the waveguide weakly guided so that longer wavelength has a lower index.

10. **Distinguish between multimode, material and waveguide dispersion.**
Multimode group delay/dispersion refers to change in the group velocity among the propagation modes at a single frequency. Material Dispersion is due to variation in the refractive index of the core material as a function of wavelength. Waveguide dispersion depends upon the fiber design. The propagation constant is the function of the ratio of fiber dimension (i.e., core radius) to the wavelength.

11. **Which types of dispersions cause spreading of an optical pulse in an optical fiber?**
Chromatic dispersion as well as inter-modal dispersion mainly causes dispersion (i.e., spreading of the optical pulse) in an optical fiber. Dispersion is typically measured as a time spread per distance traveled (ns/km). Single-mode fiber has only one mode, so inter-modal dispersion does not occur in it. But in multimode fiber, inter-modal dispersion is the dominant cause of dispersion, and chromatic dispersion can also occur at 850 nm wavelength.

12. **How does chromatic dispersion create pulse spreading?**
The speed of light is dependent on the refractive index, i.e., $c = c_0/n$; where c_0 represents the speed of light in a vacuum. The index of refraction, n, varies with the light transmission wavelength. All optical sources (such as LEDs and LDs) have some color, or variation in wavelength output. The low wavelength portion of the pulse travels slower than the high wavelength one, creating pulse spreading.

13. **Mention some fundamental properties of chromatic dispersion.**
Chromatic dispersion is measured in units of time divided by distance and transmitter source spectral width (ps/nm–km). It is zero near 1310 nm in silica optical fibers and also zero near 1550 nm in dispersion

shifted optical fibers. Even at the dispersion zero, there is some pulse spreading due to the spectral width of the optical source.

14. Draw a suitable diagram to depict the concept of pulse spreading arising due to multimode propagation of optical signal in an optical fiber.

Fig. 6.21 shows the concept of pulse spreading due to multimode propagation.

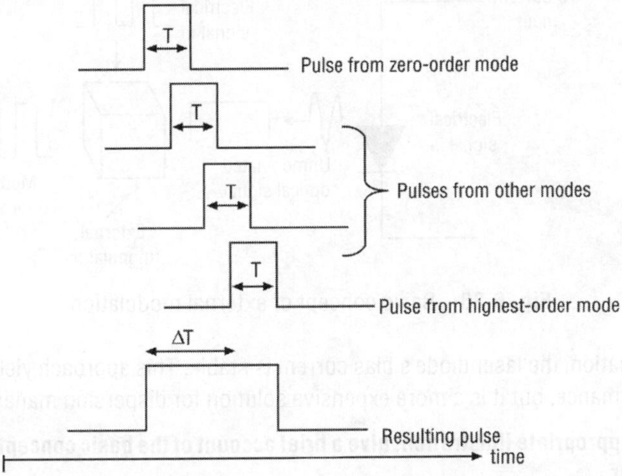

Fig. 6.21 Pulse spreading due to multimode propagation

15. What are the effects of pulse spreading due to chromatic dispersion?

As the optical pulse propagates down the fiber, it tends to spread and generate Inter Symbol Interference (ISI). As a result, it limits either the bit rate or the maximum achievable distance at a specified bit rate. This is due to the fact that the refractive index of the fiber core has a wavelength dependent factor. This makes different frequency-components of the optical pulses to travel at different speeds. Higher bit rates experience higher signal degradation due to chromatic dispersion.

16. Between direct modulation and external modulation, which approach would you prefer as a dispersion management solution and why?

Fig. 6.22 shows the basic concept of direct modulation of laser diode being used as an optical source.

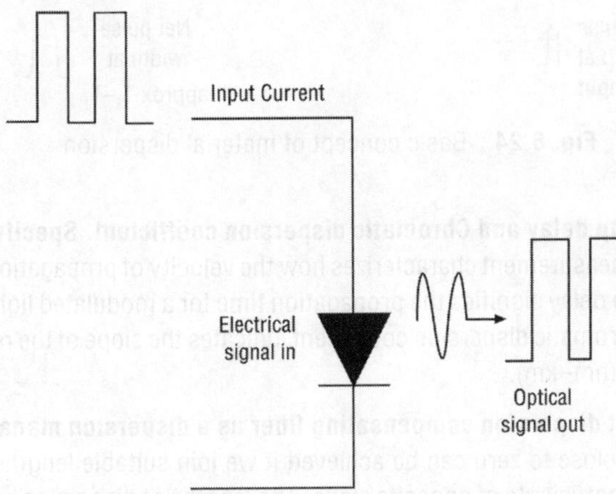

Fig. 6.22 Basic concept of direct modulation

In direct modulation, the laser diode's bias current is modulated with signal input to produce modulated optical output. This approach is straightforward and low cost, but is susceptible to chirp (spectral broadening) thus exposing the signal to higher dispersion.

Fig. 6.23 shows the basic concept of external modulation of laser diode being used as an optical source.

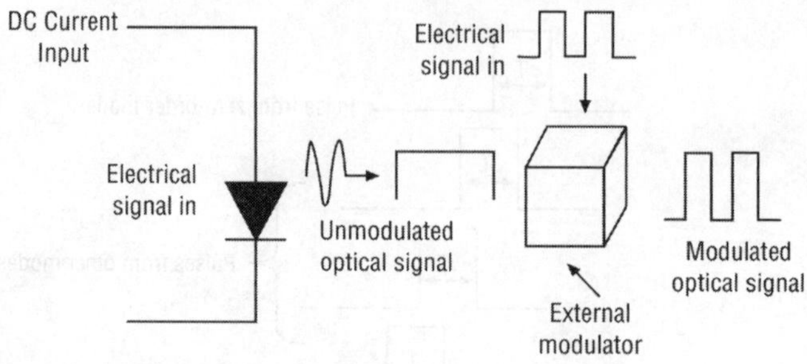

Fig. 6.23 Basic concept of external modulation

In external modulation, the laser diode's bias current is stable. This approach yields low chirp and better dispersion performance, but it is a more expensive solution for dispersion management.

17. With the help of appropriate illustration, give a brief account of the basic concept of material dispersion in an optical fiber.

We know that in an optical fiber the propagation velocity varies with operating wavelength. Thus an optical pulse made up of many wavelengths will be spread out in time as it propagates within an optical fiber. Fig. 6.24 illustrates the basic concept of material dispersion with two wavelength example.

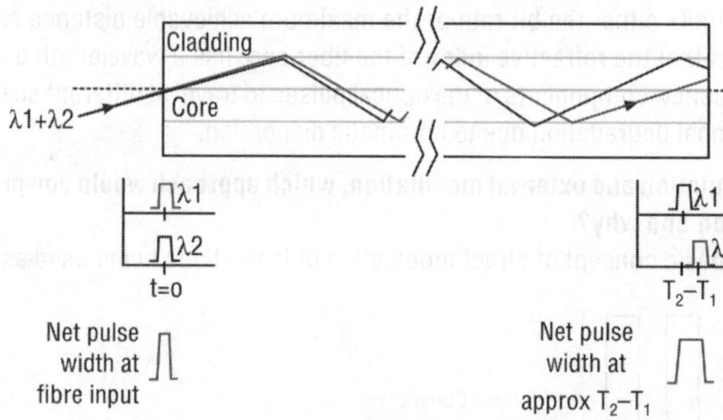

Fig. 6.24 Basic concept of material dispersion

18. Define the terms: Group delay and Chromatic dispersion coefficient. Specify their units.

Chromatic dispersion measurement characterizes how the velocity of propagation of a light pulse changes with wavelength. Group delay signifies the propagation time for a modulated light wave, and is expressed in picoseconds (ps). Chromatic dispersion coefficient indicates the slope of the relative group delay curve, and is expressed in ps/(nm–km).

19. Comment on the use of dispersion compensating fiber as a dispersion management solution.

An average dispersion close to zero can be achieved if we join suitable lengths of optical fibers having chromatic dispersion coefficients of opposite signs. The lengths of dispersion compensating fiber can be of the order of several kilometers. These can be inserted at any point in the fiber–optic communication link, either at the transmitter or at the receiver. Although the total dispersion is close to zero, this technique can also be employed to manage FWM and CPM since at every point we have dispersion which translates in decoupling different channels limiting the mutual interaction.

20. What is the relationship of chromatic dispersion to bit rate?

Chromatic dispersion becomes a serious problem at 10 Gbps and beyond. This leads to a higher bit error rate (BER). In fact, the acceptable extent of chromatic dispersion depends on the bit rate and is related as inversely proportional to its square.

21. With the help of suitable functional block diagrams, differentiate between pre-compensation and post-compensation dispersion management techniques.

Fig. 6.25 depicts a typical functional block schematic diagram of pre-compensation dispersion management technique.

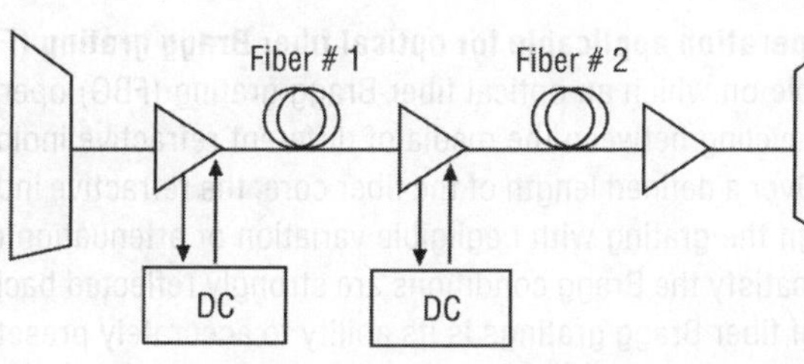

Fig. 6.25 A typical pre-dispersion compensation (DC) block diagram

Fig. 6.26 depicts a typical functional block schematic diagram of post-compensation dispersion management technique.

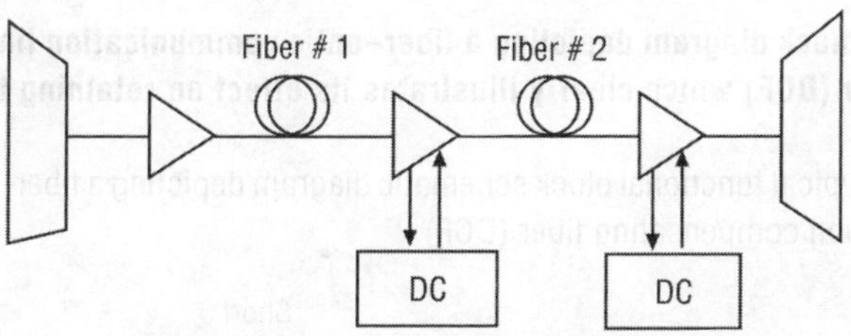

Fig. 6.26 A typical post-dispersion compensation (DC) block diagram

22. Would you prefer electronic dispersion compensation (EDC) technique of dispersion management? Give reasons to support your argument.

Electronic dispersion compensation technique is widely used in dispersion management in fiber–optic communications. Due to direct detection at the optical receiver, chromatic dispersion (which is considered as a linear distortion in the optical domain) is transformed into non-linear distortions after the received optical signal is converted to an electrical signal. Therefore, the concept of non-linear channel modeling is realized using decision feedback equalizers (DFE) and feed forward equalizer (FFE) structures are employed. But the use of electronic dispersion compensation reduces the bit rate.

23. Highlight the salient features of Optical Fiber Bragg Grating (FBG).

In long-haul fiber–optic communication links, optical fiber Bragg grating (FBG) has found widespread practical applications in compensation of dispersion-broadening optical pulses. Fiber Bragg gratings are constructed by laterally exposing a single-mode fiber's core to a periodic pattern of an intense ultraviolet light. This results in a permanent increase in the refractive index value of the fiber's core. It is equivalent to the creation of a fixed-index modulation, known as grating, in proportion to the pattern of the exposure. At each periodic fractional change in the core's refractive index, a small amount of light is reflected. All the reflected light signals combine coherently to one large reflected light signal at a particular wavelength when the grating period is approximately half of the wavelength of the input light signal. This is referred to as the Bragg condition, and the wavelength at which this reflection occurs is called the Bragg wavelength. Light signals at wavelengths other than the Bragg wavelength are essentially transparent because these are not phase matched.

24. List typical dispersion related parameters of conventional single-mode fiber.

- Dispersion parameter = 17.0 ps/(nm–km) at 1550 nm
- Slope = 0.057 ps/(nm^2–km) at 1550 nm
- Dispersion parameter range at 1530–1565 nm = 15.9–17.8 ps/(nm–km)
- Dispersion parameter range at 1570–1620 nm = 18.1–21.0 ps/(nm–km)

25. State the principle of operation applicable for optical fiber Bragg grating (FBG) and elaborate it.

The fundamental principle on which an optical fiber Bragg grating (FBG) operates is Fresnel reflection. It states that the light traveling between the media of different refractive indices may refract as well as reflect at the interface. Over a defined length of the fiber core, the refractive index will typically alternate. Light propagates through the grating with negligible variation or attenuation of the optical signal. Only those wavelengths that satisfy the Bragg conditions are strongly reflected back. One of the fundamental feature and advantage of fiber Bragg gratings is its ability to accurately preset and maintain the grating wavelength. The central wavelength of the reflected optical signal satisfies the Bragg condition, $\lambda_B = 2\Lambda n_{eff}$; where λ_B is the Bragg wavelength, Λ represents the grating period (i.e., the distance between two adjacent maximum points of the periodic refractive index), and n_{eff} is the fiber core's effective refractive index. It is also called modal index because it depends on the wavelength as well as on the mode in which the light propagates.

26. Draw a functional block diagram depicting a fiber–optic communication link using short dispersion compensating fiber (DCF) which clearly illustrates its effect on retaining the original shape of the transmitted pulse.

Fig. 6.27 depicts a typical functional block schematic diagram depicting a fiber–optic communication link using short dispersion compensating fiber (DCF).

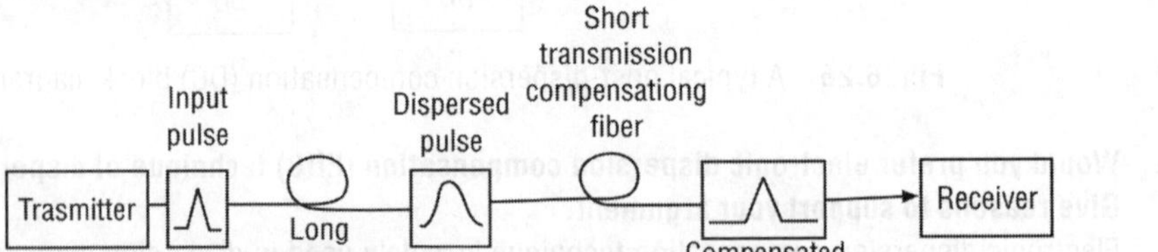

Fig. 6.27 A typical fiber–optic link using short DCF

It can be seen that the transmitted input pulse is broadened due to the dispersion in long transmission fiber. This is called dispersed pulse. A short dispersion compensating fiber is introduced in transmission fiber and compensated pulse is retrieved which is identical to transmitted input pulse.

Multiple Choice Questions

1. Fiber–optic communication systems and networks are primarily limited by
Statement I: the dispersive effects
Statement II: the non-linear effects
Statement III: the fiber losses
A. Only Statement I is correct. B. Only statements I and II are correct.
C. Only statements I and III are correct. D. All statements are correct.

2. Which statement is not correct?
A. It is possible to compensate for fiber losses with proper use of optical amplifiers but it further worsens the dispersion problem.
B. Dispersion-induced broadening of optical pulse imposes the most serious limitations on the system performance.
C. The bandwidth of the optical fiber is a direct measure of the dispersion-limited transmitted bit rate.
D. The 3-dB fiber bandwidth is directly related with the length of the fiber.

3. The 3-dB bandwidth of an optical fiber is expressed as (D is the dispersion parameter in ps/(km–nm), L represents the length of the fiber in km, and σ_λ represents the RMS spectral width of the optical source in nanometers)
A. $f_{3dB} \approx \frac{0.188}{|D| L\sigma_\lambda}$. B. $f_{3dB} \approx \frac{0.188|D|}{L\sigma_\lambda}$
C. $f_{3dB} \approx \frac{0.188L}{|D|\sigma_\lambda}$ D. $f_{3dB} \approx \frac{0.188\sigma_\lambda}{|D|L}$

4. The bit rate R_B is related with 3-dB fiber bandwidth by the relationship
A. $R_B \le 1.33 f_{3dB}$ B. $R_B \ge 1.33 f_{3dB}$
C. $R_B \le \frac{1.33}{f_{3dB}}$ D. $R_B \le \frac{1}{1.33 f_{3dB}}$

5. The effects of Group-Velocity Dispersion (GVD) can be reduced by
Statement I: employing an optical source having narrow linewidth
Statement II: operating near zero-dispersion wavelength λ_{ZD} of the optical fiber
A. Only Statement I is true. B. Only Statement II is true.
C. Both statements are true. D. None of the statements is true.

6. For a directly-modulated distributed feedback (DFB) laser, the maximum transmission distance can be estimated using the following expression:
A. $L < \frac{1}{4R_B|D|\sigma_\lambda}$. B. $L > \frac{1}{4R_B|D|\sigma_\lambda}$.
C. $L < \frac{4|D|}{R_B\sigma_\lambda}$. D. $L \ge \frac{1}{4R_B|D|\sigma_\lambda}$.

7. The maximum estimated transmission distance for a directly modulated DFB laser having $\sigma_\lambda = 0.15$ nm and operating at 2.5 Gbps bit rate is [$D = 10$ ps/(nm–km)]
A. 10.5 km B. 42 km
C. 168 km D. 672 km

8. The limiting transmission distance for an externally modulated DFB laser having $\sigma_\lambda = 0$ and operating at 2.5 Gbps bit rate is [D = 10 ps/(nm–km), $\beta_2 = -20$ ps^2/km at λ = 1550 nm]

A. 500 km B. 1000 km
C. 250 km D. 100 km

9. Actual implementation of dispersion management techniques can be carried out by using
Statement I: pre-compensation techniques at the optical transmitter.
Statement II: post-compensation techniques at the optical receiver.
Statement III: dispersion-compensating fibers along the fiber.

A. Only Statement I is true B. Only statements I and II are true
C. Only statements I and III are true D. All statements are true

10. In a pre-chirp pre-compensation dispersion management,
Statement I: For unchirped Gaussian pulses ($C = 0$), $L = L_D$.
Statement II: For chirped Gaussian pulses ($C = 1$), $L = 1.366L_D$.
Statement III: For chirped Gaussian pulses ($C = 1/\sqrt{2}$), $L = \sqrt{2}L_D$.

A. Only Statement I is true B. Only statements I and II are true
C. Only statements I and III are true D. All statements are true

11. *Statement I:* Directly-modulated semiconductor lasers used as optical source chirp the optical pulse automatically but the value of C is negative and the required condition $\beta_2 C < 0$ is not satisfied.
Statement II: In the case of external modulation of semiconductor lasers, optical pulses are nearly chirp-free and the required condition $\beta_2 C < 0$ is fully satisfied.

A. Only Statement I is true B. Only Statement II is true
C. Both statements are true D. None of the statements is true

12. Using an electro-absorption or a Mach–Zehnder (MZ) modulator, optical pulse can be chirped with

A. $C > 0$. B. $C < 0$.
C. $C = 0$. D. $C = ½$.

13. Which statement is true?

A. In a simple non-linear pre-chirp pre-compensation dispersion management technique, the output of optical transmitter is amplified with a SOA operating in the gain saturation region.
B. In a simple non-linear pre-chirp pre-compensation dispersion management technique, the output of optical transmitter is amplified using a Raman fiber amplifier.
C. In a simple non-linear pre-chirp pre-compensation dispersion management technique, the output of optical transmitter is amplified using an EDFA.
D. In a simple non-linear pre-chirp pre-compensation dispersion management technique, the output of optical transmitter is not amplified.

14. *Statement I:* Electronic equalization is the most practical technique for dispersion compensation in coherent fiber–optic communication systems.
Statement II: A linear electronic circuit cannot compensate GVD since all phase information is lost as a photodetector responds to optical intensity only.
Statement III: The non-linear equalization technique allows the recovery of the dispersion-induced optical pulse.

A. Only statements I and III are true B. Only Statement II is true
C. Only statements II and III are true D. All statements are true

15. An opto–electronic equalization technique for dispersion management is based on a

A. power splitter. B. Mach–Zehnder (MZ) modulator.
C. transversal filter. D. $LiNbO_3$ modulator.

16. MZ interferometer optical filter

A. has a bandwidth ~ 10 GHz
B. is insensitive to input polarization
C. cannot act a programmable optical filter
D. has fixed operating wavelength

17. In dispersion-shifted fibers, the wavelength of zero dispersion is shifted to the region of lowest attenuation which mostly lies in the ______ wavelength region.

A. 850 nm
B. 980 nm
C. 1300 nm
D. 1550 nm

18. Which statement is not correct?

A. DCF is a fiber loop of relatively shorter length which has dispersion equal and opposite to that of dispersion in the transmitting fiber.
B. DCF can be introduced either in the beginning (i.e., pre-compensation), or in the end (i.e., post-compensation) between two in-line optical amplifiers.
C. DCF cannot provide continuous compensation over a wide range of optical wavelengths.
D. DCF has a small core size which may make it prone to certain types of nonlinearities.

19. The condition for perfect dispersion compensation in DCF is

A. $L_2 = \left(\frac{D_1}{D_2}\right)L_1$.
B. $L_2 = -\left(\frac{D_1}{D_2}\right)L_1$.
C. $L_2 = -\left(\frac{D_2}{D_1}\right)L_1$.
D. $L_1 = -\left(\frac{D_1}{D_2}\right)L_2$.

20. Statement I: Chirped fiber Bragg gratings possess a comparatively narrower stop band.
Statement II: The Bragg wavelength $\lambda_B = 2\Lambda n_{eff}$ changes along the length of the grating.
Statement III: For WDM systems, we are required to employ a different type of chirped fiber Bragg grating to enable compensation for each channel.
Statement IV: Apodization method cannot also be used for dispersion compensation for non-uniform Bragg grating period.

A. Only statements I and II are true
B. Only Statement II is true
C. Only Statement II and III are true
D. All statements are true

21. Statement I: The most developed dispersion-compensating gratings (DCGs) are chirped fiber Bragg gratings (FBGs).
Statement II: Two or more fiber gratings can be combined to form a transmission fiber, which provides dispersion compensation with relatively high insertion loss.
Statement III: A chirped mode coupler is an all-optical fiber device which is designed using the principle of chirped distributed resonant coupling.

A. Only Statement I and II are true
B. Only Statement I and III are true
C. Only statements II and III are true
D. All statements are true

22. Which method(s) is used to reduce the impact of dispersion?

A. External modulation (to reduce chirping).
B. Small dispersion fiber.
C. Dispersion compensating fiber.
D. Any one of these.

Keys to Multiple Choice Questions

1. B	**2.** D	**3.** A	**4.** A	**5.** C	**6.** A	**7.** B	**8.** A	**9.** D	**10.** B
11. C	**12.** A	**13.** A	**14.** D	**15.** C	**16.** A	**17.** D	**18.** C	**19.** B	**20.** C
21. B	**22.** D								

Review Questions

1. Dispersion-induced pulse broadening imposes serious limitations on the system performance. What are these limitations? How are they related to each other?
2. How can the effects of Group-Velocity Dispersion (GVD) be reduced? What is the practical difficulty to make the operating wavelength equal to the specified zero-dispersion wavelength for a particular optical fiber cable?
3. Justify with suitable example that using an external modulator with DFB laser enhances the limiting transmission distance significantly as compared to that of a directly modulated DFB laser, for 2.5 Gbps bit rate.
4. What is meant by dispersion management? Why is it necessary? List various techniques of practical implementation of dispersion management.
5. Write the expression which relates the transmission distance (L) with dispersion length (L_D) in case of pre-chirp dispersion compensation method. Also specify the condition that should be satisfied by a chirped optical pulse that may travel for longer distances before it exceeds the allocated bit duration.
6. Draw a functional block schematic of the pre-chirp pre-compensation dispersion management technique. Illustrate the waveforms at frequency-modulated (FM) output of the DFB laser, shape of the optical pulse available at the output of an external modulator, and pre-chirped optical pulse that is finally transmitted.
7. Explain novel coding pre-compensation dispersion management techniques. Comment on the bit rate versus transmission distance achieved in each technique.
8. Discuss the following post-compensation dispersion management techniques:
   (a) Electronic Equalization
   (b) Opto–electronic Equalization
   (c) Optical Equalization

   Compare and contrast their advantages and limitations.
9. How do dispersion compensating fibers offer different approach for dispersion management in comparison to pre- and post-compensation techniques?
10. With the help of dispersion versus wavelength characteristics of standard, dispersion-shifted fibers (DSFs) as well as dispersion-flattened fibers (DFFs), describe the pros and cons of their usage in fiber–optic systems.
11. Define the characteristics figure of merit (FOM) used to describe the quality of DCF. For dispersion length of DCF to be small, what is the constraint on the dispersion coefficient?
12. How can the dispersion coefficient of DCF be made negative and attenuation be minimized? What is the major drawback of using DCF for dispersion management in long-haul applications?
13. What is the significance of Bragg condition in fiber Bragg gratings as dispersion-compensation scheme? Illustrate the principle on which the operation of a fiber Bragg grating is based.
14. List two basic fabrication methods employed for fiber Bragg gratings (FBG). Differentiate between uniform-period and non-uniform period FBGs.
15. There are two methods for dispersion compensation when the grating is made to function as a reflection filter and the wavelength of the optical signal falls within its stop band. Describe them briefly for uniform period FBGs.

16. Chirped fiber Bragg gratings are the most developed dispersion-compensating gratings (DCGs). Illustrate the basic operation of chirped fiber Bragg grating for chromatic dispersion compensation of (a) one wavelength; (b) three wavelengths as used in WDM.

17. What are the parameters on which the dispersion coefficient of a chirped grating depends? List various methods of fabrication of chirped fiber gratings.

18. Write short note on the followings:
(i) Chirped dual-mode coupler
(ii) Tapered dual-core fiber

Numerical Problems

1. For a specified dispersion parameter of the fiber = 17 ps/(nm–km), find the dispersion for 80 km length of the fiber. *[Ans.: 1360 ps/nm]*

2. Determine the insertion loss of a 13.6 km long DCF having specified fiber loss parameter as 0.5 dB/km. *[Ans.: 6.8 dB]*

3. Compute the figure-of-merit for a DCF having dispersion parameter as 100 ps/(nm–km). The fiber loss parameter is 0.5 dB/km. Comment on the acceptance of the computed value of figure-of-merit. *[Ans.: 200 ps/(nm–dB); No, a large FOM is desirable]*

4. Determine the dispersion for chirped fiber Bragg grating having refractive index = 1.45 and the difference between Bragg wavelengths at the ends of grating = 0.2 nm. [Use $c = 3 \times 10^5$ km/s] *[Ans.: 4.8×10^7 ps/(km–nm)]*

5. Calculate the material dispersion effect for LED with line width of 100 nm for an optical fiber cable having dispersion coefficient parameter, D_m = 22 ps/(km–nm) at 1310 nm. *[Ans.: 2.2 ns]*

6. Calculate the material dispersion effect for a laser with a line width of 2 nm for an optical fiber cable having dispersion coefficient parameter, D_m = 22 ps/(km–nm) at 1310 nm. *[Ans.: 44 ps]*

7. Determine the waveguide dispersion coefficient at 1310 nm for refractive index n_2 = 1.48 and percent change in refractive index Δn = 0.2%. *[Ans.: -1.9 ps/(nm–km)]*

8. An optical fiber has chromatic dispersion coefficient = 8ps/(nm–km) and linewidth of 2 nm. Compute (a) the bandwidth and length product; (b) the optical bandwidth for 10 km of this kind of fiber. *[Ans.: (a) 36.9 Gbps–km; (b) 2.8 GHz]*

9. Find the fiber loss for 10 km length which has insertion loss specification as 0.25 dB per km. *[Ans.: 2.5 dB]*

10. Consider the effect of frequency chirping as broadening of the Gaussian-shaped optical pulse by a factor of 6 as compared to that of its transform-limited optical pulse width. Determine the dispersion-limited transmission distance if the light wave system operates at 1550 nm and uses direct-modulation at bit rate of 10 Gbps. Given D = 17ps/(km–nm). *[Ans.: 12 km]*

11. Calculate the intermodal dispersion and Bandwidth length product of a grade index fiber of 50 μm core with refractive index of n_1 = 1.480 and n_2 = 1.460 used at 1300 nm wavelength. *[Ans.: 0.026 ns; 9.6 Gbps/km]*

12. Chromatic dispersion that causes the shorter wavelength of the optical signal to travel relatively faster than the longer wavelength limits the maximum transmission distance. Determine the transmission distance for 2.5 Gbps data rate for a directly modulated high-chirp laser diode source whose specified line width = 0.5 nm. The dispersion coefficient of the fiber used is 17 ps/(nm–km) at λ = 1550 nm. *[Ans.: 47 km]*

13. Calculate the transmission distance for 2.5 Gbps and 10 Gbps data rates for an externally modulated very low-chirp laser diode source used as dispersion management solution for chromatic dispersion. Its specified line width is 1.2 times the transmitted bit rate. The dispersion coefficient of the optical fiber used is specified as 17 ps/(nm–km) at 1550 nm wavelength.
[Ans.: 1000 km at 2.5 Gbps and 61 km at 10 Gbps]

14. For a directly-modulated DFB laser having σ_λ = 0.15 nm, estimate the maximum transmission distance of an optical fiber system operating at 2.5 Gbps bit rate. [Use D = 10 ps/(nm–km)] *[Ans.: 42 km]*

15. Find the limiting transmission distance for an externally modulated DFB laser having σ_λ = 0 and operating at 2.5 Gbps bit rate. [Use D = 10 ps/(nm–km); β_2 = –20 ps^2/km at λ = 1550 nm] *[Ans.: 500 km]*

16. In pre-chirp dispersion compensation technique, write the expression for dispersion which relates with dispersion length (L_D). Also specify the condition that should be satisfied by a chirped optical pulse that is capable of propagating for longer transmission distance before it broadens beyond its allocated duration of the transmitted bit.

(a) For unchirped Gaussian pulses, the transmission distance (L) is equal to the dispersion length (L_D).

(b) For chirped Gaussian pulses having C = 1, L increases L_D by 36.6 %.

(c) For chirped Gaussian pulses having $C = 1/\sqrt{2}$, L increases L_D by 60 %.

(d) At what value of C, the maximum improvement occurs? *[Ans.: $C = 1/\sqrt{2}$]*

17. A 13.6 km DCF fiber having dispersion parameter = 100 ps/(nm–km) is used for GVD compensation of a 80-km fiber having its specified dispersion parameter = 17 ps/(nm–km). Can this DCF reduce the dispersion to zero? *[Ans.: YES]*

WDM Concepts and Components

CHAPTER **7**

Chapter Objectives

After studying this chapter, you should be able to

- ✓ describe the principle of wavelength division multiplexing (WDM)
- ✓ understand WDM system configuration
- ✓ know about the applications of WDM-based systems
- ✓ explain different types of WDM components including transmitters and receivers
- ✓ analyze system performance issues and WDM soliton systems

Wavelength division multiplexing (WDM) is the second major fiber–optic revolution in the field of telecommunications. WDM is a technology which combines many different segments of wavelength range, called different independent optical channels, into the same optical fiber. The best feature of an optical fiber is that it has a wide spectral region which ranges from 1260 nm to 1675 nm. The light source used in high-capacity optical fiber communication systems emits a narrow wavelength band of less than 1 nm, thus enabling simultaneous transmission of many optical channels using the same optical fiber. WDM allows a huge increase in capacity of an optical fiber as compared to point-to-point link that carries only a single optical channel. Another big advantage of WDM is that different transmission formats can be supported by various optical channels. It means that without the need of common signal format, any data rate can be transmitted simultaneously and independently using the common optical fiber.

This chapter focuses on WDM concepts and components used in high-capacity optic–fiber communication networks. The discussion begins with the principle of wavelength division multiplexing which contains an orthogonal set of optical carriers with a suitable guard band, which a single-mode fiber can propagate. This is followed by a brief discussion on WDM system configuration involving a number of optical devices. An account of applications of WDM systems is presented next. The discussion is carried forward by describing various types of WDM components, including transmitters and receivers. Finally, an analysis of system performance issues and WDM soliton systems are covered.

7.1 Principle of Wavelength Division Multiplexing

Wavelength division multiplexing (WDM) is based on the fundamental physical principle which states that many optical rays having different wavelengths can be propagated together over a common optical channel with no interference. The concept of WDM is analogous to the basic concept of frequency division multiplexing (FDM) in which the available bandwidth of a communications channel in its frequency domain is divided into multiple sub-bands (called user channels). It implies that each user channel occupies only a part of the wide frequency spectrum. In WDM, each user channel is recognized by an optical wavelength. Remember the relationship between the wavelength and frequency as $\lambda = c/f$, which implies that shorter the wavelength of the signal, higher will be its frequency, and vice-versa. We can say that optical FDM is WDM. In optical fiber communications, each wavelength serves as a separate channel (i.e., an optical fiber). Different wavelengths are properly spaced (similar to guard band in FDM) so as to avoid any possibility of inter-channel interference. Fig. 7.1 depicts the fundamental concept of WDM.

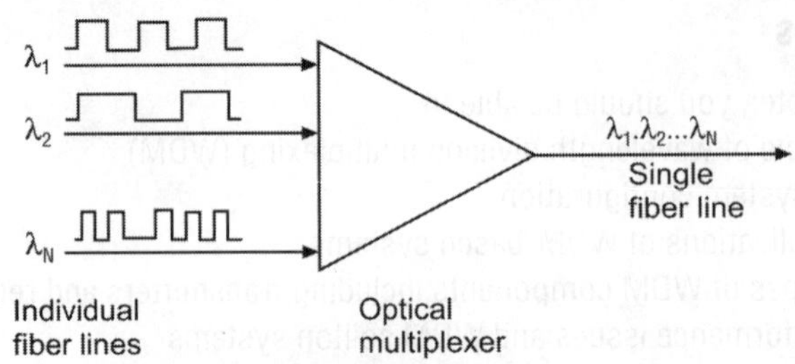

Fig. 7.1 Fundamental concept of WDM

WDM technology uses multiple wavelengths on individual fiber lines to transmit information over a single fiber line using optical multiplexer. Fig. 7.2 illustrates a conceptual difference between time-division multiplexing (TDM) which uses TDM–MUX or optical TDM (which uses OTDM–MUX) and WDM that uses WDM–MUX.

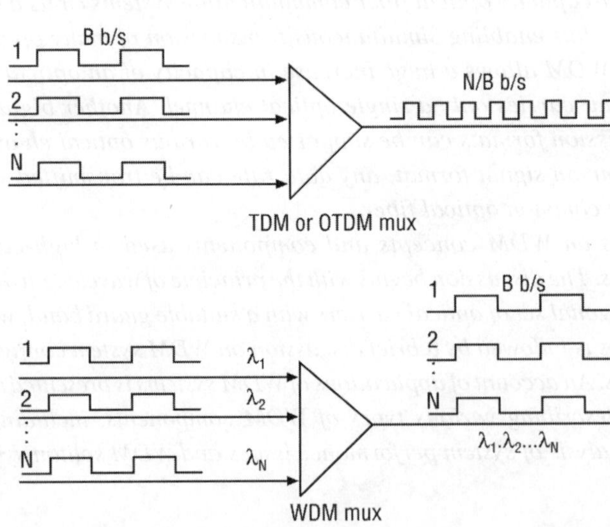

Fig. 7.2 TDM vs WDM

Most optical networks use a combination of TDM and WDM. In this, fixed time-slots are time-division multiplexed onto a specific wavelength and then employ optical TDM multiplexer (OTDM

MUX). It allows multiple users to share the entire capacity of one WDM wavelength. In most of the cases, a single wavelength capacity is much more than that needed by an individual user. The basic concept of WDM is to use the optical channels (frequency slots in terms of wavelength channels) to carry user data. Data transmission formats may have different analog or digital asynchronous bit rates. An optical channel is capable of carrying any type of data format. In general, WDM enables the upgradation or enhancement of information-carrying capacity of present optical networks without any addition of optical fibers. Fig. 7.3 illustrates the principle of operation of a typical WDM system.

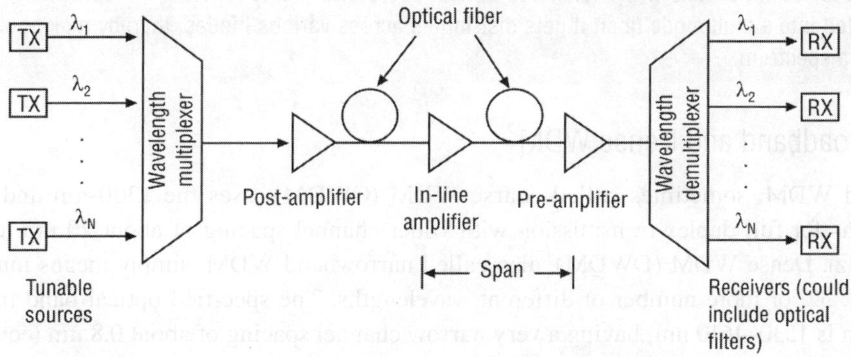

Fig. 7.3 Principle of operation of a WDM system

As shown, WDM contains an orthogonal set of optical carriers ($\lambda_1, \lambda_2, \ldots \lambda_N$) emitted by corresponding tunable optical sources (Lasers or LEDs) generating a data stream. All these optical signals are combined by an optical MUX, known as wavelength multiplexer. The power of the multiplexed optical signal can be boosted by using post optical amplifier and then transmitted simultaneously over the common optical fiber. The main function of wavelength multiplexer is to combine different optical signals (varying in wavelengths) into a continuous wavelength spectrum and launch them over the same optical fiber. A chain of optical amplifiers as post optical amplifier which is used to boost the optical power, an in-line optical amplifier which is used to compensate for the fiber attenuation, as well as a pre-amplifier for increasing the sensitivity of optical receiver are used along the optical fiber link. An optical DEMUX, known as wavelength demultiplexer, at the receiver end separates the signals having different wavelengths and directs them to appropriate optical receivers comprising of optical filters and photodetectors. Thus, we see that there may be a requirement of several types of optical components/devices (both passive and active) for the purpose of amplification, combining, isolating different wavelength signals, and distributing to respective optical receivers.

Similar to a guard band, there is a small spacing in between the wavelengths, which reduces the non-linear effects and possible inter-channel interference in WDM. The relationship to define wavelength separation between adjacent wavelengths is given as

$$\text{Wavelength separation} = \frac{\text{Frequency separation} \times (\text{wavelength})^2}{\text{speed of light}} \tag{7.1}$$

The range of wavelengths carried in the optical fiber varies. A common set of wavelengths used is mostly in the 1550-nm region (called C-band).

It may be noted that WDM happens because a single-mode optical fiber can support many different wavelengths at the same time. As we know that in a single-mode fiber (SMF), only the fundamental mode of propagation exists that occupies a very narrow wavelength spectrum. The entire coupled energy will be in the fundamental mode. This is suitable for Wavelength Division Multiplexing. Another advantage of using WDM is that the effective bandwidth of an optic–fiber link is multiplied several times.

> ***Note:*** Is WDM possible with multimode optical fibers? Not exactly! When the optical signal is injected into a multimode fiber, it gets distributed across various modes, thereby making it too wide a spectrum.

7.1.1 Broadband and Dense WDM

Broadband WDM, sometimes called coarse WDM (CWDM), uses the 1300-nm and 1550-nm wavelengths for full-duplex transmission with wider channel spacing of about 20 nm (equivalent to 100 GHz). Dense WDM (DWDM), also called narrowband WDM, simply means multiplexing of 4, 8, 16, 32, or more number of different wavelengths. The specified optical band in terms of wavelength is 1530–1610 nm, having a very narrow channel spacing of about 0.8 nm (equivalent to 25 GHz). This has been possible due to low attenuation of optical signals in silica fiber at 1300-nm and 1550-nm, as shown in Fig. 7.4.

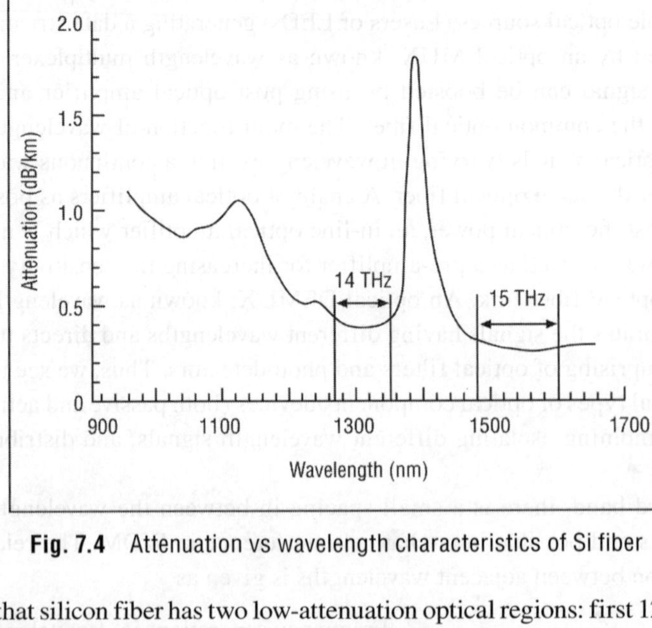

Fig. 7.4 Attenuation vs wavelength characteristics of Si fiber

The plot shows that silicon fiber has two low-attenuation optical regions: first 1270–1350 nm (nearly around 1300 nm) and second 1480–1600 nm (nearly around 1550 nm). A high quality optical source usually has a narrow linewidth. It implies that the available low-attenuation regions can provide several operating wavelength windows. In order to realize dense WDM system, we can use a number of such optical sources, each one transmitting a different peak wavelength (having very narrow spectral width) while maintaining sufficient space among their operating wavelengths. But it is very necessary that inter-channel optical signal interference is not created while maintaining the message integrities from each of the independent optical sources throughout their transmission up to the receiving end for final conversion to electrical signals. For example, it is possible to propagate 50 different optical channels in the 1530–1560 nm optical band on a common optical fiber cable by using a narrow-linewidth laser having 0.8 nm spectral band (equivalent to 100 GHz bandwidth). This is the fundamental principle of DWDM operation. Multiple data signals are transmitted using different optical wavelengths through a single fiber. Input optical signals are assigned specified frequencies within a designated frequency spectrum. The fiber capacity is increased several times when these optical signals are multiplexed and transmitted using a single fiber. If we use erbium doped optical amplifier, then the transmission capabilities are increased by 4–8 times that of equivalent TDM Systems.

7.1.2 Salient Features of WDM

- *Capacity Upgrade:* WDM is capable of enhancing the capacity of optical fiber communications networks significantly. DWDM offers almost unlimited transmission capacity.
- *Transparency and Scalability:* In WDM, each optical channel can carry any type of transmission signal format and scalable.
- *Wavelength Switching:* Wavelength-switched architecture allows reconfiguration of optical layer. Wavelength-routed telecom networks are based on rigid optical fiber infrastructure.
- *Wavelength Routing and Dynamic Provisioning:* The use of wavelength-sensitive optical routing devices can use wavelength as one additional dimension in designing tele-communication switches and networks.

7.1.3 Advantages of WDM

- In WDM, the wavelength of the channel is used for optical switching, optical routing, or dispensing each optical channel to its designated receiver. As a result, wavelength division multiple access (WDMA) network is an all-optical network.
- WDM enables leasing out an individual wavelength instead of leasing an entire fiber–optic link, thus enhancing the scope to a larger scale of its usage.
- WDM allows optical networks to increase their capacity manifold (i.e., 6000 times). For example, in DWDM, it is possible to operate one optical fiber in bit rate of the order of Tbps.
- With the deployment of tunable lasers as optical sources, and dynamic wavelength/spectrum management, WDM enables to support 'bandwidth and quality of service (QoS) on demand' requirements.
- Due to increase in traffic-carrying capacity with the use of WDM in optical networks, it is possible to deliver audio, video, and data channels together at high speed and at a reasonably low cost.

Facts to Know

From the transmission point of view, WDM divides the entire fiber–optic bandwidth into many segments and each signal (wavelength) uses its individual bandwidth segment. WDM uses parallel transmission as in case of frequency-division multiplexing (FDM), not serial transmission as in case of time-division multiplexing (TDM).

7.1.4 Typical Applications of WDM Systems

In WDM systems, various optical signals having different wavelengths are transmitted simultaneously using the same optical fiber after each one is modulated by independent electrical bit streams. Frequency division multiplexing (FDM) for analog channels, or time-division multiplexing (TDM) for digital channels may be used by electrical bit streams in the electrical domain before these are subjected to wavelength-division multiplexing (WDM). Each wavelength channel can carry several time-division multiplexed channels which can be further multiplexed in WDM–MUX for transmission over the common optical fiber medium. An optical demultiplexer is used at the optical receiver end of WDM communications link for segregating individual channels of the received multiplexed optical signal.

Usually, optical fibers offer very large bandwidth. WDM simply exploits this capability of optical fibers. As an instance, WDM channels carrying 100s of 10-Gbps data rate can be propagated successfully using the same fiber by reducing the inter-channel separation to be less than 100 GHz. We can say that the capacity of fiber–optic communication links can exceed 10 Tbps. Due to the availability of various optical WDM components, WDM-based systems have many applications, such as

1. *Point-to-point high-capacity optic–fiber links:* WDM helps to increase the overall bit rate (high capacity) for long-haul (point-to-point) optical fiber communication links which serves the core or backbone of a wide area telecommunication network. If *N* channels having equal bit rates are multiplexed using optical multiplexers and then transmitted over the same fiber, then the overall capacity increases (or is multiplied) by a factor of N.
2. *LAN using WDM:* A broadcast star topology may be generally employed to multiplex several optical channels at the LAN level, e.g., Lambdanet.
3. *MAN and WAN using WDM:* A MAN is formed by connecting several LANs in a ring topology using passive wavelength routers. A WAN can be formed by connecting several MANs in which different nodes are usually interconnected following a mesh topology that uses dynamically configurable optical cross-connect switches and wavelength converters extensively.
4. *Multi-Access WDM Networks:* The use of channel wavelength switching, optical signals routing, or even dispensing each optical channel to its designated optical receiver in WDM results in all-optical multi-access WDM networks. In this, each user can transmit data to as well as receive data from any other user who is always connected to the WDM network. They provide a random bi-directional data access between them.

Facts to Know

DWDM technology enables meeting growing demands of users by expanding fiber network rapidly. DWDM along with ATM provides new user services while simplifying the network configuration at reduced costs. Even current and new TDM systems can be added to existing DWDM technology, which can virtually create a system providing almost endless capacity.

Example 7.1 Optical Bandwidth

In a fiber–optic system, the usable spectral band $\Delta\lambda$ is specified as 120 nm in the standard 1550-nm optical band. Find the optical bandwidth.

Solution:

We know that optical bandwidth is $|\Delta \upsilon| = \left(\frac{c}{\lambda^2}\right)|\Delta\lambda|$

Given the central wavelength, $\lambda = 1550$ nm and the wavelength deviation, $|\Delta\lambda| = 120$ nm

$$\therefore \quad |\Delta\upsilon| = \left(\frac{3\times10^8\, m/s}{\left(1550\times10^{-9}\, m\right)^2}\right)\times\left(120\times10^{-9}\, m\right) = 15\times10^{12}\, Hz,\ or\ 15\ THz \qquad \textbf{Ans.}$$

Example 7.2 Number of Wavelength Channels

A fiber–optic transmission system is required to operate in the spectral band of 1536–1556 nm. If the maximum channel spacing is constrained to have 500 GHz, then, how many wavelength channels can be multiplexed in the system?

Solution:

We know that optical bandwidth is $|\Delta \upsilon| = \left(\frac{c}{\lambda^2}\right)|\Delta\lambda|$

Given the spectral band = 1536–1556 nm

Therefore, the central wavelength, $\lambda = \frac{1536 + 1556}{2} = 1546$ nm

The wavelength deviation around central wavelength, $|\Delta\lambda| = 1556 - 1536 = 20$ nm

$$\therefore \quad |\Delta\upsilon| = \left(\frac{3\times10^8\, m/s}{\left(1546\times10^{-9}\, m\right)^2}\right)\times\left(20\times10^{-9}\, m\right) = 2.15\times10^{12}\, Hz$$

Given channel spacing = 500 GHz

Hence, number of wavelength channels $= \frac{2.5\times10^{12}}{500\times10^{9}} = 5$ **Ans.**

Section Practice Problems

1. Determine the spectral band in the 1310-nm wavelength region if the usable optical bandwidth is 14 × 10^{12} Hz. *[Ans.: 80 nm]*
2. How many independent signals can be sent on a single fiber in the 1525–1565 nm spectral band if the channel spacing is 50 GHz? *[Ans.: 100]*
3. The WDM system operates in the standard C and L optical bands, covering the whole 1530–1610 nm wavelength range. Assuming channel spacing to be 25 GHz, determine the number of channels that can be transmitted through this system? *[Ans.: 389]*

7.2 WDM System Configurations

WDM systems and networks can be configured by using a wide range of well-designed optical devices and components which are required to perform various functions such as generation and combination of multi-wavelength optical signals, transporting and amplifying these signals during propagation through the optical network, followed by separating as well as receiving these signals at their respective destinations. Broadly speaking, there are active optical networks and passive optical networks.

- An *active optical network* (AON) is the one that uses electrically-powered optical switching components. Examples of such devices include a wavelength router, a cross-connect switch required to manage signal distribution as well as to direct signals to desired destinations.
- A *passive optical network* (PON) is the one that does not include electrically-powered switching components. It uses optical splitters to separate and collect optical signals as they propagate through the fiber–optic network. However, electrically-powered devices are required at the transmitting and receiving ends of the link. Passive optical network based on a typical hybrid WDM–TDM is able to combine a very large capacity provided by WDM, whereas TDM enables to share the dynamic bandwidth. Optical burst-mode receiver (BM–RX) is a sub-system used in advanced PON that can offer guaranteed QoS. It requires extensive additional circuitry to provide automatic phase alignment at the beginning of each optical burst signal, fast AGC, DC offset compensation, automatic threshold setting and common mode rejection improvement. Optical BM–RX must have high sensitivity, fast response time and greater dynamic range.

WDM network components can be broadly classified into two main categories:

- *Active Components*– Active optical components or devices used in WDM network require external power to be functional. For example, optical tunable sources/transmitters, optical receivers, optical amplifiers, optical switches, active MUXs/DEMUXs, tunable optical filters, etc.
- *Passive Components*– Passive optical components or devices used in WDM network do not require any type of external power for their operation. For example, wavelength selective couplers, wavelength selective splitters, optical isolators, optical circulators, passive MUXs/ DEMUXs, optical attenuators, fixed optical filters, etc.

There are some other WDM network optical components which can be either active or passive ones, i.e., hybrid ones. Examples of such optical components include wavelength converters, add–drop MUXs (ADMs), cross-connects, etc.

> ***Note:*** Optical amplifiers are key in DWDM systems. They can be used as power (booster) amplifier, in-line amplifier, pre-amplifier, or/and LAN booster amplifier along with star coupler.

7.2.1 Classifications of WDM Systems

Broadly speaking, there are two distinct categories of WDM systems.

- Single-hop all-optical WDM systems and networks
- Multi-hop all-optical WDM systems and networks

A single-hop all-optical WDM network is the one in which all available nodes are directly connected to one another, resulting in a fully-connected network. Lambdanet, Rainbow, Starnet are typical examples of broadcast-and-select type WDM network.

- In Lambdanet, there is one optical transmitter and *N* number of optical receivers in each of N nodes in the network. An optical transmitter emits at a specified wavelength, whereas *N* number of optical receivers operate at *N* number of different wavelengths. The output of various optical transmitters is combined by a WDM device, known as passive star coupler. This device also distributes received optical signals to all optical receivers uniformly. This implies that each node in the network receives all the available traffic. The whole network is transparent to the modulation format or even the bit rate.
- The Rainbow network is similar to lambdanet except the use of tunable optical filters in place of bank of optical receivers. Up to 32 nodes, each one of them capable of transmitting 1 Gbps signals over 20–30 km, can be supported by this network. Its main disadvantage is relatively slow process of tunable filters.
- Starnet uses packet switching technique in which up to 1.25 Gbps bit rates per node can be supported over a distance of 10 km with SNR close to 24 dB.

> ***Note:*** In a multi-hop all-optical WDM network, nodes are partially connected to each other. An optical signal emitted by one mode in the network usually passes through several intermediate nodes and require number of hops before reaching its destination. Examples of such networks include the Banyan network, the Shuffle network, and the deBruijn network.

A WDM Network with a Feeder Ring

This is an example of multiple-hop network. The backbone of the WDM network is connected with a feeder ring through an egress node. Through access node, the feeder ring is capable of delivering data to many other feeder rings. In order to ensure robustness, four fibers are typically employed– two fibers carry data in the anticlockwise and clockwise direction, respectively, whereas other set of two fibers serve as back-up fibers and carry data whenever there is a failure detected in a point-to-point optical fiber communication link. This is also known as a self-healing feeder ring. Another WDM device, known as add–drop MUX (ADM), is included at all nodes of a WDM network which has a feeder ring so as to enable add or drop any individual channel.

> ***Note:*** Fiber–optic communication networks are nowadays mainly deployed to provide services to a very large number of potential users which are usually spread over a relatively large service area. Therefore, WDM lightwave systems can be classified into LAN, MAN and WAN.

(a) *Local Area Network (LAN)*– It is a broadcast star topology. With an objective of serving a relatively small geographical area, number of different optical channels are combined in LAN.

(b) *Metropolitan Area Network (MAN)*– A MAN consists of several LANs which are connected together with the help of passive wavelength routers. A ring topology is mostly used for MANs.

(c) *Wide Area Network (WAN)*– A WAN comprises of several MANs which are connected using a mesh topology. Optical switches and wavelength converter devices are extensively used to form a WAN so as to make it dynamically configurable. Optical cross-connects are used in the nodes.

As one of the physical layers of the LAN communications protocol, Optical Ethernet provides transmission of data over optical fiber cable. Various switches and internet servers are connected using optical ethernet which can provide data rate up to 100 Gbps (40-Gbps as an interim data rate between 10- and 100-Gbps). Multimode fiber cables are specified for distances up to 100 m, and single mode fiber cables are specified up to 10 km (or, even up to 40 km) by IEEE 802.3ba Task Force. Single mode optical ethernet transceivers are based on the use of photonic integrated circuits using a quad laser driver and quad direct-modulation laser array as optical source. Multimode optical ethernet transceivers at 100 Gbps use 850-nm vertical-cavity surface-emitting lasers as optical source.

As mentioned earlier, DWDM components mainly include an optical transmitter (laser with precise stable wavelength), the erbium-doped fiber optical amplifiers, a high-performance optical fiber having low attenuation, an optical receiver comprising of photo detectors and optical demultiplexers, and other WDM devices such as optical ADMs and cross-connect switches.

7.2.2 High Capacity Point-to-Point WDM Links

We have studied that the main function of WDM links is to form the backbone network with enhanced bit rate capacity. In a typical WDM link, the output optical signals of several optical transmitters, each one of them operating at its allocated wavelength, is multiplexed, and then coupled to an optical fiber for propagation to the receiving end of the optic–fiber communication link. An optical demultiplexer (DEMUX) used at its receiver end distributes each received channel to the desired destination (i.e., an optical receiver). When N number of individual optical channels having their respective bit rates as R_{B1}, R_{B2}, ……, R_{BN} are multiplexed and propagated together over a fiber length L, then the product of net bit rate and the length (i.e., $R_B \times L$) is given as

$$R_B \times L = (R_{B1} + R_{B2} + \ldots\ldots + R_{BN}) \times L \tag{7.2}$$

For equal bit rates (i.e., $R_{B1} = R_{B2} = \ldots\ldots = R_{BN}$), we see that the overall system capacity is multiplied by N times. However, inter-channel cross-talk limits the minimum channel spacing.

Table 7.1 gives typical capacity for different high-capacity WDM systems.

Table 7.1 Capacity of WDM systems

No. of Channels, N	*Bit Rate per Channel, B*	*Capacity of WDM System, N × B*
120	20 Gbps	2.40 Tbps
132	20 Gbps	2.64 Tbps
160	20 Gbps	3.20 Tbps
82	40 Gbps	3.28 Tbps
256	40 Gbps	10.24 Tbps
273	40 Gbps	10.92 Tbps

Factors limiting the number of channels in WDM

- At the optical transmitter end, the stability as well as the tunability of semiconductor lasers such as distributed feedback (DFB) lasers.

- Finite bandwidth with uniform gain provided by optical amplifiers, e.g., 100 nm in Raman fiber amplifiers and 40 nm in EDFAs.
- Signal degradation during propagation over the fiber because of non-linear effects.
- Inter-channel cross-talk during demultiplexing at the optical receiver end.

Analog fiber optic links are primarily used for transmission of analog signals and are capable to operate in an industrial noisy environment. The standard analog input signal levels are ± 10 V DC. However, in some applications that require lower input range of the order of only 100 mV, the analog links can be re-configured using internal differential analog operational amplifiers. In order to increase the signal bandwidth or permit transmission of complex signals, additional accessories may be included in the analog fiber optic links.

Facts to Know

The initial WDM systems utilized mainly 1310 nm and 1550 nm wavelength ranges. Due to advancement of technology, nowadays Dense WDM systems can utilize more number of wavelength ranges (16, 32, 64, 128, or even more) in the 1550 nm region. Each one of these wavelengths is capable of transmitting data rates as high as 10 Gbps per channel. Typical channel spacing of the order of 50, 100, 200 and 1000 GHz are possible depending on the laser linewidth and bandwidth of the optical filter.

7.2.3 SONET/SDH Standards

Synchronous Optical Network (SONET), also known as Synchronous Digital Hierarchy (SDH), is based on a fiber ring or mesh topology, and describes the standards and specifications for synchronous, high-speed, point-to-point optical TDM networks. In SONET/SDH, the signals are switched and routed electronically, whereas transmission of signals are carried out optically using optical fiber as the medium between nodes. The conversion from optical-to-electronic signals take place at each node. SONET defines the standard signal formats which have been adopted only in North America. SDH defines the standard signal formats which have been adopted in continents other than North America. The basic SONET signal, known as synchronous transport signal (STS) – 1, has a transmission bit rate of 51.84 Mbps. Higher-level SONET signals are typically integer multiples of transmission bit rate of STS-1. This means that an STS-N signal has a transmission bit rate which is exactly equal to N (where N = 1, 3, 12, 24, 48, and 192) multiplied by the basic transmission bit rate of STS-1 which is specified as 51.84 Mbps. In practice, SONET links are referred to as physical layer Optical Carrier (OC) links after converting electrical signals to optical signals, followed by efficient encoding for transmission.

Let us now turn our attention toward SDH in which the basic transmission bit rate is 155.52 Mbps. It is designated as Synchronous Transport Module (STM)– 1, and happens to be equivalent to STS-3 (3 × 51.84 Mbps). Higher transmission bit rates in SDH are accordingly designated as STM-M (where M can have values such as 1, 4, 16, and 64). If we compare SDH with SONET in terms of the nature of signals, then SDH does not differentiate between a logical electronic signal (for example, STS-N signal in SONET) and an equivalent physical optical signal (for example, OC-N signal in SDH), and hence both types of signals are labeled STM-M. The most used levels of SONET (OC-N) and electronic (STM-M) along with transmission line rates are depicted in Table 7.2.

Table 7.2 SONET level and equivalent SDH level

S. No.	SONET Level	Electronic Level	Transmission Line Rate	Equivalent SDH Level
1.	OC Level # 1	STS Level # 1	@ 51.84 Mbps	-
2.	OC Level # 3	STS Level # 3	@ 155.52 Mbps	STM Level # 1
3.	OC Level # 12	STS Level # 12	@ 622.08 Mbps	STM Level # 4
4.	OC Level # 24	STS Level # 24	@ 1244.16 Mbps	STM Level # 8
5.	OC Level # 48	STS Level # 48	@ 2488.32 Mbps	STM Level # 16
6.	OC Level # 96	STS Level # 96	@ 4976.64 Mbps	STM Level # 32
7.	OC Level # 192	STS Level # 192	@ 9953.28 Mbps	STM Level # 64

In a standard one-way (simplex) point-to-point optical fiber communication link, only a single optical source such as a laser diode is used at its transmitting end and only one photodetector such as p–i–n photodiode is used at its receiving end with a single optical fiber cable serving as the optical signal propagating medium between them. Let this fiber cable has a capacity of 2.5 Gbps, then we can easily determine the number of OC-3 SONET or STM-1 SDH data streams. From the tabulated data, we know that each OC-3 SONET level or STM-1 level can carry data at 155.52 Mbps. Thus, 2.5 Gbps/155.52 Mbps = 16 number of OC-3 SONET or STM-1 SDH data streams can be electrically multiplexed together and then it can modulate a laser diode for sending optical signal over this fiber. If we wish to upgrade the number of OC-3 (or, equivalently/STM-1) channels to 64 (say) over the same fiber, then four number of identical and independent optical channels with adequate separation between wavelengths can be used by WDM system.

Facts to Know

Plesiochronous digital hierarchy (PDH) is another type of optical signal hierarchy standard used mostly in North America, Europe, and Japan. It refers to three quasi-synchronous types of optical digital networks. It was used before SONET/SDH standards came into existence.

Example 7.6 Transmission Distance

A WDM system uses a fiber having a specified attenuation @ 0.2 dB/km. If the system is designed to have a power margin of 30 dB, then find the transmission distance.

Solution:

Given fiber loss = 0.2 dB/km

Power margin = 30 dB

To utilize complete power margin, the fiber loss can be tolerated up to 30 dB over a transmission distance of $\frac{30 \text{ dB}}{0.2 \text{ dB/km}} = 150$ km **Ans.**

Example 7.7 Distance-Bit Rate Product

Let 50 channels be carried by a WDM system. If the capacity of each channel is 2.5 Gbps, then find the distance-bit rate product for the transmission distance of 100 km.

Solution:

Given number of channels = 50 and the capacity of each channel = 2.5 Gbps

Therefore, total capacity of WDM system = 50 × 2.5 Gbps = 125 Gbps

Given transmission distance = 100 km

Hence, distance-bit rate product = 100 km × 125 Gbps = 12.5 Tbps–km **Ans.**

Section Practice Problems

1. A WDM system has fiber loss of 0.3 dB/km. For a specified frequency band of 7.5×10^{12} Hz and channel spacing of 75×10^{9} Hz between channels, compute the transmission distance for specified power margin of 30 dB. *[Ans.: 100 km]*
2. Let 100 channels be carried by a WDM system. If the capacity of each channel is 10 Gbps, then find the distance-bit rate product for the transmission distance of 50 km. *[Ans.: 50 Tbps–km]*

7.3 Tunable Optical Filters

Tunable optical filters are active components which are mostly used at the receiver end of a WDM system. As the name suggests, its main function is to get tuned and select a desired optical channel from the multiplexed received signal. Tunable optical filters have the ability to change the wavelength they select dynamically. The bandwidth of the tunable optical filter should be properly designed for a particular channel being selected. It should be sufficiently large so that the desired optical channel can pass through without much loss of the signal. On the other hand, it should be relatively small so that it can adequately reject the adjacent channel. The primary operation of tunable optical filter is illustrated in Fig. 7.5.

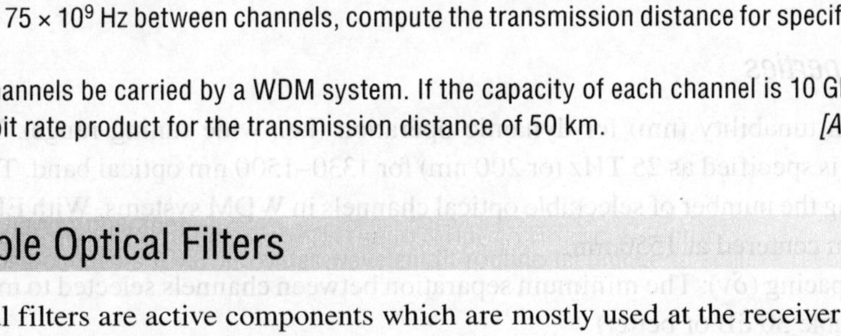

Fig. 7.5 Primary operation of a typical tunable optical filter

Tunable filters are made by changing at least one branch of an interferometry optical filter in terms of either its refractive index or length of the propagation path using some control mechanism. When these parameters vary, the phase as well as the intensity of the propagating light wave changes as a function of wavelength. In this way, the wavelength selectivity is achieved.

To increase the number of channels, tunable optical filters can be cascaded having different values of free spectral range (FSR). For example, four optical channels of a high-resolution optical filter can be cascaded with four optical channels of a low-resolution optical filter within FSR, which results in 16 unique channels. It is shown in Fig. 7.6.

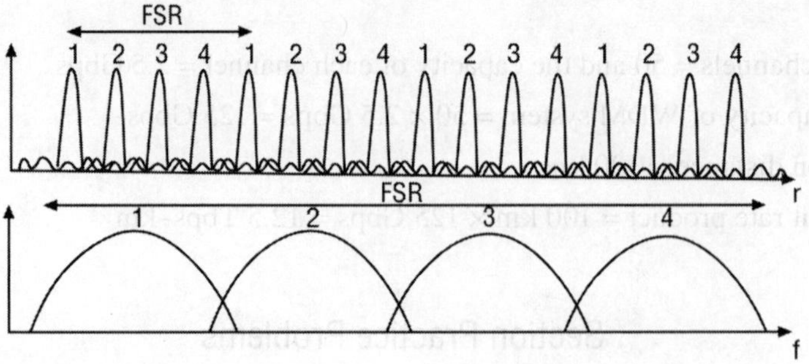

Fig. 7.6 Cascading filters with different FSRs

Desirable Properties

- Wavelength tunability (nm) for dynamic operation (i.e., wide tuning range). Typical tuning range ($\Delta\nu$) is specified as 25 THz (or 200 nm) for 1330–1500 nm optical band. This enables for maximizing the number of selectable optical channels in WDM systems. With EDFA, normally $\Delta\lambda = 35$ nm centered at 1550 nm.
- Channel Spacing ($\delta\nu$): The minimum separation between channels selected to minimize cross-talk (desirable 30 dB or better)
- Maximum number of channels ($N = \Delta\nu / \delta\nu$)
- Flat passband and steep slopes
- Relatively faster tuning speed (usually in milliseconds) that enables to minimize the access time
- Low insertion loss
- Negligible cross-talk
- Insensitive to polarization of optical signals
- Stable operation
- Small size and low cost

Classification of Tunable Optical Filters

(a) Tunable optical filters based on optical interference
  (i) Fabry–Perot Interferometer (FPI)
  (ii) Mach–Zehnder Interferometer (MZI)
(b) Tunable optical filters based on optical diffraction
  (i) Grating-based Michelson
  (ii) Acousto–optic
  (iii) Electro–optic

7.3.1 Fabry–Perot Interferometer Filters

An FPI tunable optical filter basically comprises of a cavity that is constructed by using two mirrors at its either ends. The length of the cavity is electronically adjusted with the help of a piezoelectric transducer. Fig. 7.7 shows a Fabry–Perot interferometer tunable optical filter and its transfer function.

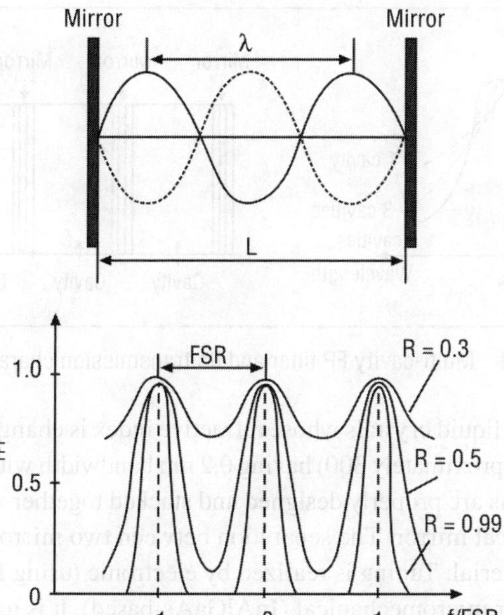

Fig. 7.7 Fabry–Perot filter and its transfer function

Free spectral range (FSR) is an important parameter of a Fabry–Perot filter which is used to define the separation between periodic transmission peaks. It is given by

$$FSR = \Delta v_L = \frac{c}{2n_g L} \tag{7.3}$$

where, n_g represents the group refractive index, and L represents the cavity length.

The FPI filter bandwidth Δv_{FP}, or approximate bit rate R_B should be very large, so that all the frequency components of the selected channel are passed. That is,

$$\Delta v_{FP} \sim R_B = \frac{\Delta v_L}{F} \tag{7.4}$$

where, F represents the finesse of FP filter which is expressed as $F = \dfrac{\Delta v_L}{\Delta v_{FP}}$ (7.5)

Neglecting the internal losses, the parameter 'finesse' of FP filter is given by

$$F = \frac{\pi\sqrt{R}}{\sqrt{(1-R)}} \tag{7.6}$$

where, R represents the mirror reflectivity.

An FPI filter uses two different optical fibers having an air gap in between. The two ends of the optical fiber are coated in such a way that they can act as highly reflective mirrors that can reflect optical signals. Then, the whole assembly is encapsulated within a chamber made of piezoelectric material. The length of the air gap length is varied electronically to enable tuning of a specified optical channel. Typical transmission characteristics along with an arrangement of multi-cavity thin-film optical filter is depicted in Fig. 7.8.

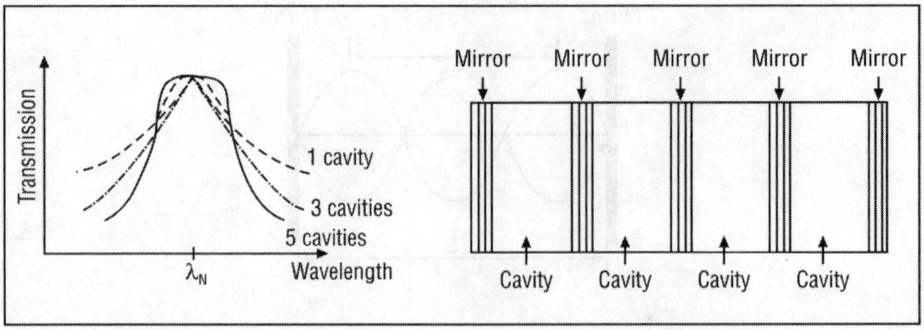

Fig. 7.8 Multi-cavity FP filter and its transmission characteristics

Tunable FP filter using liquid crystals whose refractive index is changed electronically for tuning, provide high value of F (approximately 300) having 0.2 nm bandwidth with switching time of 10 μs – 1 ms. A number of thin films are properly designed and stacked together which can serve the purpose of a highly reflective optical mirror. The separation between two mirrors is obtained by placing an appropriate dielectric material. Tuning is realized by electronic (using InGaAsP/InP waveguide) or thermo-optic (Si-based) or micromechanical (InAlGaAs-based). It is used for narrow-band filter.

Advantages of FP Interferometer Filters

- Wide dynamic range
- Narrow bandwidth
- High tuning speed
- Low polarization dependent loss (PDL)

Disadvantages of FP Interferometer Filters

- Limited number of channels, typically less than 100 (F ≈155)
- Relatively slow tuning due to mechanical nature
- Poor stability
- Low side–lobe suppression ratio (SSR)

Facts to Know

It is possible to integrate the Fabry–Perot (FP) interferometer tunable optical filters within the system and that too without incurring any coupling losses. These filters are widely employed in commercial WDM optical fiber communication links. However, fixed optical filters can be used to construct optical receivers, multiplexers and demultiplexers.

7.3.2 Mach–Zehnder Interferometer Filters

MZI tunable optical filters can be realized when two ports of a 3-dB optical coupler at its output are connected to two output ports of a second 3-dB optical coupler. Such an arrangement is depicted in Fig. 7.9.

Tuning over the desired optical range can be obtained by varying the refractive index of one of the arm of MZI either by heating it, or by placing Electro–optic material (e.g., lithium niobate, $LiNbO_3$).

By applying drive voltage to one of the two waveguides, an electric field is created. As a result, the signals that appear at the output of two waveguides will have either in-phase (0° phase shift), or 180° phase shift. Accordingly, the optical signal will be either passed or blocked.

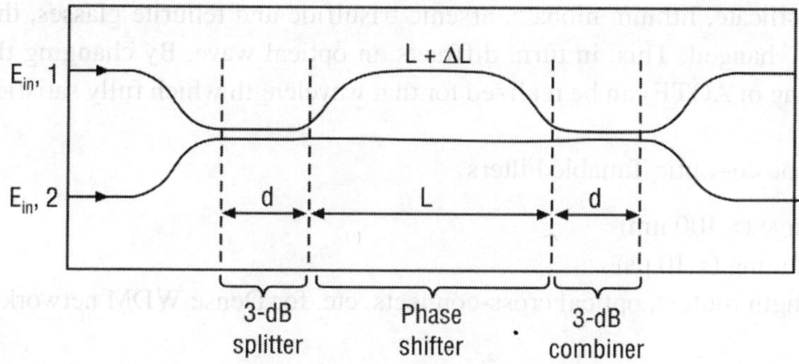

Fig. 7.9 Basic MZ interferometer (MZI)

A cascaded chain of MZ interferometer tunable optical filters with their relative delays is adjusted suitably. Tuning is carried out by varying the length of arms slightly that can achieve Finesse value of 1600. MZ chain comprises of a splitter, a combiner and a delay. The adjustable delay controls one of the path lengths that may result in a phase difference when combined. Wavelengths with 180° phase difference are filtered out. It is capable of selecting closed spaced channels (suitable for DWDM applications). However, it exhibits a slow response (~ 1 ms) because of thermal tuning mechanism, therefore, low tuning speed.

7.3.3 Grating-based Michelson Filters

We know that a fiber Bragg grating functions like a reflection optical filter. We can control its central wavelength by varying the grating period. Also, we can manipulate its bandwidth either by changing the strength of the fiber Bragg grating (FBG), or by chirping the period of the grating slightly. An optical circulator is used for fiber gratings. How can a fiber grating be made to function as a narrowband tunable optical filter? It can be done simply by creating a shift in the phase in the center of the grating. FBG-based Michelson interferometer filters can be designed using a distributed Bragg reflector (DBR) structure. In order to obtain fast tuning (of the order of nanoseconds), gratings can also be used with FP and MZ interferometer tunable optical filters. It can be integrated with optical amplifiers and optical receivers used in WDM.

Grating filters consist of a flat layer of plastic or glass material having a number of parallel grooves that can reflect light at all angles. However, at a particular angle, only a certain wavelength adds constructively. It is desirable to place the filter at the proper angle to select a certain wavelength. Fiber Bragg gratings are used for low insertion loss which varies with temperature. A grating is directly induced into the fiber core. There exists a great similarity between thin-film interference filters and fiber Bragg gratings, with the exception that the materials placed onto a substrate layer are either of low index or of high index. But it results in high insertion loss, thermal instability, as well as poor spectral profile.

7.3.4 Acousto–optic Tunable Filters

In an AOTF, acoustic waves are normally used to form the gratings dynamically. It is based on the principle of photo–elastic effect. When acoustic waves propagate within acousto–optic material such as fused silica, lead silicate, lithium niobate, arsenic trisulfide and tellurite glasses, the refractive index is periodically changed. This, in turn, diffracts an optical wave. By changing the acousto–wave frequency, tuning of AOTF can be realized for that wavelength which fully satisfies the Bragg condition.

Advantages of Acousto–optic Tunable Filters

- Wide tuning range (> 100 nm)
- Relatively fast tuning (< 10 μs)
- Used in wavelength routers, optical cross-connects, etc. for Dense WDM networks

7.3.5 Electro–optic Tunable Filters

The principle of operation as well as the structure of Electro–optic tunable filter (EOTF) is similar to AOTF except that an Electro–optic effect using lithium niobate ($LiNbO_3$) material creates the fiber Bragg grating. Finger-like electrodes are used to induce this grating. Tuning is obtained by changing the voltage applied to the electrodes in order to vary refractive index. Tuning speed of EOTFs is quite high, may be of the order of nanoseconds. But their dynamic range is less (~ 10 nm) and low side-lobe suppression ratio. Table 7.3 shows a comparison of key parameters of various tunable optical filters.

Table 7.3 Tunable optical filters – a comparison of key parameters

S. No.	*Type of Tunable Optical Filter*	*Tuning Range (nm)*	*Tuning Time*
1.	Fabry–Perot	≈ 500	1–10 ms
2.	Acousto–optic	≈ 250	10 μs
3.	Electro–optic	≈ 16	1–10 ns
4.	Liquid Crystal Fabry–Perot	≈ 30	0.5–10 μs

Facts to Know

Tunable optical filters can also be implemented using an optical amplifier which has smaller gain-bandwidth product than channel separation. The peak–gain wavelength is changed for tuning. Other methods of realizing tunable optical filters include the use of SBS for selective amplification of a channel, SOAs with a DFB structure for narrow gain-bandwidth, and FWM in SOAs.

Example 7.10 Free Spectral Range of FP Filter

Determine the free spectral range (FSR) of a Fabry–Perot (FP) interferometer tunable optical filter which is designed with the group index $n_g = 1.5$ of the intra-cavity material used in the construction of a FP filter having L = 100 μm.

Solution:

We know that in FP filter, $FSR = \Delta\nu_L = \dfrac{c}{2n_g L}$

For the given n_g = 1.5 and L = 100 μm, we have

$$FSR = \Delta\nu_L = \frac{3\times10^8 m/s}{2\times1.5\times\left(100\times10^{-6} m\right)} = 1 \text{ THz} \qquad \textbf{Ans.}$$

Example 7.11 Finesse of FP Filter

Determine the finesse of a Fabry–Perot (FP) interferometer tunable optical filter to transmit bit rate of 2.5 Gbps if the bandwidth of the filter is approximately equal to the transmission bit rate. Its FSR is specified as 1 THz.

Solution:

We know that in FP filter,
$$F = \frac{\Delta\nu_L}{\Delta\nu_{FP}}$$

For given FSR $\Delta\nu_L$ = 1 THz and filter bandwidth $\Delta\nu_{FP}$ ~ bit rate = 2.5 Gbps, we have

$$F = \frac{1\times10^{12}}{2.5\times10^{9}} = 400 \qquad \textbf{Ans.}$$

Example 7.12 Finesse and Number of Channels

Neglecting the internal losses in a Fabry–Perot interferometer (FPI) tunable filter, determine the following:

(a) finesse of the FP filter if it is designed with 99% reflecting mirrors.
(b) the maximum number of channels that this FP filter can select if the spectral efficiency is 1/3.

Solution:

(a) We know that in FP filter, $F = \dfrac{\pi\sqrt{R}}{\sqrt{(1-R)}}$ where R represents the mirror reflectivity.

For given R = 0.99, we have

$$F = \frac{\pi\sqrt{0.99}}{\sqrt{(1-0.99)}} = 312.6 \qquad \textbf{Ans.}$$

(b) In a FP filter, the number of channels that can be selected is given by

$$N < \eta_s \times \left(\frac{\Delta\upsilon_L}{\Delta\upsilon_{FP}}\right) = \eta_s \times F$$

For given spectral efficiency η_s = 1/3 and calculated F = 312.6, we have

$$N = \frac{1}{3}\times312.6 \approx 104 \qquad \textbf{Ans.}$$

Section Practice Problem

1. A Fabry–Perot (FP) optical filter is designed to operate at λ = 1550 nm and selects 100 optical channels having channel separation = 0.2 nm. What is the length of the FP filter? Take refractive index = 1.5.

[Ans.: 40 μm]

7.4 WDM MUX/DEMUX

A WDM multiplexer (MUX) is an optical device that couples or combines a number of optical signals having different wavelengths. On the other hand, a WDM demultiplexer (DEMUX) is an optical device that splits or separates a number of optical signals having different wavelengths. In other words, a WDM–MUX combines several wavelength channels into one common fiber, whereas a WDM–DEMUX separates several wavelengths available in one fiber into individual wavelengths. Fig. 7.10 shows the basic concept of wavelength multiplexer (MUX) and demultiplexer (DEMUX).

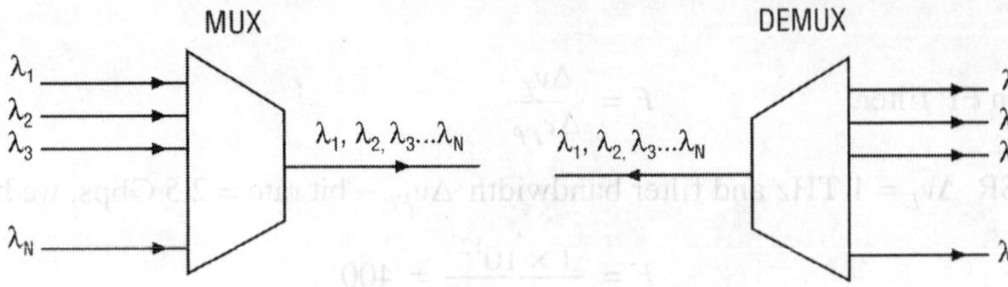

Fig. 7.10 Basic concept of wavelength multiplexer (MUX)

Fig. 7.11 shows the characteristics of a WDM MUX/ DEMUX.

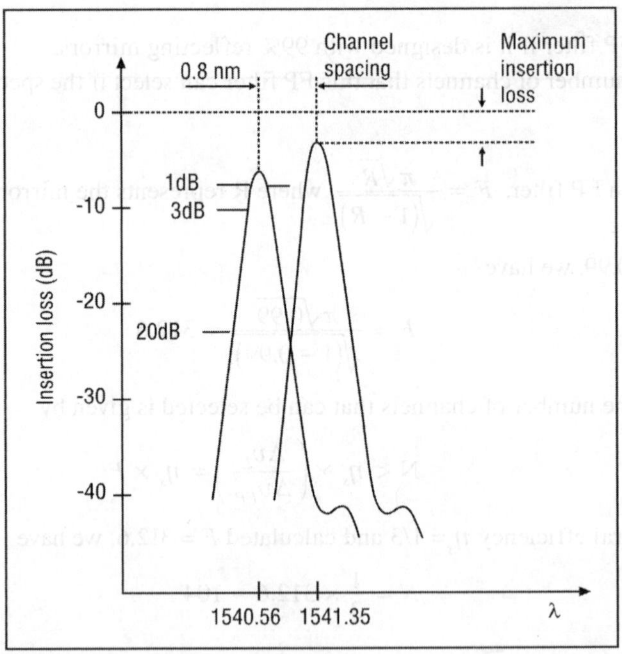

Fig. 7.11 Basic concept of wavelength demultiplexer (DEMUX)

A bidirectional WDM MUX/ DEMUX can perform both multiplexing and demultiplexing in a single device.

Desirable properties of Optical MUX/ DEMUX

For MUX– Insertion loss per channel should be low.

For DEMUX– Each channel should be separable without any leakage from the adjacent channels. Optical DEMUX must be insensitive to the incident optical signal's polarization. Moreover, inter-channel cross-talk must be quite small (<-20 dB).

Classification of WDM MUX/ DEMUX

Depending on a wavelength-selective mechanism employed to realize a WDM MUX/ DEMUX, we have the following two types:

(i) Diffraction-based MUX/ DEMUX– It uses diffraction grating which disperses incident multiplexed optical signal into its different wavelength constituents (DEMUX operation).

(ii) Interference-based MUX/ DEMUX– It makes use of different WDM devices that include optical 3-dB directional couplers and optical filters.

7.4.1 Grating-based DEMUX

It is based on the principle of Bragg diffraction produced by an optical Fiber Bragg grating. The basic function of a simple optical demultiplexer using fiber Bragg grating (FBG) is depicted in Fig. 7.12.

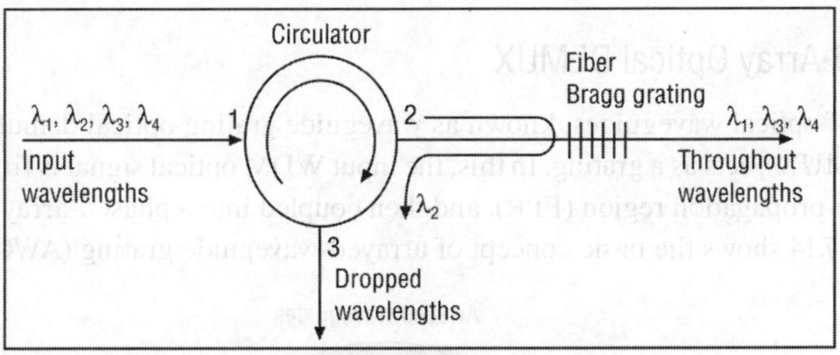

Fig. 7.12 Basic concept of DEMUX function using FBG

The grating period of grating-based optical MUX can be controlled to a specified wavelength by following the Bragg condition given as $2\Lambda n_{eff} = \lambda_B$, where, Λ represents the grating period, n_{eff} represents the effective refractive index, and λ_B represents the Bragg wavelength. It may be noted that each grating is responsible for reflection of only Bragg wavelength. This implies that there is requirement of etching of several gratings for multiple wavelengths. Fig. 7.13 illustrates an example of wavelength selective DEMUX.

In another structure, an optical DEMUX uses a graded-index lens and a conventional lens which enables the focusing of the input WDM signal onto a reflection grating. Different wavelengths are separated spatially. They can be focused by another lens onto individual graded-index profile fibers which not only simplifies the alignment procedure but also leads to a relatively compact device.

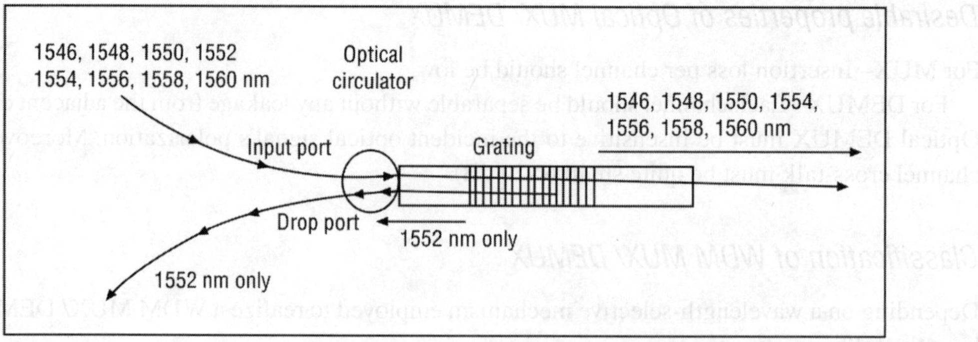

Fig. 7.13 An example of optical DEMUX using FBG

7.4.2 Interference-based MUX/ DEMUX

Multiplexers based on the Mach–Zehnder (MZ) filter use several MZ interferometers (MZI). For example, a four-channel optical MUX may comprise of three MZ interferometers. It is required that the phase shift between two arms of MZI should be wavelength-dependent. For this purpose, its one arm is deliberately kept relatively longer as compared to the other arm. Multiple directional fiber couplers are used to construct multiplexers for the coarse WDM.

7.4.3 Phased-Array Optical DEMUX

A phased-array of optical waveguides, known as waveguide grating optical demultiplexer (phased-array optical DEMUX), acts as a grating. In this, the input WDM optical signal is first passed through a lens, called free-propagation region (FPR), and then coupled into a phased-array of optical planar waveguides. Fig. 7.14 shows the basic concept of arrayed-waveguide grating (AWG).

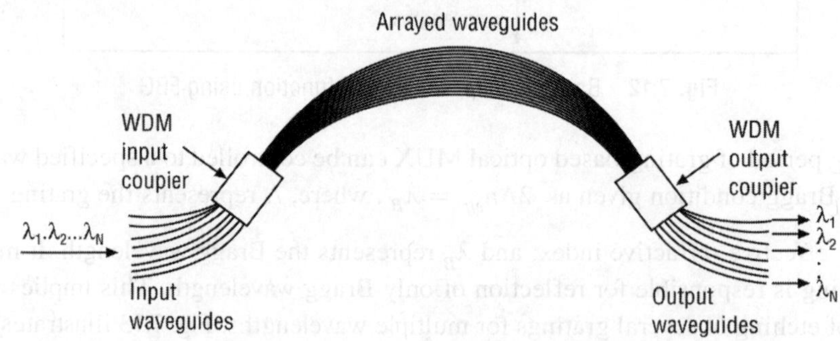

Fig. 7.14 Basic concept of Arrayed–Waveguide Grating (AWG)

Since waveguides have different lengths in arrayed–waveguide grating, the WDM signal experiences a different phase shift in each waveguide. Each input waveguide corresponds to a different center wavelength and channel spacing. Several waveguides around the central one will correspond to the correct channel spacing within the tolerance, and the peak wavelengths will vary from one waveguide to another. Phase-shifts depend on wavelength too. Different wavelength channels concentrate on various output waveguides. It is possible to fabricate such arrayed waveguide gratings by using semiconductor material such as Si, InP, or $LiNbO_3$.

Example 7.14 Number of Channels in DEMUX

Calculate the number of channels that are demultiplexed by a phased-array optical demultiplexer if its free spectral range is 1600 GHz and channels are spaced at 50 GHz at central wavelength of 1550 nm.

Solution:

Given free spectral range =1600 GHz and channel spacing = 50 GHz

We know that the number of channels demultiplexed by a phased-array optical DEMUX is given by

$$N = \frac{\text{Free Spectral Range}}{\text{Channel Spacing}}$$

Hence, $N = \dfrac{1600\text{ GHz}}{50\text{ GHz}} = 32$ **Ans.**

Section Practice Problem

1. An optical DEMUX is designed based on arrayed waveguide gratings. Show that the order of the arrayed waveguides is 121 if 16 channels are required to be demultiplexed with channel spacing of 100 GHz at 1550 nm central wavelength.

7.5 Add–Drop Multiplexer (ADM)

WDM add–drop multiplexer (ADM) is needed for optical metropolitan area networks (MANs) and wide area networks (WANs) for adding or dropping one or more wavelength channels without disturbing the integrity of other wavelength channels. ADM can be configured using two 3-port optical circulators with fiber Bragg gratings. Fig. 7.15 illustrates the basic concept of ADM using FBG by taking example data.

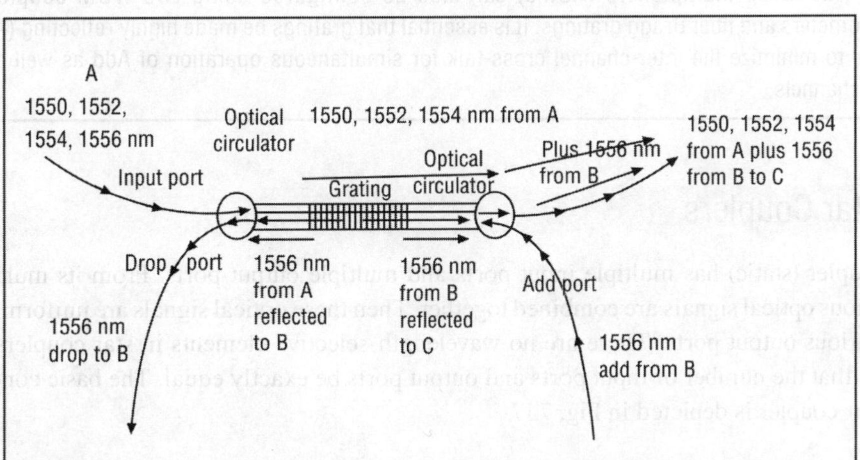

Fig. 7.15 The basic concept of Add–Drop MUX (ADM) using FBG

As shown, FBG is bidirectional, i.e., it reflects in both directions. The WDM signal having different wavelengths such as 1550 nm, 1552 nm, 1554 nm, and 1556 nm are incident on input port of the optical circulator from site A. The wavelength 1556 nm happens to be within the stop band. Therefore, this wavelength is completely reflected and appears at site B through the DROP port of the same optical circulator. Similarly, it is possible to add the 1556 nm wavelength through the ADD port of second optical circulator. The remaining wavelengths are not affected by it and appears at port C (i.e., 1550 nm, 1552 nm and 1554 nm from site A as well as 1556 nm from site B). Thus, this arrangement is quite simple in design.

Fig. 7.16 shows the extended Add–Drop MUX arrangement which uses two three-port circulators with their respective tunable fiber gratings and optical multiplexers and demultiplexers.

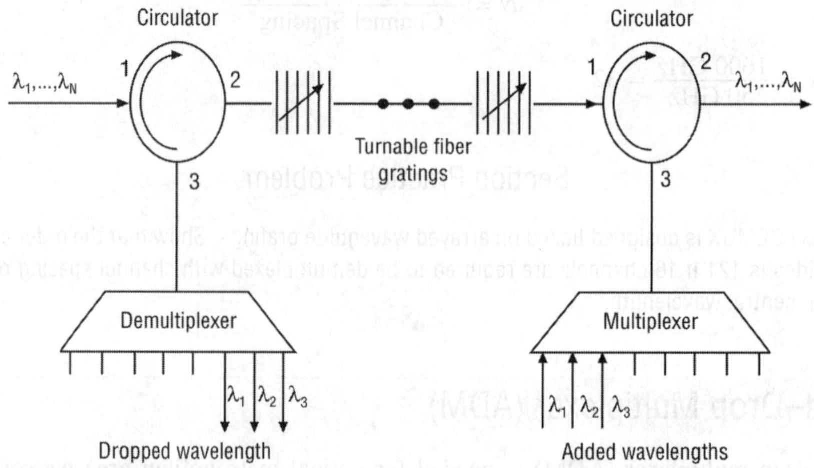

Fig. 7.16 An example of extended Add–Drop MUX (ADM)

Facts to Know

Optical Add–Drop Multiplexers (ADMs) can also be configured using two WDM couplers, MZ interferometers and fiber Bragg gratings. It is essential that gratings be made highly reflecting (~ 100%) in order to minimize the inter-channel cross-talk for simultaneous operation of Add as well as Drop optical channels.

7.6 Star Couplers

A star coupler (static) has multiple input ports and multiple output ports. From its multiple input ports, various optical signals are combined together. Then these optical signals are uniformly divided among various output ports. There are no wavelength-selective elements in star couplers. It is not necessary that the number of input ports and output ports be exactly equal. The basic concept of an optical star coupler is depicted in Fig. 7.17.

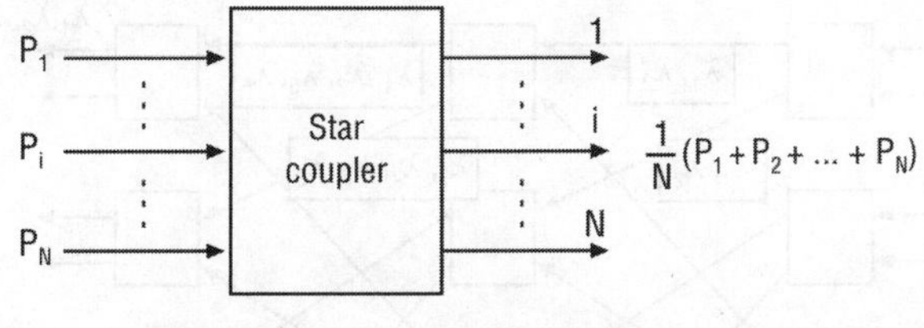

Fig. 7.17 A basic optical star coupler

Star couplers are usually used in WDM LAN applications. There are two kinds of star couplers:

- *Using 3-dB single-mode fiber couplers*– A basic 3-dB 2 × 2 single-mode optical fiber coupler divides both of its input optical signals between its two output ports. It is possible to form higher-order N × N single-mode optical fiber couplers, also known as star couplers, by simply combining multiple 2 × 2 3-dB single-mode optical fiber couplers so far N is an integral multiple of 2 . The complexity of such optical star couplers increases significantly when number of ports increases.
- *Using the fused bi-conical tapering method* – In this method, a very large number of individual optical fibers are fused together. Then, the fused portion is elongated till a bi-conically tapered structure is formed. In this tapered section of fibers, optical signals from individual fibers interact with one another and are mixed together in a uniform manner, and then appear at its various output ports. This arrangement and principle of operation is also applicable for multimode optical fibers. Fig. 7.18 shows the concept of fused bi-conical coupler, also called directional coupler.

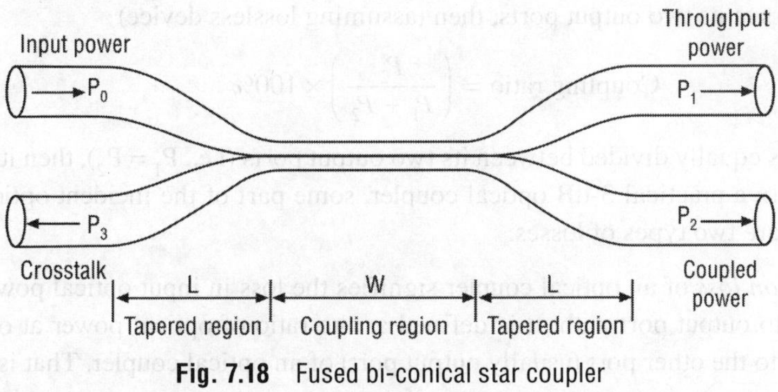

Fig. 7.18 Fused bi-conical star coupler

In fused bi-conical star couplers, two single mode fibers are twisted, melted, pulled and fused together over length *W*, called coupling region; tapered section of length *L*, called tapered region; total draw length = *L*+*W*. There is a significant decrease in *V*-number in the coupling region; energy in the fiber core leak out and gradually couples into the second fiber. The length is adjusted for changing the power ratio between both outputs. It is possible to design fused bi-conical star couplers which can operate satisfactorily over a relatively large range of wavelengths.

Fig. 7.19 shows a 8 × 8 bi-directional star coupler which is obtained by cascading three similar stages of 3-dB optical couplers.

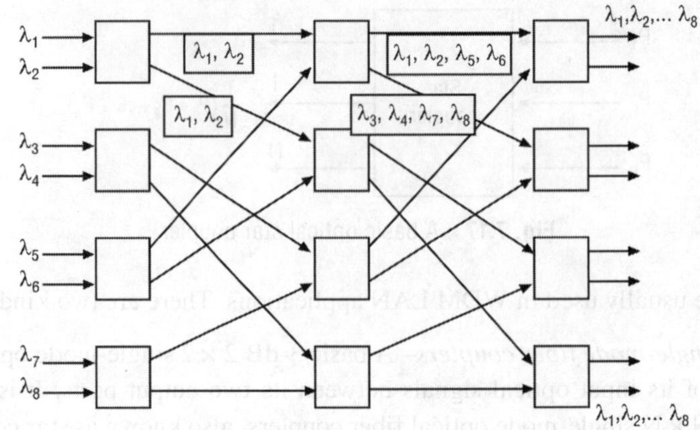

Fig. 7.19 8 × 8 bi-directional star coupler

Number of 3-dB optical couplers required, $N_e = \frac{N}{2}\log_2 N$ (7.7)

For example, for a 8 × 8 bi-directional star coupler i.e., N = 8, we have

$$N_e = \frac{8}{2}\log_2 8 = 12$$

The performance of an optical coupler can be specified by means of the coupling ratio, also known as the splitting ratio. It is defined as the ratio of percent division of optical power between the output ports of an optical coupler. If P_0 represents the input power at the input port, and P_1 and P_2 represent the output powers at its two output ports, then (assuming lossless device)

$$\text{Coupling ratio} = \left(\frac{P_2}{P_1 + P_2}\right) \times 100\% \tag{7.8}$$

If the power is equally divided between its two output ports (i.e., $P_1 = P_2$), then it is called a 3-dB optical coupler. In a practical 3-dB optical coupler, some part of the incident optical power is lost within it. There are two types of losses:

- The *insertion loss* of an optical coupler signifies the loss in input optical power for a specific input port to output port path. It is defined as the ratio of optical power at one port (usually input port) to the other port (usually output port) of an optical coupler. That is,

$$\text{Insertion loss (dB)} = 10\log\left(\frac{P_i}{P_j}\right) \tag{7.9}$$

where, P_i is the optical power level at input port *i* and P_j is the optical power level at particular output port *j* (*i* and *j* are integers 1, 2, 3, ….).

- The *excess loss* of an optical coupler signifies the ratio of the input (incident) optical power to the total output optical power at both output ports. For example, for a 2 × 2 optical star coupler,

$$\text{Excess loss (dB)} = 10\log\left(\frac{P_0}{P_1 + P_2}\right) \tag{7.10}$$

where, P_0 is the input optical power level and P_1, P_2 are the optical power levels at output port 1 and 2, respectively.

Cross-talk is another performance parameter of an optical coupler. Cross-talk is a measure of the amount of isolation between the input optical power applied at one port and the optical power reflected back (or, scattered) into the other input port of a 2 × 2 optical coupler. That is,

$$\text{Crosstalk (dB)} = 10\log\left(\frac{P_3}{P_0}\right) \tag{7.11}$$

where, P_3 is the reflected power at other input port and P_0 is the input power at the required input port.

Example 7.16 Number of 3-dB Couplers

How many 3-dB couplers are needed to design a 4 × 4 bi-directional star coupler?

Solution:

We know that the number of 3-dB star couplers needed to form an N × N star coupler is given by

$$N_c = \frac{N}{2}\log_2 N$$

For a 4 × 4 bi-directional star coupler, N = 4.

Hence, number of 3-dB couplers needed $N_c = \frac{4}{2}\log_2 4 = 4$ **Ans.**

Example 7.17 Coupling ratio

An optical power $P_0 = 100$ μW is applied at the input port of a 2 × 2 bi-conical tapered optical fiber coupler. The output optical power levels at three ports are $P_1 = 45$ μW at output port 1, $P_2 = 42.5$ μW at output port 2, and $P_3 = 3.15$ nW at other input port 3. Determine the percent coupling ratio. Assume that the device is lossless.

Solution:

The *coupling ratio* refers to the percent splitting of input optical power among the output ports of an optical coupler.

We know that coupling ratio (%) = $\left(\frac{P_2}{P_1 + P_2}\right) \times 100$

For the given $P_1 = 45$ μW and $P_2 = 42.5$ μW, we have

$$\text{Coupling ratio (\%)} = \left(\frac{42.5\ \mu\text{W}}{45\ \mu\text{W} + 42.5\ \mu\text{W}}\right) \times 100\% = 48.57\%$$ **Ans.**

Example 7.18 Insertion Loss of 2 × 2 Coupler

An optical power $P_0 = 100$ μW is applied at the input port of a 2 × 2 bi-conical tapered fiber coupler. The optical power levels at three ports are $P_1 = 45$ μW at output port 1, $P_2 = 42.5$ μW at output port 2, and $P_3 = 3.15$ nW at other input port 3. Determine the insertion loss (dB) in the following situations:

(i) between input port and output port 1
(ii) between input port and output port 2

Solution:

The insertion loss is defined as the loss of optical power for an input port P_i to a particular output port P_j. That is,

$$\text{Insertion loss (dB)} = 10\log\left(\frac{P_i}{P_j}\right)$$

(a) For the given $P_0 = 100$ μW, $P_1 = 45$ μW, we have

$$\text{Insertion loss (dB)} = 10\log\left(\frac{100\ \mu\text{W}}{45\ \mu\text{W}}\right) = 3.47 \text{ dB} \qquad \textbf{Ans.}$$

(b) For the given $P_0 = 100$ μW, $P_2 = 42.5$ μW, we have

$$\text{Insertion loss (dB)} = 10\log\left(\frac{100\ \mu\text{W}}{42.5\ \mu\text{W}}\right) = 3.716 \text{ dB} \qquad \textbf{Ans.}$$

Example 7.19 Excess Loss of 2 × 2 Coupler

Let an optical power $P_0 = 400$ μW be applied at the input port of a 2 × 2 bi-conical tapered fiber coupler. The optical powers at three output ports are $P_1 = 180$ μW at output port 1, $P_2 = 170$ μW at output port 2, and $P_3 = 12.6$ nW at other input port 3. The device is not lossless. Determine the excess loss (dB) for this 2 × 2 3-dB optical coupler.

Solution:

We know that the excess loss of an optical coupler is defined as the ratio between the input optical power at the input port and the total output optical power at both output ports. That is,

$$\text{Excess loss (dB)} = 10\log\left(\frac{P_0}{P_1 + P_2}\right)$$

For given $P_0 = 400$ μW, $P_1 = 180$ μW *and* $P_2 = 170$ μW, we have

$$\text{Excess loss (dB)} = 10\log\left(\frac{400\ \mu\text{W}}{180\ \mu\text{W} + 170\ \mu\text{W}}\right) = 0.58 \text{ dB} \qquad \textbf{Ans.}$$

Example 7.20 Cross-talk Level in 2 × 2 Coupler

Consider an optical power level of $P_0 = 400$ μW applied at the input port of a 2 × 2 bi-conical tapered optical fiber coupler. The optical powers are 180 μW, 170 μW, and 12.6 nW at three ports P_1, P_2, and P_3, respectively. How much would be the cross-talk level (dB)?

Solution:

We know that cross-talk in an 2 × 2 optical coupler is a measure of the amount of the isolation between the input optical power at its input port (Port 1) where the input optical signal is applied and the optical power that is reflected back into the other input port (Port 3). It is expressed as

$$\text{Crosstalk (dB)} = 10\log\left(\frac{P_3}{P_0}\right)$$

For given $P_0 = 400$ μW, and P_3 = *12.6 nW*, we have

$$\text{Crosstalk (dB)} = 10\log\left(\frac{12.6\times10^{-9}\,W}{400\times10^{-6}\,W}\right) = -45 \text{ dB} \qquad \textbf{Ans.}$$

Section Practice Problems

1. An optical power level of 20 μW is applied at the input port of a 2 × 2 bi-conical tapered fiber coupler. The output optical power at throughput port is 9 μW and at coupled port is 8.5 μW. Determine the percent coupling ratio. *[Ans.: 48.6%]*
2. An optical power level of 20 μW is specified at the input of a 2 × 2 3-dB bi-conical tapered optical fiber coupler. The output optical powers at two output ports are 9 μW and 8.5 μW, respectively. Determine the insertion loss (dB) from its input port to output ports 1 and 2. *[Ans.: 3.47 dB; 3.72 dB]*
3. If an optical power level of 20 μW is applied at the input port of a 2 × 2 bi-conical tapered fiber coupler and the output optical power at the throughput port is 9 μW, and the output optical power at the coupled port is 8.5 μW, then what would be the excess loss (dB)? *[Ans.: ~0.6 dB]*
4. A 2 × 2 single-mode bi-conical tapered fiber coupler has been designed with a splitting ratio of 40:60. The specified insertion loss from input port of coupler to output port 1 (i.e., 60% channel) is 2.7 dB and that from input port of coupler to output port 2 is 4.7 dB (40% channel). If an optical power level of 100 μW is applied at its input port, then what would be the output optical power levels at (i) port 1; (ii) port 2. *[Ans.: a) 53.7 μW; b) 33.9 μW]*
5. Consider an optical power P_0 = 800 μW applied at the input port of a 2 × 2 3-dB bi-conical tapered optical coupler. The optical powers at the other three ports of the coupler are P_1 = 360 μW (output port 1), P_2 = 340 μW (output port 2), and P_3 = 25.2 nW (other input port). How much will be the cross-talk level (dB)? *[Ans.: -45 dB]*

7.7 Wavelength Converters

A wavelength converter is primarily used to change the wavelength of the incident optical signal to a new wavelength while maintaining the integrity of the data content. There are various schemes for realizing wavelength converters which are briefly discussed below:

(a) *Opto–electronic Regenerators:* Opto–electronic regenerators convert optical data into electric domain and use an optical transmitter at the desired wavelength, as shown. It uses standard components like optical transmitters and receivers. So, it is relatively easy to implement. There is a possibility of net amplification and insensitivity to input polarization. However, there is a limited transparency to that data format as well as to the bit rate. Moreover, speed is limited due to the use of electronic circuitry.

(b) *Cross-gain Saturation:* Whenever a weak field along with a strong field is amplified inside a semiconductor optical amplifier (SOA), cross-gain saturation occurs. For example, the optical pulsed signal at a specific wavelength (say, λ_1) is launched into the semiconductor optical amplifier (SOA) along with a low-power continuous wave light beam at another wavelength (say, λ_2). The optical gain of SOA is mostly saturated at the wavelength λ_1. The continuous wave light beam is amplified by a small amount during 1 bit duration, and by a large amount during 0 bit duration, resulting in reverse polarity bit pattern at λ_2 through optical filter tuned at λ_2. Fig. 7.20 shows a wavelength converter that uses co-propagation mode which is based on the cross-gain modulation as used in SOA.

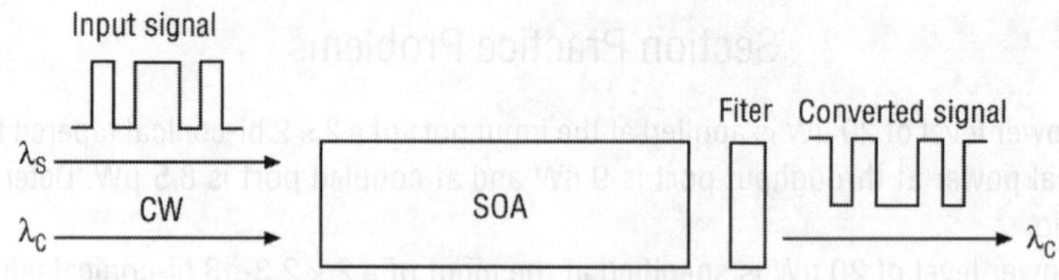

Fig. 7.20 Wavelength converter using SOA

Cross-gain saturation type wavelength converter has a distinct advantage- it can work up to 40 Gbps bit rate. An overall optical gain can be provided to λ-converted signal. It is possible to make it insensitive to polarization. But it offers relatively low on-off contrast. The performance is degraded due to amplified spontaneous emission (ASE) noise, phase reversal of information data, and phase distortion because of frequency chirping.

(c) *Cross-absorption Saturation*: The SOA is used as an absorption medium which enables to resolve the problem of phase reversal. An electro-absorption modulator is used for wavelength conversion. Fig. 7.21 shows a wavelength converter using cross-absorption saturation in a semiconductor laser.

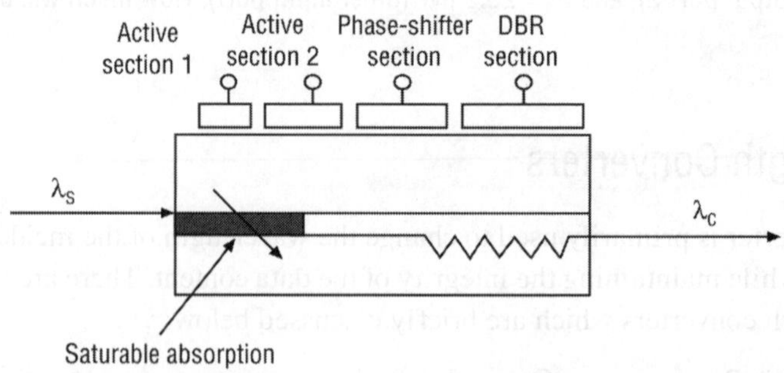

Fig. 7.21 Wavelength converter using cross-absorption saturation

(d) *XPM-based MZ Interferometer (MZI):* In cross-phase modulation (XPM)-based MZI, each arm contains a semiconductor optical amplifier (SOA). The optical pulsed signal at one wavelength (say, λ_1) is divided in the first optical coupler with most of the optical power passing through one arm of MZ Interferometer filter. Simultaneously, the optical continuous wave signal at a different wavelength (say, λ_2) is equally divided in two parts by the optical coupler. These two optical signals propagate together in both arms of the MZI filter. Since the first wavelength λ_1 is not present, the CW wavelength is present at the cross port. Because of cross-phase modulation (XPM), the CW wavelength at λ_2 is directed towards desired port. An optical filter blocks the λ_1 signal. Fig. 7.22 shows an interferometer wavelength converter using XPM.

Why is the XPM-based MZ interferometer wavelength converter preferred over cross-gain saturation wavelength converter? The reason for this is that wavelength converter using XPM-based SOA is not able to reverse the bit sequence. Moreover, it yields a relatively higher ON–OFF contrast and operates at a higher transmission bit rate (up to 80 Gbps) and the signal degradation is less due to ASE noise. But it has a narrow input optical power dynamic range.

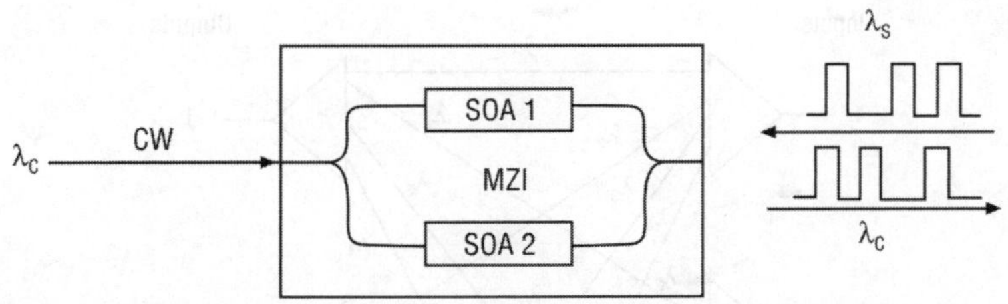

Fig. 7.22 Wavelength converter using SOA based on XPM

(e) *FWM-based SOA scheme:* In FWM-based wavelength converter, a high-intensity continuous wave pump signal is propagated into the semiconductor optical amplifier along with optical signal whose wavelength is required to be converted. The pump frequency is chosen as $(\lambda_1 + \lambda_2)/2$. At the output port of SOA, the input optical signal is available at λ_2. In this scheme, wavelength conversion over 80 nm are possible. It can offer high transmission bit rate (~ 100 Gbps) operation irrespective of data format and bit rate. The conversion efficiency too can be very high. It exhibits net gain with reversal of frequency chirp. It necessitates a tunable optical source, resulting in large coupling loss.

7.8 Wavelength Routers

Wavelength router is an optical device used in WDM applications which combines the functions of an optical star coupler with optical MUX and DEMUX operation. The WDM optical signals that enter from *N* input ports of a wavelength router are demultiplexed into individual optical channels having different wavelengths, and are routed directly towards the *N* output ports. Then, the WDM signal present at each port of wavelength router consists of wavelength channels at various input ports. It appears to be equivalent to a cyclic form of WDM demultiplexing operation. Fig. 7.23 shows the basic concept of wavelength router.

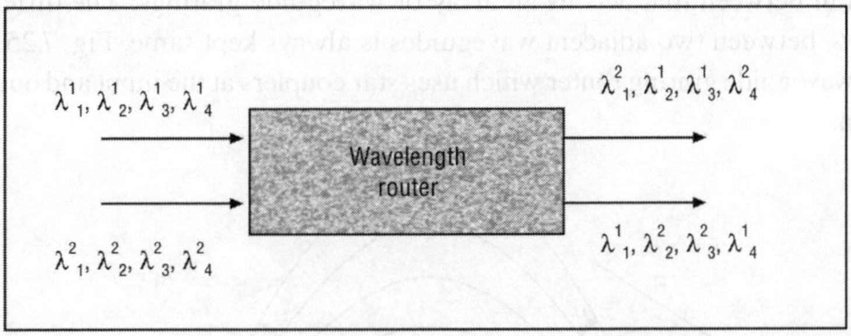

Fig. 7.23 Basic concept of wavelength router

Due to non-usage of any external electrical power, this device is commonly known as a passive wavelength router. It is also known as a static wavelength router due to non-dynamic reconfiguration capability of the routing topology. Fig. 7.24 shows a 4 × 4 hardwired (i.e., non-reconfigurable) wavelength router.

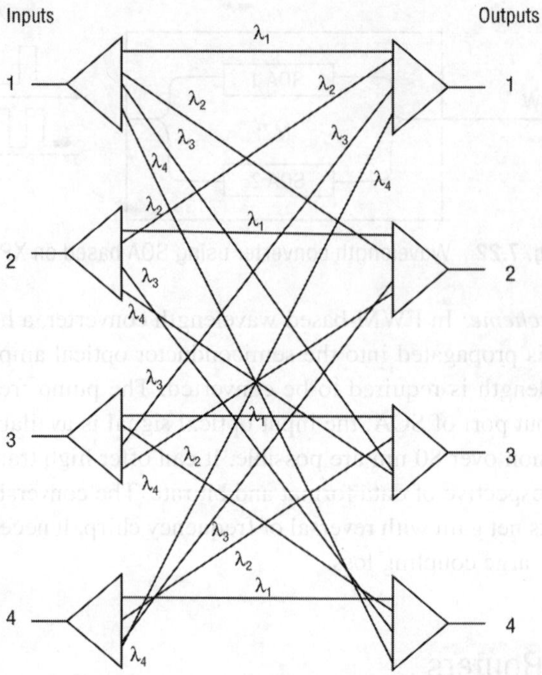

Fig. 7.24 A non-reconfigurable wavelength router (4 × 4)

Waveguide Grating Router (WGR)

Waveguide grating router (WGR) is based on arrayed waveguide grating (AWG). It mainly comprises of ($N \times N'$) optical star couplers, where N stands for number of ports on one side and N' stands for number of ports on the other side. The N' number of ports of one optical star coupler are directly connected with N' ports of another optical star coupler by using an array of N' number of waveguides (Si/SiO_2 or InP) in between that acts as an array of waveguide gratings. The difference in length, designated by ΔL, between two adjacent waveguides is always kept same. Fig. 7.25 shows a typical arrangement of waveguide grating router which uses star couplers at the input and output with grating array in between.

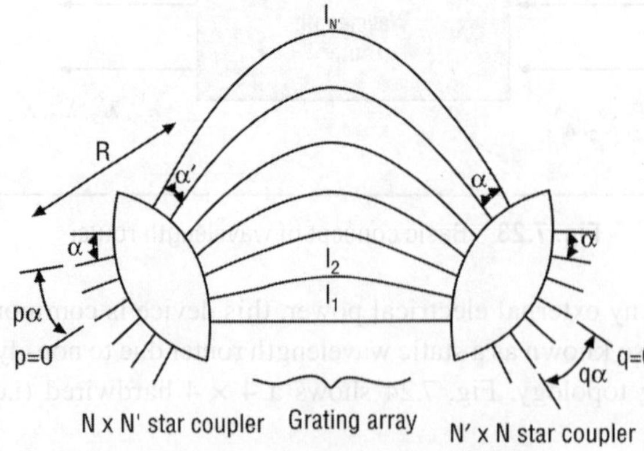

Fig. 7.25 Waveguide grating router

Where do waveguide grating routers find application? They are widely employed as multichannel optical transmitters and optical receivers, add-drop optical multiplexers, waveguide routers as well as tunable optical filters. Small channel spacing with less cross-talk can be achieved.

7.9 Optical Cross-Connects (OXC)

Optical cross-connects (OXC) provides the basic function of wavelength routing which can be reconfigured as required in the wide-area WDM networks while maintaining its transparency in operation. The OXC device has *N* number of input ports, each one is capable of receiving a WDM signal that consists of *M* number of different wavelengths. The issue of a limited number of available wavelengths can be resolved by employing wavelength-reuse technique that use dynamic routing technique. The generic structure of an OXC is based on optical MUX/DEMUX, optical switches and optical Add–Drop MUX/DEMUX.

Optical DEMUX splits the input signal into different wavelengths contained in it. Each individual wavelength is then directed toward the stack of *M* number of optical switches, each one being easily reconfigured for routing the optical signal in desired way. Each optical switching unit receives *N* input signals having identical wavelengths. With the purpose of adding or dropping of a specified optical channel, an additional input and output port is included in every optical switch. The output of optical switches is then routed to *N* number of optical MUXs to form the WDM signal. Such an optical cross-connect requires *N* number of optical MUXs, *N* number of optical DEMUXs, as well as $[M \times (N+1)^2]$ number of optical switches. Fig. 7.26 shows the basic concept of heterogeneous WDM cross-connect.

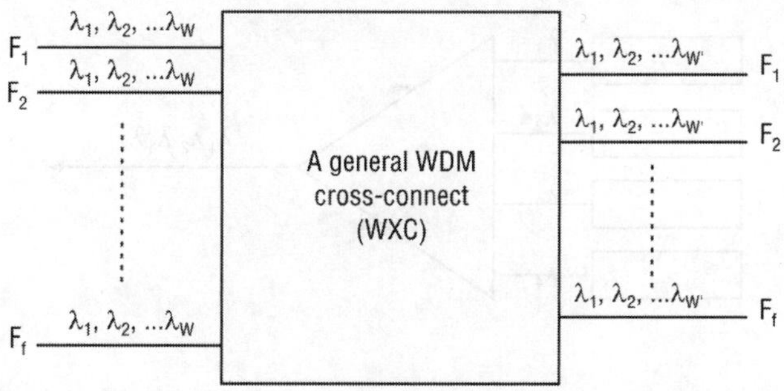

Fig. 7.26 Basic concept of WDM cross-connect

Fig. 7.27 shows a $P \times P$ fully reconfigurable as well as tunable wavelength-routing switches having *M* number of different wavelengths that uses 2×2 optical cross-point devices and photonic switches.

Several technologies can be used for making optical switches such as

- Micro-electro–mechanical system (MEMS) mirrors
- Liquid crystals
- Thermo–optic MZ switches
- Y-junction switches using semiconductor optical amplifiers
- Directional–coupler switches ($LiNbO_3$ waveguides)
- Electrographic switches ($LiNbO_3$ crystals)

- Optical fibers along with optical circulators and fiber gratings
- Air bubble jet technology

The major disadvantage of optical cross-connect is the exponential growth of optical devices and their interconnections with an increase in number of nodes and wavelengths in WDM system.

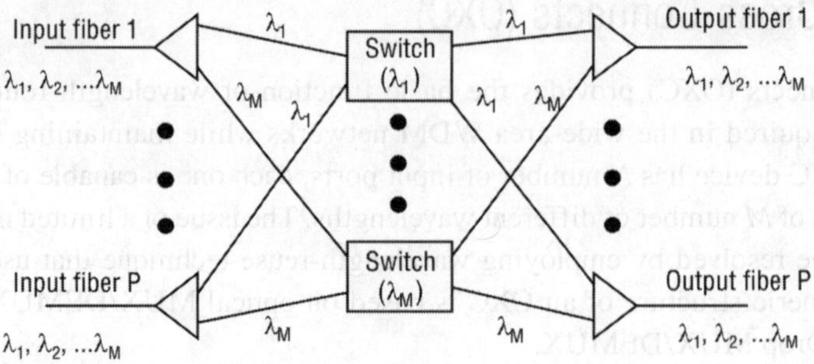

Fig. 7.27 Wavelength-routing switch

7.10 WDM Transmitters

WDM transmitter generally comprises of laser array (each laser is a tunable laser which can be tuned to desired fixed wavelength) and an optical multiplexer which is tunable across a range of wavelengths. Fig. 7.28 shows a simplified structure of WDM transmitter.

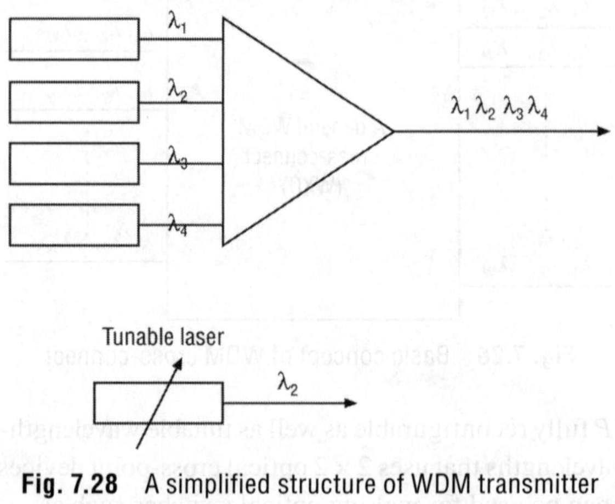

Fig. 7.28 A simplified structure of WDM transmitter

7.10.1 Essential Requirements

1. Quality of generated optical signal
    - The linewidth has to be as narrow as possible, e.g., DFB lasers have linewidth of 10–40 MHz.
    - The side-mode suppression ratio (SMSR) has to be as high as possible, e.g., DFB lasers have SMSR of 40 dB.

- A laser has to operate in a single longitudinal mode, e.g., DFB lasers with fiber Bragg gratings.
- Frequency chirp has to be eliminated.

All these requirements lead to the use of external modulation. So direct modulation is not recommended in WDM networks.

2. Stability of generated optical signal
   - Variations in output power results in variations in linewidth and non-linear optical effects in single-mode fibers as four-wave mixing (FWM) and SBS. But variations in linewidth are not acceptable in WDM operation.
   - Variations in peak wavelengths are unacceptable because it causes channel cross-talk variations resulting in an increase in BER.
   - The relative intensity noise (RIN) needs to be minimized.
3. Reliability should be quite high.
4. Power consumption should be minimum so as to reduce the heat produced by a laser.
5. Tunability of lasers to all WDM wavelengths including tuning speed, ability to emit several wavelengths simultaneously.

7.10.2 Types of Tunable Lasers

Large number of distributed feedback lasers operating at different wavelengths are used as optical sources in most of the WDM systems. This arrangement is almost impractical with a large number of wavelength channels. Mainly, there are two alternatives available to realize WDM transmitters:

1. *Tunable Lasers*– It is always preferred to have a single laser that can be tuned over the complete range of desired wavelengths instead of using an array of lasers. Only one (DFB or DBR) laser that has fiber Bragg grating filter built within the lasing cavity is usually used. Wavelength is tuned by either changing the operating temperature of the grating at the rate of 0.1 nm per degree centigrade, or by varying the injection electrical current into its passive section (~ 0.006 nm/mA). The tuning range decreases with the optical output power. There are four kinds of tunable lasers available:
   - Temperature tuning lasers: They exploit the dependence on temperature of the refractive index of an active area. This characteristics results in the drifting of the peak frequency radiated by a laser diode. Both the tuning range and tuning speed of these lasers are not very good.
   - External cavity tunable lasers: Wavelengths are changed by varying the resonance condition for the lasing wavelength. It is based on using diffraction gratings as one of the laser's reflectors. It comprises of MQW laser diode active medium with anti-reflecting coating on one of its two ends, collimating lens, diffraction grating device capable of rotating with coarse tuning of >50 nm as well as lateral-motion for fine tuning (in GHz). A diffraction grating works as a highly reflective mirror, reflecting a wavelength λ_i such that $d \sin \theta_i = m\lambda_i$; where, d represents the diffraction grating period, θ_i represents the angle of incidence (i.e., tilting angle) and $m = 0, \pm1, \pm2, \pm3$ …. Changing the values of d and θ_i, wavelength can be tuned by a laser.
   - Sectional distributed Bragg reflection (DBR) tunable lasers: It basically comprises of three different sections, namely, MQW InGaAsP active section, phase-alignment section, and DBR section. Incident light actually obtains gain in the first section and desired power can be set.

The second section provides a phase shift of a reflected wave. The third section provides the limited tuning range (~ 10 nm).

- Integrated Cavity Lasers: A number of active media (i.e., optical amplifiers) are terminated by a common cleaved-mirror facet at one end, and optically connected to an optical MUX and filter at the other end. The MUX has a single output port optically connected to the second cleaved mirror facet. These facets form a laser cavity. It represents many individual lasers within a single cavity and a MUX combines all beams into a single output.

Fig. 7.29 shows tunable laser characteristics having a tuning range of 10–15 nm typical.

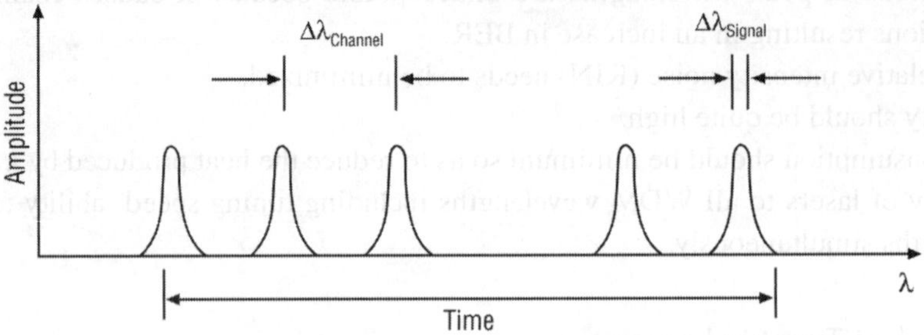

Fig. 7.29 Tunable laser characteristics

2. *The Vertical-Cavity Surface-emitting Laser (VCSEL)*– VCSEL basically offers a two-dimensional laser array. It can cover a relatively wider optical band. It provides an economical solution for data-transfer applications such as LAN.

Tunable lasers can also be categorized based on the tuning mechanisms employed. These are as follows:

- *Mechanically Tuned Lasers:* This category of lasers basically uses Fabry–Perot cavity which is placed quite close to the external cavity that serves as a lasing medium. This arrangement helps to filter out undesired wavelengths. The distance between two mirrors is physically adjusted for the purpose of tuning.
- *Acousto–optic Tuned Lasers:* In this type of lasers, the required refractive index is changed in an external cavity with acoustic waves.
- *Electro–optic Tuned Lasers:* In this type of lasers, the required refractive index is changed in an external cavity by applying an external variable current.
- *Injection-Current Tunable Lasers:* This type of tunable laser is based on a diffraction grating inside or outside of the lasing medium. The lasing medium is, in fact, an optical waveguide where the refractive index is made to vary periodically between two different values of the wavelengths. Only those wavelengths that exactly match with that produced with the refractive indices of the grating as well as the grating period will only be amplified. Tuning is possible by injecting an electric current which varies the refractive index of the grating.

  There are two types of injection-current based tuned lasers - DFB and DBR. The grating exists in both DFB and DBR type of tunable lasers but the difference lies in its location with respect to the lasing medium. In DFB, the grating is present within it, whereas in DBR, it is moved outside it.
- *Laser Arrays:* It is a set of pre-tuned lasers, each one tuned at its designated wavelength only.

Table 7.4 gives a comparison of various tunable lasers in terms of key parameters such as tuning range and time taken for tuning.

Table 7.4 Tunable lasers – tuning range and time comparison

S. No.	*Tunable Laser*	*Tuning Range (nm)*	*Tuning Time*
1.	Mechanical (external cavity) laser	≈ 500	1–10 ms
2.	Acousto–optic laser	≈ 83	10 μs
3.	Electro–optic laser	≈ 7	~ 1–10 ns
4.	Injection–Current (DFB and DBR) laser	≈ 10	1–10 ns

7.10.3 Optical Modulation Types

There are two types of optical modulation techniques used in WDM transmitters.

- Analog Modulation Techniques– Amplitude Modulation (AM), Frequency Modulation (FM), and Phase Modulation (PM)
- Digital Modulation Techniques– Amplitude Shift Keying (ASK), Frequency Shift Keying (FSK), and Phase Shift Keying (PSK)

Binary Amplitude Shift Keying (BASK) is also called ON–OFF keying (OOK). This is the most preferred choice in optical fiber communications. It is simple to implement for propagation of optical (light) signals.

There are two ways in which the optical modulation can be applied.

- Direct Modulation– for switching the optical source (such as laser) ON and OFF
- External Modulation– for modulating the output light from the optical source (such as laser)

Fig. 7.30 shows the basic concept of direct modulation of laser diode being used as an optical source.

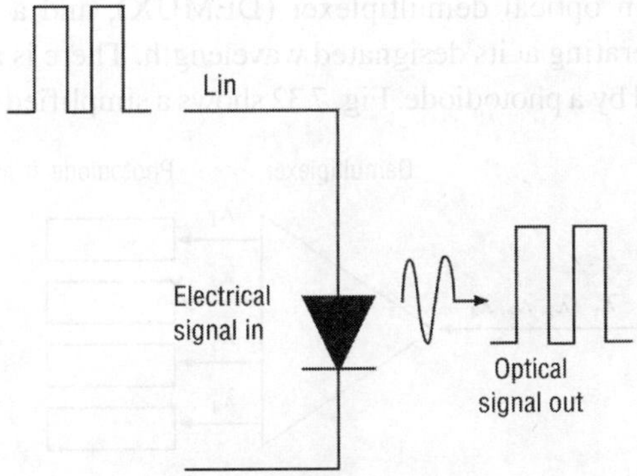

Fig. 7.30 Basic concept of direct modulation

In direct modulation, the laser diode's bias current is modulated with signal input to produce modulated optical output. This approach is straightforward and low cost, but is susceptible to chirp (spectral broadening) thus exposing the signal to higher dispersion.

Fig. 7.31 shows the basic concept of external modulation of laser diode being used as an optical source.

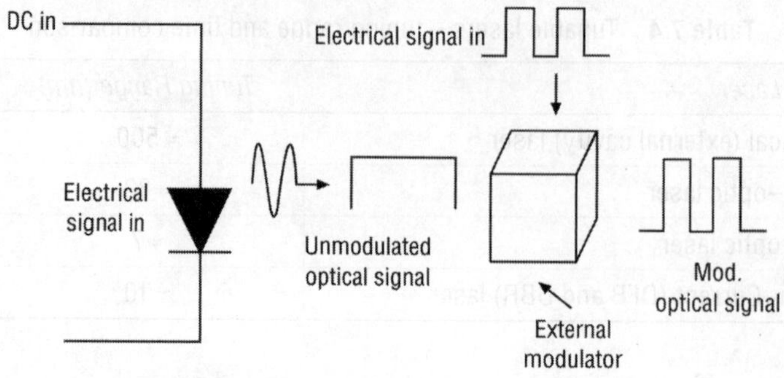

Fig. 7.31 Basic concept of external modulation

In external modulation, the laser diode's bias current is stable. This approach yields low chirp and better dispersion performance, but it is a more expensive solution for dispersion management. Integrated laser and modulation is cost effective.

> **Facts to Know**
>
> Mechanical tunable laser has a wider tuning range than others, whereas electro-optic and DFR/DBR tunable lasers have minimum tuning time.

7.11 WDM Receivers

Tunable receivers usually involve optical components and devices such as passive DEMUXs, tunable filters, and arrayed waveguide gratings (AWGs). In a simplified arrangement, a WDM receiver essentially comprises of an optical demultiplexer (DEMUX), and a photodetector (also called photodiode) array, each operating at its designated wavelength. There is an alternate arrangement of using tunable filter followed by a photodiode. Fig. 7.32 shows a simplified structure of WDM receiver.

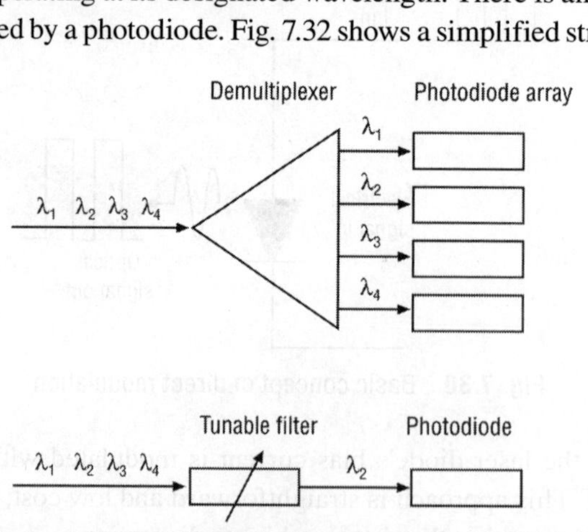

Fig. 7.32 A simplified structure of WDM receiver

Photodetectors can be used either for direct detection or coherent detection of incoming light signal.

- *Direct Detection:* In this method, a photodetector converts a stream of light into a continuous flow of electrons, followed by amplification. Then, it is passed through a threshold device to determine a sequence of binary 0s or 1s. For example, in a *p-n* junction (known as *p-n* photodiode), or a *p-i-n* photodiode (consisting of an intrinsic semiconductor material between *p*-type and *n*-type semiconductor materials), the light beam strike at the *p-n* junction which creates more number of electron-hole pairs (EHPs) in both *p* and *n* semiconductor regions, resulting in a current flow.
- *Coherent detection:* In this method, the phase information is used in detecting the incoming signal. The incident light is added to the local oscillator (a monochromatic laser), then, it is detected by a photodetector. This method is more elaborate and it is quite difficult to maintain phase information

Essential Requirements of WDM Receivers

1. Spectral width or wavelength range- Sufficient and compatible to that of EDFAs (up to 80 nm)
2. Receiver sensitivity- High enough to overcome SNR degradation due to channel cross-talk
3. Tuning time ~ nanoseconds
4. Operating temperature sensitivity (-40°C to +85°C)
5. Polarization independence
6. Power consumption and packaging– minimum power consumption, compactness and reliability
7. Immunity to internal noise and EMI

There are two basic approaches to selecting a desired wavelength for WDM network.

- Tunable optical transmitter, and fixed optical receiver (TTFR)
- Fixed optical transmitter, and tunable optical receiver (FTTR)

There are two basic methods to achieve wavelength selectivity of a receiver: active and passive. In active method, a tunable filter actively seeks a desired channel in optical domain. In passive method, optical DEMUX splits the received wavelengths into individual wavelength before these are directed to their respective photodiode in photodetector array. Selection of the desired channel is carried out by electronic components in the electrical domain. Passive optical DEMUXs are more reliable and high switching speed (~ ns) can be achieved.

Facts to Know

WDM networks require transceivers (transmitters + receivers) which are able to radiate and accurately detect closely spaced wavelength channels with unprecedented requirements in terms of wavelength stability. WDM network transceivers have been developed as opto–electronic integrated circuits (OEIC).

7.12 System Performance Issues

Inter-channel cross-talk is the most important system performance issue in the design of WDM networks. *Inter-channel cross-talk* refers to the transfer of optical power from one wavelength channel to another wavelength channel. As a result, the system performance degrades considerably. Several

WDM components including optical switches, optical filters, and optical DEMUX, their behavior is not perfect. Therefore, inter-channel cross-talk is more likely to occur even in a perfect linear optical channel (i.e., optical fiber cable. It may be noted that inter-channel cross-talk also occurs due to non-linear effects in optical fibers.

Inter-band cross-talk refers to another type of interference that may occur from optical signals having different wavelengths, and it mainly affects channel spacing. With the use of proper narrow-band optical filters, Inter-band cross-talk can be minimized.

Similarly, *Intra-band cross-talk* refers to another type of interference that may occur from optical signals having the same wavelength on an adjacent fiber. It usually occur in switching nodes and can accumulate when the optical signal propagates from one node to another node. It is not easy to minimize its effect by using optical filters.

There are mainly two main mechanisms: Linear cross-talk and non-linear cross-talk.

7.12.1 Linear Cross-talk

Depending on its origin, linear cross-talk is of two types as given below:

(a) Hetero-wavelength Linear Cross-talk

Hetero-wavelength means 'out-of-band wavelength'. Hetero-wavelength linear cross-talk occurs when optical tunable filters and optical DEMUXs allow leakage of some part of the optical power from adjacent optical channels which may lead to interference with the detection process. Because of its incoherent nature, it poses less problems.

Consider a tunable optical filter which is required to allow the mth optical channel. Because of linear cross-talk, its output optical power that is available at the input of the photodetector at the optical receiver is expressed as

$$P = P_m + \sum_{n \neq m}^{N} T_{mn} P_n \tag{7.12}$$

where, P_m represents the optical power in the desired m^{th} channel, T_{mn} denotes the transmissivity of the optical filter for n^{th} channel on selection of the mth channel, and N is total number of channels incident on filter.

If $T_{mn} \neq 0$ for $m \neq n$, then inter-channel cross-talk occurs. This means that this type of cross-talk is related to the channels other than the detected channel and exists outside its occupied spectral band. That is why it is called *out-of-bound cross-talk*. It mainly depends on the optical power available in the adjacent channels. How can we counteract the effect of this type of linear cross-talk on the overall system performance? The following may be considered.

- Requirement of the additional power required at the optical receiver, known as the *power penalty*.
- To determine the amount of power penalty, we need to estimate the corresponding increase in electrical current that is required to maintain desired BER value.

(b) Homo-wavelength Linear Cross-talk

The word 'Homo-wavelength' simply means 'in-band wavelength'. As the name suggests, homo-wavelength linear cross-talk mainly results from WDM components and devices that are used for switching and routing in an optical communication network. For example, for an $(N \times N)$ configuration of the wavelength router, there exists N^2 different possibilities of dividing the available N-wavelength WDM signal. Now, let us consider that the output signal is available at wavelength designated as λ_m. Then, there will be total (N^2-1) interfering signals which can co-exist with the desired signal. Out of these, there will be $(N-1)$ number of different optical signals which will have identical wavelength λ_m. The remaining $[(N^2-1) - (N-1) = N \times (N-1)]$ number of optical signals have different wavelengths which may be eliminated by other WDM devices. The $(N-1)$ number of cross-talk signals having identical wavelength, also known as *in-band cross-talk*, mainly occur due to imperfect filtering operation because of partially overlapping transmission peaks by a waveguide grating router.

7.12.2 Non-Linear Cross-talk

As discussed previously, non-linear effects mainly occur when optical power is extremely high. Non-linear cross-talk are more significant in dense WDM. They can be broadly classified into two groups:

- *Scattering phenomena:* There are two different types of scattering phenomena. These are Stimulated Brillouin Scattering (SBS) and Stimulated Raman Scattering (SRS). In order to minimize the impact of non-linear cross-talk because of SBS and SRS, we need to use moderate optical powers in the channels. In addition, an optimum plan for densely packed channels can be worked out in order to minimize the overall spectrum width.
- *Refractive index phenomena:* There are three different types of refractive index phenomena. These are (a) self-phase modulation (SPM), (b) cross-phase modulation (XPM), and (c) four-wave mixing (FWM). Due to the impact of SPM phenomenon, the spectrum of the optical signal is increased and leads to non-linear cross-talk or an unexpected dispersion penalty. The impact of FWM is degradation in signal-to-noise ratio in dispersion shifted fiber that results in non-linear cross-talk. Due to this, the channel capacity of a DWDM system gets limited.

Thus, several non-linear effects as described above results in inter-channel cross-talk and the system performance is considerably degraded.

(i) Non-linear SRS Cross-talk

It is based on the phenomenon that light incident with molecules creates scattered light. Moreover, the wavelength of scatter light is longer than that of the corresponding incident light. What does it mean? It means that the frequency of a part of the light propagating in a Raman-active fiber is decreased, i.e., the Stokes wave. How can we estimate the frequency range covered by the Stokes wave? The frequency range can be estimated by well-defined Raman gain spectrum that may be about 40 THz less than the frequency of the incident optical signal. As an instance, the maximum Raman gain occurs at 13 THz in silica fiber, as shown in Fig. 7.33.

In WDM networks, the optical fiber itself functions as a Raman optical amplifier. So far the difference in wavelength remains within the specified optical bandwidth of the Raman optical gain, the shorter frequency (i.e., longer wavelength) channels can be amplified by the shorter wavelength (longer frequency) channels. It is quite obvious that the shortest wavelength channel can pump many channels simultaneously and hence is the most depleted one. This is equivalent to transfer of energy among channels. It is, in fact, known as Raman-induced cross-talk which is responsible for degradation of the system performance.

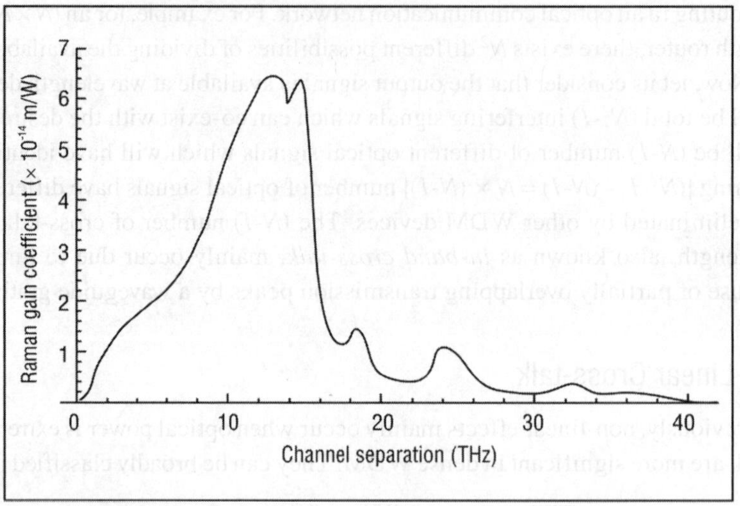

Fig. 7.33 Raman gain coefficient vs channel separation

Now the question arises, how we can avoid Raman cross-talk from occuring. It is possible, provided we make the output power of optical channels quite small. As a result, the SRS-induced optical gain may be almost nil for the whole fiber length under consideration. Also it can be further minimized if we insert suitable in-line optical filters which should be able to eliminate the low-frequency noise that may occur due to the longest used optical channel.

(ii) Non-linear SBS Cross-talk

First of all, we have to understand the basic similarities and dissimilarities between SBS and SRS. The phenomenon of Stimulated Brillouin scattering (SBS) is quite similar to stimulated Raman scattering (SRS) except that the shift in frequency is caused by acoustic waves instead of molecular vibrations. In this case the direction of propagation of the Stokes wave is just opposite to the direction of the incident light. If the optical pulse width is greater than 1 µs, then the SBS takes place at a relatively lower level of the input optical power. However, for short duration optical pulses (having width less than 1 µs), the SBS has almost no effect. There is a possibility that there is an energy transfer from a short-wavelength channel to a long-wavelength channel by SBS provided the channel separation is equal to the so-called Brillouin shift, which is about 10 GHz in 1550-nm region. So, it can be easily avoided provided two channels can counter-propagate for Brillouin amplification to occur. It may be noted that when all optical signals in various channels travel only in the forward direction, then SBS will not induce any type of inter-channel cross-talk. But rather, it severely limits the power handling capability of the channel. If we investigate the reason behind this, we find that a fraction of the optical channel power is transferred toward a backward-propagating Stokes wave which might have been produced due to noise under the threshold condition given by $g_B P_{th} L_{eff}/A_{eff} \approx 21$ is met. It does not depend on the number of other wavelength channels. But it is possible to satisfy this threshold condition for each wavelength channel even at lower levels of optical power. There is another way to suppress the SBS that is called the cross-phase modulation (XPM) which is usually produced by the adjacent channels. But it happens to be a source of cross-talk itself.

(iii) SPM- and XPM-based Non-linear Cross-talk

The performance of WDM system is degraded by both the self-phase modulation (SPM) as well as the cross-phase modulation (XPM). As mentioned earlier, there are two types of non-linear refraction-based phenomena - SPM and XPM. The main reason for SPM to occur is due to fluctuations in the optical signal's power. This, in turn, causes changes in the non-linear phase of the optical pulse signal, the extent of which can be determined by

$$\phi_{NL} = n_2 k_0 L |E|^2 \tag{7.13}$$

where, n_2 represents non-linear coefficient for refractive index, $k_0 = \frac{2\pi}{\lambda}$, L denotes the fiber length, and $|E|^2$ represents the optical intensity.

When the WDM network uses a number of optical amplifiers in a long-haul optic–fiber communication link, then the signal quality is degraded due to SPM-induced non-linear effects (i.e., chirping).

The primary reason for the occurrence of cross-phase modulation (XPM) is that the refractive-index depends on the light intensity. This results in phase shift in an optical signal which depends on the intensity as it travels through the optical fiber cable. Of course, the phase shift in the optical signal also depends on the power of adjacent channels. The second reason for the occurrence of XPM is due to a change in the intensity level of an optical signal which propagates at a different wavelength. In addition to GVD, XPM leads to reduction in SNR at the receiver. The usage of low-GVD fibers may lead to reduction in the XPM-induced cross-talk in WDM systems. However, this arrangement may result in four-wave mixing (FWM) phenomenon. The XPM has an advantage too. It finds application in wavelength convertors in which a pump signal is modulated at a specific wavelength from an intensity-modulated signal at a different wavelength.

(iv) FWM-induced Non-linear Cross-talk

Now let us turn our attention to another type of non-linear cross-talk that degrades the system performance. Consider two different wavelengths having frequencies f_1 and f_2. These two frequencies can mix together to generate sidebands having frequency components as $2f_1 - f_2$ and $2f_2 - f_1$. How can these sidebands cause interference? Interference occurs due to overlapping with those frequency components that are actually employed for transmission of data. Is it possible to minimize them? Yes, it can be minimized by using non-uniform spacing of the wavelength. Thus, four-way mixing (FWM) phenomenon usually happens whenever different wavelength channels travel with the same phase for longer duration. This results in the generation of new signals at the same frequency spacing as the original. The effect of FWM will be more if the channels are closer to each other or if each channel has more power. The phase-matching condition is satisfied for relatively smaller channel spacing. This can happen only in case a WDM system is made to operate quite nearer to λ_{ZD} of dispersion-shifted fiber. FWM-induced non-linear cross-talk can result in

- Coherent in-band cross-talks for equally spaced channels.
- Incoherent out-of-band cross-talks when channels are not spaced equally.

So, clearly there are two main reasons for the degradation in the system performance. One, obviously, is the attenuation in the channel power. And the other one, which tends to more severe, is due to presence of coherent non-linear cross-talk.

The impact of non-linear processes such as SRS, SBS, SPM, XPM, and FWM depend on three factors:

- the optical power level in the fiber
- the effective length of the fiber
- the effective cross-sectional area of the fiber core

The optical power level decreases along the fiber length due to attenuation. The effective length of the fiber takes into account optical power absorption along the fiber length and depends on the attenuation per unit fiber length α (dB/km). The effective length of the fiber L_{eff} is given by

$$L_{eff} = \frac{1 - e^{-\alpha L}}{\alpha}$$

We know that for a specified optical power level, the light intensity in a fiber is inversely proportional to the cross-sectional area of the fiber core. Due to increase in the effects of non-linearities with the light intensity in the fiber, the non-linearities decrease with the cross-sectional area of the fiber core. Moreover, there is non-uniform distribution of the optical power across the cross-sectional area of the fiber core, an effective cross-sectional area of the fiber core A_{eff} can be used. Table 7.5 shows standard values of effective cross-sectional areas of the fiber core for different types of fibers.

Table 7.5 Standard cross-sectional areas of fiber core

S. No.	*Type of Fibers*	*Standard Cross-Sectional Area of Fiber Core (μm^2)*
1.	Non-dispersion-shifted single-mode fibers	80
2.	Dispersion-shifted fibers	55
3.	Dispersion-compensating fibers	20

Let us turn back our attention to finding ways to minimize the FWM-induced degradation in the performance of WDM systems. If WDM systems are designed with reduced channel power and unequal channel spacing, then it is possible. But there are many WDM components and devices such as waveguide-grating routers and tunable optical filters that require equal channel spacings. A practical solution to this issue has been addressed by using non-zero dispersion shifted fibers (NZDSFs) in WDM systems.

7.12.3 Link Budget Design

A simplex point-to-point optic–fiber link comprises of an optical source, optical fiber, and an optical detector. To analyze a link, it is necessary to have the desired transmission distance, the specified bit-error-rate (BER), and the channel bandwidth or data rate. The characteristics of optical source (LED or laser diode) include emission wavelength and pattern, spectral line width and number of propagation modes, output optical power, and effective emission area. An optical fiber can be either single-mode or multimode having parameters such as core size and refractive-index profile, numerical aperture or mode-field diameter, attenuation, and bandwidth or dispersion. An optical detector can be p–i–n or avalanche photodiode type with designated operating wavelength, sensitivity, responsivity and speed.

In order to ensure the desired fiber–optic link performance, it is essential to carry out the budget analysis of the link power as well as the system rise-time. For a point-to-point optical fiber communication link, an optical *power-loss model* states that the actual optical power received at the input port of the photodetector is dependent upon several factors. These include the amount of optical power launched into the optical fiber by an optical source, the fiber loss, the losses due to connectors and fiber splices, etc. In addition, a link margin of about 6–8 dB is generally used for systems to include losses due to aging of various components used in the fiber–optic link, variations in operating temperature, and transmission losses due to any other component added in the link at a later stage. Hence, total loss in optical power permissible between the output of an optical source and the input of the photodetector of an optic–fiber communication link is given as

$$P_T = P_S - P_R = 2l_c + \alpha_f L + \text{link margin} \tag{7.14}$$

where, P_T represents the total loss in optical power in an optic–fiber link, P_S represents the optical power at the transmitting end of the fiber cable connected with the optical source, P_R represents the sensitivity level of the optical receiver, l_c is the loss due to fiber connectors (one each at transmitting and receiving end of the optical fiber cable), α_f represents the specified fiber attenuation in dB/km, and L represents the transmission distance of the optic–fiber link.

A *rise–time budget analysis* is used to determine the limitation of an optic–fiber communication link due to dispersion. By definition, total rise time of the optic–fiber link is given as the square root of the sum of square of various rise times produced by optical transmitter, the group-velocity dispersion (GVD), the modal dispersion in case of multimode fiber, and the optical receiver to the degradation in original rise-time of the transmitted optical pulse. That is,

$$t_{link} = \sqrt{\sum_{i=1}^{N} t_i^2} = \sqrt{t_{tx}^2 + t_{GVD}^2 + t_{\text{mod}}^2 + t_{rx}^2} \tag{7.15}$$

where, $t_{GVD} \approx |D|\sigma_\lambda L$; D represents the dispersion parameter ns/(nm–km), σ_λ represents the half-power spectral width of the optical source, and L represents the fiber length.

$$\Rightarrow \qquad t_{GVD} = \frac{440 \times L^q}{B_0};$$

q is the parameter which ranges between 0.5 and 1, and B_0 represents the bandwidth (in MHz) of a 1-km length of the optical fiber cable.

$$\Rightarrow \qquad t_{rx} = \frac{350}{B_{rx}};$$

B_{rx} being the 3-dB electrical bandwidth of the receiver in MHz.

It may be noted that all the times are specified in nanoseconds.

7.12.4 Power Penalty

In general, power penalty represents an increase in optical signal power (usually expressed in dB) which is required to maintain the specified value of BER even in the presence of optical signal impairments such as linear cross-talk in WDM. Due to a specific impairment in the received signal, the signal-to-noise ratio (SNR) decreases.

It may be recalled that for p-i-n photodiode based optical receiver with Gaussian noise,

$$BER = Q\left(\frac{I_1 - I_0}{\sigma_0 + \sigma_1}\right) = Q\left(\frac{R(P_1 - P_0)}{\sigma_0 + \sigma_1}\right)$$

$$\because I_0 = RP_0,\ I_1 = RP_1 \quad (7.16)$$

where, the decision threshold is optimal, i.e. $I_{th} = \dfrac{R(\sigma_0 P_1 + \sigma_1 P_0)}{\sigma_0 + \sigma_1}$ (7.17)

Let us denote a factor $r = \dfrac{R(P_1 - P_0)}{\sigma_0 + \sigma_1}$ (7.18)

In the presence of linear cross-talk, consider that P_1', P_0', σ_0', and σ_1' denote the respective received optical power levels and standard deviations of noise power levels. Then, at the same SNR, we can write

$$r' = \frac{R(P_1' - P_0')}{\sigma_0' + \sigma_1'} \quad (7.19)$$

The power penalty (PP) in dB can be expressed as

$$PP = 10\log(r) - 10\log(r') = -10\log\left(\frac{r'}{r}\right) \quad (7.20)$$

$$\Rightarrow \quad PP = -10\log\left(\frac{\dfrac{R(P_1' - P_0')}{\sigma_0' + \sigma_1'}}{\dfrac{R(P_1 - P_0)}{\sigma_0 + \sigma_1}}\right) \quad (7.21)$$

When the thermal noise is dominant, then $\sigma_0 = \sigma_1 = \sigma_{th}$

Thus, noise is independent of the signal power. Therefore, $\sigma_0' = \sigma_1' = \sigma_{th}$

In this case, we can write

$$PP_{PIN-rec} = -10\log\left(\frac{P_1' - P_0'}{P_1 - P_0}\right) \quad (7.22)$$

Recall that for the APD receiver (which is shot noise dominant),

$$\sigma^2_{shot} = 2e(G_m)^2 F_A(G_m) RPB_e \quad (7.23)$$

Assume $\sigma_1 \infty \sqrt{P_1}$; $\Rightarrow$ $\sigma_1 = \alpha\sqrt{P_1}$, where α is constant

Let $P_0 \ll P_1$

If the threshold is zero, then $\sigma_0 \ll \sigma_1$

Therefore, $$PP = -10\log\left(\frac{\dfrac{RP_1'}{\sigma_1'}}{\dfrac{RP_1}{\sigma_1}}\right) = -10\log\left(\frac{\dfrac{P_1'}{\sigma_1'}}{\dfrac{P_1}{\sigma_1}}\right) \quad (7.24)$$

We have $\frac{P_1'}{\sigma_1'} \approx \frac{P_1'}{\alpha\sqrt{P_1'}} = \frac{\sqrt{P_1'}}{\alpha}$; $\frac{P_1}{\sigma_1} \approx \frac{P_1}{\alpha\sqrt{P_1}} = \frac{\sqrt{P_1}}{\alpha}$

$$\Rightarrow \qquad PP = -10\log\sqrt{\left(\frac{P_1'}{P_1}\right)} \tag{7.25}$$

Fig. 7.34 shows inter-channel and intra-channel cross-talk power penalty (dB) limited by thermal–noise as a function of cross-talk level (dB).

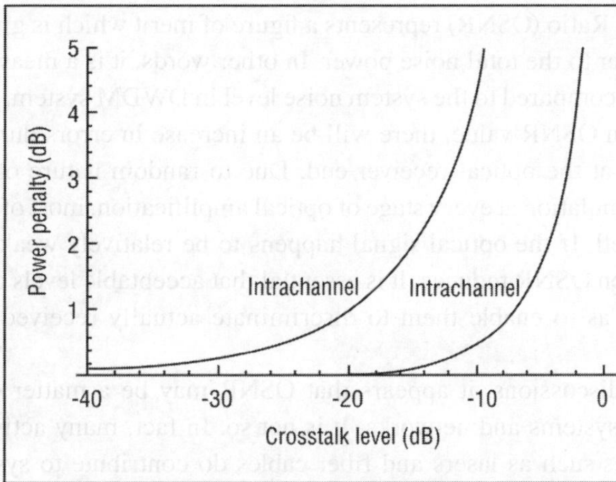

Fig. 7.34 Power penalty vs cross-talk level

It may be noted that signal-spontaneous noise limited power penalties for intra-channel as well as inter-channel cross-talk would be reduced by 50%.

Overall Design Considerations

- Single channel high speed systems use dispersion shifted fibers which is hard to use for WDM systems with an objective to upgrade the optic–fiber link capacity in the future because of system performance issues due to FWM. Hence, WDM systems use standard single-mode fibers or NZDSF.
- Need of chromatic dispersion compensation.
- Most systems use NRZ and ultra-long-haul systems use chirped RZ modulation.
- Reducing power or having large effective area can reduce the effect of non-linearities.
- Inter-channel spacing of ≈ 0.8 nm (equivalent to 100 GHz at 1550 nm) is common.
- For loop application, coarse WDM is used.
- We have to consider the bandwidth of optical amplifiers because the output power of amplifiers is limited to 20~ 25 dBm.
- Due to an increase in the number of wavelength channels, the input power for each channel decreases such that there is overall reduction in the total system span.
- We can use all-optical networks that are transparent in terms of bit rate, protocols, and modulation formats.

Facts to Know

The polarization of propagating optical signals through the optical fiber too plays a vital role to determine the amount of required power penalty. The worst condition occurs in case the interfering optical signals possess identical polarization state. In other words, we can say that the power penalty is high when the polarization is matched or out of phase.

7.12.5 OSNR-Based DWDM Design

Optical Signal-to-Noise Ratio (OSNR) represents a figure of merit which is given by the ratio of the total optical signal power to the total noise power. In other words, it is a measure of how strong the optical signal level is as compared to the system noise level in DWDM system. It is always expressed in dB. With reduction in OSNR value, there will be an increase in errors during the process of bit detection and recovery at the optical receiver end. Due to random nature of the noise as well as possibility of their accumulation at every stage of optical amplification, most of the optical amplifiers amplify the noise as well. If the optical signal happens to be relatively weaker (or, the noise level happens to be more), then OSNR reduces. It is essential that acceptable levels of OSNR are required by optical receivers so as to enable them to discriminate actually received optical signals from probable system noise.

From the on-going discussions, it appears that OSNR may be a matter of concern in optical amplifier-based WDM systems and networks. It is not so. In fact, many active and passive WDM components and devices such as lasers and fiber cables do contribute to system noise. This may severely create design issue in an OSNR-limited WDM system. We know that GVD poses a serious problem (in the form of inter-symbol interference due to optical pulse broadening, and consequently requirement of power penalty) for high transmission bit rate (greater than 2.5 Gbps) requirements in a single-mode WDM system. This may degrade the system's OSNR. In system design calculations, however, ASE noise due to optical amplifiers may still be considered as the most prominent source for degradation in OSNR and increase in power penalty. When the desired optical as well as the system noise are amplified simultaneously, the OSNR value signifies the quality of the received optical signal. Therefore, we can say that the optical system designs which are based on OSNR remain a vital design tool.

The design of an optic–fiber communication system must meet the specified BER requirements. The Q-factor of the received optical signal provides a qualitative measure of the receiver's overall performance because it is a function of the OSNR. The Q-factor describes the minimum required SNR value in order to achieve a specified BER figure for a given optical signal. Higher the value of Q-factor, better will be the BER. But it is rather difficult to calculate the BER value. Whereas the analysis based on Q-factor is relatively easier and convenient. The Q-factor is usually expressed in dB. It is related with OSNR as given by the following expression:

$$Q_{bB} = OSNR_{dB} + 10\log\left(\frac{B_0}{B_c}\right) \tag{7.26}$$

where, B_0 represents the optical bandwidth (in MHz) of the photodetector used at the optical receiver, and B_c represents the electrical bandwidth (In MHz) of the optical filter used at the receiver.

It can be seen that if $B_0 > B_c$, then $OSNR_{dB} > Q_{dB}$. For practical designs, $OSNR_{dB} > Q_{dB}$ by minimum 1–2 dB. For a relatively higher transmission bit rate system design, the receiver margin is normally kept as 2 dB. Hence, we can say that the Q-factor is directly proportional to the OSNR value to some extent. Optical spectrum analyzers are often used for performing noise calculations.

Facts to Know

It is highly recommended to consider OSNR for the channel having worst impairment (usually on either end of the given optical band) in a typical multichannel WDM network. If we use a cascaded arrangement of EDFAs, then there will be a continuous degradation in OSNR value due to its ASE and with transmission distance as well. In order to minimize OSNR degradation, distributed Raman optical amplifiers should be used.

For a particular optic–fiber link, we must calculate OSNR value and design the link considering both OSNR and limitations due to dispersion. We have earlier discussed that dispersion can be compensated to a large extent by employing dispersion management techniques. But the compensation for degraded OSNR can be achieved only by adopting 3R optical signal regeneration technique (i.e., Optical–Electrical–Optical, O–E–O). Due to its higher cost, this scheme is not recommended for the design of a multi-channel WDM system. For a WDM link, we first consider system limitations due to OSNR. OSNR-based WDM system design necessitates confirming that the achieved value of OSNR is quite near to the required value for an acceptable BER at the final receiver.

OSNR have a critical impact in DWDM link when multiple in-line optical amplifiers are deployed in the network due to their amplified spontaneous emission noise. Optical receivers have OSNR tolerance limit. If OSNR degrades, noise level increases, and the receiver has greater difficulty in decoding signal information. Consequently, it results in higher BER. In optical links with multiple amplification nodes, OSNR calculations need to be carried out along the way for each node in order to identify the point where OSNR degrades to the critical value. OSNR can be improved by introducing O–E–O regenerator. We can say that OSNR is more crucial for DWDM systems as compared with SDH systems.

Note: Internet Protocol (IP) over DWDM deals with transmission of data packets using an optical layer in an all-optical DWDM network for its operation and capacity. It has the capability to support bit-rates of Optical Carrier OC-192 and higher standards. In IP over DWDM system, the transport layer (also called the open architecture) is all-optical with protocol transparency. This helps in increasing bandwidth, maintaining a high data rate with reduced latency. It offers a new era in an optical networking.

7.12.6 Other Design Issues

In all-optical WDM networks, transmission impairments at the physical layer is accumulated and poses a challenge to provide reliable fiber–optic link. It is necessary to maintain the value of signal-to-noise ratio above the specified threshold limit at the optical receiver end. For mitigation of transmission impairments such as four-wave mixing components, wavelength-routing algorithm and physical layer impairment-aware routing algorithm are required to be implemented. In WDM systems, there are number of other design issues related to transmitter and receiver characteristics. These are given below:

- Stability of wavelength associated with each channel (hence the channel spacing) against variations in temperature, ageing, etc.
- Loss of signal power due to various reasons such as transmission loss, insertion loss, and distribution loss. Use of optical amplifiers, however, can compensate for these losses
- Requirement of gain flatness of optical amplifiers over the whole optical spectrum of the WDM signals
- Power management in WDM networks
- Build-up of amplifier noise when WDM signal is processed by number of optical amplifiers
- Dispersion (i.e., pulse broadening) as the WDM signal propagates along the length of the optical fiber. It severely limits the spacing between bits, maximum bit rate, and hence the maximum transmission distance for a specified bit rate. So there is need of effective dispersion-management techniques for WDM networks
- Number of wavelengths to use that decide total optical band occupied and the optical bandwidth of various WDM devices
- Optical fiber as the medium of optical signal propagation at 1300 nm and 1550 nm with 200 nm bandwidth each
- Optical amplifier with 35–40 nm bandwidth
- Injection–current laser with 10 nm tuning range
- Fabry–Perot optical filter having entire low-attenuation region and tuning range, e.g., Electro–optic filter with 16 nm tuning range
- Channel spacing and transmission bit rates
- Link budget and rise-time budget
- Non-linearities in the optical fiber
- Resolution of optical transmitters and optical receivers used in WDM systems
- Higher network capacity by using more number of channels may lead to higher network costs and more complex protocols
- Power considerations including signal-to-noise ratio

Facts to Know

For next generation wide-area backbone communication networks, all-optical WDM optic–fiber networks that employ wavelength routers are under consideration. They involve a number of inter-connected wavelength routers (each being able to support several wavelength channels).

7.13 WDM Soliton Systems

We have seen that the phenomenon of dispersion imposes a serious concern for long-haul optic–fiber communications networks operating at high transmission bit rate. The GVD limits its information-carrying capacity. With an objective to overcome this problem, we can select a highly stable optical pulse having suitable shape for transmission through optical fiber communication systems. Such type of optical pulse is known as a soliton. Thus, *an optical soliton* is a pulse that preserves its shape and other characteristics over a long distance. In soliton-based optic–fiber systems, the effect of GVD is counteracted by a phenomenon known as self-phase modulation (SPM). WDM soliton systems are, therefore, capable of operating at high bit rate with large information-carrying capacity over several thousands of kilometers distances by employing required number of optical amplifiers.

It may be noted that an ideal soliton can maintain a constant dispersion in an ideal (lossless) optical fiber cable. Practically, due to varying dispersion and fiber loss, the quality of the soliton pulse is severely degraded with the transmission distance. However, with an introduction of suitable dispersion compensation schemes in soliton-based systems, known as dispersion-managed soliton technique, these problems have been overcome to a large extent. This has resulted in an enhancement of dispersion tolerance as well as the system power margin.

WDM soliton systems essential need an optical source at the transmitting end which can emit frequency chirp-free ultrashort pulses (having duration of the order of picoseconds) at a relatively higher repetition rate > the shape of soliton pulse should be such that it resembles the *sech* (a squared hyperbolic secant function) waveform. It is also required that such an optical must generate solitons in the 1550 nm optical band. One of the technique employed to generate soliton pulses having 20–30 ps width is the use of gain switching which is obtained when the laser is biased below its specified threshold level and pumped regularly at higher threshold level. However, due to variations in the refractive index (determined by laser's specified linewidth enhancement factor), each optical pulse at the output of the optical source happens to be frequency-chirped. One of the possible solution to this problem is to use mode-locked semiconductor lasers in which the sequence of nearly chirp-free optical pulses is generated. In order to allow mode-locking of the laser, the grating provides a self-tuning mechanism for operation over a wider modulation frequency range. With this arrangement, soliton like pulses having 12–18 ps duration at 40 Gbps repetition rate can be produced.

A tunable Raman fiber optical amplifiers in 1620–1660 nm optical band can be used with solitons having femtosecond duration. This enhances the system capacity significantly. But during the femtosecond pulse duration region, the stimulated Raman scattering contributes to the higher-order non-linear distortion. This may result in a unstable propagation of such solitons along the length of the optical fiber. Therefore, it is recommended to use practical solitons having about 1 ps duration. The intensity and duration of soliton can be preserved by using an adaptive feedback that can control the Raman frequency shift. Erbium-doped fiber lasers can generate solitons having 30 ps pulse duration.

It is quite obvious that in long-haul optical solitons communication links, the soliton energy reduces due to fiber losses. A reduced peak power of the soliton, in turn, may weaken the SPM effect which would have counteracted the impact of GVD. As a result, there would be broadening of the soliton. Therefore, in order to maintain the intensity of the soliton as it moves along the fiber length, it must be amplified regularly with the help of either distributed or lumped optical amplification mechanisms. Lumped optical amplification mechanism is employed when the physical separation between in-line optical amplifiers is less than dispersion fiber length. However, this scheme is not recommended for systems operating at relatively higher transmission bit rates (> 10 Gbps). For such systems, a distributed amplification scheme is preferred which is based on fiber loss compensation locally at regular spacing. It uses Raman fiber optical amplifiers to achieve distributed gain with pumping the fiber carrying the signal at $\lambda \sim 1480$ nm. An alternative approach to obtain distributed gain is by doping the fiber lightly with Er ions and regular pumping. This makes solitons to propagate over long distances within these active fibers.

It must be remembered that the dispersion-managed solitons do not undergo a large change in its non-linear phase. In such situations, the solitons behave just like a linear optical pulse of shorter duration. It is possible to compensate for the soliton broadening, if any, due to positive GVD by employing equivalent negative GVD. Over the average soliton period, however, the soliton regains its basic characteristics. For small dispersion problems, non-linear Schrödinger equations fully describe the system; whereas for large dispersion problems, the soliton almost resembles with a Gaussian pulse having frequency chirp.

With the conventional non-return-to-zero (NRZ) signaling format, high bit rate transmissions are not possible for longer distances. But dispersion-managed solitons provide a viable solution due to reduction in distortion which may arise because of various non-linear effects.

One of the serious concerns with soliton transmissions in long-haul high bit rate optic–fiber systems is soliton–soliton interaction which may change GVD along the length of the fiber. Due to this, Gordon–Haus jitter occurs, which has a cubic dependence on the propagation distance that limits soliton transmission distances. However, it is possible to overcome these problems by implementing appropriate dispersion management schemes in the region of normal dispersion, and applying synchronous modulation for optical pulse retiming and shaping. This is known as soliton control. It also enables to retime the position of the soliton pulse which has been affected by the jitter due to ASE noise of optical amplifiers.

Facts to Know

Nowadays soliton-based systems along with EDFAs are being deployed in applications that are required to transmit data at a very high bit rate over much longer distances. They can provide extremely high information carrying capacity (of the order of multi-Gbps) without the use of regenerators. Soliton-based optical switches are used for optical computation purpose. In soliton-based WDM systems along with optical amplifiers, ultrahigh speed communication superhighways with much higher transmission bit rates (of the order of several Tbps) can be obtained.

Points to Remember

- In fiber–optic communications, transmitting number of wavelengths (optical channels) simultaneously on to the same optical fiber is termed as wavelength division multiplexing (WDM).
- WDM enables capacity upgrade of existing optical network without adding optical fibers.
- WDM systems and networks require a wide range of optical components to generate and combine the multi-wavelength optical signals, transport and amplify these signals as they traverse the optical fiber network, and then separate and receive these signals as they reach their respective destinations.
- In a WDM system, a tunable optical filter is mostly used for selection of a desired wavelength channel.
- A Fabry–Perot (FP) interferometer tunable optical filter comprises of a cavity that is formed by using two mirrors having its length electronically controlled with the help of a piezoelectric transducer.
- A Mach–Zehnder (MZ) interferometer tunable optical filter can be formed by connection two output ports of the first optical coupler with output ports of second optical coupler
- Fiber Bragg grating based Michelson interferometer optical tunable filters can be designed using a distributed Bragg reflector (DBR) structure.
- In Acousto–optic tunable filters (AOTFs), acoustic waves are used to form the grating dynamically and the acousto-wave frequency is changed to realize tuning.

- In Electro–optic tunable filter (EOTF), the fiber Bragg grating is created by an Electro–optic effect using Lithium Niobate $LiNbO_3$ material.
- A WDM MUX combines several wavelength channels into one fiber, whereas a WDM DEMUX separates several wavelengths available in one fiber into individual wavelengths.
- Grating-based DEMUX is based on the principle of Bragg diffraction. When the grating period is adjusted by using the Bragg condition, $2\Lambda n_{eff} = \lambda_B$ to a specific wavelength, the optical gratings is formed.
- WDM add–drop multiplexer (ADM) is used to add or drop one or more wavelength channels while maintaining the integrity of existing wavelength channels. It can be configured using two 3-port optical circulators with fiber Bragg gratings.
- A star coupler combines number of wavelength channels available at its multiple input ports, and then splits them accordingly at the multiple output ports. It does not contain any wavelength-selective elements.
- A wavelength converter is used to change the wavelength of the optical signal available at its input port while maintaining the data integrity.
- Wavelength converters can be realized by several mechanisms such as Opto–electronic regenerators, cross-gain saturation and cross-absorption in SOAs, XPM-based MZ Interferometers, and FWM-based SOAs.
- Wavelength routers is a WDM device that is used to combine the function of an optical star coupler with MUX and DEMUX operation.
- Optical cross-connects (OXC) provides wavelength routing technique to achieve reconfiguration of the wide-area WDM network ensuring transparent operation.
- WDM transmitter generally comprises of laser array (each laser is a tunable laser which can be tuned to desired fixed wavelength) and an optical multiplexer which is tunable across a range of wavelengths.
- WDM transmitters can employ analog modulation or digital modulation schemes by employing either direct or external modulation.
- A WDM receiver basically comprises of an optical demultiplexer (DEMUX), and an array of photodiodes, each operating at its designated wavelength.
- In the design of WDM networks, the most important system performance issue is the inter-channel cross-talk which consists of linear and non-linear cross-talk.
- There are several non-linear effects such as SRS, SBS, XPM, SPM, and FWM that occur in optical fibers which can degrade the system performance due to inter-channel cross-talk caused by them.

Important Equations

The wavelength separation in WDM = $\dfrac{\text{Frequency separation} \times (\text{wavelength})^2}{\text{speed of light}}$.

The optical bandwidth or the deviation in frequency, $|\Delta\upsilon| = \left(\dfrac{c}{\lambda^2}\right)|\Delta\lambda|$; *where $\Delta\lambda$ represents the wavelength deviation around central wavelength λ.*

Free spectral range in FP tunable optical filter, $FSR = \Delta v_L = \dfrac{c}{2n_g L}$; *where n_g represents the group index of intra-cavity material and L represents the length of FP filter.*

The finesse of FP filter, $F = \dfrac{\pi\sqrt{R}}{\sqrt{(1-R)}}$, *where R represents the mirror reflectivity.*

The guide length of waveguide coupler, $L = \frac{\pi}{2\kappa}(m+1)$ with $m = 0,1,2,.....$; *where κ is the coupling coefficient which is almost monotonically proportional to wavelength.*

Number of 3-dB optical couplers needed to form an N × N star coupler,

$$N_c = \frac{N}{2}\log_2 N = \frac{N}{2} \times \frac{\log N}{\log 2}$$; *where log 2 = 0.301.*

Splitting loss of star coupler $= -10\log\left(\frac{1}{N}\right) = 10\log N$; *where N is number of input ports.*

Excess loss of the $N \times N$ star coupler $= -10\log\left(F_T^{\log_2 N}\right)$; *where F_T is the fractional power passing through each 3-dB optical coupler element with $0 \le F_T \le 1$ (i.e., a fraction $1 - F_T$ of power is lost in each 2 × 2 element).*

Total loss of the $N \times N$ star coupler $= 10(1 - 3.322\log F_T)\log N$; *where F_T is the fraction of power traversing each 3-dB coupler element with $0 \le F_T \le 1$.*

The length difference in MZI multiplexer arms, $\Delta L = \frac{1}{2n_{eff}\left(\frac{1}{\lambda_1} - \frac{1}{\lambda_2}\right)} = \frac{c}{2n_{eff}\Delta\upsilon}$; *where n_{eff} represents the effective refractive index, λ_1 and λ_2 are two wavelengths at two input ports of a basic 2 × 2 MZI MUX, and $\Delta\upsilon$ represents the frequency separation of two wavelengths.*

The difference in path-lengths of adjacent waveguides, $\Delta L = m\frac{\lambda_c}{n_c}$; *where the integer m represents the diffraction grating order, λ_c represents the central wavelength for the propagation path from the center of the input waveguide to the center of the output waveguide, and n_c represents the refractive index of the grating array waveguides in phase–array based WDM devices.*

Output optical power at the photodetector, $P = P_m + \sum_{n \ne m}^{N} T_{mn}P_n$; *where P_m represents the optical power in the desired m^{th} channel, T_{mn} represents the optical filter transmissivity for channel# n when channel# m is chosen, and N is total number of channels incident on filter.*

Amount of non-linear phase-shift contributed by SPM, $\phi_{NL} = n_2 k_0 L|E|^2$; *where n_2 is non-linear coefficient for the refractive index, $k_0 = \frac{2\pi}{\lambda}$, L represents the fiber length, and $|E|^2$ represents the optical signal intensity.*

The Bragg wavelength $\lambda_B = 2\Lambda n_{eff}$; *Λ represents the grating period (i.e., the distance between two adjacent maximum points of the periodic refractive index), and n_{eff} represents the effective refractive index of the fiber core.*

Key Terms with Definitions

Active optical components Devices used in WDM network require external power to be functional.

Active optical network Uses electrically powered switching equipment to manage signal distribution and direct signals to specific destinations.

ADM	Add–Drop Multiplexer used in WDM for adding or dropping one or more wavelength channels are dropped or added while maintaining the integrity of other wavelength channels.
AOTF	Acousto–optic Tunable Filter in which acoustic waves are used to form the grating dynamically.
Broadband WDM	WDM that uses the 1300-nm and 1550-nm wavelengths for full-duplex transmission with wider channel spacing of about 20 nm (equivalent to 100 GHz).
Coarse WDM	Same as broadband WDM.
DEMUX	A WDM demultiplexer that separates several wavelengths available in one fiber into individual wavelengths.
DWDM	Dense WDM, or narrowband WDM that can multiplex 4, 8, 16, 32, or more number of different wavelengths in the 1530 nm to 1610 nm optical band having a very narrow channel spacing of about 0.8 nm (equivalent to 25 GHz).
EOTF	Electro–optic Tunable Filter in which an Electro–optic effect using lithium niobate $LiNbO_3$ is used to create the fiber Bragg grating.
FBG	Fiber Bragg grating works as a mirror, selectively reflecting Bragg wavelength only, and thus transmitting all the other wavelengths of the optical signal.
FP interferometer filter	A Fabry–Perot interferometer tunable optical filter that comprises of a cavity that is constructed by using two optical mirrors at its ends, with its length being controlled electronically with the help of an external piezoelectric transducer.
Grating-based Michelson Filter	A fiber Bragg grating functions like a reflection filter in which its middle wavelength is adjusted by varying the grating period. Its bandwidth is controlled either by varying the grating strength or by small chirping of the grating period.
Hetero-wavelength	Out-of-band wavelength.
Homo-wavelength	In-band wavelength.
Inter-band cross–talk	The interference from signals on different wavelengths that affects channel spacing.
Interferometry optical filter	Optical filters using an interferometer which have frequency-dependent transmission characteristics and is quite sensitive to the input wavelength.
Intra-band cross-talk	A type of interference which arises from optical signals having same wavelengths but propagating on adjacent fibers in WDM.
LAN	Local Area Network – A broadcast star topology that is usually employed to combine multiple channels over a relatively small geographical area.
MAN	Metropolitan Area Network – Formed by connecting several LANs with the help of passive wavelength routers.
MUX	A WDM multiplexer that combines several wavelength channels into one optical fiber.
MZ interferometer filter	A Mach–Zehnder interferometer tunable optical filter that can be formed by connecting both ports of an optical coupler to both ports of another optical coupler.
OXC	Optical cross-connects that provides wavelength routing scheme in order to reconfigure wide-area WDM network, retaining its transparent nature.
Passive optical components	Devices used in WDM network that do not require any type of external power for their operation.
Passive optical network	Networks that primarily use optical splitters for the purpose of separating and collecting optical signals throughout the network.

Power penalty Additional power that is required at the optical receiver of a WDM network in order to mitigate the degradation due to non-linear cross-talk.

Single-hop all-optical WDM network A fully-connected network, also known as mesh technology or broadcast star technology, in which there is direct connection among all nodes.

Star coupler Combines the optical signals available at its numerous input ports and then divides the optical signal in equal proportions among the output ports.

Tunable optical filter Active WDM component that selects a desired channel at the receiver.

WAN Wide Area Network– Several MANs are connected in a WAN employing mesh topology to connect all nodes.

Wavelength converter A WDM component used for conversion of wavelength available at its input port to another wavelength at its output port while ensuring that data integrity is preserved.

Wavelength router A WDM component which can combine the functions of an optical star coupler with optical MUX and DEMUX operations.

WDM Wavelength Division Multiplexing which refers to simultaneous propagation of several optical signals having different wavelengths using the common optical fiber cable, without causing any interference among them.

WDMA Wavelength Division Multiple Access in an all-optical WDM network which permits the use of channel wavelength for optical routing, optical switching, or dividing each wavelength channel to desired optical receiver.

WDM receiver Comprises of an optical demultiplexer (DEMUX) and an array of photodiodes, each one operating at its designated wavelength.

WDM transmitter Comprises of laser array (each laser is a tunable laser which can be tuned to desired fixed wavelength) and an optical multiplexer which is tunable across a range of wavelengths.

Short Answer Type Questions

1. **What are the various types of network mediums used in WDM transmission system? Give a simplified functional block schematic diagram of a WDM system.**

The network medium may be an optical fiber link in WDM transmission system. For broadcast application, it may be a passive star coupler (PSC). It can also be a combination of fiber links and a network of optical or electronic switch. A simplified functional block schematic diagram is depicted in Fig. 7.35.

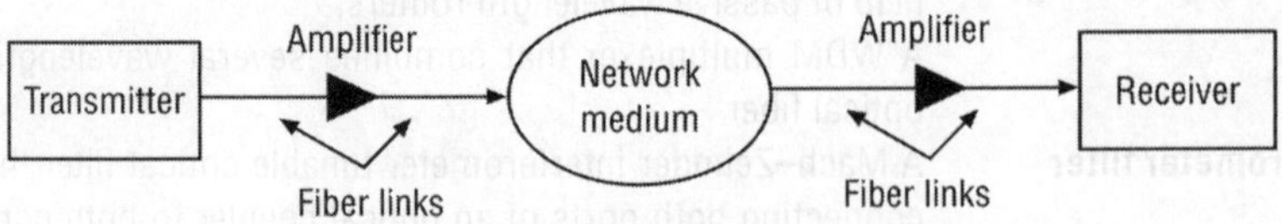

Fig. 7.35 A typical WDM transmission system

The transmitter comprises of one or many optical transmitters. It can operate either using a fixed wavelength, or tunable wavelengths over a wide range of wavelengths. An optical transmitter basically comprises of an optical source such as a laser, an external modulator, or/and a tunable optical filter. An optical MUX is normally used to combine several optical if multiple optical transmitters are used. The optical receiver comprises of a combination of a tunable optical filter, photodetector, and an optical demultiplexer (if needed with an photodetector array). Optical amplifiers are needed in various locations along the fiber link in order to compensate for the fiber losses.

2. Draw a functional block schematic of a typical WDM system with brief description of each functional block.

Fig. 7.36 shows a functional block schematic of a typical WDM-based system.

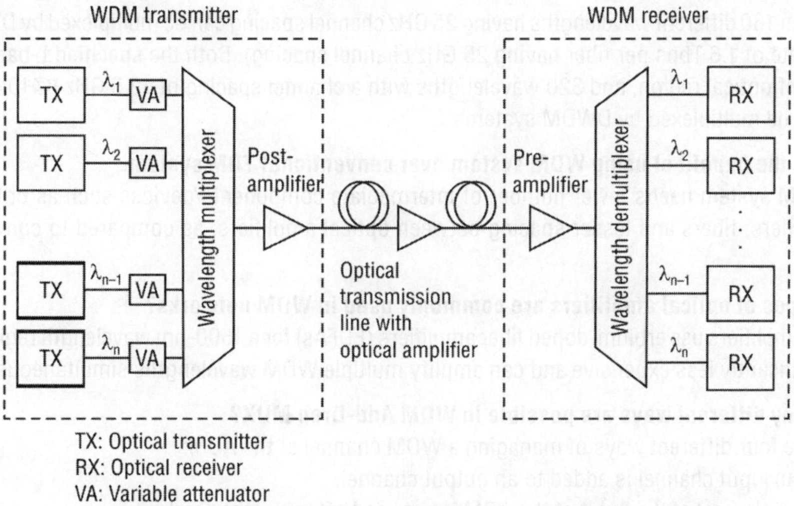

Fig. 7.36 Functional block schematic of a WDM-based system

At the transmitter end of the WDM, many optical channels are multiplexed using an optical MUX unit. The resultant WDM signal is then amplified by an optical amplifier, and then coupled with the optical fiber cable. At the WDM receiver end, the received WDM signal is again amplified by a pre-optical amplifier, followed by demultiplexing by optical DEMUX unit and then sent to their respective receivers. Also, the optical amplifiers are deployed as in-line amplifiers to amplify the optical signal which helps to compensate for the fiber loss. It also boosts the optical power at transmitting and receiving ends.

3. Summarize the basic principles of Dense WDM (DWDM).

(1) Bandwidth of a modulated laser: 10–50 MHz (i.e., 0.001 nm)
(2) Guard band: 0.4–1.6 nm (typical)
(3) Typical spectral band, for example, 120 nm @ 1550 nm and 80 nm @1300 nm band
(4) Discrete wavelengths form different channels for modulation, routing and switching
(5) Requirement of various types of active and passive devices

4. What are benefits and limitations of DWDM in optical networks?

One of the major benefit of DWDM is to provide very high capacity optical networks. Theoretically a very large number of individual channels can be propagated simultaneously in an optical fiber. DWDM networks can be physically realized using precise wavelength selective devices. Practically, wavelength selective (optical signal processing) components and non-linear effects limit the system performance. There are certain passive signal processing devices such as fiber Bragg gratings which are effectively employed in WDM networks. Moreover, optical amplifiers are must to provide long transmission distances without repeaters.

5. **List the range of wavelengths in optical S, C, L and U-band.**
Optical S-band ≈ 1480-nm wavelength region
Optical C-band ≈ 1550-nm wavelength region (1540–1570 nm)
Optical L-band ≈ 1574–1608 nm wavelength region
Optical U-band ≈ 1650-nm wavelength region

6. **How many wavelengths are multiplexed in DWDM systems? What is the typical bit rates that can be achieved in an optical fiber?**
More than 160 different wavelengths having 25 GHz channel spacing can be multiplexed by DWDM systems (at the rate of 1.6 Tbps per fiber having 25 GHz channel spacing). Both the specified L-band and C-band spectra of optical region, and 320 wavelengths with a channel spacing of 12.5 GHz (@10 Gbps per fiber can also be multiplexed by DWDM systems.

7. **Mention the benefit of using WDM system over conventional TDM system.**
The WDM system needs fewer number of intermediate components/devices such as optical repeaters or amplifiers, fibers and lesser spacing between optical amplifiers, as compared to conventional TDM system.

8. **What types of optical amplifiers are commonly used in WDM networks?**
Optical amplifiers use erbium-doped fiber amplifiers (EDFAs) for a 1500-nm wavelength region. The EDFAs are significantly less expensive and can amplify multiple WDM wavelengths simultaneously.

9. **How many different ways are possible in WDM Add-Drop MUX?**
There are four different ways of managing a WDM channel at the ADM:
   - Add: An input channel is added to an output channel.
   - Drop: A channel at the input of the ADM is dropped off to another node.
   - Through: This channel is a straight pass-through WADM.
   - Drop-and-Continue or Bridge: This configuration allows payload to be dropped off and also passed downstream.

10. **Mention some special requirements of a WADM device.**
Due to the diverse customer mix, each WADM channel should be capable of carrying a different data rate and channel mix. The ADM must be able to demodulate each wavelength from the composite signal, and drop, pass through, or add the wavelengths, as required.

11. **What is the utility of WADM in optical networks?**
As the capacity of optical systems increases and the system's users are located in different geographical regions, WADM provides great flexibility in bandwidth management in a fast, efficient and cost-effective manner to meet the customers' requirements.

12. **State the basic function of wavelength converters.**
Wavelength converters converts the input wavelengths available on different fibers connected to its input and can be programmed to modify the wavelength and output modified wavelength. There are different types of wavelength converters available such as Full Range Wavelength converters (FWC), Limited Range Wavelength converters (LWC). To reduce cost, one can share wavelength converters among fiber links.

13. **How do arrayed waveguide grating routers function?**
Arrayed Waveguide Grating Routers (AWGRs) are primarily passive optical devices which can reroute wavelength channels within optical fibers. They have *M* number of inputs and *M* number of outputs. They can process different wavelengths ranging from 0 to *M - 1*. The wavelength channel *i* at input port *j* is routed to the same wavelength at output port (*i - j*) mod *M*. They are easily available and inexpensive.

14. How can two light paths having identical wavelengths share a common optical fiber link?

In WDM, we experience a constraint of wavelength continuity that makes optical networks different from circuit-switched telephone networks. This means that two light paths that share a common fiber link should not be assigned the same wavelength. But there is a solution. By using wavelength converters, we can two light paths having same wavelength share a common fiber link.

15. Switching speed is the main problem at the heart of the optic–fiber communication network infrastructure. Suggest an appropriate solution.

The solution lies in the design of inexpensive WDM cross-connect (WXC) which are fast and easily scalable. Fig. 7.37 illustrates the basic concept of WDM cross-connect.

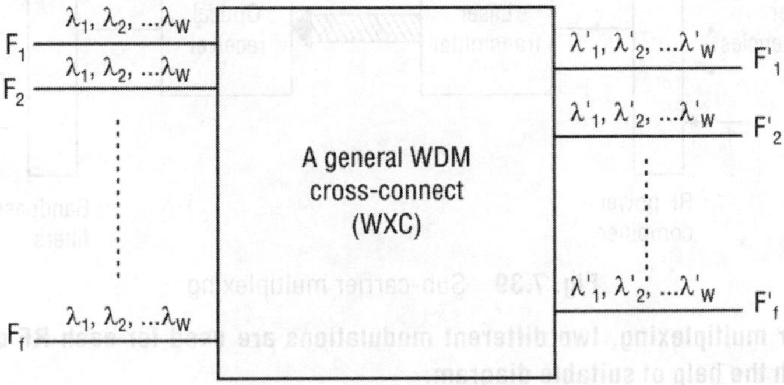

Fig. 7.37 Basic concept of WDM cross-connect

16. What is meant by sub-carrier multiplexing (MUX)?

In sub-carrier multiplexing technique, each modulated RF carrier signal may serve as an independent sub-carrier. The unmodulated optical signal acts as the main carrier signal for modulation purpose. Frequency division multiplexed (FDM) multichannel systems is known as sub-carrier modulation. Fig. 7.38 shows the basic concept of sub-carrier multiplexing.

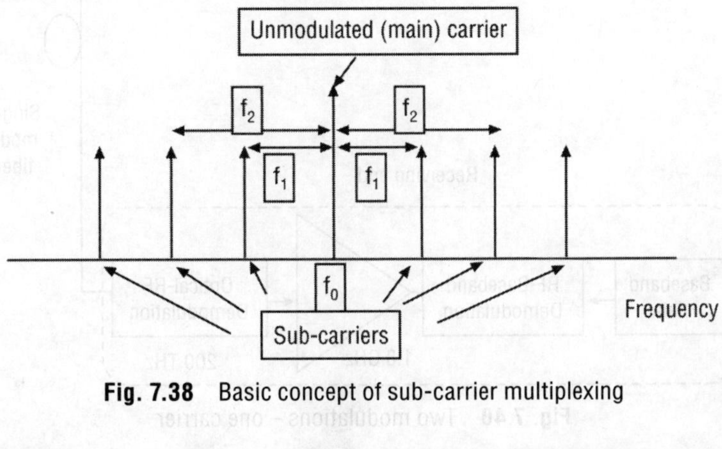

Fig. 7.38 Basic concept of sub-carrier multiplexing

17. Draw the functional block schematic of sub-carrier multiplexing used in CATV distribution.

Fig. 7.39 illustrates the functional block schematic of sub-carrier multiplexing over a common optical channel.

A number of modulated carrier frequencies (f_1, f_2, ….., f_N) are first combined in RF power combiner, called frequency division multiplexing (FDM). The composite FDM signal is then applied at the input of laser transmitter in which the signal in electrical domain is converted to the optical signal. It is transmitted onto a common fiber–optic channel which is received by optical receiver. The output of optical receiver is composite FDM signal in electrical domain. The individual signal is then separated using bandpass filters.

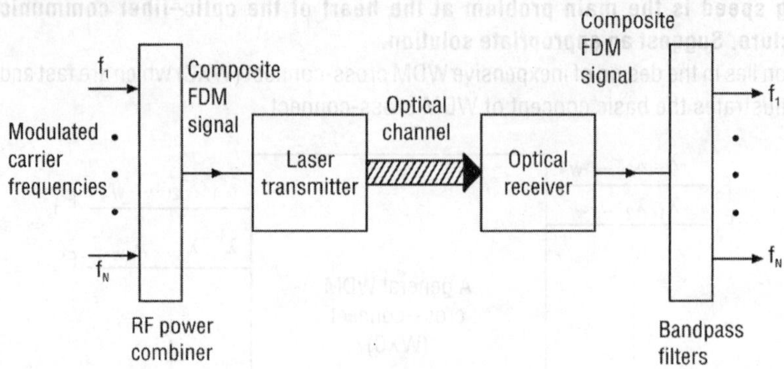

Fig. 7.39 Sub-carrier multiplexing

18. 'In sub-carrier multiplexing, two different modulations are used for each RF carrier'. Justify this statement with the help of suitable diagram.

Fig. 7.40 shows the concept of using two different modulations (Baseband-RF modulation and RF-Optical modulation) in sub-carrier multiplexing in optic–fiber communications.

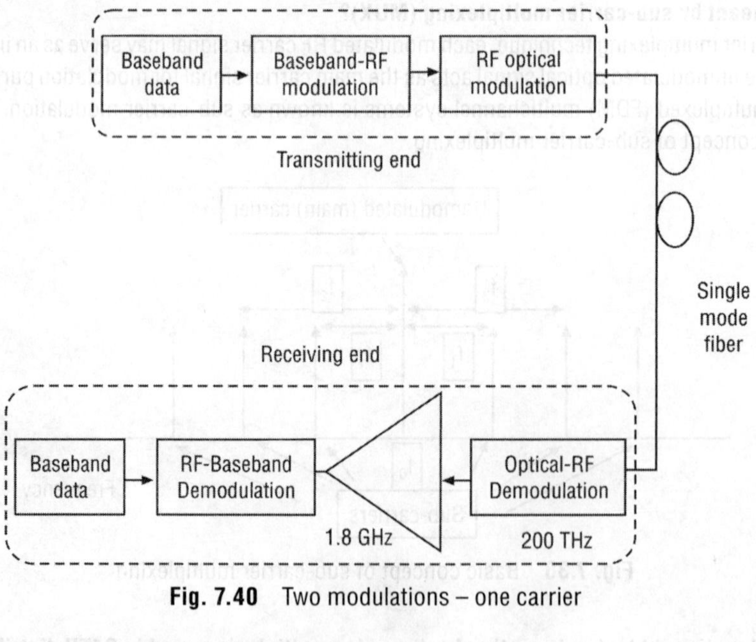

Fig. 7.40 Two modulations – one carrier

19. List the pros and cons of sub-carrier multiplexing.

- Both analog as well as digitally-modulated sub-carriers can be used.

- Voice, digital audio, high-definition video or data can be carried by each RF carrier.
- Different modulation techniques can be used on RF carriers.
- It is not easy to analyze the performance of sub-carrier multiplexing.

20. What are the types of analog modulation techniques used in CATV distribution?

The frequency spectrum of 50–88 MHz as well as 120–550 MHz is assigned for CATV distribution application. The information is either amplitude modulated (AM) or frequency modulated (FM) on RF carrier which is followed by intensity modulation by the laser. That is, it uses either AM or FM (analog modulation techniques) for RF electromagnetic signal to optical signal conversion. Use of AM leads to simpler implementation but SNR > 40 dB for each channel is needed in addition to high linearity. FM offers better SNR and less linearity requirement.

21. Why is multimode fiber not suitable for WDM systems?

Many different electromagnetic modes like TE01, TM01 that remain quite stable during propagation of light within the optical fiber cable. This makes too wide spectrum, as depicted in Fig. 7.41.

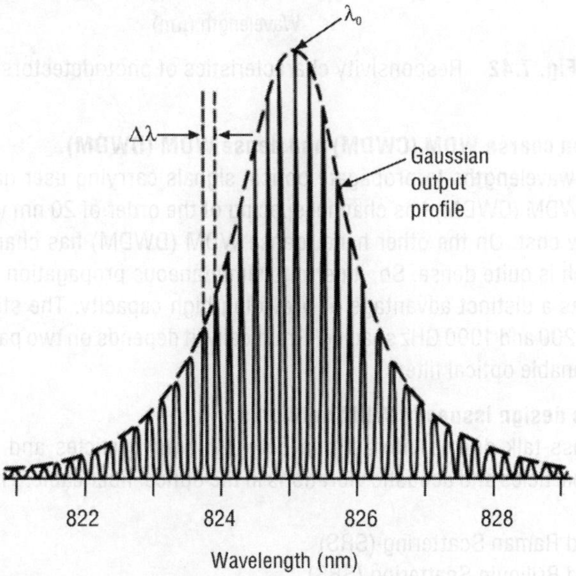

Fig. 7.41 Multimode laser spectrum

This is the reason that multimode laser as an optical source is not generally used in a DWDM system.

22. With the help of responsivity versus wavelength characteristics of photo detectors, show that narrow band optical filters are necessary for separation of different wavelength channels prior to applying them to photodetectors used in DWDM receivers.

Fig. 7.42 shows the responsivity characteristics curves for photodetectors over 600–1800 nm optical bands for three types of semiconductor materials Si, Ge, InGaAs used for photodetectors along with quantum efficiencies from 10% to 90%.

As we can see, photodetectors are quite sensitive over wide spectrum (600 nm). This necessitates the requirement of narrowband optical filters for separation of different wavelength channels prior to applying them to photodetectors used in DWDM receivers.

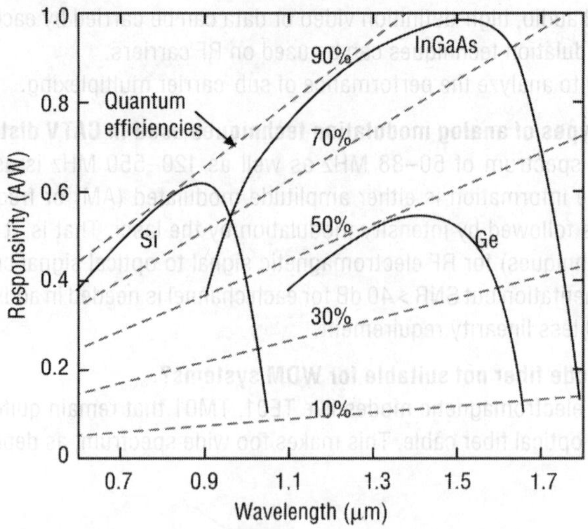

Fig. 7.42 Responsivity characteristics of photodetectors

23. Distinguish between coarse WDM (CWDM) and dense WDM (DWDM).

WDM uses several wavelengths to propagate optical signals carrying user data using a single optical fiber cable. Coarse WDM (CWDM) has channel spacing of the order of 20 nm which is quite wide. It has an advantage of low cost. On the other hand, dense WDM (DWDM) has channel spacing of the order of 0.8 nm only which is quite dense. So, it permits simultaneous propagation of more than 16 different wavelengths and has a distinct advantage of providing high capacity. The standard channel grids are available in 50, 100, 200 and 1000 GHz spacing. Practically, it depends on two parameters - laser linewidth and bandwidth of tunable optical filters.

24. Summarize various design issues in WDM networks.

(1) Non-linear cross-talk due to interactions between light particles and molecular vibrations, or between light particles and acoustic vibrations in the optical fiber cable. There are two types of such mechanisms
  (i) Stimulated Raman Scattering (SRS)
  (ii) Stimulated Brillouin Scattering (SBS)

(2) Non-linear variations in the refractive index due to varying light intensity such as
  (i) Self-Phase Modulation (SPM)
  (ii) Cross-Phase Modulation (XPM)
  (iii) Four Wave Mixing (FWM)

25. Is DWDM flexible? Give sufficient reasons to support your answer.

DWDM is a protocol which does not depend on the transmission bit rate. It implies that data signals such as IP, SONET and ATM can be propagated through the same data stream irrespective of differences in their data rates. The signals are never terminated within the optical layer which allows the independence of protocol and bit rate. This also permits DWDM technology to be employed along with installed equipment in the existing telecommunication network. So, there is a lot of flexibility to enhance system capacity within any part of the networks.

26. What are specific advantages of point-to-point DWDM systems?

As far as the point-to-point architecture of DWDM is concerned, it is fairly simple to build and maintain. It also enables transparency in protocol and modulation formats, incremental expansion, and capacity increase with time. On the other hand, there is every possibility of significant reduction in initial installation costs. Point-to-point solutions are extremely efficient without need of any optical amplifiers or additional equipment.

27. With the help of conceptual diagrams, differentiate between 1 × 2 configuration optical splitter, 2 × 1 configuration optical combiner, and 2 × 2 configuration optical coupler.

The basic concept of 1 × 2 configuration optical splitter, 2 × 1 configuration optical combiner, and 2 × 2 configuration optical coupler is shown in Fig. 7.43.

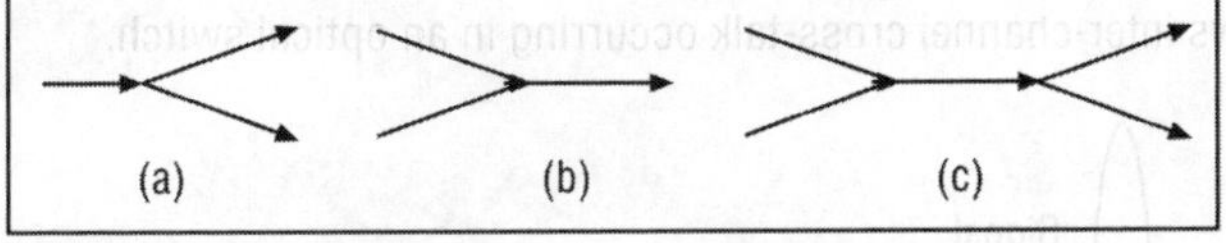

Fig. 7.43 (a) 1 × 2 configuration optical splitter; (b) 2 × 1 configuration optical combiner; (c) 2 × 2 configuration optical coupler

28. With the help of functional diagram, give the concept of passive star coupler (PSC).

In an optical star coupler, the optical signal present at any one port is broadcasted to every other port. The output power is then simply given as the input power divided by number of ports, ignoring the excess loss, if any. Fig. 7.44 shows the functional diagram of a 16 × 16 configuration passive optical star coupler which uses combiners, couplers, and splitters.

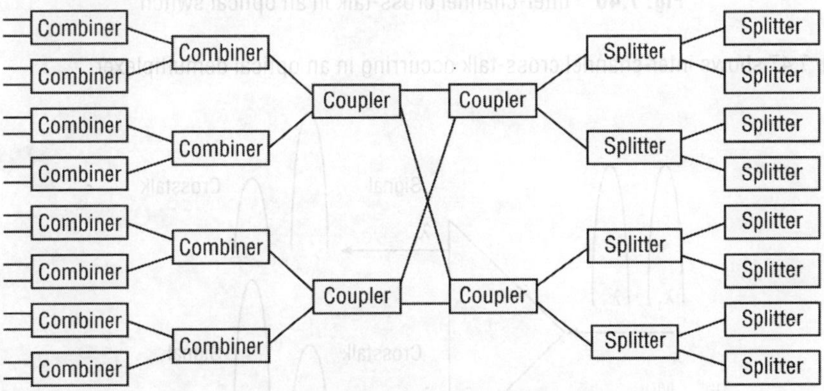

Fig. 7.44 A 16 × 16 configuration passive optical star coupler

29. Show the effect of stimulated Raman scattering (SRS) in WDM.

Fig. 7.45 shows the effect of SRS, depicting the transfer of optical power available at a smaller wavelength channel to the higher wavelength channel.

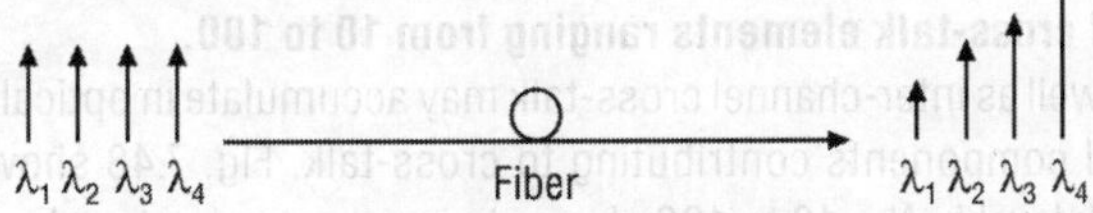

Fig. 7.45 Effect of SRS

To estimate the net effect of SRS in a given WDM system, Raman gain coefficient is given by

$$g(\Delta) = \begin{cases} g_R \dfrac{\Delta\lambda}{\Delta\lambda_c}; & \text{if } 0 \le \Delta\lambda \le \Delta\lambda_c \\ 0; & \text{otherwise} \end{cases}$$

For $\Delta\lambda_c$ = 125 nm, $g_R \approx 6 \times 10^{-14}$ m/w (at λ = 1550 nm) represents the peak value of Raman gain coefficient.

30. With the help of suitable illustration, show the phenomenon of inter-channel cross-talk in (a) an optical switch with inputs of different wavelengths; (b) an optical demultiplexer.

(a) Fig. 7.46 shows inter-channel cross-talk occurring in an optical switch.

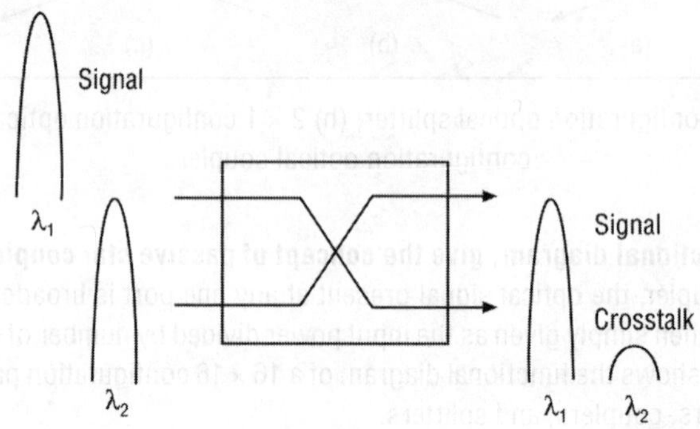

Fig. 7.46 Inter-channel cross-talk in an optical switch

(b) Fig. 7.47 shows inter-channel cross-talk occurring in an optical demultiplexer.

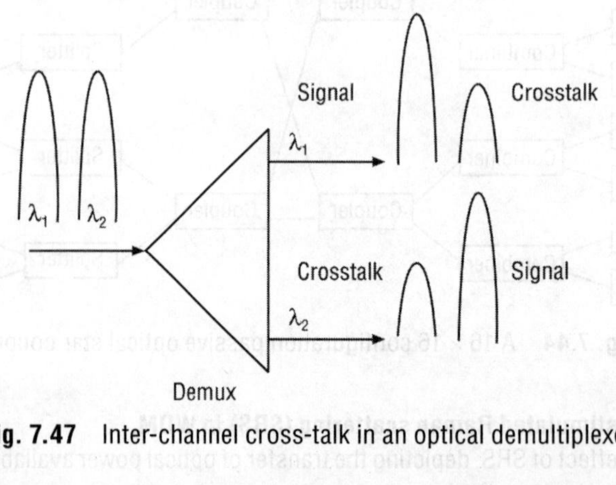

Fig. 7.47 Inter-channel cross-talk in an optical demultiplexer

31. Can intra-channel and inter-channel cross-talk accumulate in optical networks? Illustrate the effect of cross-talk level on power penalty with the help of plot between power penalties versus cross-talk level for a number of cross-talk elements ranging from 10 to 100.

Yes, intra-channel as well as inter-channel cross-talk may accumulate in optical networks, depending upon the number of optical components contributing to cross-talk. Fig. 7.48 shows the plot between power penalty and cross-talk level in N = 10 to 100 elements in an optical network.

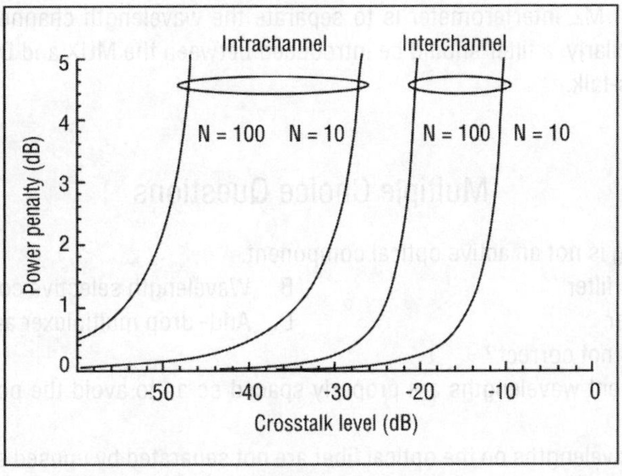

Fig. 7.48 Power penalty vs cross-talk level in a network

32. 'The near end cross-talk is more severe than the far end cross-talk in a bidirectional optical system.' Justify this statement with the help of suitable diagrams.

Fig. 7.49 illustrates the functional diagrams of using either optical MUX/DEMUX or combination of MUX, Optical Circulator and DEMUX in two different approaches in a bidirectional optical system.

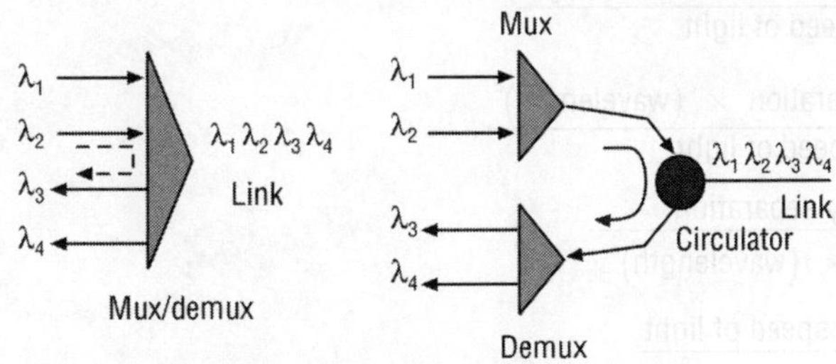

Fig. 7.49 Bidirectional optical systems

It may be noted that the near end cross-talk is more severe than the far end cross-talk. Hence, it is recommended that one use an optical circulator with independent MUX and DEMUX in a bidirectional transmission system rather than an integrated device MUX/DEMUX.

33. Suggest a technique to reduce cross-talk occurring in an optical switch and MUX/DEMUX.

Cross-talk in an optical switch can be reduced using wavelength dilation technique in which four Mach–Zehnder interferometer (MZI) are used with two optical switches, as shown in Fig. 7.50.

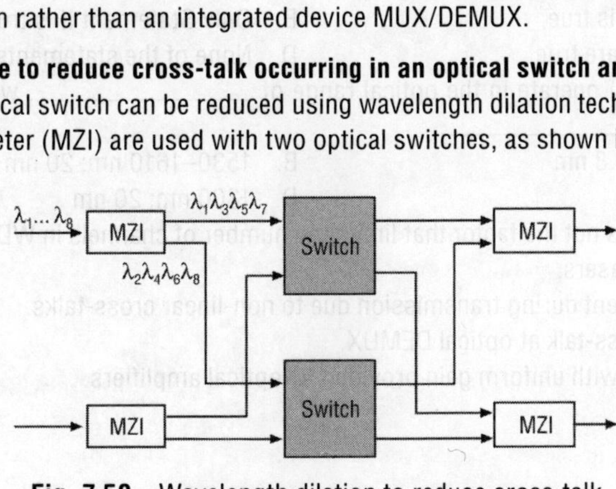

Fig. 7.50 Wavelength dilation to reduce cross-talk

The function of MZ interferometer is to separate the wavelength channels into two groups or to combine them. Similarly, a filter should be introduced between the MUX and DEMUX in order to reduce intra-channel cross-talk.

Multiple Choice Questions

1. One of the following is not an active optical component.
- A. Tunable optical filter
- B. Wavelength selective coupler
- C. Optical amplifier
- D. Add–drop multiplexer and demultiplexer

2. Which statement is not correct?
- A. In WDM, different wavelengths are properly spaced so as to avoid the possibility of inter-channel interference.
- B. In WDM, the wavelengths on the optical fiber are not separated by unused spectrum which may help to prevent their interference with each other.
- C. WDM technology uses multiple wavelengths on individual fiber lines to transmit information over a single fiber line using optical multiplexer.
- D. Each optical channel in a WDM system can have any type of data signaling format (i.e., analog or digital asynchronous bit rates).

3. The relationship to define wavelength separation between adjacent wavelengths is given as
- A. $\frac{\text{Frequency separation} \times \text{wavelength}}{\text{speed of light}}$.
- B. $\frac{\text{Frequency separation} \times (\text{wavelength})^2}{\text{speed of light}}$.
- C. $\frac{\text{Frequency separation}}{\text{speed of light} \times (\text{wavelength})^2}$
- D. $\frac{\text{Wavelength} \times \text{speed of light}}{(\text{Frequency separation})^2}$

4. Statement I: WDM happens because a single-mode optical fiber can support many different wavelengths at the same time.
Statement II: WDM is not possible with multimode fibers.
- A. Only Statement I is true.
- B. Only Statement II is true.
- C. Both statements are true.
- D. None of the statements is true.

5. Dense WDM (DWDM) operate in the optical range of ________, with a very narrow channel spacing of about ________
- A. 1530–1610 nm; 0.8 nm
- B. 1530–1610 nm; 20 nm
- C. 1300 nm; 0.8 nm
- D. 1300 nm; 20 nm

6. One of the following is not the factor that limits the number of channels in WDM.
- A. Stability of DFB lasers.
- B. Signal enhancement during transmission due to non-linear cross-talks.
- C. Inter-channel cross-talk at optical DEMUX.
- D. Finite bandwidth with uniform gain provided by optical amplifiers.

7. Which type of tunable optical filter is depicted in Fig. 7.51?

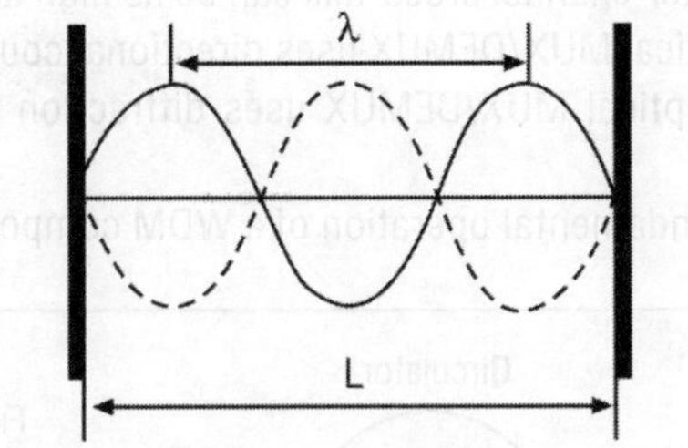

Fig. 7.51 For MCQ 7

A. Mach–Zehnder Interferometer
B. FBG-based Michelson
C. Electro–optic
D. Fabry–Perot Interferometer

8. In Fabry–Perot interferometer tunable optical filter, periodic transmission peaks are separated by free spectral range (FSR) which is given as __________; where, n_g represents the group index of intra-cavity material used in FPI, and L represents the length of FP filter.

A. $\Delta v_L = \frac{1}{2cn_gL}$
B. $\Delta v_L = \frac{c}{n_gL}$
C. $\Delta v_L = \frac{c}{2n_gL}$
D. $\Delta v_L = \frac{2c}{n_gL}$

9. Tunable FP filter using liquid crystals whose refractive index is changed electronically for tuning, provide
Statement I: high value of F (~ 300).
Statement II: bandwidth of about 0.2 nm.
Statement III: switching time of 10 μs–1 ms.
A. Only Statement I is true.
B. Only statements I and II are true.
C. Only statements I and III are true.
D. All the statements are true.

10. A cascaded chain of MZ interferometer tunable active filter
Statement I: comprises of a splitter, a combiner and a delay.
Statement II: cannot achieve Finesse value of 1600.
A. Only Statement I is true.
B. Only Statement II is true.
C. Both statements are true.
D. None of the statements is true.

11. Acousto–optic tunable filters (AOTFs) can provide
Statement I: wide tuning range (> 100 nm).
Statement II: relatively fast tuning (< 10 μs).
A. Only Statement I is true.
B. Only Statement II is true.
C. Both statements are true.
D. None of the statements is true.

12. Which one provides fastest tuning time?
A. Fabry–Perot Tunable Optical Filter
B. Liquid Crystal Fabry–Perot Tunable Optical Filter
C. Tunable Acousto–optic Filter
D. Tunable Electro–optic Filter

13. Which statement is true?
A. For WDM MUX, insertion loss per channel should be low.
B. For WDM DEMUX, inter-channel cross-talk can be as high as 20 dB.
C. Diffraction-based optical MUX/DEMUX uses directional couplers and optical filters.
D. Interference-based optical MUX/DEMUX uses diffraction grating dispersing input optical signal spatially.

14. Fig. 7.52 illustrates the fundamental operation of a WDM component. Identify it.

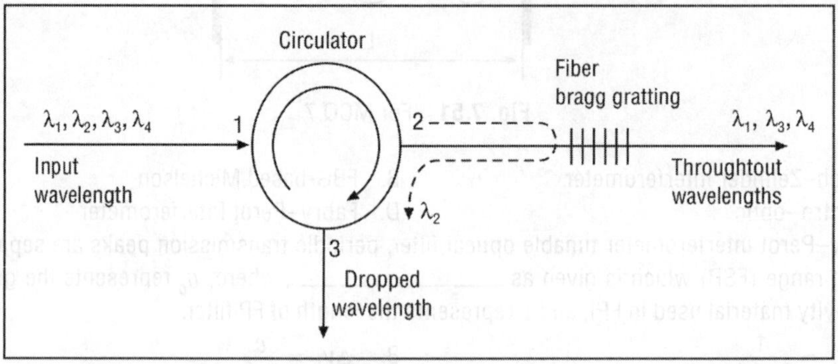

Fig. 7.52 For MCQ 14

A. Interference-based Optical MUX/DEMUX
B. Phased–Array Optical DEMUX
C. Grating-based Optical MUX
D. Grating-based Optical DEMUX

15. WDM add-drop multiplexer (ADM) is needed for optical ________ for adding or dropping one or more optical channels while maintaining the integrity of other optical channels.
A. MANs as well as WANs
B. LANs as well as WANs
C. LANs as well as MANs
D. LANs only

16. For a 8 × 8 bi-directional star coupler, how many 3-dB couplers are needed?
A. 8
B. 12
C. 24
D. 64

17. Statement I: A wavelength converter is a WDM component that can convert the input wavelength at its input port to a new wavelength without modifying the data content of the optical signal.
Statement II: Cross-gain saturation type wavelength converter has a distinct advantage that it can work up to 40 Gbps bit rate
Statement III: An electro-absorption modulator can be used for wavelength conversion.
Statement IV: The cross-gain saturation wavelength converter is preferred over XPM-based MZ interferometer wavelength converter.
A. Only statements I, II are true.
B. Only statements I, III are true.
C. Only statements I, II, and III are true.
D. All statements are true.

18. ________ is a WDM component which is capable of combining the functions of an optical star coupler with optical MUX and DEMUX operations.
A. Wavelength Router
B. Wavelength Converter
C. Optical Cross-Connect
D. Optical Modulator

19. Fig. 7.53 illustrates the function of ________ type of WDM component.
A. Wavelength Grating Converter
B. Wavelength Grating Router
C. Star Coupler
D. WDM Add–drop Multiplexer.

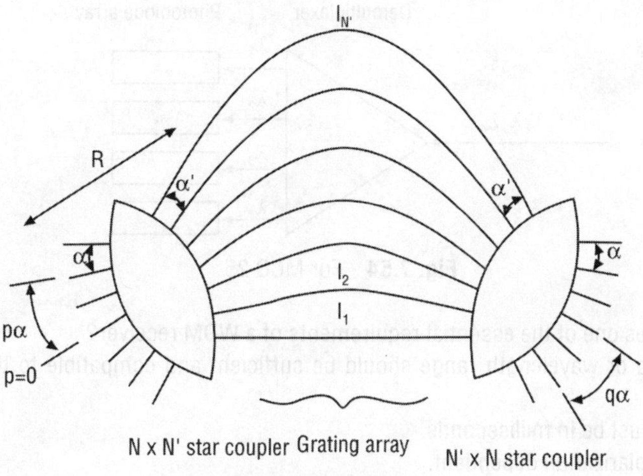

Fig. 7.53 For MCQ 19

20. Statement I: Optical cross-connects (OXC) provides a wavelength routing scheme that can reconfigure the local-area WDM network.

Statement II: Optical cross-connects (OXC) provides a wavelength routing scheme that can reconfigure the metropolitan-area WDM network.

Statement III: Optical cross-connects (OXC) provides a wavelength routing scheme that can reconfigure the wide-area WDM network

A. Only statements I and II are true.
B. Only Statement III is true.
C. Only statements II and III are true.
D. All the statements are true.

21. Statement I: WDM transmitter generally comprises of laser array and an optical demultiplexer.

Statement II: Direct modulation is not recommended in WDM transmitter.

Statement III: Only DFB or DBR laser that has grating filter in the lasing cavity is usually used in WDM transmitter.

A. Only statements I and II are true.
B. Only statements I and III are true.
C. Only statements II and III are true.
D. All the statements are true.

22. One of the following is not the kind of tunable lasers.

A. Internal cavity tunable laser
B. Integrated cavity laser
C. Temperature tuning laser
D. Sectional distributed Bragg reflection (DBR) tunable laser

23. ________________ type of tunable laser has the widest tuning range but requires more tuning time.

A. Mechanical
B. Acousto–optic
C. Electro–optic
D. Injection–Current (DFB and DBR)

24. Analog as well as digital types of optical modulation techniques can be used in WDM transmitters. But ________________ modulation scheme is preferred over others.

A. Amplitude Modulation (AM)
B. Frequency Modulation (FM)
C. Amplitude Shift Keying (ASK)
D. Phase Shift Keying (PSK)

25. Fig. 7.54 shows a simplified structure of

A. WDM Demultiplexer
B. Wavelength Router
C. WDM Receiver
D. Wavelength Converter

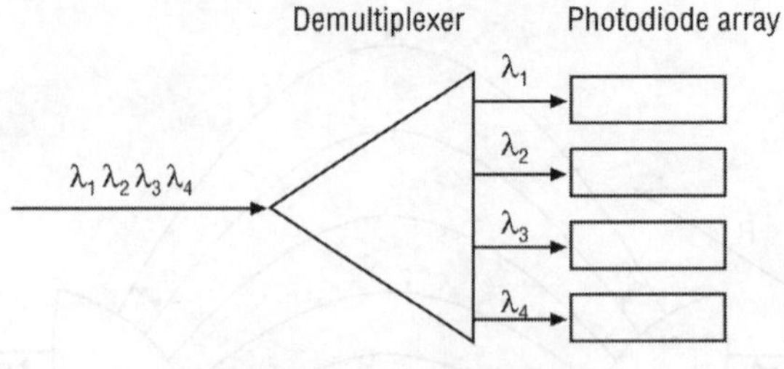

Fig. 7.54 For MCQ 25

26. Which one specifies one of the essential requirements of a WDM receiver?
   A. Spectral width or wavelength range should be sufficient and compatible to that of EDFAs (up to 80 nm).
   B. Tuning time must be in milliseconds
   C. It should be polarization dependent.
   D. Immunity to external noise.

27. Statement I: Cross-talk occurs due to non-linear effects in fibers.
   Statement II: Cross-talk does not occur in a perfectly linear fiber channel.
   Statement III: Intra-band cross-talk usually occurs in switching nodes and can accumulate over a number of nodes.
   A. Only statements I and II are true. B. Only statements I and III are true.
   C. Only statements II and III are true. D. All the statements are true.

28. Statement I: Power penalty is the add-on optical power that is needed at the optical receiver to counteract the effect of cross-talk.
   Statement II: The cross-talk power penalty cannot be calculated by finding the increase in current needed to maintain a certain value of BER.
   A. Statements I and II are true. B. Only Statement I is true.
   C. Only Statement II is true. D. None of the statements is true.

29. It is possible to avoid Raman inter-channel cross-talk provided
   A. The output optical channel power is kept low enough so as to have almost negligible SRS-induced amplification over the length of the fiber.
   B. The output optical channel power is kept high enough so as to have almost negligible SRS-induced amplification over the length of the fiber.
   C. The output optical channel power is kept low enough so as to have almost negligible SBS-induced amplification over the length of the fiber.
   D. The output optical channel power is kept high enough so as to have almost negligible SBS-induced amplification over the length of the fiber.

30. Stimulated Brillouin Scattering mechanism happens at
   A. Relatively higher input optical power levels for wider optical pulses (> 1 μs) but not for shorter optical pulses (< 1 μs).
   B. Relatively lower input optical power levels for shorter optical pulses (< 1 μs) but not for wider optical pulses (> 1 μs).
   C. Relatively lower input optical power levels for wider optical pulses (> 1 μs) but not for shorter optical pulses (< 1 μs).
   D. Relatively higher input optical power levels for shorter optical pulses (< 1 μs) but not for wider optical pulses (> 1 μs).

Keys to Multiple Choice Questions

1. B	**2.** B	**3.** B	**4.** C	**5.** A	**6.** B	**7.** D	**8.** C	**9.** D	**10.** A
11. C	**12.** D	**13.** A	**14.** D	**15.** A	**16.** B	**17.** C	**18.** A	**19.** B	**20.** B
21. C	**22.** A	**23.** A	**24.** C	**25.** C	**26.** A	**27.** B	**28.** B	**29.** A	**30.** C

Review Questions

1. State the underlying principles of the WDM techniques.

2. How is WDM technique different from FDM technique? List various advantages of WDM technique.

3. Distinguish between WDM and DWDM.

4. What is the standard base frequency and channel spacing specified by ITU for DWDM?

5. Compare the salient features of an optical Fabry–Perot interferometer filter with (i) a grating based optical filter, and (ii) Acousto–optic filter.

6. How can we change the coupling ratio of a 2 × 2 coupler?

7. Describe the principles of operation of (i) a 2 × 2 configuration optical directional coupler, and (ii) an N × N configuration optical star coupler.

8. List various types of devise for optical multiplexing/demultiplexing used in WDM. Compare their merits and demerits.

9. The role of a tunable optical filter in a WDM system is to choose a desired wavelength channel. Are tunable optical filters passive or active components? With the help of suitable illustration, show its basic operation.

10. What are the desirable properties of tunable optical filters? List their types and discuss the principle of operation of any one of them.

11. Why are tunable sources needed? Explain the principle of operation of at least two types of tunable lasers.

12. Illustrate the basic concept of optical DEMUX function using fiber Bragg grating? Specify the necessary Bragg condition.

13. WDM add-drop multiplexer (ADM) can be configured using two 3-port optical circulators with fiber Bragg gratings. Give an example of extended ADM using tunable fiber gratings.

14. The function of an optical static star coupler is to combine the wavelength channels available from its many input ports and divide it equally among the output ports. Should the number of input and output ports be equal? Show an 8 × 8 configuration bi-directional star coupler by cascading three stages of 3-dB optical couplers.

15. The basic function of a wavelength converter is to change the incident wavelength at its input port to a new wavelength at its output port while preserving the data integrity of the optical signal. Discuss any two techniques of realizing wavelength converters.

16. Wavelength router combines the basic functions of an optical star coupler with that of optical MUX and DEMUX. Draw a 4 × 4 non-reconfigurable architecture of a wavelength router. How is it different from waveguide grating router?

17. Show a $P \times P$ reconfigurable architecture of a tunable wavelength-routing switch having M number of different wavelengths that uses a basic 2 × 2 configuration of an optical crosspoint devices and photonic switches. List different technologies that can be used for making optical switches.

18. Give a brief account of essential requirements of WDM transmitters and receivers.

19. Write short notes on the following:
(a) Optical modulation methods
(b) Tunable lasers

20. How do you specify the performance of an optical coupler?

Numerical Problems

1. How many independent signals will be transmitted through a single optical fiber cable in the 1525–1565-nm optical band if the narrow-linewidth lasers are used at optical source having 0.8 nm spectral band? *[Ans.: 50]*
2. Find the optical bandwidth for a spectral band specified as $\Delta\lambda = 80$ nm in the wavelength region of 1310 nm. *[Ans.: 14 THz]*
3. What would be the approximate optical bandwidth if the usable spectral band $\Delta\lambda$ is 120 nm in the wavelength region of 1550 nm? *[Ans.: 15 THz]*
4. The transmission bandwidths in the 1310-nm and 1550-nm regions allow the use of many simultaneous channels with narrow spectral widths. Calculate the total available bandwidth if $\Delta\lambda = 80$ nm and 120 nm, respectively in these two low-loss wavelength regions. *[Ans.: ~30 THz]*
5. Show that the optical bandwidth is 100 GHz for channel spacing of 0.8 nm (as specified in ITU-T standards) for the wavelength region of 1535 to 1562 nm.
6. Consider the low-loss region of a silica fiber optical communication system to be 1520–1580 nm. How many channels can be multiplexed if the channel spacing = 75 GHz? *[Ans.: 100]*
7. A WDM system has fiber loss specification as 0.25 dB/km. For a given frequency band of 7.5×10^{12} Hz and channel spacing of 75×10^9 Hz between channels, compute the transmission distance for given 30 dB power margin. *[Ans.: 120 km]*
8. Consider 100 channels carried by a WDM system. If the capacity of each channel is 2.5 Gbps, then what would be the distance-bit rate product for the transmission distance of 120 km? *[Ans.: 30 Tbps-km]*
9. Calculate the free spectral range (FSR) of a phased-array optical demultiplexer with 32 channels spaced at 50 GHz at central wavelength of 1550 nm. *[Ans.: 1600 GHz]*
10. Determine the order of the arrayed waveguides of an optical DEMUX which is designed based on arrayed waveguide gratings if 16 channels are required to be demultiplexed spaced at 100 GHz at 1550 nm central wavelength. *[Ans.: 121]*
11. How many 3-dB optical couplers will be sufficient for designing a typical 16 × 16 configuration bi-directional optical star coupler? *[Ans.: 32]*
12. A 2 × 2 lossless optical coupler is using identical single-mode fibers. What would be the interaction length required to achieve a splitting ratio of 10:90? *[Ans.: 1.25/κ]*
13. Determine the waveguide dispersion coefficient at 1310 nm for refractive index $n_2 = 1.48$ and percent change in refractive index $\Delta n = 0.2\%$. *[Ans.: -1.9 ps/(nm-km)]*
14. An optical power level of 200 μW is applied at the input port of a 2 × 2 bi-conical tapered fiber coupler. The output optical power at throughput port is 90 μW and at coupled port is 85 μW. Compute the percent coupling ratio. *[Ans.: 48.6%]*
15. A 2 × 2 bi-conical tapered optical fiber coupler is applied with an input optical power level = 200 μW. It is observed that the output optical powers at two output ports of optical coupler are 90 μW and 85 μW, respectively. Determine the insertion loss (dB) from
    (i) input port to output port # 1.
    (ii) input port to output port # 2. *[Ans.: i) 3.47 dB; ii) 3.72 dB]*
16. If an optical power level of 200 μW is applied at the input port of a 2 × 2 bi-conical tapered fiber coupler and the output optical power at its throughput port = 90 μW and at its coupled port = 85 μW, then what would be the excess loss (dB)? *[Ans.: ~0.6 dB]*
17. A 2 × 2 single-mode bi-conical tapered fiber coupler has been designed with a splitting ratio specified as 40:60. The measured insertion loss from its input port to its output port # 1 (i.e., 60% channel) is 2.7 dB and that from its input port to its output port # 2 is 4.7 dB (40% channel). If an optical power level of 200 μW is applied at its input port, then what would be the output optical power levels at (a) port 1; (b) port 2. *[Ans.: a) 107.4 μW; b) 67.8 μW]*

18. In the data sheet of a 2 × 2 single-mode bi-conical tapered fiber, the following specification are given:
- splitting ratio: 0.67
- insertion loss for its output port # 1 = 2.7 dB
- insertion loss for its output port # 2 = 4.7 dB

If the input power = 200 μW, then compute the excess loss (dB). *[Ans.: 0.575 dB]*

19. An optical power level, $P_0 = 200$ μW is applied at the input port of a 2 × 2 bi-conical tapered fiber coupler. The output optical powers at the other three ports are $P_1 = 107.4$ μW, $P_2 = 67.8$ μW, and $P_3 = 6.3$ nW. Show that the coupling ratio is 40/60.

20. Cross-talk is one of the key performance parameters for an optical 3-dB coupler which measures the degree of isolation between the input at one port and the optical power reflected back into the other input port. Consider an optical power level of $P_0 = 200$ μW applied at the input port of a 2 × 2 bi-conical tapered fiber coupler. The output optical powers at its three different ports are $P_1 = 90$ μW, $P_2 = 85$ μW, and $P_3 =$ 6.3 nW. How much will be the cross-talk level (dB)? *[Ans.: -45 dB]*

21. A symmetrical waveguide coupler has a coupling coefficient of 0.6/mm. Find the coupling length for m = 0, 1, and 2. *[Ans.: 2.62 mm; 5.24 mm; 7.86 mm]*

22. A 32 × 32 single-mode coupler is required to be designed using a cascade of 3-dB fused-fiber a typical 2 × 2 configuration optical couplers. Show that the number of 3-dB optical couplers needed would be 80.

23. A 32 × 32 single-mode coupler is made from a cascade of 3-dB fused-fiber 2 × 2 couplers. Assuming that 5% of the power is lost in each element, determine

(a) the splitting loss
(b) the excess loss
(c) overall loss

Express your answers in dB. *[Ans.: a) 15 dB; b) 1.1 dB; c) 16.1 dB]*

24. Let the input wavelengths of a typical 2 × 2 configuration Mach–Zehnder Interferometer Multiplexer be separated by 10 GHz. Find the difference in waveguide length. Use $n_{eff} = 1.5$ for a silicon waveguide. *[Ans.: 10 mm]*

25. The frequency separation in the input wavelengths of a 2 × 2 silicon MZI Multiplexer is 130 GHz. Show that the waveguide length difference is 0.77 mm.

26. In an N × N waveguide grating multiplexer, a central design wavelength $\lambda_c = 1550$ nm and the refractive index of the grating array waveguides $n_c = 1.45$. Determine the waveguide length difference for I[st] order waveguide grating MUX. *[Ans.: 1.07 μm]*

27. Let the maximum change in the refractive index of a tunable distributed Bragg reflector laser be 0.65% at $\lambda = 1550$ nm. If the spectral width of this optical source is 0.02 nm at transmission bit rate of 2.5 Gbps, then determine the tuning range of the laser and number of channels within it. Assume that the channel spacing is 10 times the spectral width of this optical source so that inter-channel cross-talk can be completely avoided. *[Ans.: 10 nm; 50]*

28. Design a broadband WDM 3-dB coupler which splits two wavelengths. The two step-index single-mode fibers used to make the coupler are identical. The coupling coefficient for $\lambda_1 = 1.0483$/mm and for $\lambda_2 =$ 1.2839/mm. Find the position of output ports with respect to input ports for given two wavelengths. *[Ans.: 0.749 mm; 0.611 mm]*

CHAPTER 8

Optical Measurements

Chapter Objectives

After studying this chapter, you should be able to

- ✓ know the requirements and benefits of optical fiber measurements
- ✓ describe techniques and procedures for measurement of optical fiber parameters
- ✓ understand measuring equipment for laboratory and field tests used in optical fiber communications

Optical measurements are necessary to verify the operational characteristics of the optical fiber communication link. Various measurement techniques and special-purpose test equipments are employed for determining key performance parameters of the constituent components and devices including the optical fiber. It is quite obvious that optical measurements are needed at different levels of research and design, manufacturing and production of optical components and devices, installation and commissioning of optical fiber communication systems in the field. There is wide variety of optical measurement and test equipments used. These include optical power meter, optical oscilloscope and spectrum analyzer, optical time-domain reflectometer (OTDR), optical waveform analyzer, connector inspection microscope, dispersion analyzer, live fiber detector, talk-set, optical test set (combined source and power meter), etc. All these measurements are wavelength specific. Fiber attenuation and occurrence of faults in the optical fiber link is the main concern in ensuring the desired performance. There are several challenges involved with optical measurements like multiple wavelengths/channels, high optical power levels, need to carry tests remotely along with a high degree of automation.

Optical power and insertion loss measurements are among the easiest yet the most important optical measurements in optical fiber communications. An OTDR has several uses such as loss measurements as well as fault detection. Live fiber detectors and talk-sets are useful portable test equipment for the purpose of installation, maintenance and repair. Software prediction of an OTDR trace is a recent development in optical measurements. This chapter focuses on optical measurements of transmission properties of major constituents of optical fiber communication system such as optical source power output, optical amplifier noise characteristics, modulation response, insertion loss, fiber attenuation, dispersion parameters, and link fault detection.

8.1 Requirements of Optical Fiber Measurements

Optical fiber communication systems are evolving with innovations and numerous applications. Existing copper cables are being replaced with optical fibers everywhere in all accessible areas.

Advanced telecommunication systems comprise of complex optical fiber systems, i.e., passive and active all optical networks. This necessitates more reliable and accurate techniques of optical measurements and tests. There are a variety of performance determining parameters associated at component and system level. For example,

- optical gain and noise figure of optical amplifier that enable WDM systems
- bandwidth response, spectrum width and dispersion for high data rate (> 10 Gbps) applications that require compatible characteristics of optical devices
- wavelength, power and signal-to-noise ratio in WDM systems with 100 GHz having narrow wavelength spacing

Before we proceed to optical fiber measurement techniques, we must have thorough knowledge of the basic features of an optical fiber communication link. The primary objective of measurement is to determine whether the system complies with its desired design goals. In order to guarantee overall system performance, all the associated components and devices within the optical communication link must be properly specified and characterized.

The optical fiber measurements have been standardized by several organizations. The International Telephone and Telegraph Consultative Committee (CCITT) has made recommendations for single-mode fiber measurements. In the US, the Electronics Industries Association (EIA) has published numerous fiber optic test procedures (FOTPs).

8.2 Optical Transmitter Measurements

There are mainly three optical sources– the LED, the Fabry–Perot laser and the DFB laser. Each one of these is having a completely different output versus wavelength characteristics. An optical source could be a wavelength tunable laser or a broadband laser. An optical power meter can measure the gain of the optical amplifiers, whereas an optical spectrum analyzer (OSA) can automatically measure and display the following parameters:

(i) Total power output
(ii) Peak and Mean wavelengths
(iii) 3-dB spectral bandwidth
(iv) Mode spacing for FP laser
(v) Side-mode suppression ratio (SSR) for DFB laser
(vi) Stop-band for DFB laser

Fig. 8.1 shows a simplified test set-up diagram for measurement of output optical power of an optical source with optical power meter.

The optical power meter contains a photodetector which converts an incident optical signal into electric current. The photodetector is characterized by the responsivity (i.e., the conversion efficiency between the input optical power and the output photocurrent, expressed in amp/watt) as a function of wavelength. It must be properly calibrated to display the optical power output of the optical source under test.

The optical power measurements are carried out to determine whether the output optical power of an optical transmitter is as specified, and to measure the output optical power of an optical fiber just prior to an optical receiver.

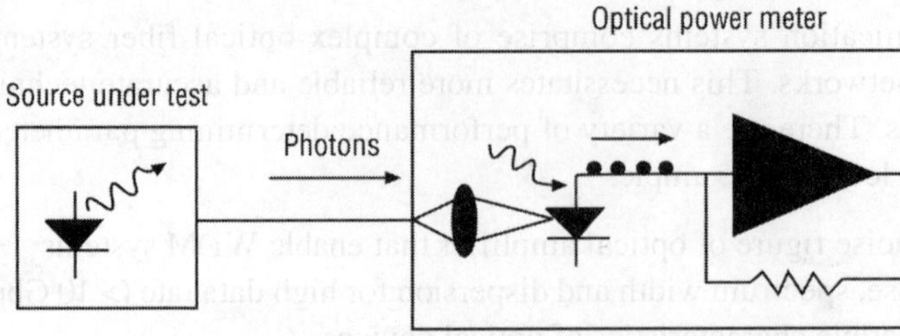

Fig. 8.1 A test set-up for measurement of optical power

Decibel Units (dBm and dBμ)

An optical power meter has the capability of displaying the measured optical power level either in mW or in dBm units. They are related by the expression

$$dBm = 10\log_{10}\left(\frac{P_o(mW)}{1\ mW}\right) \tag{8.1}$$

Similarly, for a 1 μW reference power level, the optical signal power in dBμ unit is expressed as

$$dB\mu = 10\log_{10}\left(\frac{P_o(\mu W)}{1\ \mu W}\right) \tag{8.2}$$

> ***Note:*** These two expressions are quite useful in converting the measured/specified optical power levels in dBm or dBμ units to mW or μW units, or vice versa.

Optical power meter is also used to measure the insertion loss of an optical device or length of an optical fiber by employing a stable optical source for test purpose. LED-based optical source units operate at 850 nm and 1310 nm with typical -20 dBm power output (some LED sources can be used with single mode fiber with -36 dBm typical power output). Laser-based optical source units operate at 1310 nm or/and 1550 nm with typical -7 dBm power output. Modulation with a normally 2 kHz tone is provided for use with live fiber detectors. Typically portable or handheld power meter uses ST, FC/PC, or SC type of connector adapters.

Facts to Know

Thermal detectors are used to calibrate photodetectors because they are very accurate and wavelength-independent. They measure the rise in temperature caused by optical signal absorption but suffer from poor sensitivity. Optical power meters should be made insensitive to polarization of the incident optical signal. There is a need to eliminate the reflectivity of the optical signal by the optical head of optical power meter.

For accurate measurement of the wavelength of the optical signal emitted by an optical source, Michelson interferometer configuration is used as shown in Fig. 8.2.

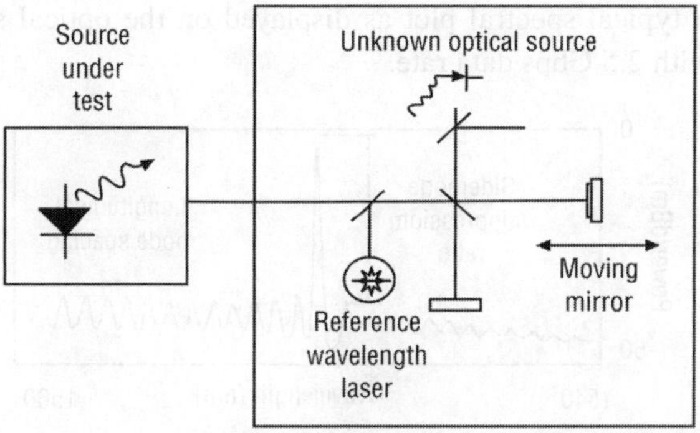

Fig. 8.2 A test set-up for measurement of wavelength

A reference laser source (helium–neon laser emitting at 632.9907 nm wavelength) with a known wavelength (reference wavelength laser) is introduced into the Michelson interferometer. The optical signal from the unknown optical source is split into two paths– one is fixed and the other is variable in length. Both signals are then recombined at a photodetector. As the variable arm is varied (i.e., moving mirror), the photodetector current varies. The wavelength meter compares the interference pattern from both lasers (reference and unknown) to determine the wavelength. This method is less sensitive to changes in operating environment.

Note: Heterodyne and homodyne analysis tools are often used to measure the unmodulated as well as modulated wave shape of the longitudinal modes in laser transmitter. This, in turn, helps to determine the linewidth and chirp of the optical signal.

An optical spectrum analyzer (that uses a diffraction grating) is generally employed to display the measured optical power versus wavelength graph. It basically consists of a tunable bandpass filter and an optical power meter. The optical rays from the optical source under test is collimated with concave mirrors and then applied to the rotating diffraction grating (for selection of the wavelength so as to reach the photodetector). It separates the incident optical ray into different angles depending on the wavelength. The grating focuses the optical rays onto an output slit. This arrangement is shown in Fig. 8.3.

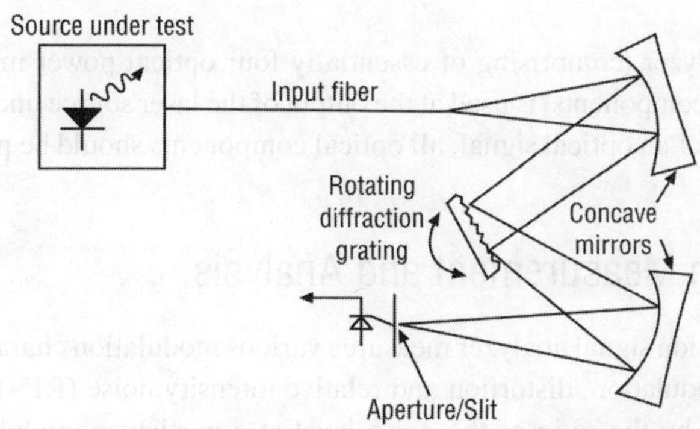

Fig. 8.3 An arrangement showing optical spectrum analysis

Fig. 8.4 illustrates a typical spectral plot as displayed on the optical spectrum analyzer for a modulated DFB laser with 2.5 Gbps data rate.

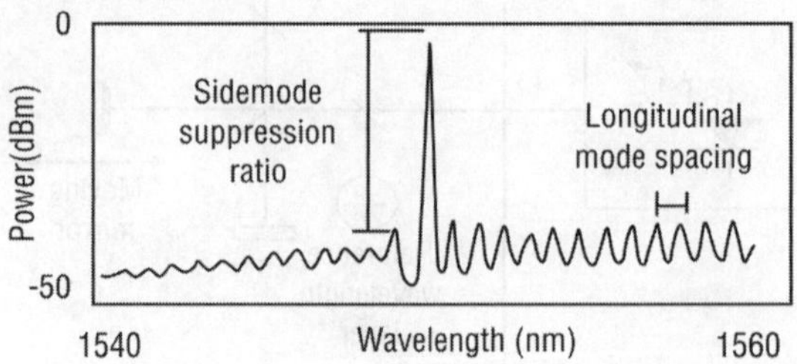

Fig. 8.4 Power vs wavelength plot on OSA

Facts to Know

The optical spectrum analyzer (OSA) can be employed to determine the very narrow bandwidth of the Fabry--Perot bandpass optical filter by measuring the diameter of the optical beam incident on the diffraction grating. For accurate spectral measurement, it is desirable that the OSA must have a very narrow passband and at least 50 dB stopband rejection.

We know that laser sources produce linearly polarized signals that influence optical gain. Therefore, it is necessary to determine the orientation of the polarized component and measure the fraction of the total light power that is polarized. Fig. 8.5 depicts a test set-up diagram using polarization analyzer instrument.

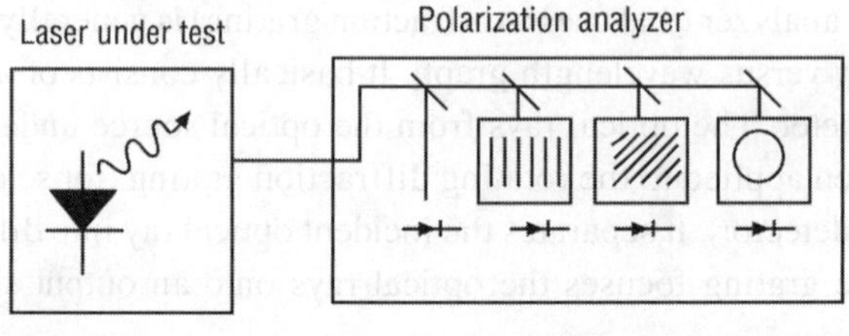

Fig. 8.5 A test set-up for measurement of polarization

A polarization analyzer (comprising of essentially four optical power meters with polarization characterizing optical components) is used at the output of the laser source under test. Due to constant changing polarization of an optical signal, all optical components should be polarization insensitive.

8.3 Modulation Measurement and Analysis

The lightwave modulation signal analyzer measures various modulation characteristics such as depth of optical intensity modulation, distortion and relative intensity noise (RIN). The relative intensity noise is characterized by the ratio of the noise level at a particular modulation frequency to the average power of the optical signal. The lightwave modulation signal analyzer basically comprises of a photodetector followed by an optical pre-amplifier and an electrical spectrum analyzer. Fig. 8.6 depicts a typical test set-up for measurement of modulation in the frequency domain.

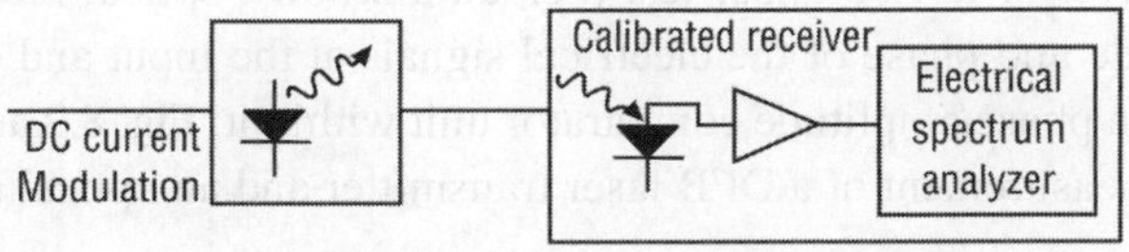

Fig. 8.6 A test set-up for modulation analysis in frequency domain

The measurement method in the frequency domain displays modulation frequency response as a function of the modulation frequency with appropriate calibration. Fig. 8.7 illustrates the power of the modulation signal as a function of the modulation frequency for a DFB laser modulated at 6 GHz.

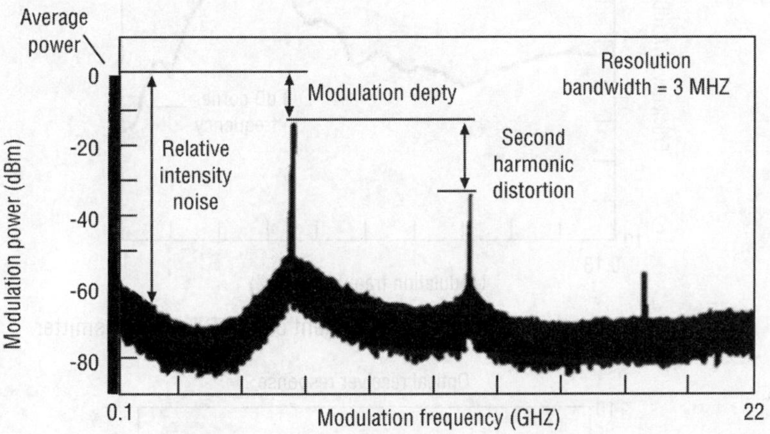

Fig. 8.7 Modulation frequency response

The measurements of RIN are normalized to a 1 Hz optical bandwidth. A DFB laser without modulation may have a RIN level as low as -145 dB/Hz.

The modulation response of an optical transmitter, an optical receiver and an optical communication links can be measured by an electrical vector network analyzer, as shown in Fig. 8.8.

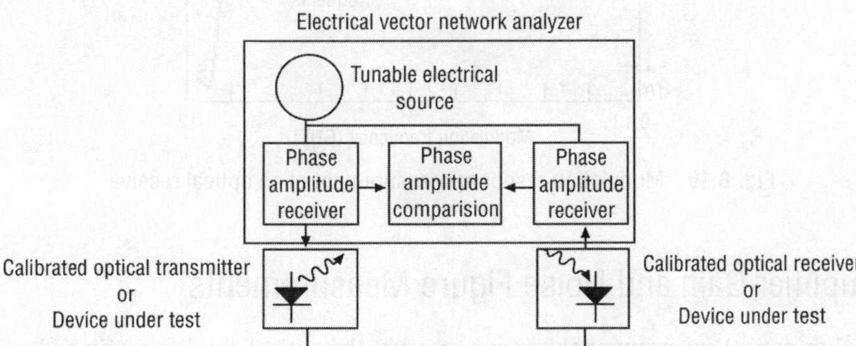

Fig. 8.8 A test set-up for modulation analysis using network analyzer

An electrical vector network analyzer basically consists of a tunable electrical source, two phase/amplitude receiver and phase/amplitude comparator. The output of its electrical source is connected to the calibrated optical transmitter or device under test (i.e., an unknown optical transmitter). A calibrated optical receiver or device under test (i.e., an unknown optical receiver) is connected to its input. The magnitude and phase of the electrical signals at the input and output of the network analyzer is compared in phase/amplitude comparator unit within it. Fig. 8.9 and Fig. 8.10 depict the modulation response measurement of a DFB laser transmitter and an optical receiver, respectively.

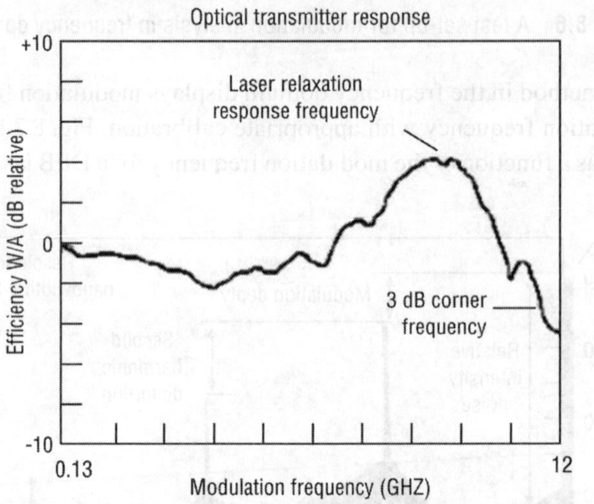

Fig. 8.9 Modulation response measurement of a DFB laser transmitter

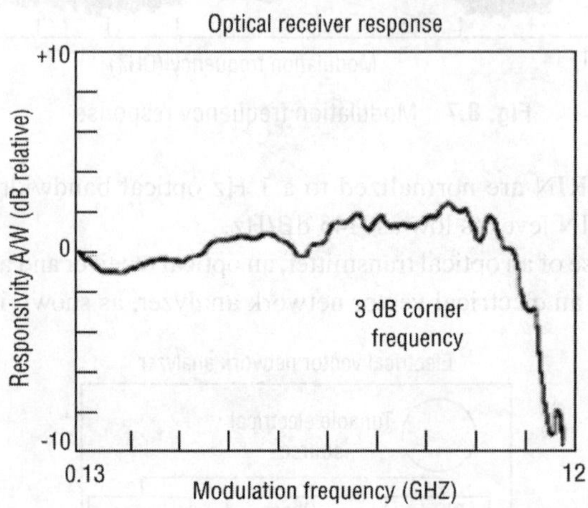

Fig. 8.10 Modulation response measurement of an optical receiver

8.4 Amplifier Gain and Noise Figure Measurements

An electrical or optical spectrum analyzer can measure the optical amplifier gain and noise figure (NF) of an EDFA. Optical amplifier noise can be measured by either optical-source-subtraction, polarization nulling, time-domain extinction, or pulse method. The measurement of amplifier gain is often carried out in large signal conditions (i.e., gain saturation) with a high-power excitation optical source. The amplifier gain G can be calculated using the following expression:

$$G = \frac{\left(P_{out} - P_{ASE}\right)}{P_{sig}} \tag{8.3}$$

where, P_{out} is the total amplifier output power which includes ASE and the amplified source spontaneous emission (SSE), P_{ASE} is the total noise spectral density from the EDFA, and P_{sig} is the input signal power entering the EDFA. The noise is characterized in optical domain with the measurement of the level of ASE at the output of the amplifier. However, the noise can be characterized in electrical domain by using a photodetector and an electrical spectrum analyzer.

Noise figure is an important parameter of the optical amplifier which represents the ratio of the signal-to-noise power ratios at its input port and output port, provided the input signal and the photodetection process have almost zero optical bandwidth and are limited by shot-noise only. The noise figure of the amplifier NF can be calculated using the expression:

$$NF = \frac{P_{ASE}}{GhcB_0} + \frac{1}{G} - \frac{P_{SSE}}{hcB_0} \tag{8.4}$$

where, h represents the Planck's constant, c is the operating frequency of the light at which the measurement is made, and B_0 denotes the optical bandwidth of the optical filter used at the optical receiver.

Fig. 8.11 shows a test set-up used to measure the gain and noise figure of an optical amplifier such as EDFA.

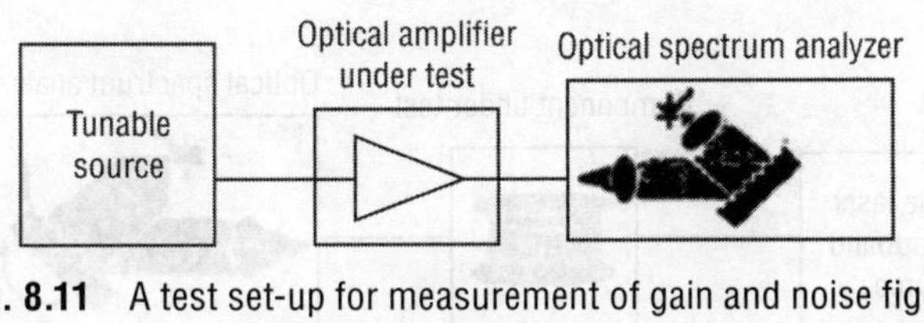

Fig. 8.11 A test set-up for measurement of gain and noise figure

Fig. 8.12 depicts a typical gain and noise figure versus wavelength measurement curve for an optical amplifier.

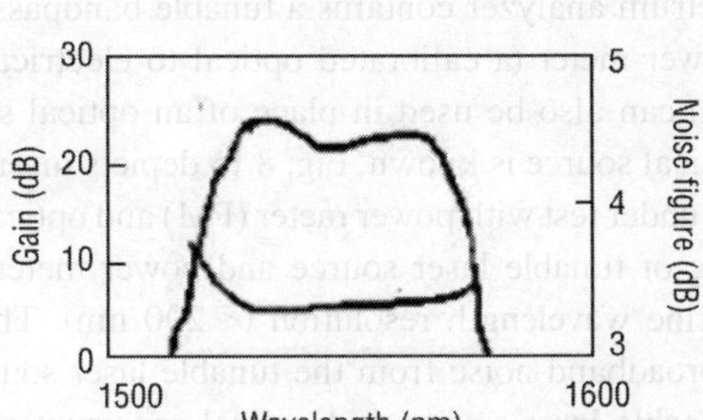

Fig. 8.12 Gain and noise figure vs wavelength measurement

8.5 Insertion–Loss Measurements

Fig. 8.13 shows the basic concept of insertion loss measurement for a fiber joint or any 2-port optical component.

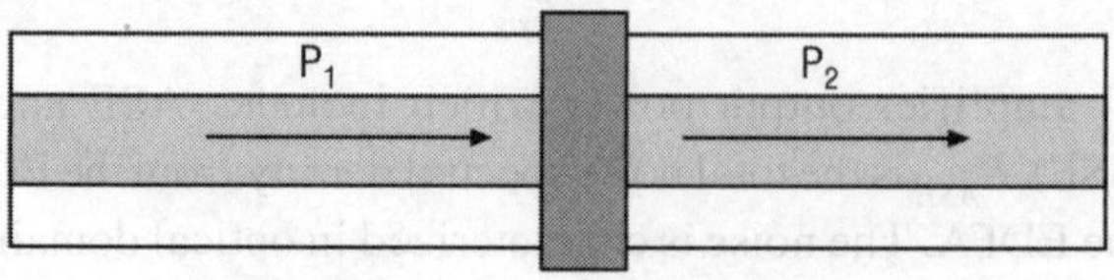

$$\text{Insertion loss in dB} = -10 \log_{10}\left(\frac{P_2}{P_1}\right)$$

Fig. 8.13 Basic concept of insertion loss

Generally, insertion loss measurement is carried out using an optical source at the input and an optical power meter at the output. Measured insertion loss values for fiber connectors are usually very small (0.1–0.5 dB). Any variations in the optical source output or/and test leads will directly affect the loss measurements. Ideally an optical source is used having a stability ten times better than the lowest value to be measured. An optical splitter may be used with a power meter to monitor the reference power output continuously for very high stability. It is recommended to use high quality clean test leads with a test-jig fixture for test lead adapters.

Fig. 8.14 shows a typical test set-up for the measurement of an insertion loss of an optical component, such as an optical fiber, 3-dB optical coupler, or any other optical device using optical spectrum analyzer.

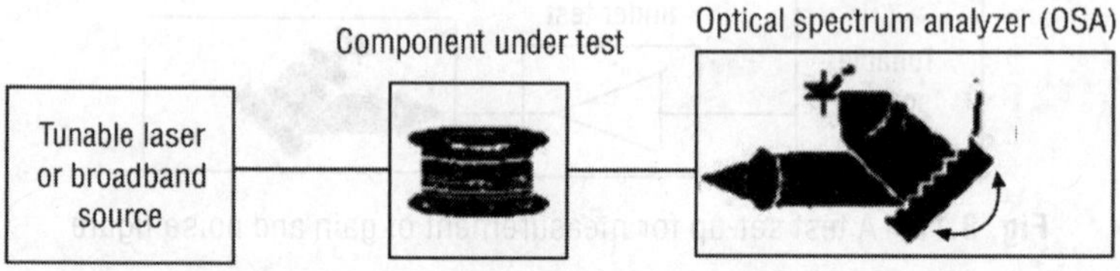

Fig. 8.14 A typical test set-up for insertion loss measurement

Basically, an optical spectrum analyzer contains a tunable bandpass filter as well as an optical power meter. An optical power meter (a calibrated optical to electrical converter without having any wavelength information) can also be used in place of an optical spectrum analyzer provided the output power level of optical source is known. Fig. 8.15 depicts insertion loss versus wavelength measurement for component under test with power meter (PM) and optical spectrum analyzer (OSA).

The test set-up consisting of tunable laser source and power meter (PM) can provide a large measurement range but of fine wavelength resolution (< 200 nm). The major limitation of such a set-up is the presence of broadband noise from the tunable laser source. On the other hand, the test set-up comprising of tunable laser source and optical spectrum analyzer (OSA) can provide additional filtering of the broadband noise emission, thereby exhibiting better performance with narrow spectral width.

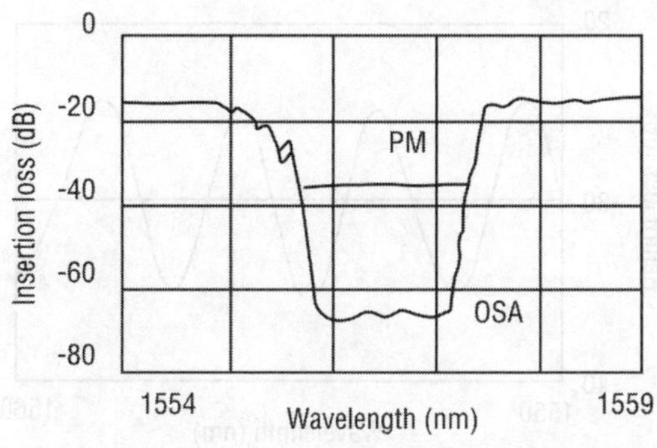

Fig. 8.15 Insertion loss vs wavelength measurement

Note: A measurement test set-up consisting of broadband optical source such as tungsten lamp emitter (that can cover entire wavelength range of optic–fiber communications) along with narrowband high power optical amplifier and optical spectrum analyzer can provide wide wavelength range coverage, fast measurement speed and moderate measurement range.

8.6 Optical Return Loss Measurements

Optical return–loss (RL) measurement is equivalent to optical reflection measurements. These measurements can be made using either a dedicated return loss test set-up (for RL ≥ 60 dB), or an Optical time-domain reflectometer (for coarse measurements). It may be noted that inherent RL of the test set-up must be at least 15–20 dB better than the best RL value to be measured. Fig. 8.16 illustrates a dedicated test set-up for measurement of optical return-loss.

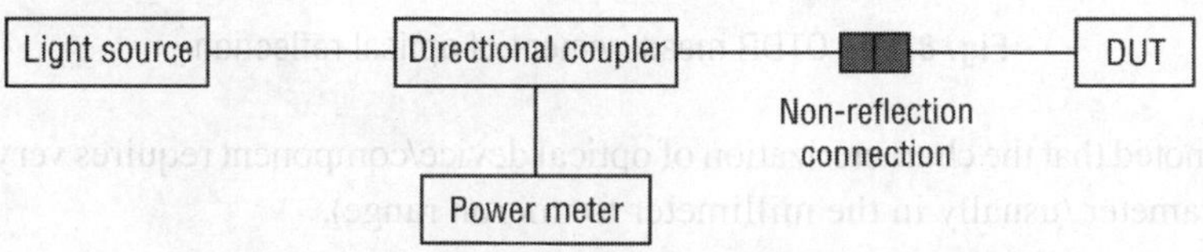

Fig. 8.16 A test set-up for measurement of return loss

The output of an optical source is first applied to a 3-dB optical directional coupler, then to an optical device under test. The directional coupler separates the reflected signal from the incident signal. The optical return loss is measured with the help of optical power meter by comparing the forward and reverse signal levels. Fig. 8.17 depicts the return–loss versus wavelength for an optical source such as a tunable laser.

For measurement of large values of optical return–loss, it is recommended to use an optical time-domain reflectometer (OTDR) technique instead of optical power meter. This is because of the fact that the locations of the reflecting surfaces become critical. Fig. 8.18 shows a test set-up using a high resolution OTDR (consisting of a Michelson interferometer and a broadband optical source to locate reflections with 20 microns accuracy) measurement technique. The resultant display to measure the return loss (dB) is also shown.

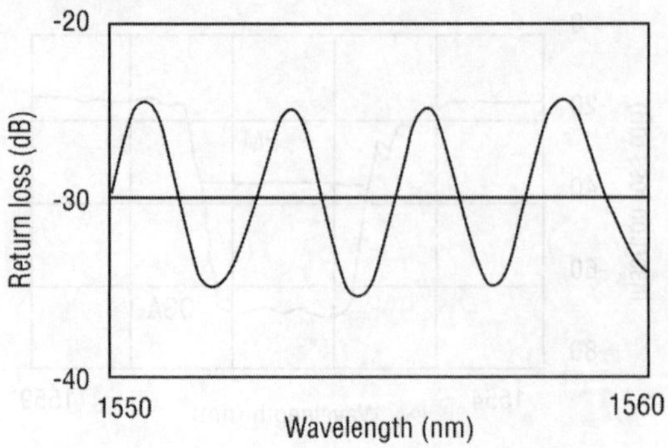

Fig. 8.17 Measurement of optical reflection

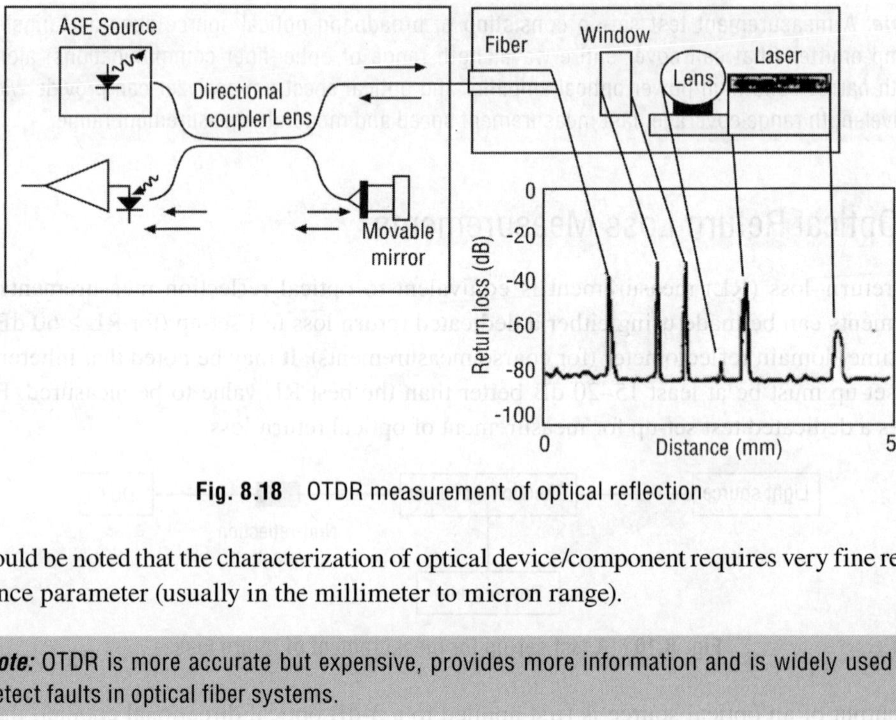

Fig. 8.18 OTDR measurement of optical reflection

It should be noted that the characterization of optical device/component requires very fine resolution in distance parameter (usually in the millimeter to micron range).

Note: OTDR is more accurate but expensive, provides more information and is widely used to detect faults in optical fiber systems.

Fig. 8.19 illustrates a typical measurement test set-up for integrated measurement of insertion–loss (IL) and return–loss (RL).

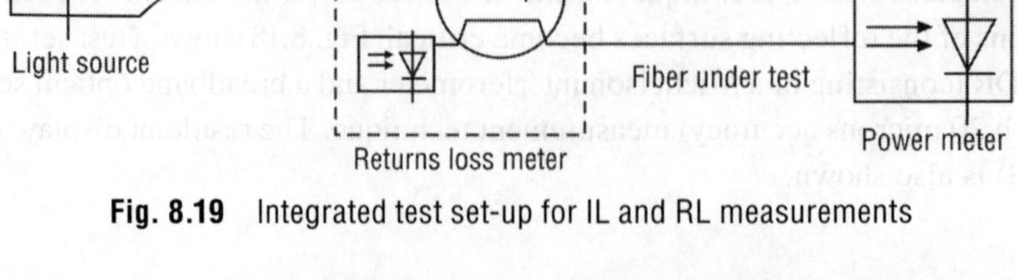

Fig. 8.19 Integrated test set-up for IL and RL measurements

8.7 Fiber Attenuation Measurements

As per TIA/EIA-568-B and ISO 11801 standards, attenuation measurement is the only required test for optical fibers. It is necessary to provide maximum attenuation limits for measured attenuation to optical fiber systems installation teams. Basically, there are three techniques employed for measuring the signal attenuation in optical fibers. These are the cut–back technique, insertion–loss method, and optical time-domain reflectometer (OTDR) method. Attenuation testing is usually carried out using either a combination of an optical source and optical power meter, or an OTDR.

(a) *The Cut–back Technique*– The cut–back technique, also known as *differential technique*, requires access to both ends of the fiber for measuring the signal attenuation in optical fibers and, therefore, is a destructive method. A known value of optical power at one or more specified wavelengths is coupled to a long length of the optical fiber under measurement and transmitted through it. The optical power is measured and noted at the other end of the cable. Now, the same cable is cut short and the procedure of measurement is repeated. Care should be taken to use identical coupling. The average fiber attenuation α in dB/km is then given by

$$\alpha(dB/km) = \frac{10}{L}\log_{10}\left(\frac{P_N}{P_F}\right) \tag{8.5}$$

where, L is the separation of two measurement points in km, the values of P_N and P_F are the measured output powers of the near (shorter length) and far (longer length) ends of the optical fiber under measurement, respectively.

If V_N and V_F represent the corresponding output voltage levels in volts from the original fiber length and the cut-back fiber length respectively, then the fiber attenuation per unit length, α (dB/km) can be expressed as

$$\alpha(dB/km) = \frac{10}{L}\log_{10}\left(\frac{V_N}{V_F}\right) \tag{8.6}$$

because electrical voltage levels are directly proportional to optical powers.

> ***Note:*** The cut-back technique for fiber attenuation measurement is regarded as the reference test method by the CCITT and EIA standards. It is also outlined in Fiber Optic Test Procedures (FOTP) for single-mode as well as multimode fibers as FOTP-78 and FOTP-46 standards, respectively.

(b) *Insertion–Loss Method*– The wavelength-tunable optical source such as laser is coupled to a small length of the optical fiber under measurement. In single-mode fiber (SMF), a cladding-mode stripper is employed so that only the fundamental mode is allowed to propagate along the fiber. For multi-mode fiber (MMF), a mode scrambler is used to ensure that the fiber core contains an equilibrium-mode distribution. If it is required to measure the fiber attenuation at different wavelengths, then a tunable optical filter can be inserted after the optical source. The attenuation of the optical fiber and the associated connectors is then given by

$$A_{dB} = 10\log\left(\frac{P_1(\lambda)}{P_2(\lambda)}\right) \tag{8.7}$$

where, $P_1(\lambda)$ and $P_2(\lambda)$ represents the launch-power level and received power level, respectively.

Fig. 8.20 shows a typical test set-up arrangement for measurement of attenuation of a certain length of the optical fiber.

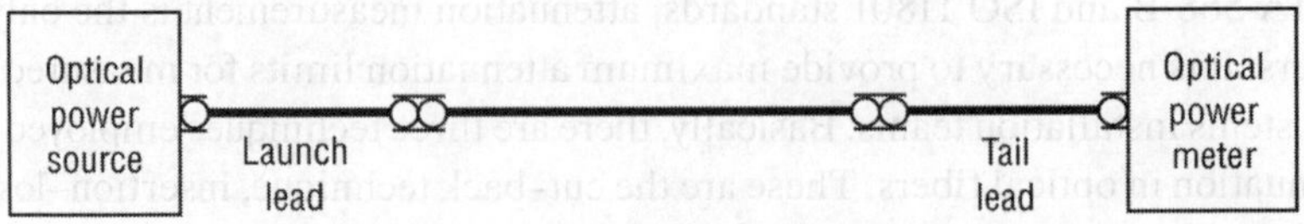

Fig. 8.20 A test set-up for fiber attenuation measurement

Initially, a short reference test lead (X) is used and the received power P_1 (dB) is noted. The test lead (X) is then replaced by the length of fiber under test and received power P_2 (dB) is noted. Then, the attenuation in the fiber length is $P_1 - P_2$ (dB). Care should be taken that the fiber length under test and the reference test lead must use identical reference connector pairs and have the same geometry from the same vendor.

(c) *OTDR Method–* A typical optical time-domain reflectometer (OTDR) comprises of an optical source such as LED or laser, a data-acquisition device, a photodetector, a central processing unit, a memory device and a visual display unit. An OTDR acts as an optical radar in which narrow-beam laser pulses are periodically launched into the optical fiber under test by using either a beam splitter or an optical directional coupler. The fiber attenuation, connector and splice losses, and a host of other specification parameters of the optical fiber communication link can be computed by simply analyzing the waveform characteristics of the backscattered light.

Note: The back-scatter measurement method using OTDR is the most popular non-destructive measurement technique for fiber attenuation, connector and splice losses as well as fault location.

Optical Measurement Standards for Fiber Connectors and Patchcords

Table 8.1 shows the international standards for measurement of optical fiber connectors and patchcords (i.e., cable assemblies).

Table 8.1 Optical measurement standards

TIA/EIA-455-34A	Interconnection Device Insertion Loss Test Set-up
TIA/EIA-455-107	System/Link Return Loss using a Loss Test Set-up
TIA/EIA-455-171	Attenuation Measurement by Substitution for • Short-length Multimode Graded Index • Single Mode Optical Fiber Cable Assemblies
IEC 60874-1	Connectors for Optical Fibers and Cables

Note: ITU-T Recommendations G.650 and G.651 standards describe the measurement techniques for total transmission loss for single-mode and graded-index multimode fibers respectively.

Facts to Know

As per TIA/EIA-568-B standards, the maximum attenuation limit for 120 m patchpanel multimode fiber under test at 1320 nm is 0.18 dB (@ 1.5 dB/km). To work out total attenuation for 120 m patchpanel to patchpanel fiber under test, one needs to add 1.5 dB for two mated connector pairs (@ 0.75 dB each) and 0.6 dB for two splices (@ 0.3 dB each). Typical attenuation for a mated pair of optical connectors is 0.35 dB.

8.8 Fiber Dispersion Measurements

The characterization of the minimum fiber dispersion wavelength is important in the design of high-speed WDM networks. For DWDM, dispersion influences cross-talk. Dispersion compensation management requires an accurate measurement of dispersion parameters. Dispersion specifications are a key differentiator for single-mode fiber. Precise compensation for chromatic dispersion needs its accurate measurement. The measurement of *chromatic dispersion* in an optical fiber is accomplished by analyzing the group delay as function of wavelength of the optical signal propagating through it. Fig. 8.21 shows a typical test set-up for measurement of the chromatic dispersion of optical fiber or any two-port optical device.

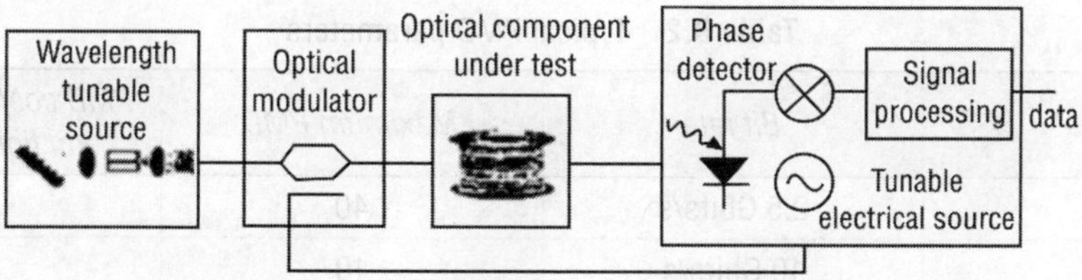

Fig. 8.21 Test set-up for chromatic dispersion measurement

The measurement procedure involves transmission of intensity-modulated signal from a wavelength tunable optical source and then comparing the phase of the detected modulation signal with that of the transmitted modulation signal. The phase comparison is repeated many times after varying the wavelength of the tunable source, resulting in the phase delay. Group delay can then be computed from the phase delay. Fig. 8.22 depicts the result for the measurement of the group delay versus wavelength of tunable laser source.

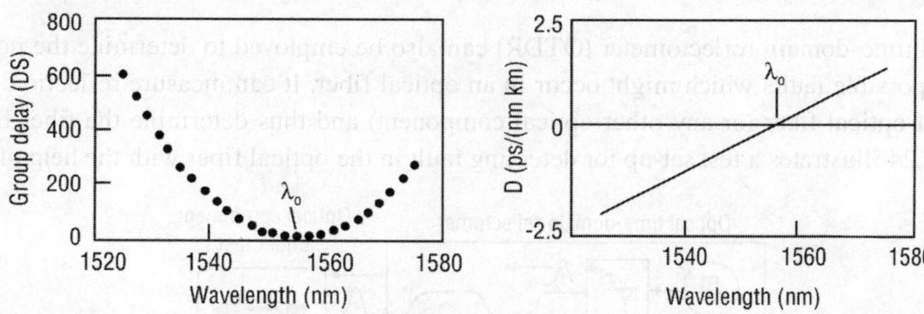

Fig. 8.22 Measurement of chromatic dispersion

Polarization of an optical signal refers to the orientation of the electric field component. Insertion loss as well as group delay of an optical fiber or any two-port optical device vary as a function of the polarization of the incident optical signal. Polarization-mode dispersion (PMD) varies randomly with time. So its measurement is relatively difficult and moreover successive measurements may differ by as much as 20%. To measure the polarization state of the optical signal propagating through the fiber or optical device, polarization analyzer is used, as shown in Fig. 8.23.

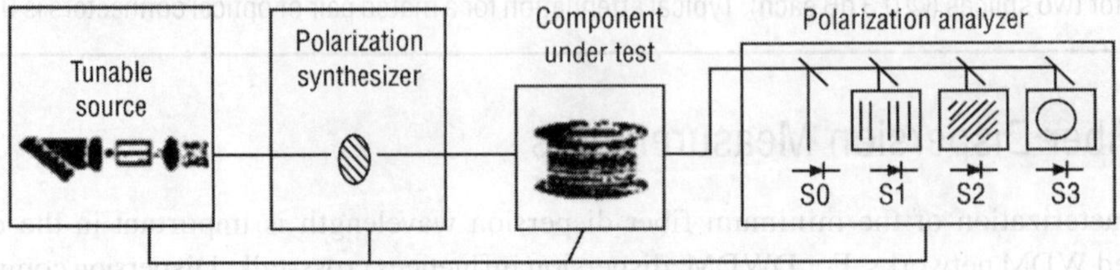

Fig. 8.23 Test set-up for polarization dispersion measurement

As shown, three well-known polarization states of the optical signal from a tunable laser source is applied to optical fiber or any other optical component under test by using polarization synthesizer. The resultant output polarization state (i.e., polarization transfer function) is characterized in the polarization analyzer. Table 8.2 gives typical PMD values for different bit rates for a 1-dB power penalty.

Table 8.2 Typical PMD parameters

	Bit rate	*Maximum PMD*	*PMD coefficient for 400 km link (ps/km$^{1/2}$)*
STM-16	2.5 Gbits/s	40	<2
STM-64	10 Gbits/s	10	<0.5
STM-256	40 Gbits/s	2.5	<0.125

This necessitates that polarization analyzer must have at least 0.05–80 ps range with less than 30 seconds measurement time, having high accuracy up to 1% and about 50 dB dynamic range. Thus, it is a special and expensive test equipment.

8.9 Optical Fiber Fault Measurements

An optical time-domain reflectometer (OTDR) can also be employed to determine the position of breaks or possible faults which might occur in an optical fiber. It can measure reflection from the surfaces of optical fiber (or any other optical component) and thus determine the fiber breaks, if any. Fig. 8.24 illustrates a test set-up for detecting fault in the optical fiber with the help of OTDR.

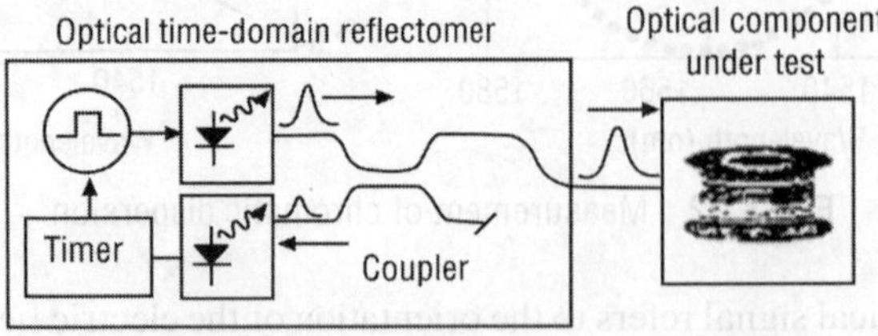

Fig. 8.24 Test set-up for fault detection

An optical pulsed signal generated by an OTDR is injected into the optical fiber cable. A small amount of the optical pulsed signal is continuously reflected back from the surface of the optical fiber under test. If the optical fiber structure has irregularities, then Rayleigh backscatter occurs and the fault in optical fiber is detected. The location of the possible fault is usually computed by using the following expression:

$$L = \frac{ct}{2n_1} \tag{8.8}$$

where, t is the time difference between the pulses reflected from the near and far ends of the fiber, and n_1 is the core refractive index of the fiber.

Fig. 8.25 shows a typical display of OTDR for measurement of reflection as return loss parameter versus distance (length of the optical fiber under test).

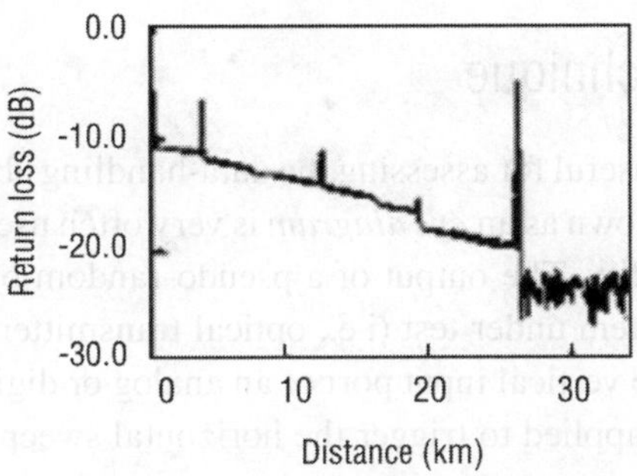

Fig. 8.25 A OTDR display for fault detection with RL

Thus, the magnitude as well as location of faults is determined by measuring the arrival time of the reflected optical signal.

Table 8.3 lists the most important single mode fiber parameters at 1300 nm and 1550 nm wavelengths.

Table 8.3 Single mode fiber parameters

S. No.	Parameter	Typical Value at 1300 nm	Typical Value at 1550 nm
1.	Refractive index	1.45	1.45
2.	Effective Area	80 μm^2	80 μm^2
3.	Attenuation	0.45 dB/km	0.25 dB/km
4.	Dispersion	~ 0 ps/nm/km	17 ps/nm/km
5.	Dispersion slope	0.08 ps^2/nm/km	0.08 ps^2/nm/km
6.	Nonlinear coefficient	~ 2 /W/km	~ 2 /W/km

Nowadays, remote fiber test systems (RFTS) are being deployed which allow defects to be identified prior to failure. Basically, it uses several remotely located plug-in OTDR units having capability to communicate directly with a central control system, as shown in Fig. 8.26.

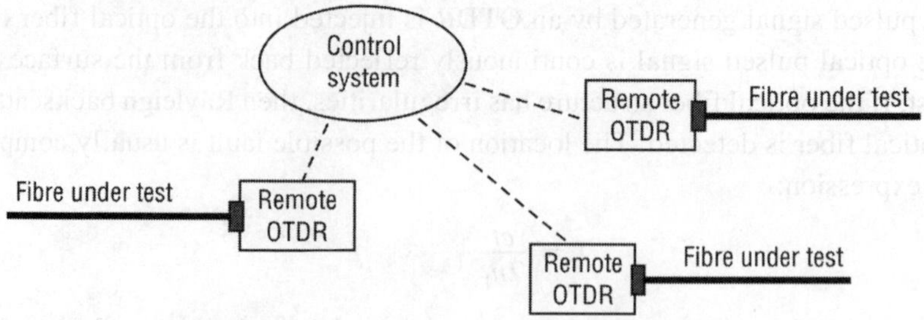

Fig. 8.26 Remote fiber integrated test set-up

Thus, RFTS can be considered as a part of an integrated maintenance and fault-detection system.

8.10 Eye–Pattern Technique

The *eye–pattern* technique is useful for assessing the data-handling ability of an optical transmission system. An eye pattern, also known as an *eye diagram* is very often used for display of the transmitted or received optical signal quality. The output of a pseudo-random bit pattern generator is directly applied to the input of the system under test (i.e., optical transmitter, fiber cable, optical receiver). Its output is then applied to the vertical input port of an analog or digital oscilloscope. The data rate of the bit pattern generator is applied to trigger the horizontal sweep of the oscillator which shows an eye pattern.

An eye pattern is created by taking the time-domain signal and overlapping the traces for a certain number of symbols. Therefore, an analog or digital storage oscilloscope (for better fidelity) is generally used to plot an eye pattern on its display. The received pulse input which may be dispersed in time is given to the vertical input of the oscilloscope. A saw-tooth type time base generator is provided to its horizontal input. It has the same time period as the incoming data, that is, sweep rate is nearly same as the symbol rate. The symbol clock is applied to the external trigger input. At the end of the fixed time interval, the signal is wrapped around to the beginning of the time axis. Thus, an eye diagram consists of many overlapping curves, as shown in Fig. 8.27.

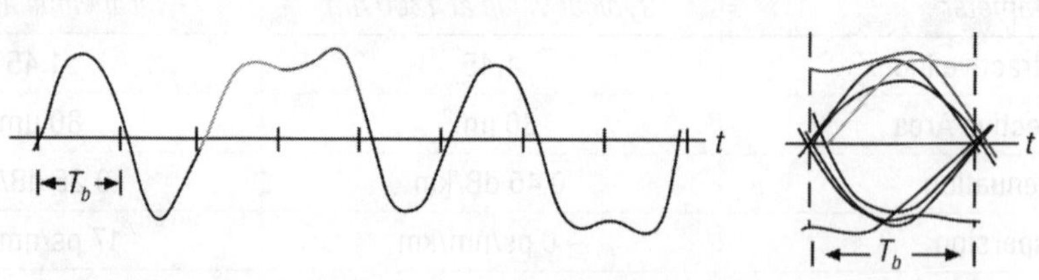

Fig. 8.27 The basic concept of an eye pattern (T_b = pulse width)

The eye opening is a useful parameter in determining the degradation of an optical fiber communication link. Fig. 8.28 illustrates an eye pattern generated by a symmetrical waveform, depicting the opening (height) and width of the eye.

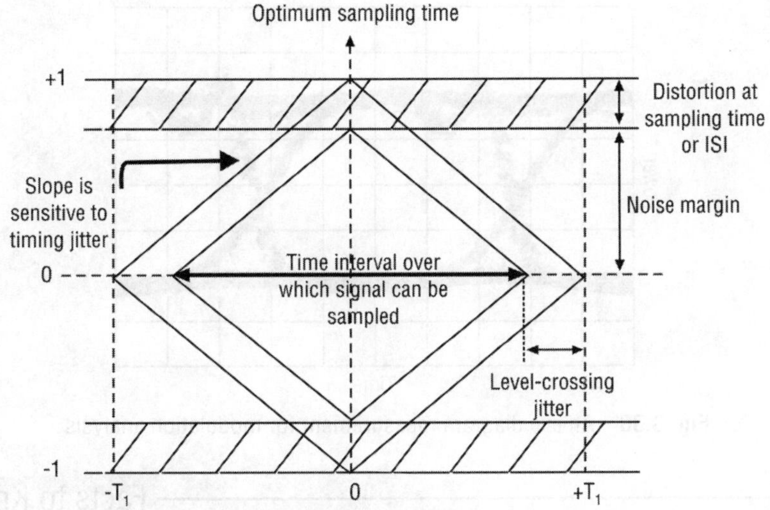

Fig. 8.28 Monitoring parameters from an eye pattern

When the eye opening is quite wide in the eye pattern, it pertains to noise-free transmission. From the typical eye pattern, we can extract several key measures regarding the signal quality. The parameters such as signal amplitude distortion, phase distortion or the timing jitter, noise margin, and system rise-time are generally extracted from the display of eye pattern.

The modulation analysis can also be carried out in time domain (display of optical power versus time) using an electrical oscilloscope. Fig. 8.29 shows a typical test set-up for modulation analysis in time domain.

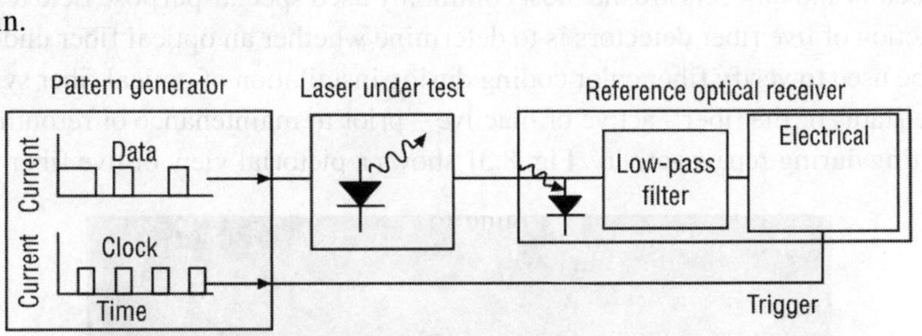

Fig. 8.29 A test set-up for modulation analysis in time domain

Note: High-speed sampling electrical oscilloscopes is often used in data communication and telecommunication systems due to involvement of Gbps data rates.

The display on the oscilloscope is known as an eye diagram, as shown in Fig. 8.30.

To generate an eye diagram on the display of an electrical oscilloscope, the output of an optical source such as laser is applied to the signal input terminal through a calibrated optical receiver and the clock signal is applied to the trigger input terminal of the oscilloscope. This results in display of all digital transitions overlaid in time.

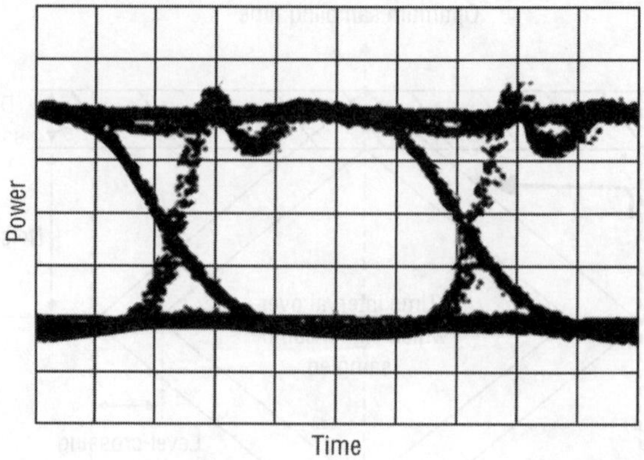

Fig. 8.30 An eye diagram measurement for modulation analysis

Facts to Know

An eye diagram can be used to troubleshoot optical fiber communication links that have poor bit-error ratio performance. In case of SDH and SONET standards, it specifies acceptable time jitter and waveform distortion.

8.11 Special-Purpose Fiber Test Equipments

Live fiber detectors and talk-sets are the most commonly used special-purpose field test equipments. The basic function of live fiber detectors is to determine whether an optical fiber under test is 'live' or not. It can be used to verify fiber color coding during installation of optical fiber systems. It helps to identify the status of the fiber - active or inactive – prior to maintenance or rerouting. It provides continuity testing during repair activity. Fig. 8.31 shows a pictorial view of live fiber detector.

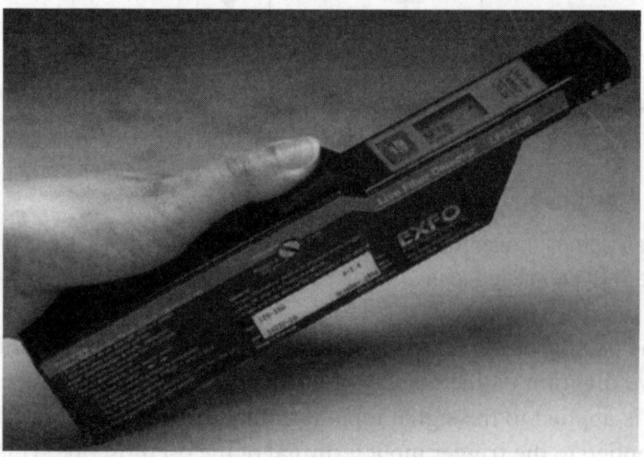

Fig. 8.31 Live Fiber Detector (*Source:* Exfo LFD-100)

It can be easily clamped onto an optical fiber under test (with 250 μm and 900 μm fibers) using a low-loss micro-bending technique so as to exhibit typically less than 0.4 dB insertion loss at 1310 nm.

A talk-set is normally used to provide live communication between technicians working on installation or repair of optical fiber communication systems. It can operate over optical fiber medium. There are two types of designs available– half-duplex and full-duplex, having provision of voice-activated operation. LED-based (at 1310 nm up to 45 km fiber length on 62.5 μm fiber) as well as laser-based (at 1310 nm up to 62 km fiber length on single mode fiber) talk-sets are available.

Facts to Know

Some live fiber detector units can display an approximate value of optical power at the point of test. There are many special test equipment which are used for carrying out various optical measurements. These include optical attenuators and optical communication analyzers.

8.12 Modelling and Simulation Tools

A number of modelling and simulation tools are available for optical fiber communications. Some of these are listed below:

1. *VPIcomponentMaker Fiber Optics*– An R&D tool for modeling, optimization, and designing various optical devices and fibers such as continuous wave and pulsed optical fiber sources, Raman optical amplifiers, doped fiber, optical signal processing for telecommunication, ultrafast and high-power optical fiber communication applications.
2. *RP Fiber Power V6*– A powerful modelling software tool for design, optimization and analysis of optical fiber devices including lasers and fiber amplifiers, optical couplers, double-clad and multi-core fibers, tapered and helical core fibers, fiber–optic telecom systems.
3. *OptSim*– RSoft's software tool with state-of-the-art simulation techniques, an easy-to-use graphical user interface and lab-like measuring test equipment. OptSim is ideally suited for computer-aided design of optical fiber communication systems at the signal propagation level including coherent systems with advanced modulation formats, DWDM systems with EDFA/Raman/SOA optical amplification, radio-over-fiber, OTDM/OCDMA, electronic dispersion compensation, free space optics, soliton transmission, etc.
4. *Gigabit Optical Link Designer (GOLD)*– It is a simulation software tool which is mainly used to analyze the impact of amplified spontaneous emission (ASE), fiber dispersion and fiber nonlinearities accumulated from optical amplifiers such as EDFAs. GOLD operates in the frequency domain for high-speed applications. It uses LabVIEW (a graphical programming language for handling data and controlling instruments) as an interface.

Example 8.1 To Express Optical Power in mW Unit

An optical power meter measures an optical signal power of +7 dBm. Express it in mW.

Solution:

We know that for a 1 mW reference power level, the optical signal power in dBm unit is expressed as

$$dBm = 10\log_{10}\left(\frac{P_o(mW)}{1\ mW}\right)$$

For given +7 dBm power level, we have

$$+7\text{ dBm} = 10\log_{10}\left(\frac{P_o(mW)}{1\ mW}\right)$$

$\Rightarrow$ $P_o(\text{mW}) = 10^{0.7} = 5\text{ mW}$ **Ans.**

Example 8.2 To Express Optical Power in dBμ Unit

An optical power meter measures an optical signal power of 800 nW. Express it in dBμ.

Solution:

We know that for a 1 μW reference power level, the optical signal power in dBμ unit is expressed as

$$dB\mu = 10\log_{10}\left(\frac{P_o(\mu W)}{1\ \mu W}\right)$$

For given power level of 800 nW or 0.8 μW, we have

$$dB\mu = 10\log_{10}\left(\frac{0.8\ \mu\text{W}}{1\ \mu W}\right) = -0.97\text{ dB}\mu$$ **Ans.**

Example 8.3 To Determine Fiber Attenuation Parameter

It is required to determine the attenuation parameter α (dB/km) of a 1.9 km long fiber. Using a photodetector as a measuring device in a cut-back attenuation measuring test set-up, a reading of 3.31 V is obtained from the photodetector at the far end of the fiber. The fiber is cut back to 2 m length from the optical source. The measurement is repeated with the photodetector at the near end of the fiber. A reading of 3.78 V is recorded now. What is the attenuation of the fiber in dB/km?

Solution:

We know that the fiber attenuation per unit length, α (dB/km) is given by

$$\alpha\ (dB/km) = \frac{10}{L}\log_{10}\left(\frac{V_N}{V_F}\right)$$

where, L = difference in the fiber lengths = 1.9 km – 2 m = 1.898 km

For the given V_N = 3.78 V and V_F = 3.31 V, we have

$$\alpha\ (dB/km) = \frac{10}{1.898}\log_{10}\left(\frac{3.78\ V}{3.31\ V}\right) = 0.3\text{ dB/km}$$ **Ans.**

Example 8.4 To calculate Fiber Attenuation

Show that the fiber attenuation per kilometer for a 2 km length of multimode fiber at an operating wavelength of 850 nm is 3.5 dB/km. The measured output voltage from the photodetector is 2.1 V originally and then increases to 10.7 V when the fiber is cut to 2 meter length in a cut-back method of fiber attenuation measurement techniques.

Solution:

We know that the fiber attenuation per unit length, α (dB/km) is given by

$$\alpha(dB/km) = \frac{10}{L}\log_{10}\left(\frac{V_N}{V_F}\right)$$

where, L = difference in the fiber lengths = 2 km – 2 m = 1.998 km

For given $V_N = 10.7$ V and $V_F = 2.1$V, we have

$$\alpha(dB/km) = \frac{10}{1.998}\log_{10}\left(\frac{10.7V}{2.1V}\right) = 3.5 \text{ dB/km}$$ **Hence Proved.**

Example 8.5 To Detect Fault in Optical Fiber

In a test set-up using optical time-domain reflectometer for locating the fault in an optical fiber, an optical power level of 200 mW is coupled from an optical power source to the optical fiber under test. The OTDR measures a reflected power level of 20 μW. The specified fiber attenuation parameter is 0.5 dB/km. Determine the distance at which the fault has occurred in the fiber length.

Solution:

We know that return loss is given by

$$RL\ (dB) = 10\log_{10}\left(\frac{P_i}{P_r}\right)$$

where, P_i is the input optical power and P_r is the reflected optical power, both having same units.

For the given $P_i = 200$ mW and $P_r = 20$ μW or 0.02 mW, we have

$$RL\ (dB) = 10\log_{10}\left(\frac{200\ mW}{0.02\ mW}\right) = 40 \text{ dB}$$

For specified fiber attenuation parameter of 0.5 dB/km, the distance at which the fault has occurred in the fiber length can be computed as (40/0.5) = 80 km **Ans.**

Points to Remember

- Optical measurement techniques are required for verifying the operational characteristics of different components and devices used in an optical fiber communication system.
- Accurate optical measurements are necessary at the design, manufacturing, installation and maintenance stages to ensure the desired performance.
- An optical signal propagating through various constituents of the system need to be measured and characterized in terms of optical power, spectral content, and polarization.
- The most common measuring and test equipment are tunable laser sources, optical power meters, spectrum analyzers, and time-domain reflectometers.
- Special-purpose optical test equipment include optical variable attenuators, polarization analyzers, optical communication analyzers, and multifunction PC-based optical test systems.
- For determining attenuations in fibers, there are three basic techniques– cut-back, insertion–loss, and using OTDR.
- OTDR is widely used for single-ended measurements of various characteristics of optical fiber communication link including fiber fault location.
- The eye pattern technique is an effective method for evaluating the performance of optical fiber data links.

Important Equations

Optical power level, $dBm = 10\log_{10}\left(\frac{P_o\,(mW)}{1\ mW}\right)$; *where, P_o is the absolute power level measured in mW.*

The optical amplifier gain, $G = \frac{(P_{out} - P_{ASE})}{P_{sig}}$; *where, P_{out} is the total amplifier output power which include ASE and the amplified source spontaneous emission (SSE), P_{ASE} is the total noise spectral density from the EDFA, and P_{sig} is the input signal power entering the EDFA.*

The noise figure of the optical amplifier, $NF = \frac{P_{ASE}}{GhcB_0} + \frac{1}{G} - \frac{P_{SSE}}{hcB_0}$; *where, h represents the Planck's constant, c is the operating frequency of the light at which the measurement is made, and B_0 denotes the optical bandwidth of the optical filter used at the optical receiver.*

The fiber attenuation, $\alpha(dB/km) = \frac{10}{L}\log_{10}\left(\frac{P_N}{P_F}\right)$; *where, L is the separation of two measurement points in km, the values of P_N and P_F are the measured output powers of the near (shorter length) and far (longer length) ends of the optical fiber under measurement, respectively.*

The fiber attenuation, $\alpha(dB/km) = \frac{10}{L}\log_{10}\left(\frac{V_N}{V_F}\right)$; *where, L is the separation of two measurement points in km, the values of V_N and V_F are the measured output voltage levels of the near (shorter length) and far (longer length) ends of the optical fiber under measurement, respectively.*

Key Terms with Definitions

Electrical vector network analyzer	Consists of a tunable electrical source, two phase/amplitude receiver and phase/amplitude comparator.
Eye pattern	Created by taking the time-domain signal and overlapping the traces for a certain number of symbols.
Modulation signal analyzer	Used to measure various modulation characteristics such as depth of optical intensity modulation, distortion and relative intensity noise.
Noise figure	The ratio of the signal-to-noise power ratios at its input port and output port provided the input signal and photodetection process have almost zero optical bandwidth and are limited by shot-noise only.
Optical power meter	Contains photodetector which converts incident optical signal into electric current and calibrated to display optical power in decibels.
Optical spectrum analyzer	Consists of a tunable bandpass filter, diffraction grating and an optical power meter to display the measured optical power versus wavelength plot.
Optical time-domain Reflectometer (OTDR)	Comprises of an optical source such as LED or laser, a data-acquisition device, a photodetector, a central processing unit, a memory device and a visual display unit.

Short Answer Type Questions

1. **Mention key parameters of different types of optical fibers which are usually specified by the manufacturers in their specification data sheets. Out of these, which parameters are not required to be measured during installation and operation phase?**

   For single-mode fiber, the manufacturers specify various important characteristics such as fiber attenuation, the effective cutoff wavelength, dispersion, and the mode-field diameter. For multimode fiber, important parameters specified include the fiber attenuation, core and cladding diameters, refractive-index profile, numerical aperture, and dispersion. Once the values of these parameters are made known by the manufacturer of the fiber, there are certain parameters which generally do not undergo significant change during cable assemblies, installation, and operation in the fields. These parameters are cutoff wavelength, refractive-index profile, mode-field diameter, numerical aperture, and fiber geometry.

2. **Do the fiber attenuation and dispersion values of an optical fiber change during the fabrication of fiber cable assemblies and their installation? How do they affect the performance of the optical fiber system?**

Yes, the specified fiber attenuation and dispersion values can change during the fabrication of fiber cable assemblies and their installation phase. Variations in chromatic and polarization-mode dispersion parameters can affect the bandwidth-distance product and thereby limit the highest achievable data rate in single-mode fiber high-speed WDM links. In multimode fibers, modal dispersion effects due to micro-bends and at fiber joints can result in additional loss. Therefore, measurement of fiber attenuation as well as dispersion parameters is of particular importance to the users, so as to locate the possible faults and breaks in optical fiber cables.

3. **After installation of an optical fiber link, which parameters are of particular interest to the testing engineers in the field?**

(1) Signal-to-noise ratio
(2) Bit-error-rate (BER)
(3) Timing jitter

These parameters are significant for measurement and performance test purposes, in order to determine the operational status of remotely located optical amplifiers and/or fault locations in the fibers. They can be easily indicated by the eye diagram .

4. **A field engineer has a typical hand-held optical tester which contains optical sources for performing more sophisticated optical power measurements. List various types of tests that are possible to be carried out.**

The specified hand-held optical tester can carry out the following tests:

- Optical power measurements as an optical power meter
- Fiber loss measurement at two different wavelengths in two directions as an optical–loss tester
- Faults and breaks detection with location in a fiber cable as a visual fault indicator
- Quality measurement of optical patch cords as an optical return–loss tester
- Full duplex communications between field engineers as an optical talk set

5. **What is the need of optical attenuators during laboratory or production tests?**

Optical attenuators are needed to provide necessary optical signal attenuation to very high output signal levels of an optical device such as an optical amplifier, in order to provide possible damage to the measuring equipment or avoid overload distortion at the output. Typical optical attenuators may provide signal attenuation up to the order of 60 dB in precise steps (1-, 2-, 10-dB) at usually 1310 or 1550 nm wavelength. The accuracy of attenuation may range from 0.001 dB to 0.5 dB from device to device.

6. **If the wavelength-dependent response of an optical fiber communication link is to be measured, what type of optical source would you recommend to be used for applying an optical signal?**

A tunable laser source such as an external-cavity semiconductor laser is the recommended optical source for the given requirement. For wavelength selection, a moving diffraction grating may be employed as a tunable optical filter along with it. Depending on the combination of the optical source and diffraction gratings, a typical test equipment can be tuned over the desired optical range of 1280–1330 nm or 1450–1565 nm with minimum optical output power level of -10 dBm. It is possible to obtain an almost flat spectral response with ±0.1 nm wavelength accuracy.

7. What is the primary function of an optical spectrum analyzer? Which is the most commonly implemented method for achieving wavelength accuracy as high as ± 0.001 nm in the measured output?

The primary function of an optical spectrum analyzer is to measure and display the output optical power as a function of specified wavelength range. For obtaining high wavelength accuracy as specified, wavelength meters based on Michelson interferometer optical filters are employed instead of general purpose diffraction-grating based optical filters (which can provide wavelength accuracy up to 0.1 nm only).

8. In a particular application, an optical spectrum analyzer (OSA) is required to be used to measure 10-MHz linewidth of a single-wavelength semiconductor laser. What type of OSA do you recommend?

For measurement of such a very narrow linewidth of an optical source, an optical spectrum analyzer which employs both heterodyne and homodyne techniques must be used.

9. Why is an optical time-domain reflectometer (OTDR) instrument considered the most useful measuring equipment in optical fiber communication systems?

An OTDR is a versatile field portable optical measurement instrument which can measure the fiber length, the fiber attenuation, the connector and splice insertion losses, return-loss, and location of faults within an optical fiber communication link. It has the capability of displaying the measured data in the form of a trace as well as of recording it in its data storage device.

10. What do you understand by a multi-functional optical test system? How can it cover a wide range of test capabilities?

A multi-functional optical test system comprises of a microprocessor-based mainframe with RS-232 interface port to connect different plug-in modules that can cover a wide range of test capabilities as per requirement. For example, plug-in module may be a tunable laser source, ASE broadband optical source, single-channel power meter, multichannel power meter, optical spectrum analyzer, variable optical attenuator, and PMD analyzer.

11. Distinguish between the cut-back technique and insertion-loss method for measurement of attenuation in optical fibers.

The cut-back technique is a destructive method which requires the optical fiber to be physically cut off a few meters from the source end after measuring the optical power at the far end of the fiber under test. The insertion-loss method is a nondestructive method and is suitable for fiber cable assemblies (i.e., fiber with connectors at its either ends). Spectral response (over a specified range of wavelengths) may be obtained with the cut-back technique, whereas an optical filter is included to determine the fiber attenuation at specified wavelength in insertion-loss method.

12. What are the two different methods for the measurement of pulse dispersion in optical fibers? Which one is preferred and why?

Time-domain and Frequency-domain intermodal dispersion measurements are the two methods for pulse-dispersion measurement in optical fibers. Time-domain method yields relative pulse amplitude-versus-time response, whereas frequency-domain method yields information on relative pulse amplitude-versus-frequency as well as phase-versus-frequency response. Data obtained from frequency-domain measurements is preferred by system designers in order to perform equalization techniques on the detected signal.

13. Why is it required to measure optical fiber attenuation and dispersion as part of field measurements?

The data sheet for the fiber provided by the manufacturer contains information on fiber attenuation and dispersion parameters. But this cannot be directly applied to installed fibers and cable assemblies connected within an optical fiber system. For instance, micro-bending affects both the fiber attenuation and dispersion. So, it is essential to measure the optical fiber attenuation and dispersion.

14. Mention key design criteria for field measurement test equipment.

The field measurement equipment should be portable and preferably low-power battery-operated compact units. They should be properly encased and sturdy. If they are ac operated by the field generators, then they should maintain reasonable accuracy of measurements even under varying supply voltages, extreme environmental conditions. The test equipment should be easily connectable to the fiber.

Multiple Choice Questions

1. The cut-back technique is used for measurement of

A. Fiber attenuation
B. Intermodal dispersion
C. Chromatic dispersion
D. Polarization-mode dispersion

2. Modulation phase-shift method is used for measurement of

A. Waveguide dispersion
B. Intermodal dispersion
C. Chromatic dispersion
D. Polarization-mode dispersion

3. Which statement is correct?

A. The measurement of polarization-mode dispersion is of high importance because it can limit the highest achievable transmission data rates for single-mode optical fiber systems.
B. If equalization techniques are required at the optical receiver end, then time-domain intermodal dispersion measurements are needed.
C. Fiber attenuation is not the results of waveguide effects, scattering mechanisms, and absorption processes.
D. Multiple wavelength optical power meters do not use photodetectors for measurement of optical signal power levels.

4. A typical OTDR consists of

A. An optical receiver, a CPU and an information-storage unit
B. An optical source, an optical receiver, a data-acquisition unit, a CPU and an information-storage unit
C. A data-acquisition unit, a CPU and an information-storage unit
D. An optical source, an optical receiver and a data-acquisition unit

5. If L_1 and L_2 represent the original and cut-back fiber lengths in km, and P_{01} and P_{02} denote the corresponding output optical powers (in same units) at a specific wavelength, then the fiber attenuation per unit length can be computed as

A. $\alpha(dB/km) = \frac{20}{L_1 - L_2}\log_{10}\left(\frac{P_{02}}{P_{01}}\right)$
B. $\alpha(dB/km) = \frac{10}{L_1 - L_2}\log_{10}\left(\frac{P_{02}}{P_{01}}\right)$
C. $\alpha(dB/km) = \frac{10}{L_1 - L_2}\log_{10}\left(\frac{P_{01}}{P_{02}}\right)$
D. $\alpha(dB/km) = \frac{10}{L_2 - L_1}\log_{10}\left(\frac{P_{02}}{P_{01}}\right)$

6. The importance of fiber dispersion measurements is governed by the type of the fiber used. For instance,

Statement I: In single-mode fibers, intra-modal (i.e., chromatic dispersion) is the dominant dispersion mechanism.

Statement II: In step-index multimode fibers, inter-modal dispersion is the dominant dispersion mechanism.

A. Only Statement I is true.
B. Only Statement II is true.
C. Both statements I and II are true.
D. None of the statements is true.

7. Dispersion effects may be characterized by

Statement I: measuring the baseband frequency response of the fiber in the time domain.

Statement II: measuring the impulse response of the fiber in the time domain.

Statement III: measuring the impulse response of the fiber in the frequency domain.
Statement IV: measuring the baseband frequency response of the fiber in the frequency domain.
A. Only statements I, II and III are true. B. Only statements I, III and IV are true.
C. Only statements III and IV are true. D. Only statements II and IV are true.

8. Frequency domain measurement is the preferred method for acquiring the
A. spectral response of the optical fibers. B. bandwidth of the multimode optical fibers.
C. bandwidth of the single-mode optical fibers. D. fiber cutoff wavelength.

9. Which statement is true?
A. The backscatter measurement method of the fiber attenuation on an optical fiber link uses the principle of optical time domain reflectometry.
B. Single-mode fiber OTDRs do not exhibit polarization noise.
C. The backscatter measurement method has the disadvantage of being destructive.
D. The OTDRs are commercially available for operation at only longer wavelength regions.

10. Statement I: The optical feedback along a fiber link can adversely affect stability of injection laser.
Statement II: Multiple reflections can contribute to the noise levels at the optical detector.
Statement III: Optical return loss measurements can be performed using an optical continuous wave reflectometer.
A. Only statements I and II are true. B. Only statements II and III are true.
C. All statements are true. D. None of the statements is true.

Keys to Multiple Choice Questions

1. A	**2.** C	**3.** A	**4.** B	**5.** B	**6.** C	**7.** D	**8.** B	**9.** A	**10.** C

Review Questions

1. With the help of suitable illustrations, show how the cut-back technique is used for the measurement of total attenuation in an optical fiber?

2. Giving the test set-up diagrams, illustarate distinctly the difference in measurement of spot attenuation and spectral loss parameters of an optical fiber.

3. Considering both frequency and time domain measurement techniques, describe the measurement of dispersion in optical fibers.

4. What is meant by optical time-domain reflectometry? How can this technique be used to carry out field measurements on optical fibers?

5. Write the expression which relates the noise figure and gain of the optical amplifier. How does optical bandwidth of the optical filter affect the overall noise figure of the optical receiver?

6. Draw a typical test set-up diagram for measurement of wavelength of an optical source. Which type of reference laser source is normally used?

7. Mention the significance of rotating diffraction grating in optical spectrum analyzer. Show a typical spectral plot (optical power versus wavelength) as displayed on it.

8. List various test equipment along with parameters that can be measured by them in optical measurements.

9. Depict a test set-up diagram using polarization analyzer instrument. How can it determine the orientation of the polarized component and measure the fraction of the total light power that is polarized?

10. What are different constituents of electrical vector network analyzer? How can it be used for analyzing the modulation response of an optical transmitter, an optical receiver and an optical communication link?

Numerical Problems

1. Convert the optical signal power level of 0.3 mW and 20 μW to dBm. *[Ans.: -5.23 dBm; -17 dBm]*
2. An optical power meter records an optical power level of 15 dBμ at the output of an optical fiber cable (just prior to an optical receiver). Express it in μW. *[Ans.: 31.6 μW]*
3. Convert the measured optical power level of +25 dBm on optical power meter at the output of an optical transmitter to numerical value of power level. *[Ans.: 316.2 mW]*
4. Show that 2.4 μW is equivalent to 3.8 dBμ, and 302 μW is equivalent to -5.2 dBm.
5. Using the cut-back technique for the measurement of fiber attenuation, the measured optical output power from the 1.5 km fiber length is 50.1 μW at 1100 nm wavelength. When the fiber is cut back to a 2 m length, the measured optical output power is 385.4 μW at the same wavelength. Show that the fiber attenuation per kilometer length at 1100 nm is computed to be approximately 5.9 dB/km.
6. An OTDR test set-up is used to find the fault in an optical fiber. An optical power level of 20 mW is coupled to the optical fiber under test. The OTDR measures a reflected power level of 2 μW. The specified fiber attenuation parameter is 0.5 dB/km. Determine the distance at which the fault has occurred in the fiber length. *[Ans.: 80 km]*

Appendix A
Fiber Optic Sensors

Recent advances in fiber optic technology and optoelectronic devices led to emergence of fiber optic sensors. With the availability of highly sensitive detectors and material loss almost approaching to negligible, it is possible to sense even slight variations in intensity level, phase shift and wavelength from external distresses on the optical fiber itself. This is the fundamental concept of fiber optic sensors. The primary function of a fiber optic sensor is to measure or monitor a physical quantity such as temperature, pressure, corrosion, humidity, and similar environmental factors. The basis of measurement is the net effect on its intensity modulation, operating wavelength, phase angle of the incident optical ray, or polarization of the light propagating through the optical fiber.

Advantages of Fiber Optic Sensors

- Multifunctional sensing capabilities including current, electric field, position, vibration, strain, viscosity, chemicals, and acoustic signals.
- Multiplexing capability to form sensing networks.
- Immune to electromagnetic and radio frequency interference.
- High sensitivity.
- Robust under worst operating conditions.
- Remote sensing capability.
- Inability to conduct electric current.
- Generally cylindrical geometry.
- Small size.
- Easy integration into composite materials and different types of natural or man-made structures.
- Lightweight.

General Structure of Fiber Optic Sensor

In its simplest form, a fiber optic sensor system comprises of an appropriate sensing element (a transducer having capability of converting the measurand quantity into an equivalent optical signal), an optical source such as an LED or an Injection Laser Diode), an optical fiber for transmission of optical signal from source to destination, an optical detector such as photodiode, and measuring instruments such as an optical spectrum analyzer (OSA) or an analog/digital oscilloscope.

The main principle of working of an optical fiber sensor is that the output of the physical transducer is used to modulate light intensity, phase angle, operating wavelength, or polarization of the optical signal. This results into a corresponding change in the operational characteristics of the optical signal received at the optical detector after traveling through optical fiber.

Classification of Fiber Optic Sensors

There are different ways of classifying fiber optic sensors. For example, these can be categorized based on their principle of operation including the process of modulation and demodulation. We know that either light intensity, frequency, phase, or polarization of the optical signal can be modulated. The variations in the physical quantity (being sensed) is used as the modulating signal. Therefore, accordingly a fiber optic sensor can be termed as intensity, frequency, phase, or polarization fiber optic sensor.

Another way of classifying fiber optic sensors is based on the location of the sensing element. For example, if the sensing device is kept external to the optical device and the resultant optical signal is carried through the optical fiber, then it is known as *extrinsic* fiber optic sensor. Whereas if the variation in the physical quantity being sensed changes some property of the optical fiber itself which in turn can change a particular characteristics of the light propagating through the optical fiber, then it is termed as an *intrinsic* fiber optic sensor.

A fiber optic sensor can also be classified based on the intended application. For example, bio-medical fiber optic sensors are used for measurement of blood circulation, glucose content in the blood, etc. Physical fiber optic sensors are used to measure physical quantities such as temperature, pressure, humidity, etc. Chemical fiber optic sensors are generally used for gas analysis, pH value determination, study of spectroscopy, etc.

Types of Fiber Optic Sensors

- The *Intensity-based* fiber optic sensors usually employ multimode large core fibers because they need more light. They are primarily based on the optical signal which undergoes through some attenuation of the signal intensity. The optical signal attenuation may be caused by bending of the optical fiber which results due to generation of force by the quantity under measurement. The optical signal attenuation can also be achieved by absorption or scattering of light by the target device. Even evanescent field and microbending loss caused by the measured quantity can also result into change in the intensity of the optical signal being carried by an optical fiber. There are numerous benefits of using intensity-based fiber optic sensors. They can perform as real distributed fiber optic sensors. They are simple to implement, feasible to multiplex several signals, and less expensive. However, they require a reference system in order to avoid false readings which might occur due to variations in the intensity of the light source as well as relative measurements.

  The *microbend fiber optic sensor* basically comprises of an optical fiber passing through two grooved plates. The periodic mechanical microbends results in attenuation of the optical signal due to coupling of the energy of the guided modes with the radiation modes. Because of exerting pressure on the upper plate, there is a slight movement which in turn can bend the fiber radius more than the critical angle. Thus, the light cannot remain confined within the core area that means it may leak into the cladding area. As a result, an intensity modulation takes place within the optical fiber.

In *evanescent wave fiber optic sensor*, the cladding is stripped from a section of the optical fiber. Then it utilizes the optical signal energy which leaks from the fiber core into the cladding to produce intensity modulation. When used as chemical sensors, an optical source is used which generates exactly a wavelength that can be easily absorbed by the chemical whose concentration is required to be determined. Any variation in the optical signal intensity measures the chemical concentration.

- The changes in the wavelength of optical signal is exploited for detection purpose in *wavelength-modulated fiber optic sensors*. Examples of such types of sensors include fiber Bragg grating (FBG), black body, and fluorescence sensors. The fiber core is usually exposed to an intense interference pattern of ultra-violet energy to form FBG. The periodic changes in refractive index of the single-mode fiber (SMF) core forms the basis of FBG optic sensor. LED that serves as a broadband optical source emits optical signal at the Bragg wavelength and is propagated through the grating of the optical fiber which reflects some part of the signal at the Bragg wavelength. Thus, FBG optic sensor acts as an effective optical filter.

  In *Fluorescent-based fiber optic sensors*, the light propagates through the optical fiber and strikes at the end tip of a probe coated with fluorescent material. This results into a fluorescent signal which is again captured by the same optical fiber. This signal is then directed towards the optical detector. Such type of sensors find wide applications in measurement of viscosity of liquids, temperature, humidity, etc., sensing of chemical materials, and medical areas.

  In the *blackbody fiber optic sensor*, a blackbody cavity which is placed at the end of an optical fiber starts to glow when the temperature is increased within the cavity. This is used to serve as an optical source. The profile of the blackbody fiber optic sensor is determined by narrow band filters used in combination with optical detectors. These sensors are widely used to measure temperature under intense RF fields with accuracy up to a few degrees centigrade.

- *Phase-modulated fiber optic sensors*, as the name suggests, makes use of variations in the phase of optical signal for sensing purpose. The electric field which is required to be sensed modulates the phase of the light propagating through the optical fiber. Interferometer such as Mach-Zehnder, Michelson, Fabry-Perot, or Fiber Bragg grating type is used to compare the modulated phase of the light in the signal fiber with the unmodulated phase of the light propagating in a reference fiber.
- *Polarization-modulated fiber optic sensors* operates on the principle that different types of polarization states of the light field are either linear, circular, or elliptical. The polarization state is same as the direction of the electric field part of the light pattern. A polarizer having a length of bi-refrigent polarization-preserving fiber is used to launch the polarized light at 45° to its preferred axis. This section of fiber functions as polarization-modulation fiber optic sensors. When external changes in stress or strain takes place, the phase difference between two polarization states is changed. Then, the output polarization state is changed according to the perturbation. Hence, by analyzing the output polarization state at the exit end of the fiber, the external perturbation can be detected.

Applications of Fiber Optic Sensors

- Measurement of temperature, humidity, pressure, displacement, velocity, stress, and strain in structures of any size or shape.
- Real-time monitoring the functional health of physical structures.
- Monitoring displacement and restoration, analyzing cracks, evaluating post-seismic damage, etc. in heritage buildings and monuments.
- Monitoring of concrete settings, concrete-steel interaction, length of cracks, pre-stress, long-term deformation, etc. in buildings and bridges.
- Measurement of spatial displacement, evaluation of post-seismic damage and neutral axis evolution.
- Monitoring of foundation, joint expansion, water leakage, distributed temperature, and measurement of spatial displacement during construction of hydraulic dams.
- Detection of joints damage, evaluation of prefabricated vaults, convergence monitoring, and multipoint optical extensometers for tunnels

Future Trends in Fiber Optic Sensors

Photonic crystal fibers have been devised which can serve as new sensor configurations and sensing mechanisms. Recent advancements in micro-fabrication technologies promise to achieve better functionality, performance, and capability to operate in all-weather conditions with increased reliability of fiber optic sensors. High-density fiber optic sensor networks can be configured by using advanced signal processing techniques.

Appendix B
Radio over Fiber

Radio over Fiber (RoF) technology can integrate advantages of both wireless networks and optical fiber networks. Recently RoF technology has found numerous applications in broadband wireless access communication networks, radio astronomy, RADAR and millimeter-wave communications, antenna array beam-forming, establishment of up- and down-converting links using photonic signals, spectroscopy, imaging, and electronic warfare.

In a typical RoF system configuration, as shown in Figure B.1, a radio signal (RF_{in} modulated signal in the millimeter-wave band) is propagated via optical fiber communication link using laser source and several electro-optical devices such as photodetector, power amplifier, circulator, Tx/Rx antenna, low-noise amplifier, etc.

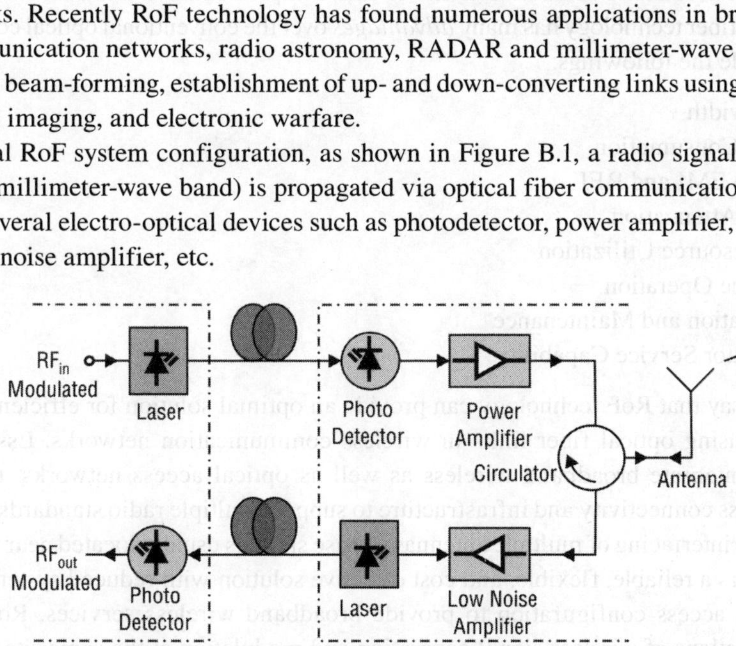

As shown, an analog or digitally modulated RF signal is carried by an analog optical fiber communication link. The RF input signal may be either baseband analog/digital data from its information source, IF-modulated signal, or the actual RF-modulated signal. It modulates the optical source such as distributed-feedback semiconductor laser diode using direct modulation technique in the central site. The output optical signal from the optical source is launched into an optical fiber between the central site and the base station. At the base station, a photodetector serves as an optical receiver that converts the optical signal to electrical (RF) signal again. The resultant signal must comply with the standard specifications as specified by the intended wireless system such as 2G/3G cellular, WLAN, WiMax, etc. In this way there is no need of generating high frequency radio carriers by the optical fiber communication link at the remote antenna site.

Usually a single optical fiber is used to transport signal in one direction (simplex). In order to enable bi-directional (duplex) communication, we require two optical fibers. It is possible to use the same fiber for two-way communication using different wavelengths in optical systems employing wavelength division multiplexing (WDM) technique. In WDM, several wavelengths carrying different information signals can be combined together and sent over a wideband optical fiber link. This also enables to maximize total data throughput necessary for broadband wireless communication networks.

The performance of the RoF system is degraded due to nonlinear distortion introduced at both modulation and photodetection process. Laser generates nonlinear distortion which can be suppressed by employing feed-forward compensation techniques. For meeting stringent requirements of linearity, more number of laser diodes, optical couplers, and photodiodes are used, which in turn may result in higher costs and system complexity. An alternative approach is to introduce predistortion device having an inverse transfer function with that of generated by nonlinear characteristics of laser diode. Another approach is to deploy a cascaded stage of lasers for realizing improved performance in terms of linearity. This offers better solution in terms of easier implementation and cost-saving.

The radio over fiber technology has many *advantages* over the conventional optical communication system that include the followings:

- Wide Bandwidth
- Less Power Consumption
- Immunity to EMI and RFI
- Low Signal Attenuation
- Dynamic Resource Utilization
- Multi-Service Operation
- Easy Installation and Maintenance
- Multi-Operator Service Capability

Thus, we can say that RoF technology can provide an optimal solution for efficient propagation of radio signals using optical fiber links in wireless communication networks. Essentially, RoF technology can integrate broadband wireless as well as optical access networks, thus enabling broadband wireless connectivity and infrastructure to support multiple radio standards and services. Moreover, remote interfacing of multiple antennas of base stations usually located near the users with a central site offers a reliable, flexible, and cost effective solution with reduced system complexity. In fixed wireless access configuration to provide broadband wireless services, RoF techniques combines the functions of wireless signal generation and modulation at the same site. This enables faster upgradation of the system with latest radio technologies. Advanced optical signal processing techniques and devices are capable of generating distortionless high-frequency analog carrier signals which are necessary for high-capacity wireless data communication links.

Appendix C
Wireless Optics

The wireless optics holds great potential for fixed wireless communications as well as for other wireless applications. In spite of all other technology developments, wireless optics has promised an economical alternative to last mile connection solution in the wireless domain. The concept of wireless optics was originally developed over few decades ago by the military. An optical wireless technology, also known as free space optics (FSO) technology, provides broadband data communication links between line-of-sight (point-to-point) locations.

Wireless optics (sometimes called fibreless optics) is an optical, wireless, line-of-sight, point-to-point, broadband technology. It basically uses infrared transmission instead of RF in the unlicensed higher frequency spectrum above 300 GHz. FSO signals are transmitted using low-power infrared beams (invisible) through free space, thereby limiting the coverage range. Another constraint on its usage is that the radiated power under any circumstances must not cross the specified limits in order to avoid any visual impairment to the human eye.

Wireless optics transceivers are generally installed in the middle or upper floors, or on an open roof of the building so that a clear line-of-sight transmission path between two stations is available. However, in some applications, these transceivers can be mounted behind a window in an existing office/home building.

In short, we can say that wireless optics is a matured technology based on line-of-sight propagation which uses optical light for transmission of user information such as voice, data, image, or video in open space. It allows optical connectivity without the use of optical fiber cables. Wireless optics system requires a light source such as LED or a LASER at the transmitter end which is capable of emitting a highly-focused light beam. The LASER beams are preferred in wireless optics because of its advantages as offered in optical fiber communications. Obviously, the only difference is the transmission medium – free space (unguided) as compared to optical fiber (guided). We know that the light travels all the way through air only at a much faster speed than through the optical fiber cable. Wireless optics, also referred to as open-air photonics or optical wireless or free-space photonics (FSP), deals with propagation of modulated optical beams in visible infrared (IR) range over air to provide broadband wireless communications. Typically, line-of-sight ranges from 100 m to a few kilometres can be obtained at throughput bandwidths up to 10 Mbps, 100 Mbps, 155 Mbps and 1250 Mbps with a possible speed up to 10 Gbps in future with the help of Wavelength Division Multiplexing (WDM) technology. Deployment cost of wireless optics infrastructure is very less as compare to fiber and it takes few days for its installation.

Issues and Challenges of Wireless Optics Links

LEDs and Infrared lights are used as a source for indoor communication link which are usually free from environmental degradations like snow, rain, fog, etc. In Indoor environment, wireless optics system is affected by only free space. There are two types of techniques used in Indoor wireless optics link: Direct line-of-sight and Diffused optical beam. The direct line-of-sight transmission requires proper alignment of transmitter and receiver without any reflection. This type of system has low multipath dispersion; higher power efficiency; less propagation path loss and higher data rate. However, it has certain drawbacks such as degradation of system performance due to shadowing effect and harmful for human eyes. Whereas diffused optical beam configuration does not necessitate any line-of-sight requirement. The optical signal is propagated from transmitter to receiver after passing through so many reflections. But it suffers from increased propagation path loss, thereby requiring more transmitting power levels. There is another limitation due to multipath dispersion phenomenon because transmitted optical signal follows different paths due to reflections from nearby objects, electrical appliances, walls, etc.

In Indoor optical wireless systems, intersymbol interference problem becomes more predominant at higher data rate. Ambient light is the main source of noise which is a combination of incandescent and fluorescent light and several other sources of noise. In indoor optical wireless systems, shadowing effect, background interference, and eye safety are the main concerns.

In Outdoor wireless optics link, absorption of transmitted optical signal takes place due to weather conditions and atmospheric factors like snow, rain, fog, presence of water droplets, etc. The scattering of light beam depends upon the operating wavelength and the size of scattering elements. The availability of channel and bandwidth vary randomly with time. So, it is required for receiver to make the possible changes so that it can cope up with the channel variations. Another solution is to use the Fractal modulation in which the spectral efficiency is kept constant over a wide range of ratio of data ret and bandwidth ratios while using fixed transmitter configurations. In order to solve the misaligned problem in between transmitter and receiver, an array of detectors can be used to achieve high signal-to-noise ratio. Adaptive equalizers can be used to reduce multipath scattering. The problem of misalignment between transmitter and receiver due to obstacles like buildings, birds, wind, earthquake etc. can be improved by using beam diverging, auto tracking capabilities and by increasing transmitted power at the input. Scintillation is a type of fluctuation which causes due to the heated air that rises from the earth, sun-rays or man-made structures including heating ducts. However, its impact can be minimized by using larger photodiode at the receiver to ensure that the signal is properly received. It also reduces the wave front distortion of optical signal which is caused by the scintillation.

Advantages of Wireless Optics

- *Lower cost.* The installation cost of wireless optics infrastructure is significantly less as compared to that of fiber-optic cables or even leasing lines from a local service provider.
- *Faster Installation.* The wireless optics infrastructure can be installed in few days or weeks, compared to months or sometimes years for installation of fiber-optics cables.
- *Flexible data transmission speed.* The data transmission speeds can be appropriately scaled as per the user's needs, ranging from 10 Mbps to 1.25 Gbps. This enables the potential users not

to pay high premium for unused capacity, if high data transmission speeds are not required for his application.

- *High data security.* Since IR transmissions cannot be as easily decoded or interpreted as RF transmissions, the wireless optics technology offers distinct advantage in providing reasonably high data security.

Disadvantages associated with Wireless Optics

- Atmospheric conditions like snow and fog can have an adverse impact on wireless optics transmissions. In regions of heavy and frequent fogs, the range is sometimes limited to 200-500 meters only.
- Signal interference caused by may be due to birds flying through IR beam and blocking it.
- Towers and tall buildings and can sway due to thunder-storms, seismic activity such as earthquakes, and so on, which may affect the direction of the IR beam.

Applications of using Wireless Optics Technology

- *Last mile connection* – The last mile connection is the communication link between a potential user and service provider. The wireless optics can be deployed in high-speed wireless communication data links between existing telecommunication networks such as the Internet service providers (ISPs) and the end-users.
- *LAN connections* – The wireless optics devices can be easily installed to offer a viable solution for interconnecting segments of LAN that are housed in different buildings of a corporate house or university campus.
- *Fiber-optic backup* – The wireless optics can be deployed as a back-up fiber-optic cables in case of temporarily break-down.
- *Backhaul* – The wireless optics can serve the purpose of carrying cell-phone traffic from cell-site antenna towers (base stations) to existing wired facilities offered by the public-switched telecommunications network.

Appendix D
Model Test Papers

Model Test Paper – Type 1

Max. Time: 3 Hours Max. Marks: 100

Note: Attempt any FIVE questions. All questions carry equal marks.

1. (a) Draw a functional block schematic of an optical fiber communication link and highlight the major functions of each block. Which is the most vital component in the link and why?

(b) Briefly describe the construction of an optical fiber cable with the help of suitable illustration, showing all parts of the cable. List at least four unique advantages of optical fiber cables.

2. (a) Using ray theory, explain the basic mechanism for the propagation of light in an optical fiber cable. In optical fibers, light travels faster in cladding as compared to core. Justify it with the help of suitable example data.

(b) The velocity of light in the fiber core of a step-index fiber is specified as 2.01×10^8 m/s. The critical angle of incidence at the intersection of the fiber core and the cladding is 80°. Estimate the values of the numerical aperture as well as the acceptance angle of incident light rays for the optical fiber placed in air. Assume that the fiber has a core diameter which is suitable to be considered by optical ray analysis.

3. (a) Compare the performance of laser diode versus LED in optical fiber communications.

(b) Find the 3-dB modulation bandwidth of a double heterojunction LED structure for a specified carrier lifetime of 6.6 ns.

4. (a) Define the terms responsivity and quantum efficiency of a photodiode, and derive expressions for the same. How are they related to each other?

(b) A laser diode with a power output of 4 mW is connected to a fiber length of 25 km having fiber loss specification of 0.2 dB/km. The receiver at the other end of the fiber has a responsivity of 0.2 A/W. How much current flows through the photodetector?

5. (a) State the principle of operation and describe the structure of a semiconductor optical amplifier.

(b) Distinguish between the amplification mechanisms in a Raman fiber amplifier (RFA) and an erbium-doped fiber amplifier (EDFA).

6. (a) Dispersion-induced pulse broadening impose the serious limitations on the system performance. What are these limitations? How are they related to each other?

(b) Draw a functional block schematic of the pre-chirp pre-compensation dispersion management technique. Illustrate the waveforms at frequency-modulated (FM) output of the DFB laser, shape of the optical pulse available at the output of an external modulator, and pre-chirped optical pulse that is finally transmitted.

7. (a) Compare the salient features of an optical Fabry-Perot interferometer filter with (i) a grating based optical filter, and (ii) acousto-optic filter.

(b) An optical power *P0* = 100 μW is applied at the input port of a 2 x 2 biconical tapered optical fiber coupler. The output optical power levels at three ports are *P1* = 45 μW at output port 1, *P2* = 42.5 μW at output port 2, and *P3* = 3.15 nW at other input port 3. Determine the percent coupling ratio. Assume that the device is lossless.

8. (a) What is meant by optical time-domain reflectometry? How can this technique be used to carry out field measurements on optical fibers?

(b) In a test set-up using optical time-domain reflectometer for locating the fault in an optical fiber, an optical power level of 200 mW is coupled from an optical power source to the optical fiber under test. The OTDR measures a reflected power level of 20 μW. The specified fiber attenuation parameter is 0.5 dB/km. Determine the distance at which the fault has occurred in the fiber length.

Model Test Paper – Type 2

Max. Time: 3 Hours Max. Marks: 100

Note: Attempt FIVE questions, taking at least TWO questions each from Part A and Part B. All questions carry equal marks.

Part-A

1. (a) Different generations of light wave systems improve the performance of an optical fiber communication system. Discuss.

(b) What are the functions of the core and cladding in an optical fiber? Why should their refractive indices be different? What would happen if the light is propagated in the fiber core without cladding?

2. (a) What is understood by index profile of an optical fiber? What are the two basic types of index profiles? Distinguish them with the help of suitable illustrations.

(b) A typical fiber is specified to have a numerical aperture of 0.1 and a fiber core radius of 3 μm. Prove that this optical fiber will operate in a single-mode configuration for a given wavelength = 0.8 μm.

3. (a) Describe the theory of semiconductor lasers and derive an expression for the threshold current.

(b) Find the external quantum efficiency assuming normal incidence at the semiconductor having refractive index of 3.5 and air interface for specified internal quantum efficiency of a double heterojunction LED of 0.75.

4. (a) Derive a general expression for power penalty (induced by intensity noise) of a p-i-n photodetector based optical receiver by considering a finite value of extinction ratio. Neglect contributions due to shot-noise and intensity-noise as compared to the thermal noise in the OFF state but not in the ON state.

(b) An optical receiver has 20 MHz bandwidth operating at a wavelength of 1100 nm. It uses an InGaAs p-i-n photodiode producing a photodiode current of 4 nA with quantum efficiency of 90%. The load resistor of the circuit is 1 kΩ. Assuming negligible surface leakage current, find the value of dark current and thermal noise current if the incident optical power is 300 nW.

Part-B

5. (a) Describe the performance parameters of semiconductor optical amplifiers in terms of noise bandwidth, optical gain, and polarisation dependence.
   (b) Illustrate the mechanism of amplification in an EDFA with a suitable energy level diagram.
6. (a) With the help of dispersion versus wavelength characteristics of standard, dispersion-shifted fibers (DSFs) as well as dispersion-flattened fibers (DFFs), describe the pros and cons of their usage in fiber-optic systems.
   (b) A 12 km DCF fiber having dispersion parameter = 100 ps/(nm-km) is used for compensating dispersion in a 80-km standard fiber having specified dispersion parameter = 17 ps/(nm-km). Can this DCF reduce the dispersion to zero? If not, what is the solution?
7. (a) WDM add-drop multiplexer (ADM) can be configured using two 3-port optical circulators with fiber Bragg gratings. Give an example of extended ADM using tunable fiber gratings.
   (b) Show a $P \times P$ reconfigurable architecture of a tunable wavelength-routing switch having M number of different wavelengths that uses a basic 2 x 2 configuration of an optical crosspoint devices and photonic switches. List different technologies that can be used for making optical switches.
8. (a) Distinguish between the cut-back technique and insertion-loss method for measurement of attenuation in optical fibers.
   (b) Show that the fiber attenuation per kilometer for a 2 km length of multimode fiber at an operating wavelength of 850 nm is 3.5 dB/km. The measured output voltage from the photodetector is 2.1 V originally and then increases to 10.7 V when the fiber is cut to 2 meter length in a cut-back method of fiber attenuation measurement techniques.

Model Test Paper – Type 3

Max. Time: 3 Hours Max. Marks: 100

Note: All questions are compulsory. There are TWO parts: Part A has 10 questions of TWO marks each, and Part B has FOUR questions of 20 marks each.

Part-A

1. Show that -20 dBm = 10 dBμ.
2. List various factors that are responsible for reduction in dispersion in graded-index multimode fibers.
3. What are the factors on which modulation bandwidth of LEDs depend? How is modulation bandwidth related to output optical power?
4. If the threshold current density for a particular laser device having an active area of 0.2×0.5 mm^2 is specified as 3×10^6 Amp/m^2, then determine the threshold current.
5. Determine the responsivity of a p-i-n photodetector at 0.85 μm operation for specified quantum efficiency of 60%.
6. Draw a suitable diagram to depict the functional properties of Erbium element.
7. Can Raman amplification provide very broadband amplification? If yes, how?
8. List the desirable properties of an ideal dispersion compensation device.
9. Can intrachannel and interchannel crosstalk accumulate in optical networks? Illustrate the effect of crosstalk level on power penalty with the help of plot between power penalties versus crosstalk level for number of crosstalk elements ranging from 10 to 100.

10. Mention key design criteria for field measurement test equipments.

Part-B

11. (a) What are essential requirements in selecting materials for optical fibers so that fiber cable can function as reliable information channel? [8]

(b) Briefly describe some practical consequences because of insulating nature of a fiber. Can optical fiber cables operate near nuclear installation? How is fiber optics useful in medical applications? [6+3+3]

(OR)

(a) What are typical benefits of graded-index profile optical fiber cables over step-index profile optical fiber cables?

(b) The core diameter of a typical step-index fiber is specified as 8 μm. The values of core refractive index and relative refractive index difference are specified as 1.46 and 0.3%, respectively. Calculate the mode field diameter (MFD) at operating wavelength of 1550 nm.

12. (a) Compare and contrast important properties of LED and LD as optical sources. Also comment on their suitability for various application of optical fiber communications.

(b) Find the threshold gain if the length of the cavity is 0.4 mm and the values of reflectivities on either ends of the cavity is 0.5. Assume loss coefficient = 3 mm^{-1}.

(OR)

(a) Define spectral response of the photodiode. What are the factors on which the spectral response and time response of a photodiode depends?

(b) Consider a silicon p-i-n photodetector and an APD to detect light at $\lambda = 850$ nm. For an incident light intensity of 0.1 mW/mm^2, the photocurrent generated by the p-i-n photodetector and APD are 10 μA and 500 μA, respectively. In both cases, the active area is 0.2 mm^2. Compute the quantum efficiency and the avalanche multiplication factor.

13. (a) How does EDFA operate? What can cause it to become a laser? Mention some promising technical characteristics of EDFA.

(b) Consider an EDFA (optical gain = 10 dB), which is used as a power amplifier after optical transmitter, is pumped at 980 nm wavelength. Assume that the amplifier input is a 0 dBm level from a laser diode transmitter. Determine the minimum required pump power for a 10 dBm output power level at 1540 nm.

(OR)

(a) With the help of plots between dispersion versus wavelength, show that dispersion compensating fiber (DCF) exhibits uniform dispersion over 1.3 μm – 1.5 μm wavelength region as compared to standard single-mode fiber and non-zero dispersion-shifted (at 1.5 μm wavelength).

(b) Out of direct modulation and external modulation, which approach would you prefer as dispersion management solution and why?

14. (a) What are various types of network medium used in WDM transmission system? Give a simplified functional block schematic diagram of a WDM system.

(b) A fiber-optic transmission system is required to operate in the spectral band of 1536 nm to 1556 nm. If the maximum channel spacing is constrained to have 500 GHz, the how many wavelength channels can be multiplexed in the syst4em? Also show the effect of stimulated Raman scattering (SRS) in WDM.

(OR)

(a) Depict a test set-up diagram using polarization analyzer instrument. How can it determine the orientation of the polarized component and measure the fraction of the total light power that is polarized?

(b) An OTDR test set-up is used to find the fault in an optical fiber. An optical power level of 20 mW is coupled to the optical fiber under test. The OTDR measures a reflected power level of 2 μW. The specified fiber attenuation parameter is 0.5 dB/km. Determine the distance at which the fault has occurred in the fiber length.

Model Test Paper – Type 4

Max. Time: 3 Hours Max. Marks: 100

Note: There are total THREE parts: Part A, Part B, and Part C.

- Part A has 10 questions of TWO marks each. Attempt ALL questions.
- Part B has FIVE questions of 10 marks each. Attempt any FOUR questions.
- Part C has THREE questions of 20 marks each. Attempt any TWO questions.

Part-A (Attempt all questions)

1. (a) What is the difference (in watts) between -60 dBm and 60 dBm?

(b) How does the light ray propagate in single mode fiber?

(c) Mention some specific aspects of an optical source to be considered for optical fiber communications.

(d) Determine the power radiated by an LED if its internal quantum efficiency is 3% and it has operating wavelength of 670 nm with a device current of 50 nm.

(e) Draw the energy-band diagram for a semiconductor p-i-n photodiode.

(f) Determine the spontaneous emission factor for an optical amplifier having optical gain = 20 dB if ASE is 1 mW for a fractional bandwidth of 5×10^{-6}.

(g) What are the effects of pulse spreading due to chromatic dispersion?

(h) Determine the free spectral range (FSR) of a Fabry-Perot (FP) interferometer tunable optical filter which is designed with the group index $n_g = 1.5$ of the intracavity material used in the construction of a FP filter having L = 100 μm.

(i) Calculate the number of channels that are demultiplexed by a phased-array optical demultiplexer if its free spectral range is 1600 GHz and channels are spaced at 50 GHz at central wavelength of 1550 nm.

(j) List various test equipments alongwith parameters that can be measured by them in optical measurements.

Part-B (Attempt any FOUR questions)

2. What are the functions of the core and cladding in an optical fiber? Why should their refractive indices be different? [6+4]

3. Derive an expression for numerical aperture. An optical fiber cable has values of refractive index of 1.46 for the fiber core. If the fractional change in the index of refraction of the optical fiber is specified as 0.01, then find its numerical aperture.

4. Illustrate a simplified block diagram of an optical receiver, depicting the essential components only.

5. Consider that an EDFA pumped at 980 nm is being used as a power amplifier with a 10-mW pump power. When optical power of an input signal is 1 mW at λs = 1550 nm, the output of the amplifier is +20 dBm. Compute a) the optical gain of the amplifier (in dB), and b) the input pump power required to achieve this gain.

6. With the help of suitable illustrations, how the cut-back technique is used for the measurement of total attenuation in an optical fiber?

Part-C (Attempt any TWO questions)

7. (a) The core refractive index and a relative refractive index difference of a multimode step-index fiber is specified as 1.5 and 2%, respectively. At operating wavelength of 1300 nm, the approximate number of propagating modes is 1000. Determine the diameter of the fiber core.

(b) Compute the threshold gain for a GaAs laser diode having the following parameters: cavity length L = 500 μm, loss coefficient a = 10 cm^{-1}, the facet reflectivities R1 = R2 = 0.32.

8. (a) State the principal noises associated with photodetectors. The responsivity of a typical photodiode is specified as 0.4 A/W for a He-Ne laser source (λ = 632.8 nm). The active area of the photodiode is 2 mm^2. What will be the output photocurrent if the incident flux is 100 $\mu W/mm^2$?

(b) Chirped fiber Bragg gratings are the most developed dispersion-compensating gratings (DCGs). Illustrate the basic operation of chirped fiber Bragg grating for chromatic dispersion compensation of (a) one wavelength; (b) three wavelengths as used in WDM.

9. (a) Elaborate specific advantages of point-to-point dense wavelength division multiplexing (DWDM) systems?

(b) Draw a typical test set-up diagram for measurement of wavelength of an optical source. Which type of reference laser source is normally used?

Appendix E
Abbreviations and Acronyms

ADM	Add-Drop Multiplexer
AGC	Automatic Gain Control
AM	Amplitude Modulation
AON	Active Optical Network
AOTF	Acousto-Optic Tunable Filter
APD	Avalanche Photodiode
AR	Anti-Reflection
ASE	Amplified Spontaneous Emission
ASK	Amplitude Shift keying
AWG	Arrayed-Waveguide Grating
BER	Bit Error Rate
BLP	Bandwidth-Length Product
BW	Bandwidth
CCITT	Consultative Committee International Telephone and Telegraph
CW	Continuous Wave
CWDM	Coarse Wavelength Division Multiplexing
dB	Decibel
DBR	Distributed Bragg Reflector
DEMUX	Demultiplexer
DFB	Distributed Feedback
DH	Double Heterojunction
DQDB	Dual Queue Distributed Bus
DWDM	Dense Wavelength Division Multiplexing
EDFA	Erbium Doped Fiber Amplifier
EDWA	Erbium Doped Waveguide Amplifier
EIA	Electronics Industries Association
EHF	Extremely High Frequency
ELED	Edge-emitting LED
ELF	Extremely Low Frequency
EMI	Electro-Magnetic Interference
EOTF	Electro-Optic Tunable Filter
FBG	Fiber Bragg Grating
FDDI	Fiber Distributed Data Interface

FDM	Frequency Division Multiplexing
FIT	Failure In Time
FM	Frequency Modulation
FOTP	Fiber Optic Test Procedures
FP	Fabry-Perot
FPA	Fabry-Perot Amplifier
FPLA	Fabry-Perot Laser Amplifier
FPR	Free-Propagation Region
FR-PVC	Flame-Retardant Polyvinyl Chloride
FSK	Frequency Shift Keying
FSR	Free Spectral Range
FTTR	Fixed Transmitter and Tunable Receiver
FWHM	Full-Width at Half-Maximum
FWM	Four Wave Mixing
GOLD	Gigabit Optical Link Designer
GVD	Group Velocity Dispersion
HF	High Frequency
Hi-OVIS	Highly Interactive Optical Visual Information System
IL	Insertion Loss
ILD	Injection Laser Diode
IR	Infrared
ISI	Intersymbol Interference
LAN	Local Area Network
LASER	Light Amplification by Stimulated Emission of Radiation
LD	Laser Diode
LED	Light Emitting Diode
LF	Low Frequency
LOS	Line-of-Sight
LPG	Long Period Grating
MAN	Metropolitan Area Network
MEMS	Micro-Electro-Mechanical System
MF	Medium Frequency
MFD	Mode Field Diameter
MMF	Multi-Mode Fiber
MQW	Multiple Quantum Well
MSM	Metal-Semiconductor-Metal
MTTF	Mean-Time-To-Failure
MUX	Multiplexer
MZ	Mach-Zehnder
MZI	Mach-Zehnder Interferometer
NA	Numerical Aperture
NEP	Noise Equivalent Power
NF	Noise Figure
NRZ	Non Return-to-Zero

NZDSF	Non-zero Dispersion Shifted Fiber
OC	Optical Carrier
OEIC	Opto Electronic Integrated Circuit
O-E-O	Optical-Electrical-Optical
OOK	On-Off Keying
OSA	Optical Spectrum Analyzer
OSNR	Optical Signal-to-Noise Ratio
OTDM	Optical Time Division Multiplexing
OTDR	Optical Time Domain Reflectometer
OXC	Optical Cross-Connects
QoS	Quality of Service
RFTS	Remote Fiber Test System
PCS	Plastic-Clad Silica
PD	Photodetector
PDH	Plesiochronous Digital Hierarchy
PDL	Polarization Dependent Loss
PISO	Parallel-in Serial-out
PM	Phase Modulation, or Power Meter
PMD	Polarization Mode Dispersion
PON	Passive Optical Network
PP	Power Penalty
PSK	Phase Shift Keying
RF	Radio Frequency
RFA	Raman Fiber Amplifier
RIN	Relative Intensity Noise
RL	Return Loss
RMS	Root Mean Square
SBS	Stimulated Brillouin Scattering
SCS	Silica-clad Silica
SDH	Synchronous Digital Hierarchy
SH	Single Heterojunction
SHF	Super High Frequency
SLD	Semiconductor Laser Diode
SLED	Surface-emitting LED
SMF	Single Mode Fiber
SMSR	Side Mode Suppression Ratio
SNR	Signal-to-Noise Ratio
SOA	Semiconductor Optical Amplifier
SONET	Synchronous Optical Networks
SPM	Self-Phase Modulation
SRS	Stimulated Raman Scattering
SSR	Side-lobe Suppression Ratio
STM	Synchronous Transport Module
STS	Synchronous Transport Signal

TDM	Time Division Multiplexing
TEC	Thermoelectric Cooler
TTFR	Tunable Transmitter and Fixed Receiver
TWA	Traveling-Wave Amplifier
TWSLA	Traveling-Wave Semiconductor Laser Amplifier
UPNRZ	Unipolar Non Return-to-Zero
UPRZ	Unipolar Return-to-Zero
VCSEL	Vertical Cavity Surface Emitting Laser
VF	Voice Frequency
VHF	Very High Frequency
VLF	Very Low Frequency
UHF	Ultra High Frequency
UV	Ultra Violet
WAN	Wide Area Network
WDM	Wavelength Division Multiplexing
WDMA	Wavelength Division Multiple Access
WGR	Waveguide Grating Router
WXC	WDM Cross-Connect
XPM	Cross-Phase Modulation

References

Agrawal, Govind P. 2010. *Fiber-Optic Communication Systems.* 4th Edition. New York: John Wiley and Sons Inc.

Aljada, M., and K. Alameh. 2005. 'Distributed Wireless Optical Communications for Humanitarian Assistance in Disasters'. *Proceedings of the Third IEEE International Workshop on Electronic Design.* Vol. 3, pp321-326 Accessed July 31, 2015. doi: 10.1109/DELTA.2006.37.

Awn, M. S., E. Leitgebi, and T. P. Marzuki. 2010. 'A Study of Fog Characteristics using Free-Space Optical Wireless Links'. *Journal of Radio Engineering.* 19(2): 213–222.

Chou, M. S., K. Parameswaran, M. M. Fejer, and I. Brener. 2000. 'Optical Frequency Mixers for WDM and TDM Applications'. *Optical Fiber Communication Conference.* 4: 16–18.

Cincotti, G., N. Wada, and K. Kitayama. 2006. 'Characterization of a full Encoder/Decocder in the AWG Configuration for Code-based Photonic Routers Part I: Modelling and Design'. *Journal of Lightwave Technology.* 24(1): 103–112.

Devaraju, M., and V. C. Ravichandran. 2014. 'Analysis of Optical CDMA System with Optical Thresholding and Decision Thresholding using Balanced Detection'. *Journal of Electronic Networks Devices and Fields.* 3(2): 81–88.

Dey, S. K., and A. Adhya. 2015. 'IP-over-WDM Network Design Methodology to Improve Efficiency in Overall Expenditure due to Cost and Energy Consumption'. *IEEE/OSA Journal of Optical Communications and Networking.* 7(6): 563–577.

Dimmick, T. E., G. Kakarantzas, T. A. Birks, and P. St. J. Russell. 2000. 'Narrow-Band Acousto–Optic Tunable Filter Fabricated from Highly Uniform Tapered Optical Fiber'. *Optical Fiber Communication Conference.* 4: 25–27.

Ganapathy, R., K. Porsezian, A. Hasegawa, and V. N. Serkin. 2008. 'Soliton Interaction under Soliton Dispersion Management'. *IEEE Journal of Quantum Electronics.* 44(4): 383-390.

Gangwar, R., S. P. Singh, and N. Singh. 2007. 'Soliton Based Optical Communication'. *Progress in Electromagnetics Research (PIER).* 74: 157–166.

Gowar, J. 2001. *Optical Communication Systems.* 2nd Edition. India: Prentice-Hall.

Hansen, K. P. 2005. 'Introduction to Nonlinear Photonic Crystal Fibers'. *Journal of Optical and Fiber Communications Research.* 2(3): 226–254.

Hasegawa, A. 2001. 'Soliton Based Ultra-High Optical Communications'. *Journal of Physics.* 57(5/6): 1097-1127.

Hauske, F. N., C. Xie, Z. Zhang, C. Li, et al. 2010. 'Frequency–domain Chromatic Dispersion Estimation'. *Optical Fiber Communication (OFC), collocated National Fiber Optic Engineers Conference, 2010 Conference on (OFC/NFOEC).* Institute of Electrical and Electronics Engineers. Accessed July 31, 2015. E-ISBN: 978-1-55752-884-1.

Jain, V. K., and J. H. Franz. 2000. *Optical Communications: Components and Systems.* Delhi: Narosa.

Juarez J. C., et al. 2006. 'Free Space Optical Communication for Next Generation Military Networks'. *IEEE Communication Magazine.* 44: 46–51.

Kaplan, A., and S. Ruschin. 2000. 'Layout for Polarization Insensitive Modulation in $LiNbO_3$ Waveguides'. *IEEE Journal of Selected Topics in Quantum Electronics*. 6(1): 83–87.

Kartalopoolos, S. V. 2000. *Introduction to DWDM Technology.* IEEE Press, and SPIE Press.

Keiser, G. 2008. *Optical Fiber Communication*. 4th Edition. New Delhi: Tata McGraw-Hill Education.

Khadir, A. A., B. F. Dhahir, and X. Fu. 2014. 'Achieving Optical Fiber Communication Experiments by OptiSystem'. *International Journal of Computer Science and Mobile Computing (IJCSMC)*. 3(6): 42–53.

Kolka, Z., O. Wilfert, D. Biolek, and V. Biolkova. 2006. 'Availability Model of Free Space Optical Data Link'. *International Journal of Microwave and Optical Technology.* 1(2): 612- 616.

Lin, Ching Fuh. 2004. *Optical Components for Communication: Principles and Applications.* Springer.

Lodro, M. M., and M. L. Joyo. 2012. '32-Channel DWDM System Design and Simulation by using EDFA with DCF and Raman Amplifiers'. *International Conference on Information and Computer Networks IPCSIT.* Singapore. 27 :216–220 .

Mahad, F. D. M., et al. 2009. 'EDFA Gain Optimization for WDM System'. *Journal of Electrical Engineering (ELEKTRIKA).* 11(1): 34–37.

Mangold, M., et al. 2014. 'Amplitude Noise and Timing Jitter Characterization of a High-Power Mode-Locked Integrated External-Cavity Surface Emitting Laser'. *IEEE Photonic Journal.* 6(1). Accessed July 31, 2015. doi: 10.1109/JPHOT.2013.2295464

Matsumoto M., et al. 2008. 'An Alternative Access Technology for Next Generation Networks Based on Full-optical Wireless Communication Links'. *Innovations in NGN: Future Network and Services, 2008. K-INGN 2008. First ITU-T Kaleidoscope Academic Conference*. Institute of Electrical and Electronics Engineers. ISBN: 978-92-61-12441-0.

Mesleh, R., et al. 2011. 'On the Performance of Different OFDM Based Optical Wireless Communication Systems'. *IEEE/OSA Journal of Optical Communications and Networking.* 3(8): 620–628.

Murthy, S. R., and M. Guruswami. 2002. *WDM Optical Networks – Concepts, Design, and Algorithms.* First Edition. Upper Saddle River, New Jersey: Prentice Hall.

Mynabaev, D. K., and L. L. Scheiner. 2001. *Fiber–Optic Communications Technology.* First Edition. Upper Saddle River, New Jersey: Prentice Hall.

Okashi, T. 2010. 'Recent Advances in Coherent Optical Fiber Communication Systems'. *Journal of Lightwave Technology.* 5(1): 44–52.

Palais, J. C. 1998. *Fiber Optic Communications.* 4th Edition. Englewood Cliffs, NJ: Prentice Hall, Inc

Pan, Z., C. Yu, and A. E. Willner. 2010. 'Optical Performance Monitoring for the Next generation Optical Communication Networks'. *Optical Fiber Technology Journal.* 16(1): 20–45.

Patel, M. G., and S. B. Khant. 2014. 'Soliton Transmission in Fiber Optics for Long Distance Communications'. *International Journal of Advanced Research in Electrical, Electronics and Instrumentation Engineering.* 3(2) : 7100–7107

Raghuvanshi, S. K., and V. Kumar. 2011. 'Analysis of Double Clad Single-Mode Step–Index Fibers'. *World Journal of Science and Technology.* 1(8): . 79–83.

Raghuvanshi, S. K., V. Palodiya, A. Kumar, and S. Kumar. 2014. 'Experimental Characterization of Fiber Optic Communication Link for Digital Transmission System'. *ICTACT Journal on Communication Technology.* 5(1): 868–876.

Rastogi, V., and K. S. Chiang. 2001. 'Propagation Characteristics of a Segmented Cladding Fiber'. *Optics Letters.* 26(8): 491–493.

Rickettes, D. S. and X. Li. 2007. 'On the Self-Generation of Electrical Soliton Pulses'. *IEEE Journal of Solid-State Circuits.* 42(8): 1657 – 1668.

Rubin, I., and J. Ling. Jan. 2000. 'Survivable All Optical Cross–Connect Meshed-Ring Communication Networks'. *Optical Networks Magazine.* 1(1): 61–72.

Sato, K., S. Kuwahara, Y. Miyamoto, and N. Shimizu. 2002. '40Gb/s Direct Modulation of Distributed Feedback Laser for Very-shortreach Optical Links'. *Electronics Letter.* 38(15): 816–817.

Senior, J. M. 2009. *Optical Fiber Communications– Principles and Practice.* 3rd Edition. Pearson Education.

Sindhi, U. J., R. B. Patel, K. A. Mehta, and V. Mishra. 2013. 'Performance Analysis of 32-Channel WDM System using Erbium Doped Fiber Amplifier'. *International Journal of Electrical and Electronics Engineering & Telecommunications.* 2(2):13–18 .

Singh, S., and R. S. Kler. 2006. 'Gain Flattening Approach to Physical EDFA for 16 × 40 Gb/s NRZ–DPSK WDM Optical Communication System'. *Fiber and Integrated Optics.* 25(5): 363–374.

Stephen, J. W., et al. 2003. 'Optical Fiber Long-Period Grating Sensors: Characteristics and Application'. *Measurement Science Technology.* 14(5):63–86 .

Wang, Xu, and Kitayama Ken-ichi. 2004.'Analysis of Beat Noise in Coherent and Incoherent Time-Spreading OCDMA'. *Journal of Lightwave Technology.* 22(10): 2226–2235.

Willaims, P. 2004. 'PMD Measurement Techniques and How to Avoid Pitfalls'. *Journal of Optical and Fiber Communications Research.* 1(1): 84–105.

Yan, M., M. Yao, H. Zhang, and Xia, Li, Zhang Y. 2008. 'Encoder/Decoder for Special Phase-Coded OCDMA System based on Amplitude Sampled FBG'. *IEEE Photon Technology Letter.* 20(10): 788–790.

Yariv, A., and P. Yeh. 2007. *Photonics– Optical Electronics in Modern Communications.* First Edition. New York: Oxford University Press.

Yeung, A., K. S. Chainge, V. Rastogi, P. L. Chu, and G. D. Peng . 2004. 'Experimental Demonstration of Single-Mode Operation of Large Core Segmented Cladding Fiber'. *Proceedings of the Optical Fiber Communication Conference.* 2: 3–6.

Zang, H., J. P. Jue, and B. Mukherjee. Jan 2000. 'Review of Routing and Wavelength Assignment Approaches for Wavelength Routed WDM Optical Networks'. *Optical Networks Magazine.* 1(1): 47–60.

Zhang, Y., S. Gao, and A. P. Zhang. 2012. 'Optically Heated Long–Period Grating as Temperature–Insensitive Fiber–Optic Refractive-Index Sensor'. *IEEE Photonic Journal.* 4(6): 2340–2345.

Zin, A. M., M. S. Bongsu, S. M. Idrus, and N. Zulkifli. 2010. 'An Overview of Radio-over-Fiber Network Technology'. *International Conference on Photonics.* 13617–13625 Institute of Electrical and Electronics Engineers. ISBN: 978-1-4244-7186-7.

Index